8th International Conference on Applied Physics and Mathematics (ICAPM 2018)

Journal of Physics: Conference Series
Volume 1039

Phuket, Thailand
27 – 29 January 2018

ISBN: 978-1-5108-6589-1
ISSN: 1742-6588

Printed from e-media with permission by:

Curran Associates, Inc.
57 Morehouse Lane
Red Hook, NY 12571

CURRAN ASSOCIATES INC. proceedings.com

Some format issues inherent in the e-media version may also appear in this print version.

Printed by Curran Associates, Inc. (2018)

Institute of Physics
Dirac House, Temple Back
Bristol BS1 6BE UK

Phone: 44 1 17 929 7481
Fax: 44 1 17 920 0979

techtracking@iop.org

Additional copies of this publication are available from:

Curran Associates, Inc.
57 Morehouse Lane
Red Hook, NY 12571 USA
Phone: 845-758-0400
Fax: 845-758-2633
Email: curran@proceedings.com
Web: www.proceedings.com

Table of contents

Volume 1039

2018 8th International Conference on Applied Physics and Mathematics (ICAPM 2018)

27–29 January 2018, Phuket, Thailand

Accepted papers received: 23 May 2018
Published online: 27 June 2018

Preface

Papers

Mathematical theory and calculation

Physical Mathematics and Applied Physics

Instrument Design and Measurement

IOP Conf. Series: Journal of Physics: Conf. Series **1039** (2018) 011001 doi:10.1088/1742-6596/1039/1/011001

PREFACE

Welcome to 2018 8th International Conference on Applied Physics and Mathematics (ICAPM 2018) held in Phuket, Thailand during January 27-29, 2018

The aim objective of ICAPM 2018 is to provide a platform for researchers, engineers, academicians as well as industrial professionals from all over the world to present their research results and development activities on Applied Physics and Mathematics. This conference provides opportunities for the delegates to exchange new ideas and application experiences face to face, to establish business or research relations and to find global partners for future collaboration.

The present volume contains the selected papers submitted to this conference and provides up-to-date, comprehensive and worldwide state-of-the art knowledge in the field of Applied Physics and Mathematics. All papers included in this collection had undergone the careful peer-review by the conference technical committee members and international reviewers before they're selected for publication. Studies presented in this volume cover the following topics: Mathematical theory and calculation, Physical Mathematics and Applied Physics, System Design and Mathematical Modeling, Applied Mathematics and Engineering, Mathematical Modeling and Analysis in Medical Field, Applied mechanics, Instrument Design and Measurement

We'd like to express our gratitude to all the committee members for their great professionalism and efforts. And thank all the sponsors and participants for their valuable contributions and support to ICAPM 2018

ICAMP 2018
Conference Committee

IOP Conf. Series: Journal of Physics: Conf. Series **1039** (2018) 011002 doi:10.1088/1742-6596/1039/1/011002

Conference Committee

International Advisory Committee Chair

Prof. I-Ming Tang, King Mongkut's University of Technology, Thonburi (KMUTT), Thailand

Conference Chair

Prof. Naoyuki Ishimura, Chuo University, Japan

Prof. Po-Liang Liu, National Chung-Hsing University, Taiwan

Program Chair

Prof. Nan-Wei Chen, Yuan Ze University, Taiwan

Assoc. Prof. Kuo-Chih Chuang, Zhejiang University, China

Prof. Jong-Shenq Guo, Tamkang University, Taiwan

Technical Committee

Prof. Praveen Agarwal, Anand Intl. College of Engg, India

Prof. Dongsun Seo, Myongji University, South Korea

Prof. Xiaodi Li, Shandong Normal University, China

Prof. Dilbaj Singh , SGTB Institute of Mgt. & IT, India

Assoc. Prof. Wael Zaki, Khalifa University, UAE

Assoc. Prof. Dr. Abdul Ghaffar Jaffar, Balochistan University of Information Technology, Pakistan

Assoc. Prof. Jiann Lin Chen, I-Shou University, Taiwan

Assoc. Prof. Dr. S. Joseph Antony, University of Leeds, UK

Assoc. Prof. Sixing Tao,Shangqiu Normal University, China

Assoc. Prof. Montip Tiensuwan, Mahidol University, Thailand

Dr. Zailan Siri, University of Malaya, Malaysia

Dr. M. Premjit Singh, Manipur University, India

Dr. Ayman Abd-Elhamed, Helwan University, Egypt

Dr. Wooi-Nee Tan, Multimedia University, Malaysia

Dr. Yan-Gu Lin, National Synchrotron Radiation Research Center, Taiwan

Dr. Vikas Malik, Jaypee Institute of Information Technology, India

Dr. Lokendra K. Balyan, Indian Institute of Information Technology Design and Manufacturing Jabalpur, India

Asst. Prof. Yong-Yuk Won, Myongji University, South Korea

Asst. Prof. Thamir Abdul Ameer Hassan, Al -Karkh University of Science Iraq, Iraq

Asst. Prof. Abdalla Mansur, Al Ain University of Science and Technology, United Arab Emirates

Dr. Berhanu Assaye Alaba, Bahir Dar University, Ethiopia

Dr. Mitali Konwar, Moran College, India

Dr. Khushboo Verma, UNIVERSITY OF LUCKNOW, India

Dr. Yang Xi, Diodes Inc., USA

Dr. Ibrahim Rose Irnawaty, UNIVERSITI SAINS ISLAM MALAYSIA, MALAYSIA

Dr. Suntaree Unhapipat, Mahidol University, Thailand

Dr. Siti Suzilliana Putri Mohamed Isa, Universiti Putra Malaysia, Malaysia

Prof. Ilona Tregub, Financial University under the Government of the Russian Federation, Russia

Peer review statement

All papers published in this volume of *Journal of Physics: Conference Series* have been peer reviewed through processes administered by the proceedings Editors. Reviews were conducted by expert referees to the professional and scientific standards expected of a proceedings journal published by IOP Publishing.

IOP Conf. Series: Journal of Physics: Conf. Series **1039** (2018) 012001 doi:10.1088/1742-6596/1039/1/012001

On varying coefficients of spatial inhomogeneous nonlinear Schrödinger equation

N. Karjanto[1] **and J. Tan**[2,3]

[1] Department of Mathematics, University College, Sungkyunkwan University, Natural Science Campus, 2066 Seobu-ro, Jangan-gu, Suwon 16419, Gyeonggi-do, Republic of Korea
[2] Department of Mechanical Engineering, Faculty of Engineering, The University of Nottingham Malaysia Campus, Semenyih 43500, Selangor, Malaysia

E-mail: natanael@skku.edu

Abstract. A nonlinear evolution equation for wave packet surface gravity waves with variation in topography is revisited in this article. The equation is modeled by a spatial inhomogeneous nonlinear Schrödinger (NLS) equation with varying coefficients, derived by Djordjević and Redekopp (1978) and the nonlinear coefficient is later corrected by Dingemans (1997). We show analytically and qualitatively that the nonlinear coefficient and the corresponding averaging value, stated but not derived, by Benilov, Flanagan and Howlin (2005) and Benilov and Howlin (2006) are inaccurate. For a particular choice of topography and wave characteristics, the NLS equation alternates between focusing and defocusing case and hence, it does not admit the formation of a classical soliton, neither bright nor dark one.

1. Introduction

In this article, we revisit a nonlinear evolution equation for wave packet surface gravity waves with variation in topography. The equation is described by the spatial nonlinear Schrödinger (NLS) equation with varying coefficients. It was derived in [1] and the correction to the nonlinear coefficient was given in [2, 3, 4]. The equation has been studied in the context of smooth topography in [5] and strong smooth topography in [6]. However, it can be verified that the nonlinear coefficients used in these studies are inaccurate. In this article, we compare analytically and qualitatively between the accurate and inaccurate nonlinear coefficients and a consequence of choosing particular wave characteristics and fluid depth variation to the model equation.

2. On an evolution equation

A nonlinear evolution for the packet of surface gravity waves propagating along uneven bottom is modeled by the spatial inhomogeneous nonlinear Schrödinger (NLS) equation, also known as the NLS equation with varying, non-constant coefficients [1]. We notice that the NLS equations with non-constant coefficients adopted in [5] and [6] presented in different expressions even though both model equations refer to [1]. Equation (2.2) in [5] is written with negative signs in front of the dispersive and nonlinear coefficients, α and β, respectively. It reads

$$i\left(\partial_x A + c_g^{-1}\partial_t A + \mu A\right) - \alpha\partial_t^2 A - \beta A|A|^2 = 0. \qquad (1)$$

Meanwhile, equation (6) in [6] is written with positive signs in front of the dispersive and nonlinear coefficients, $\alpha(x)$ and $\beta(x)$, respectively. It reads

$$i\left(\partial_x A + c_g(x)^{-1}\partial_t A + \mu(x) A\right) + \alpha(x)\partial_t^2 A + \beta(x) A|A|^2 = 0. \qquad (2)$$

[3] Present address: Frost & Sullivan, 2, Jalan Kiara, Mont Kiara, Kuala Lumpur 50480, Malaysia.

For both NLS equations, $A(x,t)$ refers to the complex-valued amplitude of the wave packet for surface gravity waves propagating along topography variation, c_g is the group velocity and μ is called a dissipative coefficient; also known as an amplification/absorption coefficient in nonlinear optics; a chemical potential in Bose-Einstein condensates. It is obvious that both dispersive and nonlinear coefficients $\alpha(x)$ and $\beta(x)$ have the opposite sign from one paper to another. However, these coefficients are expressed identically in both papers, with the exception of writing h and H for the fluid depth and the term $k^2 c_g \operatorname{sech}^2 kH$ expressed in an equivalent form $-c_g(\omega^4 - k^2)$ using a hyperbolic functions relationship $\operatorname{sech}^2 kH = 1 - \tanh^2(kH)$. The dispersive and nonlinear coefficients are presented as follows, taken from [6], where later in our notation we display them using the normalized quantities $\tilde{k}$ and $\tilde{\omega}$:

$$\alpha(x) = \frac{1}{2\omega c_g}\left(1 - \frac{H}{c_g^2} + \frac{2\omega H \tanh kH}{c_g}\right) \tag{3}$$

$$\beta(x) = \frac{1}{2\omega^3 c_g}\left(3k^4 + 2\omega^4 k^2 - \omega^8 - \frac{(2k\omega + k^2 c_g \operatorname{sech}^2 kH)^2}{H - c_g^2}\right). \tag{4}$$

The dispersive coefficient $\alpha(x)$ should be the negative of expression (3), hence, the expression in [5] is correct, up to a scaling factor of the constant gravitational acceleration g. The nonlinear coefficient $\beta(x)$ (4) is found to be inaccurate and the accurate one is given explicitly in Section 3 and shown the accuracy qualitatively.

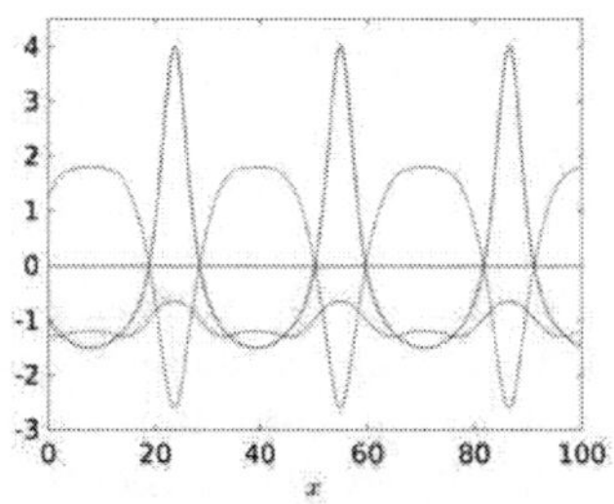

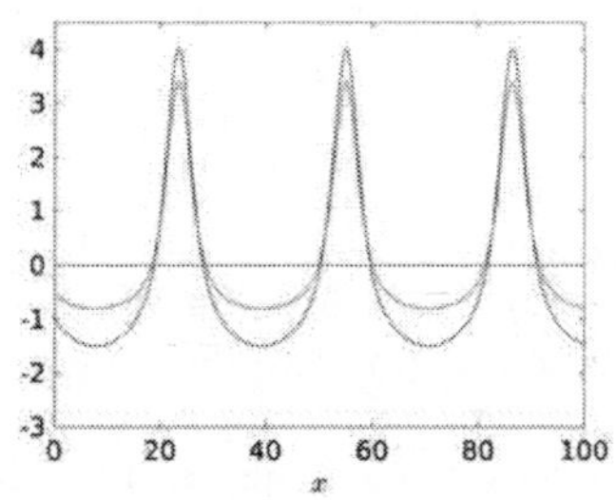

Figure 1. (Left panel) Plots of the dispersive coefficient $\alpha(x)$ (red), the (scaled) nonlinear coefficient $\beta(x)$ (blue) and the product of the dispersive and the nonlinear coefficient $\alpha(x)\beta(x)$ (green). We observe that the NLS equation alternates between focusing and defocusing case along the spatial evolution parameter x. (Right panel) Comparison plots of the (scaled) nonlinear coefficients $\beta(x)$ between an accurate one derived in [1] and corrected in [3, 4] (blue curve) and an inaccurate one (red curve), stated in [5] and [6] by referring to [1], but not derived. All cases correspond to the water depth $H(x) = 1 + 0.6\sin(x/5)$.

Applying a multiple scale method in a co-moving reference by replacing time t with $\tau = t - \int c_g(x)^{-1} dx$ and writing the coefficient μ as $c_g(x)^{-1} c_g'(x)$, the NLS equation can be written as follows, where we adopt the positive sign of the coefficients in the notation

$$i\partial_x\left(\sqrt{c_g}A\right) + \alpha(x)\partial_\tau^2\left(\sqrt{c_g}A\right) + c_g(x)^{-1}\beta(x)\left|\sqrt{c_g}A\right|^2\left(\sqrt{c_g}A\right) = 0. \tag{5}$$

There exists a relationship between the NLS equation in this form with the one expressed with negative signs in front of the dispersive and nonlinear coefficients. Employing the transformation $A = -\tilde{A}^*$, where the $*$ denotes the complex conjugate, we obtain a similar NLS equation for $\tilde{A}$, with negative signs in front of the dispersive and nonlinear coefficients, and yet these varying coefficients are identical with ones in (5), given as follows:

$$i\partial_x\left(\sqrt{c_g}\tilde{A}\right) - \alpha(x)\partial_\tau^2\left(\sqrt{c_g}\tilde{A}\right) - c_g(x)^{-1}\beta(x)\left|\sqrt{c_g}\tilde{A}\right|^2\left(\sqrt{c_g}\tilde{A}\right) = 0. \tag{6}$$

The difference in the solutions for the NLS equation (5) and (6) is only in the phase shift. Indeed, if $A(x,\tau) = a(x,\tau)e^{i\phi(x,\tau)}$ is a solution of the NLS equation (5) in its phase-amplitude form, then $\tilde{A}(x,\tau) = a(x,\tau)e^{i[\pi-\phi(x,\tau)]}$ is a solution for the NLS equation (6).

3. On dispersive and nonlinear coefficients

In the following, we adopt the spatial inhomogeneous NLS equation with positive signs of dispersive and nonlinear coefficients and hence both coefficients carry the negative sign in front of them. The original dispersive coefficient derived in [1] reads

$$\alpha(x) = \frac{-1}{2\omega c_g}\left(1 - \frac{gh}{c_g^2}(1-kh\sigma)(1-\sigma^2)\right), \qquad \text{where} \qquad \sigma = \tanh kh. \tag{7}$$

It can be shown that the normalized dispersive coefficient $\tilde{\alpha}(x)$ can be written as follows, which is in agreement with the expression (3), presented in [5] and [6], where the latter one should include the negative sign:

$$\tilde{\alpha}(x) = \frac{\alpha(x)}{g} = \frac{-1}{2\tilde{\omega}c_g}\left(1 - \frac{H}{c_g^2} + \frac{2\tilde{\omega}H\tanh\tilde{k}H}{c_g}\right). \tag{8}$$

Here, the physical water depth is scaled to normalized depth $H(x)$ using the relation $gh(x) = H(x)$. It follows that the normalized wave frequency $\tilde{\omega}$ and the normalized wavenumber $\tilde{k}$ are given by the following relations: $\omega^2 = g^2\tilde{\omega}^2$ and $k = g\tilde{k}$, respectively. The plot of the dispersive coefficient $\alpha(x)$ is displayed as the red curve on the left panel of Figure 1.

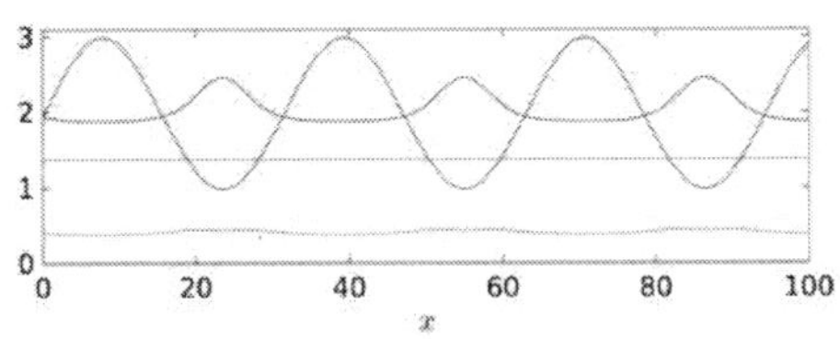

Figure 2. Plots of wave frequency $\tilde{\omega} = 1.363 = kh$, depicted as a red constant line, wavenumber $k(x)$ (black), the product of wavenumber and the water depth kh (blue) and the group velocity c_g (magenta).

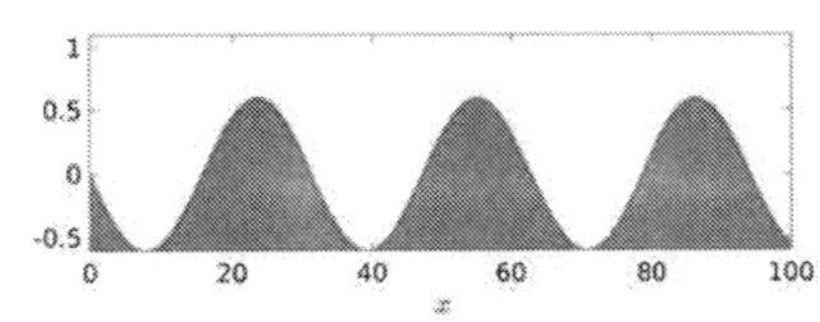

Figure 3. Sketch of the bottom topography corresponds to the water depth $H(x) = 1 + 0.6\sin(x/5)$.

It is noted in [3, 4] that there exists a tiny typographical error in the nonlinear coefficient given in [1], the expression of the final term should be $(1-\sigma^2)^2$ instead of $(1-\sigma)^2$. The original expression of the accurate nonlinear coefficient reads, where the phase velocity $c_p = \omega/k$:

$$\beta(x) = \frac{-k^4}{4\omega\sigma^2 c_g}\left[9 - 10\sigma^2 + 9\sigma^4 - \frac{2\sigma^2 c_g^2}{gh - c_g^2}\left\{4\left(\frac{c_p}{c_g}\right)^2 + 4\frac{c_p}{c_g}(1-\sigma^2) + \frac{gh}{c_g^2}(1-\sigma^2)^2\right\}\right]. \tag{9}$$

After some manipulation, the normalized nonlinear coefficient $\tilde{\beta}$ can be expressed as follows:

$$\tilde{\beta}(x) = \frac{\beta(x)}{g^3} = \frac{-1}{4\tilde{\omega}^5 c_g}\left[9\tilde{k}^6 - 12\tilde{k}^4\tilde{\omega}^4 + 13\tilde{k}^2\tilde{\omega}^8 - 2\tilde{\omega}^{12}\right] + \frac{1}{2\tilde{\omega}c_g}\frac{\left[2\tilde{k}\tilde{\omega} + \tilde{k}^2 c_g \operatorname{sech}^2\tilde{k}H\right]^2}{H - c_g^2}. \tag{10}$$

We observe that this nonlinear coefficient exhibits a different expression with the one presented and used in [5] and [6]. Compare with (4). The plot of accurate nonlinear coefficient is displayed as a blue curve on the left panel of Figure 1. The comparison between an accurate $\beta(x)$ and an inaccurate one for a particular water depth is displayed on the right panel of Figure 1. The left panel of Figure 1 also shows the plot of the product of the dispersive and the nonlinear coefficients

$\alpha(x)\beta(x)$, indicating that the NLS equation alternates between focusing and defocusing case along the spatial evolution parameter x. In the following, we show qualitatively that the nonlinear coefficient presented in [5] and [6] is indeed inaccurate.

Note that the nonlinear coefficient β changes sign when $kh \approx 1.363$ [7]. We observe in Figure 2 that for $kh > 1.363$, the carrier wave is unstable with respect to modulation and hence bright, single-soliton solution is formed. On the other hand, for $kh < 1.363$, the NLS equation does not possess bright soliton solution, but dark soliton instead since the carrier wave is modulationally stable. For this example, the scaled water depth is given by $H(x) = 1 + 0.6\sin(x/5)$ and the bottom topography is depicted in Figure 3.

The change of signs for the nonlinear coefficient corresponds to the change of the value in kh, from $kh > 1.363$ to $kh < 1.363$, or vice versa. Figure 4 shows zoom-in plots of kh and the scaled nonlinear coefficient β for the wave frequency $\tilde{\omega} = 1.363$. The left panel displays the accurate nonlinear coefficient derived in [1] and corrected in [3, 4] while the right panel displays inaccurate nonlinear coefficient given in [5] and [6] by referring to [1], with no given derivation. For both cases, the corresponding water depth is $H(x) = 1 + 0.6\sin(x/5)$. In the left panel, we observe that when $kh > 1.363$, the nonlinear coefficient $\beta < 0$, and since $\alpha < 0$, this gives a focusing case for the NLS equation and the wave train is modulationally unstable. When $kh < 1.363$, the nonlinear coefficient $\beta > 0$ and since $\alpha > 0$, the NLS equation is defocusing type and the wave train is modulationally stable. The situation does not occur for an inaccurate nonlinear coefficient, as can be observed in the right panel of Figure 4.

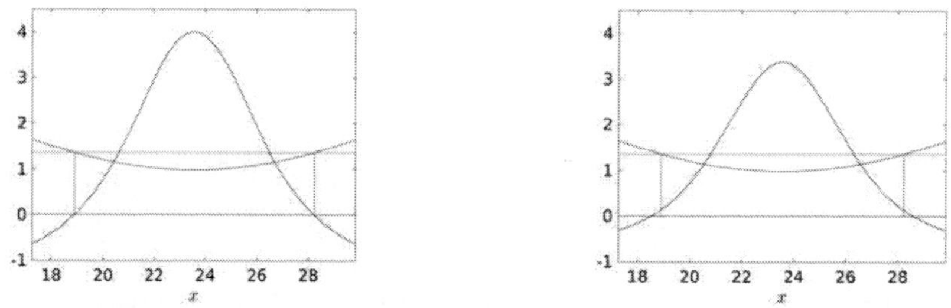

Figure 4. (Left panel) Plots of wave frequency $\tilde{\omega} = 1.363 = kh$, depicted as a red constant line, the product of wavenumber and the water depth kh (blue) and the nonlinear coefficient β (black), after some scaling. (Right panel) Similar to the left panel except for the inaccurate nonlinear coefficient β stated in [5]. We observed that when $kh \approx 1.363$, it does not correspond to the change of sign in the nonlinear coefficient. All cases correspond to the water depth $H(x) = 1 + 0.6\sin(x/5)$.

4. On averaging coefficients

For a particular choice of initial wavenumber $k(x = 0) = 2$, which corresponds to the wave frequency $\omega = 1.3885$, and the the water depth $H(x) = 1 + 0.6\sin(x/5)$, the NLS equation alternates between focusing and defocusing case along the spatial evolution parameter x. This is due to the fact that the product of the dispersive and the nonlinear coefficients $\alpha(x)\beta(x)$ is alternating between positive and negative along the horizontal position x. Hence, the formation of only bright, single-soliton solution is not admitted. Nonetheless, both asymptotic and numerical solutions can be found and a good comparison of the relative amplitude of the wave packet between the two results is presented. See Figure 2(a) in [6]. The asymptotic solution is presented in term of averaging value of the dispersive and nonlinear coefficients as well as the initial condition for the numerical computation. The averaging of a function $f(x)$ with respect to x over one period of topography T is defined as follows:

$$\langle f \rangle = \frac{1}{T}\int_0^T f(x)\,dx. \tag{11}$$

Using this definition, the averaging value of the dispersion and nonlinear coefficients are readily calculated. For the choice of wave frequency, initial wavenumber and the water depth mentioned above, the period of topography $T = 10\pi$ and we obtain $\langle\alpha\rangle = 1.0868$ and for the wave amplitude $\eta = 0.1$, the scale of dispersion $L_d = 92.0158$. However, a disagreement is found for the scale of nonlinearity L_n. Our calculation shows $L_n = 276.026$ while in [6], $L_n = 114.1$. This is due to the difference of the averaging value for the nonlinear coefficient $c_g^{-1}\beta$. Our calculation finds $\langle c_g^{-1}\beta\rangle = 2.2164$ using the inaccurate β while in [6], $\langle c_g^{-1}\beta\rangle = 11.627$. Using the accurate value of β, the averaging value $\langle c_g^{-1}\beta\rangle = -22.4010$. A comparison of the averaging values for several wave frequencies and initial wavenumbers are displayed in Table 1.

Table 1. A comparison of the averaging values of the nonlinear coefficient β/c_g. The signs in $\langle c_g^{-1}\beta\rangle$ from [6] and the calculation using inaccurate β from [6] are not adjusted (the third and the fourth columns). The sign in $\langle c_g^{-1}\beta\rangle$ has been corrected (the fifth column).

Initial wavenumber	Wave frequency	$\langle c_g^{-1}\beta\rangle$ from [6]	$\langle c_g^{-1}\beta\rangle$ (inaccurate β)	$\langle c_g^{-1}\beta\rangle$ (accurate β)
$k(0) = 2.0$	$\omega = 1.3885$	11.627	2.2164	−22.4010
$k(0) = 1.9$	$\omega = 1.3479$	3.720	−6.2451	−6.7465
$k(0) = 1.8$	$\omega = 1.3055$	−2.906	−13.4684	4.9603

5. Conclusion

We revisited the spatial nonlinear Schrödinger equation with non-constant coefficients as a nonlinear evolution equation for wave packet of surface gravity waves when it propagates over an uneven bottom, i.e. a topography with some variation. We have rewritten the dispersive and nonlinear coefficients in the form that easier to compare with the ones found in the literature. We have shown qualitatively using the change of values in kh and the change in its sign that the nonlinear coefficient used in [5] and [6] are inaccurate. A similar pattern of this inaccuracy is also carried along in the averaging value computation of the nonlinear coefficient used for an asymptotic solution of the transformed NLS equation. For a particular choice of wave characteristics and water depth, the NLS equation alternates between focusing and defocusing case along the spatial evolution parameter. Hence, the formation of the classical soliton, either bright or dark, is not admitted for this case.

Acknowledgments

The authors would like to thank Andy Chan, Siew Ling Ng and Margaret Rozario (The University of Nottingham Malaysia Campus), Teo Lee Peng (Xiamen University Malaysia Campus), Juan-Ming Yuan (Providence University, Taiwan), Mason Porter (UCLA) and Brenny van Groesen (Universiteit Twente, The Netherlands and LabMath Indonesia) for assistance, fruitful discussions and suggestions. The New Researcher Fund from The University of Nottingham University Park UK Campus and Malaysia Campus NRF 5035-A2RL20, Samsung Intramural Research Fund Project 2016-1299-000 and National Research Foundation of Korea Fund NRF-2017017743 are gratefully acknowledged.

References

[1] Djordjević V D and Redekopp L G 1978 *J. Appl. Math. Phys. (ZAMP)* **29** 950
[2] Debnath L 1994 *Nonlinear Water Waves* San Diego, CA: Academic Press
[3] Dingemans M W 1997 *Water Wave Propagation Over Uneven Bottoms* (Singapore: World Scientific)
[4] Dingemans M W and Otta A K 2001 *Advances in Coastal and Ocean Engineering* ed Liu P L-F (Singapore: World Scientific)
[5] Benilov E S, Flanagan J D and Howlin C P 2005 *J. Fluid Mech.* **533** 171
[6] Benilov E S and Howlin C P 2006 *Stud. Appl. Math.* **116** 289
[7] Benjamin T B and Feir J 1967 *J. Fluid Mech.* **27** 417.

IOP Conf. Series: Journal of Physics: Conf. Series **1039** (2018) 012002 doi:10.1088/1742-6596/1039/1/012002

Numerical solution of Benjamin-Bona-Mahony-Burger (BBMB) and regularized long-wave (RLW) equations using time-space pseudo-spectral method

A K Mittal[1], L K Balyan[1] and D Tiger [2]
[1] Discipline of Mathematics, IIITDM Jabalpur, Madhya Pradesh 482005, India.
[2] Department of Mathematics, Rajdhani College, University of Delhi 110021, India.

Email: balyan@iiitdmj.ac.in (L. K. Balyan)

Abstract. In this research article, an improved numerical scheme is developed for solving the non-linear Benjamin-Bona-Mahony-Burger (BBMB) and regularized long-wave (RLW) equations. The time-space pseudo-spectral method is developed using Chebyshev polynomials at the CGL points. The algebraic mapping is used to transform the initial-boundary value non-homogeneous problem into the homogeneous problem and further, the BBMB and RLW equations are reduced to a system of nonlinear equations. To demonstrate the performance, the method applied to two model problems and approximate results have been compared with the exact solution and other existing methods. The proposed method succeeds to obtain high order accuracy.

1. Introduction

Existences of nonlinear terms in applied mathematical physics problems are obvious. Therefore, accurate and stable numerical solutions of the nonlinear equations are important. In this paper, a high order pseudo-spectral method in time and space will be employed on BBMB and RLW equations.

Let us consider the nonlinear generalized BBMB equation in the finite space and time intervals:

$$V_t - V_{xxt} - \alpha V_{xx} + \beta\, V_x + V V_x = F(x,t), \qquad x \in [a,b], \qquad 0 \le t \le T, \tag{1}$$

with initial condition:

$$V(x,0) = V_0(x), \qquad x \in [a,b],$$

and boundary conditions:

$$V(a,t) = g(t), \qquad V(b,t) = h(t), \qquad 0 \le t \le T,$$

where $V(x,t)$ represents the fluid velocity. VV_x, nonlinear part of Eq (1), represents convection term, αV_{xx} is a dissipation term and V_{xxt} is a dispersive term. For $\alpha = 0$ and $\beta = 1$, the Eq (1) is called regularized long wave (RLW) equation. The RLW equation was first time introduced by Peregrine[1] to understand the behaviour of undular bore. For $\alpha = \beta = 1$, the Eq (1) is known as the BBMB equation and investigated by Benjamin et al.[2] in 1972. There are various extensive numerical methods to solve RLW and generalized BBMB equations, such as exp-function method[3], finite difference solution[1, 4, 5] and B-splines method[6, 7]. Kamel Al-Khaled et al.[8] obtained solitary-wave solutions of BBM-Burger equation by decomposition method. Salas et al.[9] have studied BBMB equations using the generalization of the tanh-coth method and obtained traveling wave solutions. Qinghua and Zhengzheng[10] have studied the convergence rate of the solutions of the generalized BBMB equations based upon time-space weighted energy method.

The rest of the paper is organized as follows. In section 2, a brief description of pseudo-spectral method in time and space will be discussed. In section 3, numerical results and discussion are described. Finally, the conclusion of the paper is given in section 4.

IOP Conf. Series: Journal of Physics: Conf. Series **1039** (2018) 012002 doi:10.1088/1742-6596/1039/1/012002

2. Time-space pseudo-spectral method and Discretization

We know the working domain of Chebyshev polynomials are [-1, 1], therefore, transform the problem into interval [-1, 1], using

$$x = \frac{b-a}{2}x + \frac{b+a}{2}, \quad t = \frac{T}{2}t + \frac{T}{2}.$$

Accordingly, equation (1) is transformed as follows:

$$V_t - CV_{xxt} - \alpha BV_{xx} + A\beta\, V_x + AVV_x = DF(x,t), \qquad x \in [-1,1], \qquad t \in [-1,1], \tag{2}$$

where $A = \frac{T}{b-a}$, $B = \frac{2T}{(b-a)^2}$, $C = \frac{2}{(b-a)^2}$ and $D = \frac{T}{2}$,

with initial condition:

$$V(x,-1) = \varphi(x), \qquad x \in [-1,1], \tag{3}$$

and boundary conditions:

$$V(-1,t) = \varpi_1(t), \qquad V(1,t) = \varpi_2(t), \qquad t \in [-1,1]. \tag{4}$$

The problem is non-homogeneous therefore, we consider a mapping for converting the problem from non-homogeneous to homogeneous.

$$\xi(x,t) = \frac{1-t}{2}\varphi(x) + \frac{1-x}{2}\varpi_1(t) + \frac{1+x}{2}\varpi_2(t) - \frac{(1-t)(1-x)}{4}\varpi_1(-1) - \frac{(1-t)(1+x)}{4}\varpi_2(-1). \tag{5}$$

At corner points initial and boundary conditions satisfy, $\varphi(-1) = \varpi_1(-1)$ and $\varphi(1) = \varpi_2(-1)$.

Define new variable $U(x,t)$,

$$V(x,t) = U(x,t) + \xi(x,t), \tag{6}$$

then above equation (2), is reduced as follows:

$$(U+\xi)_t - C(U+\xi)_{xxt} - \alpha B(U+\xi)_{xx} + A\beta\,(U+\xi)_x + A(U+\xi)(U+\xi)_x = DF(x,t). \tag{7}$$

Now define the residual $R(x,t)$,

$$R(x,t) = (U+\xi)_t - C(U+\xi)_{xxt} - \alpha B(U+\xi)_{xx} + A\beta\,(U+\xi)_x + A(U+\xi)(U+\xi)_x - DF(x,t). \tag{8}$$

We seek spectral approximation $U^a(x,t)$ of the form:

$$U^a(x,t) = \sum_{i=0}^{N}\sum_{j=0}^{N} \Phi_i(x)\Psi_j(t)U_{ij}^a, \tag{9}$$

where the functions $\Phi_i(x)$ and $\Psi_j(t)$ are the Nth degree Lagrange polynomials with space and time variables, respectively.

Approximation of partial derivatives at collocation points (x_p, t_q) are defined as

$$\frac{\partial^r}{\partial x^r}\left(U^a(x_p,t_q)\right) = \sum_{i=0}^{N}\sum_{j=0}^{N}\Phi_i^{(r)}(x_p)\Psi_j(t_q)U_{ij}^a = \sum_{i=0}^{N}\sum_{j=0}^{N}\Delta_{ip}^{(r)}\Psi_j(t_q)U_{ij}^a,$$

$$= \left(\left[\Delta_{[0:N,p]}^{(r)}\right] \otimes \Psi(t_q)\right)^T U^a, \tag{10}$$

and

$$\frac{\partial^s}{\partial t^s}\left(U^a(x_p,t_q)\right) = \sum_{i=0}^{N}\sum_{j=0}^{N}\Phi_i(x_p)\Psi_j^{(s)}(t_q)\,U_{ij}^a = \sum_{i=0}^{N}\sum_{j=0}^{N}\Phi_i(x_p)\Delta_{jq}^{(s)}\,U_{ij}^a,$$

$$= \left(\Phi(x_p) \otimes \left[\Delta_{[0:N,q]}^{(s)}\right]\right)^T U^a, \tag{11}$$

where

$$\Phi(x_p)=\left[\Phi_0(x_p),\ \Phi_1(x_p),\dots\dots\dots,\Phi_N(x_p)\right]^T,\quad \Psi(t_q)=\left[\Psi_0(t_q),\ \Psi_1(t_q),\dots\dots\dots,\Psi_N(t_q)\right]^T,$$

and $\otimes$ represent the kronecker product of two vectors or matrices. U^a is a vector defined by:

$$U^a=\left[U^a_{00},\dots,U_{0N,}{}^aU^a_{10},\dots,U^a_{1N},\dots,\ U_{N0}{}^a,\dots,U^a_{NN}\right]^T,$$

and $\Delta^{(r)}$ is defined r^{th} –order Chebyshev pseudo-spectral differentiation matrix. From the initial boundary conditions, the values of $\Psi_0(t)$, $\Phi_0(x)$ and $\Phi_N(x)$are zero at these CGL points. Therefore using equations (10) and (11), redefine the residual at CGL points in space and time.

$$R(x_i,t_j)=\theta_1\left[\left(I_{N-1}\otimes\Delta^1_t\right)-C\left(\Delta^2_x\otimes\Delta^1_t\right)-\alpha B\left(\Delta^2_x\otimes I_N\right)+A\left(\Delta^1_x\otimes I_N\right)\left(\beta+\xi^{ij}\right)+AI_{(N*(N-1))}\xi^{ij}_x\right]U$$
$$+\frac{\theta_1 A}{2}\left(\Delta^1_x\otimes I_N\right)U^2-DF_{ij}+\xi^{ij}_t-C\xi^{ij}_{xxt}-\alpha B\xi^{ij}_{xx}+\beta A\xi^{ij}_x+\frac{A}{2}\xi^{2,ij}_x,$$

where $\Delta^{(1)}_t=\Delta_{[1:N,1:N]}$, $\Delta^{(1)}_x=\Delta_{[1:N-1,1:N-1]}$, $\Delta^{(2)}_x=\Delta^{(2)}_{[1:N-1,1:N-1]}$,

$\theta_1=\left(e^T_{((i-1)*N)+j}\right)$ and e^T_N is a $N(N-1)$ vectors, i.e. N^{th} element is 1 and rest elements are 0. Put the residual equal to zero which leads the following system of nonlinear algebraic equation:

$$\Gamma_1U+\Gamma_2U^2=\Gamma_3, \tag{12}$$

where,

$$\Gamma_1=\left[\left(I_{N-1}\otimes\Delta^1_t\right)-C\left(\Delta^2_x\otimes\Delta^1_t\right)-\alpha B\left(\Delta^2_x\otimes I_N\right)+A\left(\Delta^1_x\otimes I_N\right)\left(\beta+I_\xi\right)+AI_{(N*(N-1))}I_{\xi_x}\right],$$

$$\Gamma_2=\frac{A}{2}\left(\Delta^1_x\otimes I_N\right)\text{ and }\Gamma_3=DF-G_1+CG_2+\alpha BG_3-\beta AG_4-\frac{A}{2}G_5,$$

$$U=\left[U_{11},\dots,U_{1N}\left|\cdots\right|U_{(N-1)1},\dots,U_{(N-1)N}\right]^T,\quad F=\left[F_{11},\dots,F_{1N}\left|\cdots\right|F_{(N-1)1},\dots,F_{(N-1)N}\right]^T,$$

$$G_1=\left[\xi^{11}_t,\dots,\xi^{1N}_t\left|\cdots\right|\xi^{(N-1)1}_t,\dots,\xi^{(N-1)N}_t\right]^T,\quad G_2=\left[\xi^{11}_{xxt},\dots,\xi^{1N}_{xxt}\left|\cdots\right|\xi^{(N-1)1}_{xxt},\dots,\xi^{(N-1)N}_{xxt}\right]^T,$$

$$G_3=\left[\xi^{11}_{xx},\dots,\xi^{1N}_{xx}\left|\cdots\right|\xi^{(N-1)1}_{xx},\dots,\xi^{(N-1)N}_{xx}\right]^T,\quad G_4=\left[\xi^{11}_x,\dots,\xi^{1N}_x\left|\cdots\right|\xi^{(N-1)1}_x,\dots,\xi^{(N-1)N}_x\right]^T,$$

$$G_5=\left[\xi^{2,11}_x,\dots,\xi^{2,1N}_x\left|\cdots\right|\xi^{2,(N-1)1}_x,\dots,\xi^{2,(N-1)N}_x\right]^T.$$

The system of nonlinear equation (12) can be solved by using Newton Raphson method.

3. Numerical results and discussion

In this section, we consider numerical examples to obtain approximate solutions using pseudo-spectral method in time and space and demonstrate the sup norm and relative errors norm, which are defined by: $L_\infty=\left\|U^a-U\right\|_\infty=Sup_{(p,q)}\left|U^a(x_p,t_q)-U(x_p,t_q)\right|$

and

$$L_2=\left(\frac{\sum_{p=0}^N\sum_{q=0}^N\left[U^a(x_p,t_q)-U(x_p,t_q)\right]^2}{\sum_{p=0}^N\sum_{q=0}^N\left[U(x_p,t_q)\right]^2}\right)^{\frac{1}{2}},$$

where U^a and U are the spectral approximations and analytical solutions, respectively.

3.1. Example 1

In this example, consider RLW equation, i.e. $\alpha=0$ and $\beta=1$ defined by Eq (1) with $F=0$.

Initial condition: $V(x,0)=3c\,\mathrm{sech}^2\left(k(x-x_0)\right),\quad x\in[a,b],$ (13)

and boundary conditions:

$$V(a,t)=3c\,\mathrm{sech}^2\left(k\left(x-va-x_0\right)\right),\quad V(b,t)=3c\,\mathrm{sech}^2\left(k\left(x-vb-x_0\right)\right),\quad 0\le t\le T, \tag{14}$$

the exact solution of this problem is [11]:

$$V(x,t)=3c\,\mathrm{sech}^2\left(k\left(x-vx-x_0\right)\right). \tag{15}$$

Here wave velocity $v=1+c$ and $k=\sqrt{\dfrac{c}{4v}}$, where c denotes solitary wave amplitude. For computation purpose, we consider the space interval $[a,b]=[-40,60]$, $x_0=0$. The results for $c=0.1$ and $c=0.03$ at a different time $T=1,5,10$ and 20 have shown in Figure 1 and 2. The exact solutions have shown by solid line and numerical solutions by blocks, which have a very good agreement with exact solutions. Figure 3 contains 3D surface plots for the numerical solution with different wave amplitude at $T=20$. The tabulated results have presented for the proposed method and other existing methods[5, 11, 12, 13] in Table 1 and 2 with different c.

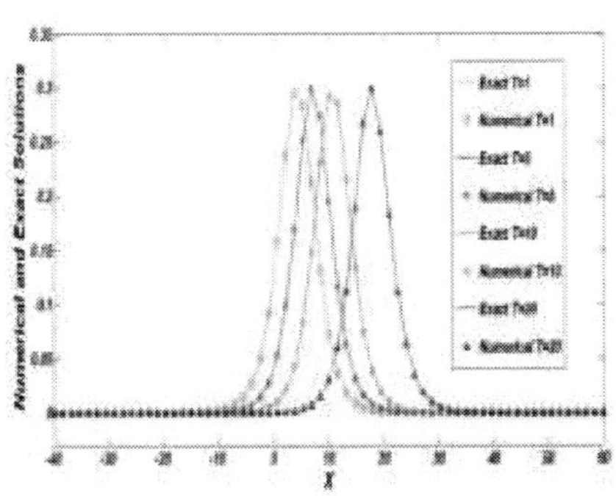

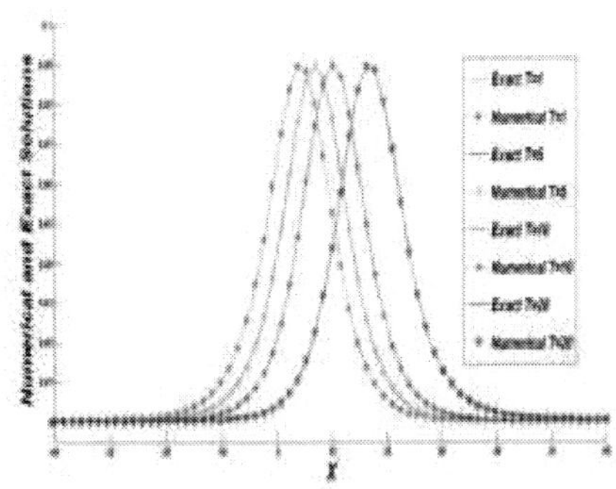

Figure 1: Exact and numerical solutions at c = 0.1. **Figure 2**: Exact and numerical solutions at c =.03

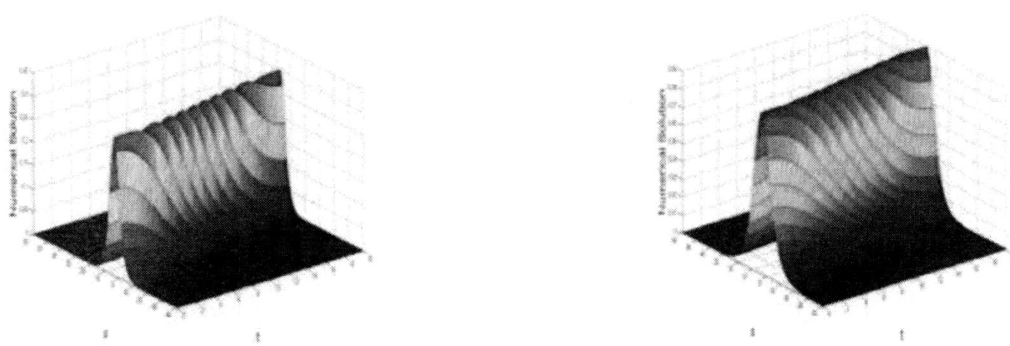

(a) (b)

Figure 3: Numerical solutions for (a) c = 0.1 and (b) c = 0.03 at $T=20$.

Table 1: Comparison of proposed method with existing method at c = 0.1.

T	Errors	Proposed Method	[12]	[11]	[5]	[13]
4	L_2	3.634e-05	3.982e-02	1.485e-05	1.2e-04	6.464e-05
	L_∞	1.314e-05	1.374e-02	5.956e-06	5.0e-05	2.496e-05
8	L_2	9.234e-04	7.946e-02	2.706e-05	2.3e-04	1.282e-04
	L_∞	4.665e-04	2.766e-02	1.051e-05	9.0e-05	5.049e-05
12	L_2	1.229e-04	1.188e-01	3.640e-05	3.4e-04	1.900e-04
	L_∞	4.141e-05	4.135e-02	1.318e-05	1.4e-04	7.470e-05

16	L_2	1.346e-04	1.577e-01	4.294e-05	4.5e-04	2.497e-04
	L_∞	1.220e-04	5.460e-02	1.448e-05	1.8e-04	9.709e-04
20	L_2	4.312e-04	1.961e-01	4.757e-05	5.5e-04	3.071e-04
	L_∞	4.237e-04	1.769e-01	1.497e-05	2.1e-04	1.177e-04

Table 2: Comparison of proposed method with existing method at c = 0.03.

T	Errors	Proposed Method	[12]	[11]	[13]
4	L_2	5.986e-05	2.928e-03	4.120e-04	4.097e-04
	L_∞	1.313e-05	7.86e-04	2.299e-04	2.299e-04
8	L_2	7.596e-05	5.816e-03	5.112e-04	5.109e-04
	L_∞	3.464e-05	1.582e-03	2.210e-04	2.210e-04
12	L_2	6.641e-04	8.698e-03	5.356e-04	5.359e-04
	L_∞	3.186e-04	2.384e-03	2.125e-04	2.125e-04
16	L_2	1.456e-04	1.158e-02	5.429e-04	5.446e-04
	L_∞	1.219e-04	3.190e-03	2.139e-04	2.120e-04
20	L_2	3.935e-04	1.445e-02	5.624e-04	5.684e-04
	L_∞	3.339e-04	3.996e-03	4.315e-04	4.278e-04

3.2. Example 2

Let us consider the BBMB equation i.e. $\alpha = 1$ and $\beta = 1$ defined by Eq (1) with

$$F(x,t) = \exp(-t)\left[\cos x - \sin x + \frac{1}{2}\exp(-t)\sin 2x\right].$$

Initial condition:

$$V(x,0) = \sin x, \qquad x \in [a,b], \tag{16}$$

and boundary conditions:

$$V(a,t) = \exp(-t)\sin a, \quad V(a,t) = \exp(-t)\sin b, \quad 0 \le t \le T, \tag{17}$$

the exact solution of this problem is[10, 14].

$$V(x,t) = \exp(-t)\sin x. \tag{18}$$

Figure 4 has shown a plot of the numerical solution by dots and exact solution by solid lines at different times intervals, namely $T = 3.0, 4.0, 5.0, 6.0$ with space interval $[a,b] = [0,\pi]$. Figure 5 contains 3D surface plots for the numerical solution at the time $T = 6$, which also depict the exactness of method. The computation results of the proposed method with existing methods [14, 15] at the time $T = 10$ have presented in Table 3, which clearly shows that the proposed method has achieved higher accuracy with minimum numbers of grid points as compared to other existing methods.

Table 3: Comparison of proposed method with existing method at time $T = 10$.

N	Proposed Method		N	[14]	[15]
	L_2	L_∞		L_2	L_∞
08	5.3251e-05	3.2597e-05	10	1.7147e- 04	2.1800e- 02
16	4.0646e-05	9.0876e-06	20	5.6341e- 05	5.3000e- 03
64	4.1282e-06	1.1177e-06	80	7.2635e- 06	3.3291e- 04
128	5.2201e-07	2.7547e-07	320	8.1631e- 07	2.0766e- 05

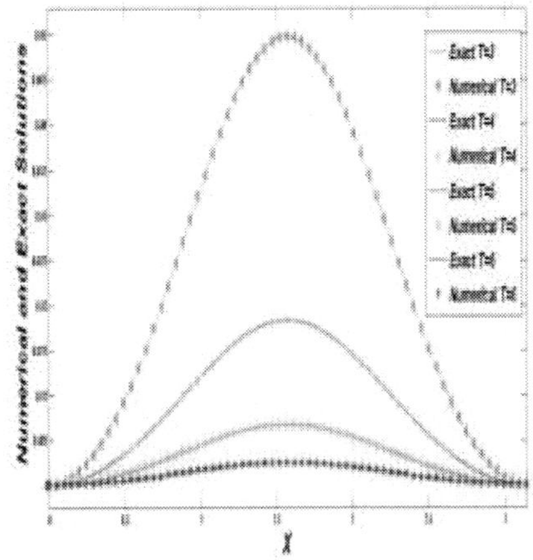

Figure 4: Numerical solutions of example 2.

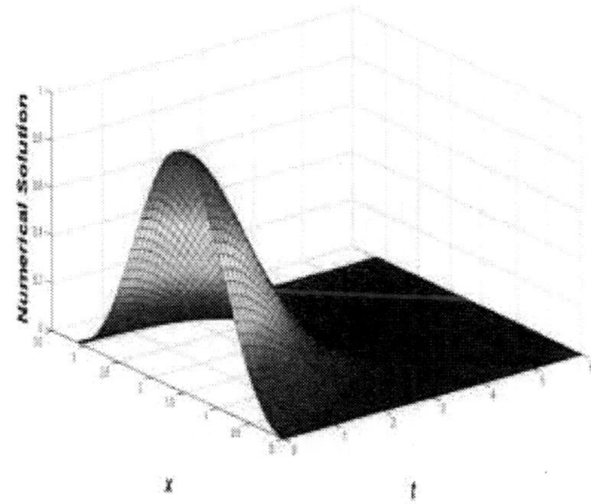

Figure 5: Numerical solutions of example 2 at $T = 6$.

4. Conclusion

In this research paper, the numerical results of non-linear Benjamin-Bona-Mahony-Burger (BBMB) and regularized long-wave (RLW) equations have obtained using time-space pseudo-spectral method with Chebyshev polynomial as basis functions and achieved spectral accuracy. The performance of the proposed method has been demonstrated by considering two examples. The computed results have shown very good agreement with the exact solutions and compete with existing results.

References

[1] Peregrine D 1966 *Journal of Fluid Mechanics*, **25**(2) pp 321- 330.
[2] Benjamin T B, Bona J L and Mahony J J 1972 *Philosophical Transactions for the Royal Society of London. Series A, Mathematical and Physical Sciences*, pp 47-78.
[3] Ganji Z, Ganji D and Bararnia H 2009 *Applied Mathematical Modelling*, **33**(4) pp 1836-1841.
[4] Eilbeck J and Mcguire G 1977 *Journal of Computational Physics*, **23**(1) pp 63-73.
[5] Kutluay S and Esen A 2006 *Mathematical Problems in Engineering*, **2006** pp 1-14.
[6] Dag I, Dogan A and Saka B 2003 *International Journal of Computer Mathematics*, **80**(6) pp 743-757.
[7] Tirmizi I A 2010 *Applied Mathematical Modelling*, **34**(12) pp 4151-4160.
[8] Al-Khaled K, Momani S and Alawneh A 2005 *Applied Mathematics and Computation*, **171**(1) pp 281-292.
[9] Salas A H and Frias B A 2010 *Applied Mathematics and Computation*, **217**(4) pp 1430-1434.
[10] Qinghua X and Zhengzheng C 2012 *Acta Mathematica Scientia*, **32**(5) pp 1743-1758.
[11] Zarebnia M and Parvaz R 2016 *Applied Mathematics and Computation*, **284** pp 79-88.
[12] Gardner L, Gardner G and Dag I 1995 *International Journal for Numerical Methods in Biomedical Engineering*, **11**(1) pp 59-68.
[13] Zarebnia M and Parvaz R 2015 *Boletim da Sociedade Paranaense de Matematica*, **35**(1) pp 127-138.
[14] Arora G, Mittal R C and Singh B K 2014 *Journal of Engineering Science and Technology, Special*, **18**(1) pp 104-116.
[15] Omrani K and Ayadi M 2008 *Numerical Methods for Partial Differential Equations*, **24**(1) pp 239-248.

Acknowledgments

The first author thankfully acknowledges the financial support provided by Ministry of Human Resource Development, India.

IOP Conf. Series: Journal of Physics: Conf. Series **1039** (2018) 012003 doi:10.1088/1742-6596/1039/1/012003

Multiple positive solutions to a class of second-order boundary value problems involving integral conditions

Bo Sun and Guiyang Jiang

School of Statistics and Mathematics, Central University of Finance and Economics, Beijing 100081, PR China.

Email: sunbo19830328@163.com

Abstract. We focus on the research of positive solutions to integral boundary value problems. By applying a fixed point theorem due to Avery and Peterson, we study the existence of at least three positive solutions to a class of second order boundary value problems with integral boundary conditions.

1. Introduction

In this paper, we will consider the positive solutions for the second-order boundary value problem with integral boundary conditions

$$u''(x)+f(x,u(x),u'(x))=0, \quad 1\le x\le 1 \tag{1.1}$$

subject to the following boundary condition:

$$u(0)=u(1)=\int_0^1 p(s)u(s)ds \tag{1.2}$$

where $p\in L^1[0,1]$, by using the fixed point theorem due to Avery and Peterson, we prove the existence of three positive solutions.

The research into the theory of boundary value problems with integral boundary conditions for ordinary differential equations has been always a very active subject in different areas of both applied mathematics and physics. Such as blood flow problems, chemical engineering, thermo elasticity, underground water flow, and population dynamics. For a detailed description of the integral boundary conditions, we refer the reader to some recent papers ([1–6]) and the references therein.

Recently, many authors have widely studied the existence of multiple solutions to the third-order boundary value problems involving ordinary differential equations. A few years ago, Guo and Ge in [6] consider the existence of positive solutions for the second order three-point boundary value problem

$$\begin{cases} x''+f(t,x,x')=0, 0<t<1, \\ x(0)=0, u(1)=\alpha u(\eta). \end{cases} \tag{1.3}$$

They obtained the existence of at least one positive solution for (1.3).

So, motivated by all the above works, this paper studies the following system

$$\begin{cases} u''(x)+f(x,u(x),u'(x)=0, 0<x<1 \\ u(0)=u(1)=\int_0^1 p(s)u(s)ds. \end{cases} \tag{1.4}$$

The specific difference between the paper above is that we use a fixed point theorem due to Avery and Peterson and the integral boundary condition.

2. Preliminaries

Definition 2.1 The map α is said to be a nonnegative, continuous, concave functional on a cone P of a real Banach space E, if $\alpha: P \to [0,\infty]$ is continuous, and $\alpha(tu+(1-t)v) \ge t(\alpha)+(1-t)\alpha(v)$, for all $u,v \in P$ and $t \in [0,1]$. Similarly, we say the map β is a nonnegative, continuous, convex functional on a cone P of a real Banach space E, if $\beta: P \to [0,\infty]$ is continuous, and $\beta(tu+(1-t)v) \ge t(\beta)+(1-t)\beta(v)$, for all $u,v \in P$ and t∈[0,1].

We consider the Banach space $E = C^1[0,1]$ equipped with the norm $\|u\| = \max\{\|u\|_{\infty,} \|u'\|_{\infty}\}$, where $\|u\|_{\infty} = \max\limits_{0\le x\le 1}|u(x)|$. Throughout this paper, we will suppose that the following assumptions hold

(H_1): $f \in C([0,1]\times[0,+\infty)\times R,[0,+\infty))$, and $f(x,u,v)>0$ for all $(x,u,v)\in[0,1]\times[0,+\infty)\times R$;

(H_2): $p \in L^1[0,1]$ is nonnegative and $0 \le \mu < 1$, where $\mu = \int_0^1 p(s)ds$.

We denote $C^+[0,1] = \{u \in E: u(x) \ge 0, x\in[0,1]\}$, $P = \{u \in E: u(x)\ge 0 \text{ is concave and } x\in[0,1]\}$. It is obvious to see that P is a cone in E.

For any $y \in C^+[0,1]$, suppose that u is a solution of the following BVP

$$\begin{cases} u''(x)+q(x)f(x,y(x),y'(x)=0, 0<x<1 \\ u(0)=u(1)=\int_0^1 p(s)u(s)ds. \end{cases}$$

Then we can easily get the solution $u(x) = \int_0^1 H(x,t)q(t)f(t,y(t),y'(t))dt$,

Where $H(x,t) = G(x,t)+\frac{1}{1-\mu}\int_0^1 G(t,s)p(s)ds,\quad G(t,s)=\begin{cases} t(1-s), 0\le t\le s\le 1 \\ s(1-t), 0\le s\le t\le 1 \end{cases},\quad \mu=\int_0^1 p(s)ds$.

Lemma 2.1 If (H_2) is satisfied, the following results are true

(1) $H(x,t)\ge 0$, for $x,t\in[0,1]$; $H(x,t)>0$, for $x,t\in[0,1]$

(2) $G(x,t)\le G(x,x)$ for $x,t\in[0,1]$.

For any $y\in C^+[0,1], T: P\to E$ is defined as

$$(Ty)(x)=\int_0^1 H(x,t)f(t,y(t),y'(t))dt \text{ for } x\in[0,1]. \tag{2.1}$$

Lemma 2.2 $T: P \to P$ is completely continuous.

Proof. For any $y \in P$, from the definition of Ty, obviously, Ty is concave. From the expression of Ty, combining with Lemma 2.2, we know that Ty is nonnegative on[0,1].

So, $TP \in P$. The continuity of T is obvious. We now prove that T is compact. Let $\Omega \in P$ be a bounded set. Then there exists R such that $\Omega = \{y\in P \| \|y\| \le R\}$. For any $y\in\Omega$, we have

$$0 \le f(s,y(s),y'(s)) \le \max\{f(s,y,y') | s\in[0,1], y\in[0,R], y'\in[-R,R]\} =: M.$$

Therefore, from (2.1) we have $\|(Ty)\|_{\infty} \le M + \frac{\mu}{1-\mu}M = \frac{M}{1-\mu},\quad \|(Ty)'\|_{\infty} \le M$.

So, $\|(Ty)\|$ is uniformly bounded .Now we prove Ty is equi-continuous. For $0\le x_1\le x_2\le\delta$, we have

$$|(Ty)(x_2)-(Ty)(x_1)| = \left|\int_0^1 (G(x_2,t)-G(x_1,t))f(t,y(t),y'(t))dt\right|$$

$$\le \begin{cases} \int_0^1 |(x_2-x_1)(1-t)| f(t,y(t),y'(t))dt, 0\le x_1\le x_2\le t\le 1 \\ \int_0^1 |(x_2-x_1)t| f(t,y(t),y'(t))dt, 0\le t\le x_1\le x_2\le 1 \\ \int_0^1 |t(1-x_2)-x_1(1-t)t| f(t,y(t),y'(t))dt, 0\le x_1\le t\le x_2\le 1 \end{cases},$$

$$|(Ty)'(x_2)-(Ty)'(x_1)| = \left|\int_{x_1}^{x_2} f(t,y(t),y'(t)dt\right| \int_{x_1}^{x_2} f(t,y(t),y'(t)dt \le M|x_2-x_1|,$$

And the similar results can easily get for $\delta \le x_1 \le x_2 \le 1$ and $0 \le x_1 \le \delta \le x_2 \le 1$. The Arzelà-Ascoli theorem makes sure that $T\Omega$ is relatively compact, which means T is compact. Then, $T: P \to P$ is completely continuous, and each fixed point of T in P is a solution of (1.1) and (1.2).

Theorem 2.1. Let P be a cone in a real Banach space E. Let γ and θ be nonnegative continuous convex functionals on P, α be a nonnegative continuous concave functional on P, and ψ be a nonnegative continuous functional on P Satisfying $\psi(\lambda u) \le \lambda \psi(u)$ for $0 \le \lambda \le 1$, such that for some positive numbers M and d, $\alpha(u) \le \psi(u)$ and $\|u\| \le M\lambda(u)$, for all $u \in P(\gamma, d)$. Suppose $T: P(\gamma, d) \to P(\gamma, d)$, is completely continuous and there exist positive numbers a, b, and c with $a < b$ such that

(S_1) : $\{u \in P(\gamma,\theta,\alpha,b,c,d) \mid \alpha(u) > b\} \ne \varnothing$ and $\alpha(Tu) > b$ for $u \in P(\gamma,\theta,\alpha,b,c,d)$;

(S_2) : $\alpha(Tu) > b$, for $u \in P(\gamma,\theta,\alpha,b,d)$ with $\theta(Tu) > c$;

(S_3) : $0 \notin R(\gamma,\psi,a,d)$, and $\psi(Tu) < a$ for $u \in R(\gamma,\psi,a,d)$ with $\psi(u) = a$.

Then T has at least three fixed points, u_1, u_2, and $u_3 \in P(\lambda, d)$, such that

$$\gamma(u_i) \le d, \text{ for } i = 1, 2, 3,\ b < \alpha(u_1),\ a < \psi(u_2), \text{ with } \alpha(u_2) < b \text{ and } \psi(u_3) < a.$$

3. Existence of three positive solutions to (1.1) and (1.2)

For the convenience of notation, we denote

$$L = \int_0^1 G(0,t)dt + \frac{1}{1-\mu}\int_0^1\int_0^1 G(t,s)p(s)dsdt,\quad Q = \int_0^1 G(\delta,t)dt + \frac{1}{1-u}\int_0^1\int_0^1 G(t,s)p(s)dsdt]\cdot$$

We define the nonnegative, continuous, concave functional α, the nonnegative, continuous convex functional θ, γ, and the nonnegative, continuous functional ψ on the cone P by

$$\gamma(u) = \max_{0\le x\le 1}|u'(x)|,\quad \psi(u) = \theta(u) = \max_{0\le x\le 1}|u(x)|,\quad \alpha(u) = \min_{0\le t\le 1}|u(x)|$$

By the concavity of u, for all $u \in P$, we have $N\theta(u) \le \alpha(u) \le \theta(u) = \psi(u)$,

Where $N = \max\left\{\dfrac{\min_{0\le s\le 1} p(s)\delta}{2-\min_{0\le s\le 1} p(s)\delta}, \dfrac{\min_{0\le s\le 1} p(s)(1-\delta)}{2-\min_{0\le s\le 1} p(s)(1-\delta)}\right\}$, $\max u(x) = u(\delta)$, where $0 \le \delta \le 1$.

Lemma 3.1. For $u \in p$,there must be a constant $M = (1-\int_0^1 p(s)ds)^{-1}$, such that $\|u\| \le M\gamma(u)$

Proof. For $u \in p$, by the concavity of u and Lemma 2.2, there is $u(x) - u(1) \le \max_{0\le x\le 1}|u'(x)|, x \in [0,1]$,

Taking $u(1) = \int_0^1 p(s)u(s)ds$ into consideration, we then have

$$\max_{0\le x\le 1}|u(x)| \le \max_{0\le x\le 1}|u(x)|\int_0^1 p(s)ds + \max_{0\le x\le 1}|u'(x)|,$$

Thus, $(1-\int_0^1 p(s)ds)\max_{0\le x\le 1}|u(x)| \le \max_{0\le x\le 1}|u'(x)|$.

Then, we get $\|u\| \le \dfrac{1}{1-\int_0^1 p(s)ds}\max_{0\le x\le 1}|u'(x)| = M\gamma(u)$. Thus, the proof is completed.

Lemma 3.2 Let $0 < p(s) < 1$. For $u \in P$, There will be a real number $0 < N < 1$, such that

$$\min_{0\le x\le 1}|u(x)| \ge N\|u\|_\infty.$$

Proof. For any $u \in P$ and the the concavity of u, we can get

$$\min_{0\le x\le 1}|u(x)| = u(0) \text{ or } \min_{0\le x\le 1}|u(x)| = u(1),\quad \|u\|_\infty = \max_{0\le x\le 1}|u(x)| = u(\delta).$$

And because of the concavity of u, we now divide the proof into two cases.

Case 1. If $\min_{0\le x\le 1}|u(x)| = u(0)$, then using the cancavity and positivity of u ,we can get

$$\int_0^1 p(s)u(s)ds \ge \min_{0\le s\le 1} p(s)\frac{\delta(u(\delta)+u(0))}{2},$$

then from(1.2),we have $u(0) \geq \frac{\min_{0\le s\le 1} p(s)\delta}{2-\min_{0\le s\le 1} p(s)\delta} u(\delta)$, that is $\min_{0\le x\le 1} |u(x)| \geq \frac{\min_{0\le s\le 1} p(s)\delta}{2-\min_{0\le s\le 1} p(s)\delta} \|u\|_\infty$.

Case 2. If $\min_{0\le x\le 1} |u(x)| = u(1)$, then using the cancavity and positivity of u ,we can get

$$\int_0^1 p(s)u(s)ds \geq \min_{0\le s\le 1} p(s) \frac{(1-\delta)(u(\delta)+u(1))}{2},$$

then from(1.2), we have $u(1) \geq \frac{\min_{0\le s\le 1} p(s)(1-\delta)}{2-\min_{0\le s\le 1} p(s)(1-\delta)} u(\delta)$, that is $\min_{0\le x\le 1} |u(x)| \geq \frac{\min_{0\le s\le 1} p(s)(1-\delta)}{2-\min_{0\le s\le 1} p(s)(1-\delta)} \|u\|_\infty$.

For $0 < p(s) < 1$, we can easily see that $0 < N < 1$. These equations complete the proof.

The main results of our paper is as follows:

Theorem 3.1. Let $0 < a < b \le d / 2k$ be given, and suppose that f satisfies the following conditions:

(l_1): $f(t,w,v) \le 2d$ for $(t,w,v) \in [0,1]\times[0,Md]\times[-d,d]$;

(l_2): $f(t,w,v) > b/L$ for $(t,w,v) \in [0,1]\times[b,b/N]\times[-d,d]$;

(l_3): $f(t,w,v) < a/Q$ for $(t,w,v) \in [0,1]\times[0,a]\times[-d,d]$.

Then the boundary value problem (1.1)(1.2) has at least three positive solutions u_1,u_2,u_3, such that

$$\max_{0\le x\le 1} |u'_i(x)| \le d, for\ i=1,2,3,\quad b < \min_{1/k\le x\le 1-(1/k)} |u_1(x)| \le d,$$

$$a < \max_{0\le t\le 1} |u_2(x)|, \text{ with } \min_{1/k\le x\le 1-(1/k)} |u_2(x)| < b,\quad \max_{0\le x\le 1} |u_3(x)| < a.$$

The Arzelà-Ascoli theorem guarantees that $T\Omega$ is relatively compact, which means T is compact. Then, $T:P\to P$ is completely continuous, and each fixed point of T in P is a solution of (1.1)(1.2). We now show that the conditions of Theorem 2.1 are satisfied. The proof is divided into four steps.

(1) We will show that condition (l_1) implies that $T:\overline{P(r,d)} \to \overline{P(r,d)}$.

In fact, for $u \in \overline{P(r,d)}$, there is $r(u) = \max_{0\le x\le 1} |u'(x)| \le d$. By lemma 2.1 and 2.1, we have

$$\gamma(Tu) = \max_{0\le x\le 1} |(Tu)'(x)| = |(Tu)'(\delta)| = \left|\int_0^1 (1-t)f(t,u(t),u'(t))dt - \int_0^\delta f(t,u(t),u'(t))dt\right|$$

$$= f(t,u(t),u'(t)) \left|\int_0^1 (1-t)dt + \int_0^\delta dt\right| \le 2d\cdot\frac{1}{2} = d.$$

Thus, we have $\gamma(Tu) \le d$.

(2) We show that the condition (S1) in Theorem 2.1 holds, we choose $u(x) = \frac{b}{N}$. Obviously

$$\alpha(u) = \min_{0\le x\le 1} |u(x)| = \frac{b}{N} > b,\quad \theta(u) = \max_{0\le x\le 1} |u(x)| = u(\delta) = \frac{b}{N},\quad \gamma(u) = \max_{0\le x\le 1} |u'(x)| = 0 \le d.$$

Hence, $u \in P(\gamma,\theta,\alpha,b,\frac{b}{N},d)$ and $\{u \in P(\gamma,\theta,\alpha,b,\frac{b}{N},d) \mid \alpha(u) > b\} \ne \varnothing$.

If $u \in P(\gamma,\theta,\alpha,b,\frac{b}{N},d)$, then from condition ($l_2$), we have $f(t,w,v) > b/L$, and it follows that

$$\alpha(T(u)) = \min_{0\le x\le 1} |Tu(x)| = (Tu(0)) = \int_0^1 G(0,t)f(t,y(t),y'(t))dt + \frac{1}{1-\mu}\int_0^1\int_0^1 G(t,s)p(s)f(t,y(t),y'(t))dsdt$$

$$> \frac{b}{L}\left[\int_0^1 G(0,t)dt + \frac{1}{1-u}\int_0^1\int_0^1 G(t,s)p(s)dsdt\right] = b.$$

Hence, we can get that $\alpha(Tu) > b$, for all $u \in P(r,\theta,\alpha,b,kb,d)$.

Thus, condition (S1) in Theorem 2.1 holds.

We point out that both $u(0)$ and $u(1)$ can be the minimum. But as we can see from the previous proof, if $\min_{0\le t\le 1} |u(x)| = u(1)$, we can use the same strategy to get the same outcome, so we omit it here.

(3) Now, we will show that condition (S_2) of Theorem 2.1 is satisfied. Take $u \in P(r,\theta,\alpha,b,kb,d)$

with $\theta(Tu) > \frac{b}{N}$, for all $u \in P(r,\alpha,b,d)$, we have $\alpha(Tu) = \min_{0\le x\le 1}|Tu(x)| \ge N \max_{0\le x\le 1}(Tu(x)) = N\theta(Tu) > b$.
Consequently, condition (S2) in Theorem 2.1 satisfied.
(4) At last, we prove (S3) of Theorem 2.1 is also satisfied.
Obviously, as $\psi(0) = 0 < a$, $0 \notin R(r,\psi,a,d)$. With $\psi(0) = 0 < a$, for $u \in R(r,\psi,a,b)$, we have

$$\psi(Tu) = \max_{0\le x\le 1}|(Tu)(x)| = Tu(\delta) = \int_0^1 G(\delta,t)f(t,u(t),u'(t))dt + \frac{1}{1-\mu}\int_0^1\int_0^1 G(t,s)p(s)f(t,u(t),u'(t))dsdt$$

$$< \frac{a}{Q}[\int_0^1 G(\delta,t)dt + \frac{1}{1-u}\int_0^1\int_0^1 G(t,s)p(s)dsdt] = \frac{a}{Q}\cdot Q = a.$$ Thus we get $\psi(Tu) < a$.

Consequently, condition (S3) in Theorem 2.1 holds. So, finally the proof of Theorem 2.1 is completed by using Theorem 3.1.

4. Example

Example 4.1. We now consider the following BVP with integral boundary conditions

$$\begin{cases} u''(x) + f(x,u(x),u'(x)) = 0, x \in (0,1), \\ u(0) = u(1) = \int_0^1 p(s)u(s)ds, \end{cases}$$

Where

$$f(x,w,v) = \begin{cases} x(1-x) + w^{11} + \frac{\sqrt{|v|}}{1025}, 0 \le w \le 2, \\ x(1-x) + 2048 + \frac{\sqrt{|v|}}{1025}, w \ge 2. \end{cases}$$

Then $M = 4, N = \frac{2}{7}, L = \frac{13}{72}, Q = \frac{1}{18}$, we choose $a = 1, b = 2, d = 1025$. It is obvious that $0 \le a \le b \le 2dQ$.

What's more, f satisfies (H_1) and

(l_1): $f(t,w,v) < 2d$=2050, for $(t,w,v) \in [0,1]\times[0,4100]\times[-1025,1025]$;

(l_2): $f(t,w,v) > 2048 > b/L \approx 11.1$, for $(t,w,v) \in [0,1]\times[2,7b/2]\times[-1025,1025]$;

(l_3): $f(t,w,v) < a/Q$=18, for $(t,w,v) \in [0,1]\times[0,1]\times[-1025,1025]$.

Therefore, by Throrem 2.1 BVP in this example has at least three concave positive solutions

$$\max_{0\le x\le 1}|u_i'(t)| \le 1025, i = 1,2,3, \quad \max_{0\le x\le 1}|u_1'(t)| \le 1,$$

$$\max_{0\le x\le 1}|u_2(t)| > 1 \text{ with } \min_{0\le x\le 1}|u_2(t)| < 2, \ \min_{0\le x\le 1}|u_3(t)| > 2.$$

References

[1] B. Ahmad, A. Alsaedi and B. S. Alghamdi, Analytic approximation of solutions of the forced Duffing equation with integral boundary conditions, Nonlinear Anal., 9(2008)1727-1740, .
[2] T. Jankowski, Positive solutions for fourth-order differential equations with deviating arguments and integral boundary conditions, Nonlinear Anal., 73(2010) 1289-1299.
[3] G. Wang, G. Song and L. Zhang, Integral boundary value problems for first order integro-differential equations with deviating arguments, Journ. Comput. Appl. Math., 225(2009)602-611.
[4] G. Wang, L. Zhang and G. Song, Mixed boundary value problems for second order differential equations with different deviated arguments, Journ. Appl. Math. Infor., 29(2011)191-200.
[5] L. Kong, Second order singular boundary value problems with integral boundary conditions. Nonlinear Anal., 72(2010)2628-2638.
[6] Y. Guo, W. Ge, Positive solutions for three-point boundary value problems with dependence on the first order derivative, J. Math. Anal. Appl., 290(2004)291-301.

Positive solutions to a third-order boundary value problem with Riemann Stieltjes integral boundary conditions

Bo Sun and Jianfeng Li

School of Statistics and Mathematics, Central University of Finance and Economics, Beijing 100081, PR China.

Email: sunbo19830328@163.com

Abstract. This paper investigates the existence of concave positive solutions and establishes corresponding iterative schemes to a third-order boundary value problem with Riemann Stieltjes integral boundary conditions. The main tool is a monotone iterative technique. Meanwhile, an example is worked out to demonstrate the main results.

1. Introduction

It is well known that third-order equations arise from many branches of applied mathematics and physics. For example, in the deflection of a curved beam having a constant or varying cross section, a three layer beam, electromagnetic waves or gravity driven flows ([1-6] and the references therein).

In 2012, Jankowski [5] used a fixed point theorem to establish the existence of at least three non-negative solutions of some boundary value problems to the third-order differential equation

$$\begin{cases} x'''(t)+h(t)f(t,x(\alpha(t)))=0, & t\in(0,1), \\ x(0)=x''(0)=0,\ x(1)=\beta x(\eta)+\lambda[x], & \beta>0, \eta\in(0,1), \end{cases} \tag{1.1}$$

where λ denotes a linear functional on $C(J)$ given by $\lambda[x]=\int_0^1 x(t)\mathrm{d}\Lambda(t)$ involving a Riemann-Stieltjes integral with a suitable function Λ of bounded variation.

In [6], Sun and Zhang investigated the existence of monotone positive solution to the BVP

$$\begin{cases} u'''(t)+f(t,u(t),u'(t))=0, & 0<t<1, \\ u(0)=0,\ au'(0)-bu''(0)=a[u], & cu'(1)+du''(1)=\beta[u], \end{cases} \tag{1.2}$$

where $\alpha[u]$, $\beta[u]$ are linear functional on $C[0,1]$ given by a Riemann-Stieltjes integral.

Motivated by the work mentioned above, we apply monotone iterative techniques to obtain the existence and iteration of monotone positive solutions for the following boundary value problem:

$$\begin{cases} u'''(t)=g(t)f(t,u(t)), & t\in(0,1), \\ u(0)=\alpha[u], \quad u'(p)=0, & u''(1)+\beta[u]=\lambda[u''], \end{cases} \tag{1.3}$$

where $p>\frac{1}{2}$, $\alpha[u]=\int_0^1 u(s)\mathrm{d}A(s)$, $\beta[u]=\int_0^1 u(s)\mathrm{d}B(s)$ and $\lambda[v]=\int_0^1 v(t)\mathrm{d}\Lambda(t)$ are linear functional on $C[0,1]$ given by a Riemann-Stieltjes integral, $\alpha[u]$, $\beta[u]$, $\lambda[v]$ are suitable functions of bounded variation.

2. Preliminaries

We consider the Banach space $E=C[0,1]$ equipped with the norm $\|u\|=\max_{0\le t\le 1}|u(t)|$. Define the cone $P\subset E$ by $P=\{u\in E \mid u(t)\ge 0, u \text{ is concave and } u'(p)=0\}$.

Lemma 2.1 For the following boundary value problems

$$\gamma'''(t)=0,\quad \gamma(0)=1,\quad \gamma'(p)=0,\quad \gamma''(1)=0, \tag{2.1}$$

$$\delta'''(t)=0,\quad \delta(0)=0,\quad \delta'(p)=0,\quad \delta''(1)=1, \tag{2.2}$$

We have $\gamma(t)\equiv 1, \delta(t)=pt-\frac{t^2}{2}$, for $t\in[0,1]$.

Lemma 2.2 Suppose $\lambda[1]\ne 1$. For any $y\in C[0,1]$, the unique solution of the boundary value problem

$$\begin{cases} u'''(t)=y(t), \\ u(0)=0,\quad u'(p)=0,\quad u'(1)=\lambda[u''] \end{cases} \tag{2.3}$$

is given by $$u(t)=\left(tp-\frac{t^2}{2}\right)\int_0^1\left(1+\frac{\Lambda(s)}{1-\lambda[1]}\right)y(s)ds-t\int_0^p(p-s)y(s)ds+\int_0^t\frac{(t-s)^2}{2}y(s)ds \tag{2.4}$$

Lemma 2.3 Suppose that $\frac{1}{2}\le p\le 1, \Lambda(s)\ge 0$ and $\lambda[1]<1$, let $G(t,s)$ be the Green function

$$G(t,s):=\left(tp-\frac{t^2}{2}\right)\left(1+\frac{\Lambda(s)}{1-\lambda[1]}\right)-t(p-s)H(p-s)+\frac{(t-s)^2}{2}H(t-s),\quad 0\le t\le 1, 0\le s\le 1, \tag{2.5}$$

where $H(x-y)=\begin{cases}0, & x<y, \\ 1, & x\ge y,\end{cases}$ we have $0\le G(t,s)\le \Phi(s):=\begin{cases}\frac{p^2}{2}+\frac{p^2}{2}\left(\frac{\Lambda(s)}{1-\lambda[1]}\right), & \text{if } s\ge p, \\ \frac{p^2}{2}+\frac{p^2}{2}\left(\frac{\Lambda(s)}{1-\lambda[1]}\right), & \text{if } s\ge p.\end{cases}$

Proof. The upper bounds are obtained by finding $\max_{0\le t\le 1} G(t,s)$ for each fix s. Let $Q(s):=1+\frac{\Lambda(s)}{1-\lambda[1]}$. Since $\Lambda(s)\ge 0$, we can get $Q(s)\ge 1$. From(2.5), we have the following formula:

$$(\partial/\partial t)G(t,s)=(p-t)Q(s)-(p-s)H(p-s)+(t-s)H(t-s).$$

When $s\ge p$. The derivative is positive for $t<p$, negative for $p<t<s$ and $p\le s\le t$. When $t=p$, the derivative become zero. Therefore, under our hypothesis, the maximum of $G(t,s)$ for this fixed s occurs when $t=p$. This gives the upper half of the expression for Φ.

When $s<p$ and $t>s$ using the fact that $\Lambda(s)>0$, we have

$$G(t,s)=\left(tp-\frac{t^2}{2}\right)\left(1+\frac{\Lambda(s)}{1-\lambda[1]}\right)-t(p-s)+\frac{(t-s)^2}{2}=\left(tp-\frac{t^2}{2}\right)\frac{\Lambda(s)}{1-\lambda[1]}+\frac{s^2}{2}\le\frac{p^2}{2}\frac{\Lambda(s)}{1-\lambda[1]}+\frac{s^2}{2}.$$

When $s<p$ and $t\le s$, we have

$$G(t,s)=\left(tp-\frac{t^2}{2}\right)\left(1+\frac{\Lambda(s)}{1-\lambda[1]}\right)-t(p-s)=\left(ts-\frac{t^2}{2}\right)+\left(tp-\frac{t^2}{2}\right)\frac{\Lambda(s)}{1-\lambda[1]}\le\frac{s^2}{2}+\frac{p^2}{2}\frac{\Lambda(s)}{1-\lambda[1]}.$$

Since $Q(s)\ge 1$ and $p-\frac{t}{2}\ge 0$, for $0\le t,s\le 1$, we can easily obtain

$$0\le G(t,s)=\begin{cases} t\left(p-\frac{t}{2}\right)Q(s)+\frac{(t-s)^2}{2}, & p\le s<t, \\ t\left(p-\frac{t}{2}\right)Q(s), & p\le s, t<s, \\ t\left(p-\frac{t}{2}\right)\left(\frac{\Lambda(s)}{1-\lambda[1]}\right)+\frac{s^2}{2}, & s<p, s<t, \\ t\left(p-\frac{t}{2}\right)\left(\frac{\Lambda(s)}{1-\lambda[1]}\right)+t(s-p), & t<s<p. \end{cases}$$

We always suppose that the following assumptions hold:

(H_1) $g \in L^1[0,1], g \geq 0,$ and $\int_0^1 g(s)\mathrm{d}s > 0$;

(H_2) A, B are of bounded variation and $g_A(s), g_B(s) \geq 0$ for a.e. s, where

$$g_A(s)\int_0^1 G(t,s)\mathrm{d}A(t) \text{ and } g_B(s)\int_0^1 G(t,s)\mathrm{d}B(t)$$

(H_3) $\gamma \in C[0,1],\ \gamma(t) \geq 0,\ 0 \leq \alpha[\gamma] < 1,\ \beta[\gamma] \geq 0$;

(H_4) $\delta \in C[0,1],\ \delta(t) \geq 0,\ 0 \leq \beta[\delta] < 1,\ \alpha[\delta] \geq 0$;

(H_5) $D := (1-\alpha[\gamma])(1-\beta[\delta]) - \alpha[\delta]\beta[\gamma] > 0.$

Lemma 2.4 For any $y \in C[0,1]$, suppose that u is a solution of the following BVP:

$$\begin{cases} u'''(t) = y(t), & t \in (0,1), \\ u(0) = \alpha[u], \quad u'(p) = 0, & u''(1) + \beta[u] = \lambda[u''], \end{cases} \tag{2.6}$$

then we have

$$\begin{aligned} u(t) &= \frac{\gamma(t)}{D}\left[(1-\beta[\delta])\int_0^1 g_A(s)y(s)\mathrm{d}s + \alpha[\delta]\int_0^1 g_B(s)y(s)\mathrm{d}s\right] \\ &+ \frac{\delta[t]}{D}\left[\beta[\gamma]\int_0^1 g_A(s)y(s)ds + (1-\alpha[\gamma])\int_0^1 g_B(s)y(s)\mathrm{d}s\right] + \int_0^1 G(t,s)y(s)\mathrm{d}s. \end{aligned} \tag{2.7}$$

For any $u \in E_+, T: P \to E$ is defined

$$(Tu)(t) = \gamma(t)\alpha[gf_u] + \delta(t)\beta[gf_u] + \int_0^1 G(t,s)g(s)f_u(s)\mathrm{d}s, \tag{2.8}$$

Where $f_u(s) = f(s,u(s))$, and

$$\begin{aligned} \alpha[gf_u] &= \frac{1}{D}\left[(1-\beta[\delta])\int_0^1 g_A(s)g(s)f_u(s)\mathrm{d}s + \alpha[\delta]\int_0^1 g_B(s)g(s)f_u(s)\mathrm{d}s\right], \\ \beta[gf_u] &= \frac{1}{D}\left[\beta[\gamma]\int_0^1 g_A(s)g(s)f_u(s)\mathrm{d}s + (1-\alpha[\gamma])\int_0^1 g_B(s)g(s)f_u(s)\mathrm{d}s\right], \end{aligned} \tag{2.9}$$

Combining the expression of $\alpha[gf_u], \beta[gf_u]$ with (H_1)-(H_5), we can obtain that $\alpha[gf_u], \beta[gf_u] \geq 0$. From the definition of T, it is obvious that

$$(Tu)''(t) = \begin{cases} -\beta[gf_u] + \int_0^1 (-Q(s))g(s)f_u(s)\mathrm{d}s, & t < s, \\ -\beta[gf_u] + \int_0^1 (1-Q(s))g(s)f_u(s)\mathrm{d}s, & t > s, \end{cases}$$

$$(Tu)'(t) = \begin{cases} (p-t)\beta[gf_u] + \int_0^1 (p-t)Q(s)g(s)f_u(s)\mathrm{d}s, & t < p \leq s \text{ or } p \leq t \leq s \\ (p-t)\beta[gf_u] + \int_0^1 [(p-t)Q(s) + (s-p)]g(s)f_u(s)\mathrm{d}s, & t \leq s < p \\ (p-t)\beta[gf_u] + \int_0^1 [(p-t)Q(s) + (t-s)]g(s)f_u(s)\mathrm{d}s, & p \leq s \leq t \\ (p-t)\beta[gf_u] + \int_0^1 [(p-t)Q(s) + (t-p)]g(s)f_u(s)\mathrm{d}s, & s \leq t < p \text{ or } s < p \leq t \end{cases}$$

Lemma 2.5 If (H_1)-(H_5) are satisfied, $T: P \to P$ is completely continuous.

Proof. The proof can be obtained easily by Arzelà–Ascoli theorem, so we omit here.

3. Main results

For notational convenience, we denote

$$L_A = \int_0^1 g_A(s)g(s)\mathrm{d}s, \quad L_B = \int_0^1 g_B(s)g(s)\mathrm{d}s, \tag{3.1}$$

$$L_C = \int_0^1 \Phi(s)g(s)\mathrm{d}s, \quad L_D = \frac{(1-\beta[\delta])L_A + \alpha[\delta]L_B}{D}, \quad L_E = \frac{\beta[\gamma]L_A + (1-\alpha[\gamma])L_B}{D}. \tag{3.2}$$

Remark. From (H_1)-(H_5), we easily get that the sign of (3.1) and (3.2) are nonnegative. If $g_A(s)$ is a piecewise function related to $t\in[0,1]$, L_A is an expression of t. In this case, we denote $L_A=\max_{0\le t\le 1}\{\int_0^1 g_A(s)g(s)\mathrm{d}s\}$. The same condition is satisfied with $g_B(s)$ and L_B.

Theorem 3.1 Assume that (H_1)-(H_5) hold, if there exist two positive numbers a_1, a such that

(S_1) $f(t,u_1)\le f(t,u_2)$ for $0\le t\le 1,\ 0\le u_1\le u_2\le 1$;

(S_2) $\max_{0\le t\le 1} f(t,a)\le a_1$;

(S_3) $f(t,0)\ne 0$ for $0\le t\le 1$.

Where a_1 and a satisfy $a=\left(L_C+L_D+\frac{p^2}{2}L_E\right)a_1$. Then the bounded value problem (1.3) has positive, nondecreasing on $[0,p]$ nonincreasing on $[p,1]$ and concave solutions ω^*, ν^* such that

$$0<\omega^*\le a \text{ and } \lim_{n\to\infty}\omega_n=\lim_{n\to\infty}T^n\omega_0=\omega^*, \text{ where } \omega_0(t)=a_1(L_C+L_D)+a_1t(p-\frac{t}{2})L_E, \text{ and}$$
$$0<\nu^*\le a \text{ and } \lim_{n\to\infty}\nu_n=\lim_{n\to\infty}T^n\nu_0=\nu^*, \text{ where } \nu_0(t)=0.$$

Proof. we denote $\overline{P}_a=\{u\in P\,|\,\|u\|\le a\}$. Then, in what follows, we first prove that $T:\overline{P}_a\to\overline{P}_a$. If $u\in\overline{P}_a$, then $\|u\|\le a$, we have $0\le u(t)\le u(p)=\max_{0\le t\le 1}|u(t)|=\|u\|\le a.$

So by (S_1), (S_2) we have $0\le f(t,u(t))\le\max_{0\le t\le 1} f(t,a)\le a_1$ for $0\le t\le 1$.

In fact, $$\|Tu\|=\max_{0\le t\le 1}|Tu(t)|=(Tu)(p). \tag{3.3}$$

By (2.9),(3.1) and (S_2), we have

$$\alpha[gf_u]=\frac{1}{D}\left[(1-\beta[\delta])\int_0^1 g_A(s)g(s)f_u(s)\mathrm{d}s+\alpha[\delta]\int_0^1 g_B(s)g(s)f_u(s)\mathrm{d}s\right]\le\frac{a_1}{D}\left[(1-\beta[\delta])L_A+\alpha[\delta]L_B\right]=a_1L_H, \tag{3.4}$$

$$\beta[gf_u]=\frac{1}{D}\left[\beta[\gamma]\int_0^1 g_A(s)g(s)f_u(s)\mathrm{d}s+(1-\alpha[\gamma])\int_0^1 g_B(s)g(s)f_u(s)\mathrm{d}s\right]\le\frac{a_1}{D}\left[\beta[\gamma]L_A+(1-\alpha[\gamma])L_B\right]=aL_I. \tag{3.5}$$

From (3.4),(3.5) and Lemma 2.3, we have

$$(Tu)(p)=\max_{0\le t\le 1}(Tu)(t)=\alpha[gf_u]+\frac{p^2}{2}\beta[gf_u]+\int_0^1 G(t,s)g(s)f_u(s)\mathrm{d}s\le a_1L_D+\frac{p^2}{2}a_1L_E+a_1L_C\le a.$$

Thus, we obtain that $\|Tu\|=(Tu)(p)\le a.$ Hence, we assert that $T:\overline{P}_a\to\overline{P}_a$.

Let $\omega_0=a_1L_C+a_1L_D+a_1t(p-\frac{t}{2})L_E$, for $0\le t\le 1$, then $\omega_0(t)\in\overline{P}_a$. Let $\omega_1=T\omega_0, \omega_2=T^2\omega_0$, then $\omega_1\in\overline{P}_a$ and $\omega_2\in\overline{P}_a$. We denote $\omega_{n+1}=T\omega_n=T^n\omega_0$, $n=0,1,2,\ldots$. Since $T:\overline{P}_a\to\overline{P}_a$, we have $\omega_n\in T\overline{P}_a\subseteq\overline{P}_a$, $n=0,1,2,\ldots$. Since T is completely continuous, we assert that $\{\omega_n\}_{n=1}^{\infty}$ has a convergent subsequence $\{\omega_{n_k}\}_{k=1}^{\infty}$ and there exist $\omega^*\in\overline{P}_a$ such that $\omega_{n_k}\to\omega^*$. And we have

$$\omega_1(t)=T\omega_0(t)=\alpha[gf_{\omega_0}]+t\left(p-\frac{t}{2}\right)\beta[gf_{\omega_0}]+\int_0^1 G(t,s)g(s)f_{\omega_0}(s)\mathrm{d}s\le a_1L_D+t\left(p-\frac{t}{2}\right)a_1L_E+a_1L_C=\omega_0(t),\ 0\le t\le 1.$$

So $\omega_2(t)=T\omega_1(t)\le T\omega_0(t)=\omega_1(t),\ 0\le t\le 1$. By induction, we have $\omega_{n+1}\le\omega_n,\ 0\le t\le 1, n=0,1,2,\ldots$. Hence, we assert that $\omega_n\to\omega^*$ since T is continuous and $\omega_{n+1}=T\omega_n$. Let $n\to\infty$, we have $T\omega^*=\omega^*$. It is well known that the fixed point of the operator T is the solution of boundary value problem (1.4). Therefore, ω^* is a positive nondecreasing on $[0,p]$, nonincreasing on $[p,1]$ and concave solution of boundary value problem (1.3).

Let $\nu_0=0, 0\le t\le 1$, then $\nu_0(t)\in\overline{P}_a$. Let $\nu_1=T\nu_0, \nu_2=T^2\nu_1$, then $\nu_1\in\overline{P}_a, \nu_2\in\overline{P}_a$. We denote $\nu_{n+1}=T\nu_n=T^n\nu_0$, $n=0,1,2,\ldots$. Similarly, we assert that $\{\nu_n\}_{n=1}^{\infty}$ has a convergent subsequence $\{\nu_{n_k}\}_{k=1}^{\infty}$ and there exists $\nu^*\in\overline{P}_a$ such that $\nu_{n_k}\to\nu^*$ .We have $\nu_1(t)=T\nu_0(t)\ge 0,\ 0\le t\le 1$, So

IOP Conf. Series: Journal of Physics: Conf. Series **1039** (2018) 012004 doi:10.1088/1742-6596/1039/1/012004

$v_2(t) = Tv_1(t) \geq Tv_0(t) = v_1(t),\ 0 \leq t \leq 1.$

Hence, we assert that, $v_n \to v^*, Tv^* = v^*$. Therefore, v^* is a positive nondecreasing on $[0, p]$, nonincreasing on $[p, 1]$ and concave solution of boundary value problem (1.3).

4. Example

Let $p = \frac{2}{3}$ and $g(s) = 1$, we consider the following boundary value problem:

$$\begin{cases} u'''(t) = \frac{tu}{6}, & t \in (0,1) \\ u(0) = \alpha[u], \quad u'(\frac{2}{3}) = 0, & u''(1) + \beta[u] = \lambda[u''], \end{cases} \tag{4.1}$$

where $\alpha[u] = \int_0^1 (1-s)u(s)ds$ and $\beta[s] = \int_0^1 su(s)ds$. For these boundary conditions, we have $\gamma(t) = 1$ and $\delta(t) = \frac{2}{3}t - \frac{t^2}{2}$ corresponding to Lemma 2.1. A simple calculation shows that

$\alpha[\gamma] = \frac{1}{2}, \quad \beta[\gamma] = \frac{1}{2}, \quad \alpha[\delta] = \frac{5}{72}, \quad \beta[\delta] = \frac{7}{72}, \quad D = \frac{5}{12}, \quad L_A = \frac{355}{1296}, \quad L_B = \frac{409}{1296}, \quad L_C = \frac{7}{9}, \quad L_D = \frac{157}{243}, \quad L_E = \frac{191}{270},$

and $a = (L_C + L_D + \frac{p^2}{2} L_E)a_1 \approx 1.6a_1$. Then all the hypotheses of Theorem 3.1 are fulfilled with $a = 16$ and $a_1 = 10$. It follows from Theorem 3.1 that the boundary value problem (3.6) has two monotone positive solutions ω^* and v^* such that

$$0 < \omega^* \leq 16, \quad \lim_{n\to\infty} \omega_n = \lim_{n\to\infty} T^n \omega_0 = \omega^*, \text{ where } \omega_0(t) = -\frac{433}{27}t^2 + \frac{1732}{81}t + \frac{3460}{243},$$

$$\text{and } 0 < v^* \leq 16, \quad \lim_{n\to\infty} v_n = \lim_{n\to\infty} T^n v_0 = v^*, \text{ where } v_0(t) = 0.$$

References

[1] Graef JR and Webb JRL, 2009 *Third order boundary value problems with nonlocal boundary conditions.* (England: Nonlinear Anal)

[2] Lin X, Du Z and Liu W, 2008 *Uniqueness and existence results for a third-order nonlinear multi-point boundary value problem.* (Amsterdam: J Comput Appl Math)

[3] El-Shahed M, 2009 *Positive solutions for nonlinear singular third order boundary value problem.* (Amsterdam: Commun Nonlinear Sci)

[4] Jankowski T, 2012 *Existence of positive solutions to third order differential equations with advanced arguments and nonlocal boundary conditions.* (England: Nonlinear Anal)

[5] Webb JRL and Infante G, 2009 *Nonlocal boundary value problems of arbitrary order.* (Englan: J Lond Math Soc)

[6] Zhang H and Sun J, 2012 *Existence and iteration of monotone positive solutions for third-order nonlocal BVPs involving integral conditions.* (Hungary: Electron J Qual Theory Differ Equ)

IOP Conf. Series: Journal of Physics: Conf. Series **1039** (2018) 012005 doi:10.1088/1742-6596/1039/1/012005

Laplace Variational Method for System of Partial Differential Equations

Muhammad Aslam Noor [a]; **Ayesha Sohail** [b]; **Sarmad Arshad** [b,c*]

[a] Department of Mathematics, COMSATS Institute of Information Technology, Islamabad, Pakistan
[b] Department of Mathematics, COMSATS Institute of Information Technology, Lahore, Pakistan
[c] Department of Mathematics, Lahore Garrison University, Lahore, Pakistan

Email: m.sarmad.k@gmail.com

Abstract. The dynamics of complex biological systems can be analyzed with the aid of mathematical models. These mathematical models are mostly based on systems of coupled linear or nonlinear partial differential equations. The semi-analytic technique Laplace Variational Method has been presented in this article and is successfully applied to different mathematical models. The method simplifies the basic application procedure of Variational Iteration Method by applying the Laplace transformation. We have confirmed this advantage of this method over other methods with the help of examples and their comparative analysis.

1. Introduction

In the present era, many semi-analytical and numerical methods have been frequently used in the literature to find the solution of PDEs represents the nonlinear dynamics, such as Laplace Transform Method (LTM) [1], Parameter Expansion Method [2], Variational Iteration Method (VIM) [3], Perturbation techniques [4], Modified Homotopy Perturbation Method [5] and Homotopy Analysis Method [6, 7]. LTM is a very effective tool not only for the nonlinear differential equations with constant coefficients but also for the differential equations with variable coefficients [1]. Similarly, variational iteration method is the very strong method because it has widely been used to find the solution of various nonlinear PDEs [8, 9, 10].

Notable attention has been paid in combining two or more mathematical techniques to obtain the solution of complex nonlinear problems. In this article, a new semi-analytic technique is proposed to solve the above mentioned problem in the continuation of already available techniques such as Variational Iteration Method (VIM) [3], Laplace Transform Method (LTM) [1]. VIM was first introduced by He [8] and was optimally applied to solve different linear and nonlinear problems [9, 10]. In the present article semi-analytical technique Laplace Variational Method (LVM) obtained by combining LTM and VIM has been proposed to achieve the better approximate solution of nonlinear system of partial differential equations.

In order to obtain the recursive relations in LVM, first we take Laplace transform of every equation in the system and multiply it with the Lagrange's multipliers. Comparison have been made in between VIM and LVM. Afterwards, restrictions of variational theory are used to identify the

Lagrange's multipliers optimally. These relations neither involves the integral evaluations nor the convolutions.

We have validated the method by presenting trivial examples. After the successful implication of the technique, we have deployed it on a system of coupled nonlinear pde's. The system we have selected for analysis, governs the tumour induced angiogenesis. The results obtained are in close agreement with the results available in literature. With the aid of this new semi-analytic technique, we can solve highly nonlinear coupled systems governing complex biological phenomena very swiftly.

In this article, we have used two semi-analytical techniques LVM and VIM on advection model which have applications in flood wave propagation, simulations of advection, dispersion and chemical reactions of constraints in groundwater systems [11, 12], the nonlinear WBK equations which have applications in different nonlinear problems arises in shallow water theory [13, 14]. Again, we employ LVM and VIM on tumor angiogenesis model governed by the system of partial differential which have application in the augmentation of new blood vessels that tumors need to grow. This process is originated by the release of chemicals by the tumor and by host cells close to the tumor [15].

2. Variational Iteration Method (VIM) for System of Equations

In this section, we are using the generalised system of nonlinear partial differential equations

$$\begin{aligned} L_1\phi(x,t)+N_1(\phi(x,t),\psi(x,t)) &= f(x,t) \\ L_2\phi(x,t)+N_2(\phi(x,t),\psi(x,t)) &= g(x,t) \end{aligned} \tag{2.1}$$

where L_1 and L_2 are linear and N_1, N_2 are nonlinear operators, $f(x,t)$, $g(x,t)$ are some given functions. By using the variational iteration method, we can make the correctional functional where λ_1 and λ_2 are general Lagrange multipliers which can be evaluated optimally via variational theory. The second terms in both the equations of system Eqs.(2.1) are called correction and n is the nth order approximation. $\tilde{\phi}$ and $\tilde{\psi}$ are considered as restricted variation. We can suppose that the above correctional functionals are stationary, i.e $\delta\tilde{\phi}_n = 0$ and $\delta\tilde{\psi}_n = 0$, then the Lagrange multipliers are estimated.

Example: *1* In this example, system of nonlinear Partial Differential Equations is solved by Variational Iteration Method, i.e

$$\phi_t(x,t) = \frac{\partial}{\partial x}\{D\phi_x(x,t) - \frac{\chi\phi(x,t)\upsilon_x(x,t)}{1+\upsilon(x,t)} - \rho\phi(x,t)\psi_x(x,t)\}, \tag{2.2}$$

$$\psi_t(x,t) = \beta\phi(x,t) - \gamma\phi(x,t)\psi(x,t), \tag{2.3}$$

$$\upsilon_t(x,t) = -\eta\phi(x,t)\psi(x,t). \tag{2.4}$$

Where, $\phi = \phi(x,t)$; $\psi = \psi$ (x, t) and $\upsilon = \upsilon(x,t)$, where $\chi, \rho, \beta, \gamma, \eta$ are the constants along with the conditions

$$\phi(0,t) = 0 = \phi(\pi,t),$$

$$\psi(0,t) = 0 = \psi(\pi,t),$$

$$\upsilon(0,t) = 0 = \upsilon(\pi,t),$$

$$\phi(x,0) = \psi(x,0) = \upsilon(x,0) = \sin x.$$

To solve the system with the given initial conditions by VIM, we make the correctional functional in t-direction as

IOP Conf. Series: Journal of Physics: Conf. Series **1039** (2018) 012005 doi:10.1088/1742-6596/1039/1/012005

$$\varphi_{n+1}(x,t)=\phi_n(x,t)+\int_0^t \lambda_1 t\{(\phi_n(x,t))_t-\frac{\partial}{\partial x}\{D(\phi_n(x,t))_x\}+\frac{\partial}{\partial x}\{\frac{\chi\phi_n(x,t)(\upsilon_n(x,t))_x}{1+\upsilon_n(x,t)}\}$$
$$+\frac{\partial}{\partial x}\{\rho\phi_n(x,t)(\psi_n(x,t))_x\}\}dt, \qquad (2.5)$$

$$\psi_{n+1}(x,t)=\psi_n(x,t)+\int_0^t \lambda_2(t)\{(\psi_n(x,t))_t-\beta\phi_n(x,t)+\gamma\phi_n(x,t)\psi_n(x,t)\}dt, \qquad (2.6)$$

$$\upsilon_{n+1}(x,t)=\upsilon_n(x,t)+\int_0^t \lambda_3(t)\{(\upsilon_n(x,t))_t+\eta\phi_n(x,t)\upsilon_n(x,t)\}dt, \qquad (2.7)$$

where λ_1,λ_2 and λ_3 are general Lagrange multipliers which its optimal value is evluated via variational theory. Values of ϕ_0,ψ_0 and υ_0 are initial guess and must be taken suitably and $\tilde{\phi}_n,\tilde{\psi}_n$ and $\tilde{\upsilon}_n$ are restricted values that behaves like a constant $\delta\tilde{\phi}_n=0$, $\delta\tilde{\psi}_n=0$ and $\delta\tilde{\upsilon}_n=0$. Now we have,

$$\delta\varphi_{n+1}=\delta\varphi_n+\delta\int_0^t \lambda_1(t)\{(\phi_n(x,t))_t-\frac{\partial}{\partial x}\{D(\phi_n(x,t))_x\}+\frac{\partial}{\partial x}\{\frac{\chi\phi_n(x,t)(\upsilon_n(x,t))_x}{1+\upsilon_n(x,t)}\}$$
$$+\frac{\partial}{\partial x}\{\rho\phi_n(x,t)(\psi_n(x,t))_x\}\}dt,$$

$$\delta\psi_{n+1}=\delta\psi_n+\delta\int_0^t \lambda_2(t)\{(\psi_n(x,t))_t-\beta\phi_n(x,t)+\gamma\phi_n(x,t)\psi_n(x,t)\}dt,$$

$$\delta\upsilon_{n+1}=\delta\upsilon_n+\delta\int_0^t \lambda_3(t)\{(\upsilon_n(x,t))_t+\eta\phi_n(x,t)\upsilon_n(x,t)\}dt.$$

To find the optimal values of λ_i, we have $\delta\phi_n(0)=0$, $\delta\psi_n(0)=0$ and $\delta\upsilon_n(0)=0$, the above equations becomes

$$\delta\phi_{n+1}=(1+\lambda_1(t))\delta(\phi_n)-\int_0^t \lambda_1'(t)\delta\phi_n dt$$

$$\delta\psi_{n+1}=(1+\lambda_2(t))\delta(\psi_n)-\int_0^t \lambda_2'(t)\delta\psi_n dt$$

$$\delta\upsilon_{n+1}=(1+\lambda_3(t))\delta(\upsilon_n)-\int_0^t \lambda_3'(t)\delta\upsilon_n dt$$

which gives

$$\lambda_1(t)=\lambda_2(t)=\lambda_3(t)=-1. \qquad (2.8)$$

On substituting the values of λ_i's in Eqs (2.5), (2.6), (2.7), we obtain

$$\phi_{n+1}=\phi_n-\int_0^t\{(\phi_n(x,t))_t-\frac{\partial}{\partial x}\{D(\phi_n(x,t))_x\}+\frac{\partial}{\partial x}\{\frac{\chi\phi_n(x,t)(\upsilon_n(x,t))_x}{1+\upsilon_n(x,t)}\}$$
$$+\frac{\partial}{\partial x}\{\rho\phi_n(x,t)(\psi_n(x,t))_x\}\}dt, \qquad (2.9)$$

$$\psi_{n+1}=\psi_n-\int_0^t\{(\psi_n(x,t))_t-\beta\phi_n(x,t)+\gamma\phi_n(x,t)\psi_n(x,t)\}dt, \qquad (2.10)$$

$$\upsilon_{n+1} = \upsilon_n - \int_0^t \{(\upsilon_n(x,t))_t + \eta\phi_n(x,t)\upsilon_n(x,t)\}dt, \tag{2.11}$$

using the initial conditions

$$\phi(x,0) = \phi_0 = \psi(x,0) = \psi_0 = \upsilon(x,0) = \upsilon_0 = \sin x,$$

we get the first approximation

$$\phi_1 = \sin x - \frac{t}{2(1+\sin x)}\{\chi + 2\cos x + (2\rho + \chi)\cos 2x + \rho \sin x - 2\chi \sin x + \sin 2x + \rho \sin 3x\}$$

$$\psi_1 = \sin x - t(\gamma \sin^2 x - \beta \sin x)$$

$$\upsilon_1 = \sin x - t\eta \sin^2 x$$

3. Laplace Variational Method for System of equations

To understand the methodology, we consider the non-homogeneous system of PDEs

$$\begin{aligned} L_1(\phi) + N_1(\phi,\psi,\upsilon) &= f_1, \\ L_2(\phi) + N_2(\phi,\psi,\upsilon) &= f_2, \\ L_3(\phi) + N_3(\phi,\psi,\upsilon) &= f_3, \end{aligned} \tag{3.1}$$

subject to the initial conditions

$$\phi(x,0) = \phi_0; \quad \psi(x,0) = \psi_0; \quad \upsilon(x,0) = \upsilon_0, \tag{3.2}$$

where $L_{i'}s$ are linear and $N_{i'}s$ are nonlinear operators. $f_{i'}s$ are the known functions. The recurrence relations after applying the Laplace transform will becomes

$$\begin{aligned} \phi_{n+1}(x,s) &= \phi_n(x,s) + \lambda_1 \mathcal{L}\{L_1\phi_n(x,t) + N_1(\phi_n(x,t),\psi_n(x,t),\upsilon_n(x,t)) - f_1(x,t)\}, \\ \psi_{n+1}(x,s) &= \psi_n(x,s) + \lambda_2 \mathcal{L}\{L_2\psi_n(x,t) + N_2(\phi_n(x,t),\psi_n(x,t),\upsilon_n(x,t)) - f_2(x,t)\}, \\ \upsilon_{n+1}(x,s) &= \upsilon_n(x,s) + \lambda_3 \mathcal{L}\{L_3\upsilon_n(x,t) + N_3(\phi_n(x,t),\psi_n(x,t),\upsilon_n(x,t)) - f_3(x,t)\}. \end{aligned} \tag{3.3}$$

Now taking variation to identify the Lagrange multipliers $\lambda_{i'}s$ by using the optimality conditions

$$\frac{\delta\phi_{n+1}}{\delta\phi_n} = 0; \quad \frac{\delta\psi_{n+1}}{\delta\psi_n} = 0; \quad \frac{\delta\upsilon_{n+1}}{\delta\upsilon_n} = 0,$$

implies that $\lambda_{i'}s = \frac{-1}{s}$ by assuming that $\delta\tilde{\phi}_n = 0$, $\delta\tilde{\psi}_n = 0$ and $\delta\tilde{\upsilon}_n = 0$. Substituting the values of $\lambda_{i'}s$ and taking inverse Laplace transform on both sides of Eqs.(3.3), we get

$$\phi_{n+1} = \phi_n - \mathcal{L}^{-1}\{\frac{1}{s}\mathcal{L}\{L_1\phi_n + N_1(\phi_n,\psi_n,\upsilon_n) - f_1\}\}, \tag{3.4}$$

$$\psi_{n+1} = \psi_n - \mathcal{L}^{-1}\{\frac{1}{s}\mathcal{L}\{L_2\psi_n + N_2(\phi_n,\psi_n,\upsilon_n) - f_2\}\}, \tag{3.5}$$

$$\upsilon_{n+1} = \upsilon_n - \mathcal{L}^{-1}\{\frac{1}{s}\mathcal{L}\{L_3\upsilon_n + N_3(\phi_n,\psi_n,\upsilon_n) - f_3\}\}. \tag{3.6}$$

On substituting $n = 0,1,2,\cdots$, we get the successive pair of approximations $(\phi_1,\psi_1,\upsilon_1),(\phi_2,\psi_2,\upsilon_2),\cdots$

Example: 2
In this example, we are solving the following system of nonlinear partial differential equations by using Laplace variational method, i.e

$$\phi_t(x,t)=\frac{\partial}{\partial x}\{D\phi_x(x,t)-\frac{\chi\phi(x,t)\upsilon_x(x,t)}{1+\upsilon(x,t)}-\rho\phi(x,t)\psi_x(x,t)\}, \tag{3.7}$$

$$\psi_t(x,t)=\beta\phi(x,t)-\gamma\phi(x,t)\psi(x,t), \tag{3.8}$$

$$\upsilon_t(x,t)=-\eta\phi(x,t)\psi(x,t). \tag{3.9}$$

subject to the conditions

$$\phi(0,t)=0=\phi(\pi,t),$$

$$\psi(0,t)=0=\psi(\pi,t),$$

$$\upsilon(0,t)=0=\upsilon(\pi,t),$$

$$\phi(x,0)=\psi(x,0)=\upsilon(x,0)=\sin x.$$

The recurrence relations corresponding to the given system of partial differential equations after applying Laplace transform on both sides, we obtain

$$\phi_{n+1}(x,s)=\phi_n(x,s)+\lambda_1\{s\phi_n(x,s)-\phi_0(x,0)\}+\lambda_1\mathcal{L}\{-D\widetilde{\frac{\partial}{\partial x}(\phi_n(x,t))_x}$$

$$+\chi\frac{\partial}{\partial x}(\frac{\widetilde{\phi_n(x,t)(\upsilon_n(x,t))_x}}{1+\upsilon_n(x,t)})+\rho\frac{\partial}{\partial x}(\widetilde{\phi_n(x,t)(\psi_n(x,t))_x})\}, \tag{3.10}$$

$$\psi_{n+1}(x,s)=\psi_n(x,s)+\lambda_2\{s\psi_n(x,s)-\psi_0(x,0)\}+\lambda_2\mathcal{L}\{-\beta\widetilde{\phi_n(x,t)}+\gamma\widetilde{\phi_n(x,t)\psi_n(x,t)}\}, \tag{3.11}$$

$$\upsilon_{n+1}(x,s)=\upsilon_n(x,s)+\lambda_3\{s\upsilon_n(x,s)-\upsilon_0(x,0)\}+\lambda_3\mathcal{L}\{-\eta\widetilde{\phi_n(x,t)\upsilon_n(x,t)}\}. \tag{3.12}$$

Applying the variation, we have

$$\delta\phi_{n+1}(x,t)=\delta\phi_n(x,t)+\lambda_1\delta\{s\phi_n(x,s)-\phi_0(x,0)\}+\lambda_1\delta\mathcal{L}\{-D\widetilde{\frac{\partial}{\partial x}(\phi_n(x,t))_x}$$

$$+\chi\frac{\partial}{\partial x}(\frac{\widetilde{\phi_n(x,t)(\upsilon_n(x,t))_x}}{1+\upsilon_n(x,t)})+\rho\frac{\partial}{\partial x}(\widetilde{\phi_n(x,t)(\psi_n(x,t))_x})\},$$

$$\delta\psi_{n+1}(x,t)=\delta\psi_n(x,t)+\lambda_2\delta\{s\psi_n(x,s)-\psi_0(x,0)\}+\lambda_2\delta\mathcal{L}\{-\beta\widetilde{\phi_n(x,t)}+\gamma\widetilde{\phi_n(x,t)\psi_n(x,t)}\},$$

$$\delta\upsilon_{n+1}(x,t)=\delta\upsilon_n(x,t)+\lambda_3\delta\{s\upsilon_n(x,s)-\upsilon_0(x,0)\}+\lambda_3\delta\mathcal{L}\{-\eta\widetilde{\phi_n(x,t)\upsilon_n(x,t)}\}.$$

By using the optimality conditions, we get

$$\frac{\delta\phi_{n+1}}{\delta\phi_n}=0;\quad \frac{\delta\psi_{n+1}}{\delta\psi_n}=0;\quad \frac{\delta\upsilon_{n+1}}{\delta\upsilon_n}=0,$$

and assuming that $\delta\,\widetilde{\phi_n}=0$, $\delta\,\widetilde{\psi_n}=0$, $\delta\,\widetilde{\upsilon_n}=0$ implies that $\lambda_1=\lambda_2=\lambda_3=\frac{-1}{s}$. Substituting the values in Eq.(3.10), Eq.(3.11) & Eq.(3.12), we have

$$\phi_{n+1}(x,s)=\phi_n(x,s)-\frac{1}{s}\mathcal{L}\{\frac{\partial\phi_n(x,t)}{\partial t}-D\frac{\partial}{\partial x}(\phi_n(x,t))_x+\chi\frac{\partial}{\partial x}(\frac{\phi_n(x,t)(\upsilon_n(x,t))_x}{1+\upsilon_n(x,t)})$$
$$+\rho\frac{\partial}{\partial x}(\phi_n(x,t)(\psi_n(x,t))_x)\},\tag{3.13}$$

$$\psi_{n+1}(x,s)=\psi_n(x,s)-\frac{1}{s}\mathcal{L}\{\frac{\partial\psi_n(x,t)}{\partial t}-\beta(\phi_n(x,t))_x+\rho\phi_n(x,t)\psi_n(x,t)\},\tag{3.14}$$

$$\upsilon_{n+1}(x,s)=\upsilon_n(x,s)-\frac{1}{s}\mathcal{L}\{\frac{\partial\upsilon_n(x,t)}{\partial t}+\eta\phi_n(x,t)\upsilon_n(x,t)\}.\tag{3.15}$$

Taking inverse Laplace transform, we end up with the following recurrence relations

$$\phi_{n+1}(x,t)=\phi_n(x,t)-\mathcal{L}^{-1}\{\frac{1}{s}\mathcal{L}\{\frac{\partial\phi_n(x,t)}{\partial t}-D\frac{\partial}{\partial x}(\phi_n(x,t))_x+\chi\frac{\partial}{\partial x}(\frac{\phi_n(x,t)(\upsilon_n(x,t))_x}{1+\upsilon_n(x,t)})$$
$$+\rho\frac{\partial}{\partial x}(\phi_n(x,t)(\psi_n(x,t))_x)\}\},$$

$$\psi_{n+1}(x,t)=\psi_n(x,t)-\mathcal{L}^{-1}\{\frac{1}{s}\mathcal{L}\{\frac{\partial\psi_n(x,t)}{\partial t}-\beta\phi_n(x,t)+\gamma\phi_n(x,t)\psi_n(x,t)\}\},$$

$$\upsilon_{n+1}(x,t)=\upsilon_n(x,t)-\mathcal{L}^{-1}\{\frac{1}{s}\mathcal{L}\{\frac{\partial\upsilon_n(x,t)}{\partial t}+\eta\phi_n(x,t)\upsilon_n(x,t)\}\}.$$

For $n=0,1,2,\cdots$, we get the successive pair of approximations $(\phi_1,\psi_1,\upsilon_1),(\phi_2,\psi_2,\upsilon_2),\cdots$, where the initial approximation is $\phi_0=\psi_0=\upsilon_0=\sin x$. The first pair of approximations is

$$\phi_1=\sin x-\frac{t}{2(1+\sin x)}\{\chi+2\cos x+(2\rho+\chi)\cos 2x-\rho\sin x-2\chi\sin x+\sin 2x+\rho\sin 3x\},$$

$$\psi_1=\sin x-t(-\beta\sin x+\gamma\sin^2 x),$$

$$\upsilon_1=\sin x-t\eta\sin^2 x.$$

$$\vdots$$

4. Conclusion

The Laplace Variational Method can be easily understand with only the basic knowledge of Advanced Calculus, even the reader has no background of calculus of variations in pure mathematics. The novel method (Laplace Variational Method) is simple and easy to apply in comparison with the traditional Variational Iteration Method. The advantage of this extended variational method is the Laplace transform helps to expedite the computational cost and can easily be applied to nonlinear biological systems using user friendly softwares such as Mathematica ^{TM} and Maple ^{TM}

References

[1] Podlubny I. The Laplace transform method for linear differential equations of the fractional order. arXiv preprint funct-an/9710005. 1997 Oct 30.
[2] Taiwo OA. A parameter expansion method for two-point nonlinear singularly-perturbed boundary value problems. International journal of computer mathematics. 1995 Jan 1;55(3-4):189-96.
[3] Odibat, Momani S. Application of variational iteration method to nonlinear differential equations of fractional order. International Journal of Nonlinear Sciences and Numerical Simulation. 2006;7(1):27-34.s
[4] Kumar S. A new fractional modeling arising in engineering sciences and its analytical approximate solution. Alexandria Engineering Journal. 2013 Dec 31;52(4):813-9.
[5] Yang XJ, Srivastava HM, Cattani C. Local fractional homotopy perturbation method for solving fractal partial differential equations arising in mathematical physics. Romanian Reports in Physics. 2015 Jan 1;67(3):752-61.
[6] Ganjiani M. Solution of nonlinear fractional differential equations using homotopy analysis method. Applied Mathematical Modelling. 2010 Jun 30;34(6):1634-41.
[7] Dehghan M, Manafian J, Saadatmandi A. Solving nonlinear fractional partial differential equations using the homotopy analysis method. Numerical Methods for Partial Differential Equations. 2010 Jan 1;26(2):448-79.
[8] He J. A new approach to nonlinear partial differential equations. Communications in Nonlinear Science and Numerical Simulation. 1997 Dec 1;2(4):230-5.
[9] He JH. Approximate analytical solution for seepage flow with fractional derivatives in porous media. Computer Methods in Applied Mechanics and Engineering. 1998 Dec 1;167(1-2):57-68.
[10] He JH. Approximate solution of nonlinear differential equations with convolution product nonlinearities. Computer methods in applied mechanics and engineering. 1998 Dec 1;167(1-2):69-73.
[11] Yang Y, Endreny TA, Nowak DJ. Application of advection-diffusion routing model to flood wave propagation: A case study on Big Piney River, Missouri USA. Journal of Earth Science. 2016 Feb 1;27(1):9-14.
[12] Zheng C, Wang PP. MT3DMS: a modular three-dimensional multispecies transport model for simulation of advection, dispersion, and chemical reactions of contaminants in groundwater systems; documentation and user's guide. ALABAMA UNIV UNIVERSITY; 1999 Dec.
[13] Yan Z, Zhang H. New explicit solitary wave solutions and periodic wave solutions for Whitham Broer Kaup equation in shallow water. Physics Letters A. 2001 Jul 9;285(5):355-62.
[14] Xie F, Yan Z, Zhang H. Explicit and exact traveling wave solutions of Whitham Broer Kaup shallow water equations. Physics Letters A. 2001 Jun 25;285(1):76-80.
[15] Riabov V, Gudima A, Wang N, Mickley A, Orekhov A, Kzhyshkowska J. Role of tumor associated macrophages in tumor angiogenesis and lymphangiogenesis. The regulation of angiogenesis by tissue cell-macrophage interactions. 2014 Mar 5:63.

IOP Conf. Series: Journal of Physics: Conf. Series **1039** (2018) 012006 doi:10.1088/1742-6596/1039/1/012006

A turn based traffic allocation game

Youyuan Dong

Chongqing foreign language school, P. R China
dobgyouyuan@gmail.com

Abstract: In this paper, we consider a game in which players needing to transport goods through road A or B, make decisions in turn how much goods they transport using each road. The cost per unit goods transported on road A is constant and on road B is proportional to the total goods on that road. The objective of each player is to minimize its own cost while considering other players' choices. We present a partial solution to the n-player case and a full solution to the two player case.

1. Introduction

In this paper, we consider a version of congested transportation network with two roads where players take turns to make their decisions about how much traffic they allocate on each road. There is a wide range of literature on congested transportation network, see for example [1], [2], [3], [4], [5] and [6]. Suppose, there are n players, and for $i = 1,2,\ldots,n$, player i needs to transport T_i unit of goods using road A and B. Players make decisions in turn from 1 to n. Let a_i be the amount of goods player i allocates on road A and $b_i = T_i - a_i$ be that on road B. Assume the cost per unit of goods transported on road A is 1 and that on road B is $\sum_{i=1}^{n} b_i$. So, the total cost for player i is $a_i + b_i \sum_{i=1}^{n} b_i$. Each player in the game wants to minimize his cost while knowing that other players also want to do so. We investigate the optimal strategy of this game.

In this paper, we make two contributions. First, we show that when T_i's are sufficiently large, if everyone plays according to the optimal strategy then $b_i = 2^{-i}$ and $a_i = T_i - 2^{-i}$. In particular, player i's decision does not depend on n. Secondly, we found the full solution in the case when $n = 2$.

2. Multiplayer game with large traffic needs

In this section, we investigate the solution of the game when T_i is sufficiently large. We assume for simplicity that $T_i \geq \frac{1}{2}$ for all i.

2.1. Theorem When $T_i > \frac{1}{2}$ for all i and each player plays according to the optimal strategy, then $b_i = 2^{-i}$ and $a_i = T_i - 2^{-i}$.

Proof. We first consider player n's decision. As he makes the latest decision, he can observe all the other players' choices. He simply needs to minimize

$$a_n + b_n\left(\sum_{i=1}^{n-1} b_i + b_n\right),$$

subject to $a_n = T_n - b_n$ and $a_n, b_n \geq 0$. Substituting $a_n = T_n - b_n$ we obtain that player n's cost is

$$T_n - b_n + b_n\left(\sum_{i=1}^{n-1} b_i + b_n\right) = T_n + b_n\left(\sum_{i=1}^{n-1} b_i - 1\right) + b_n^2,$$

which is a quadratic equation in b_n and its unconditional optimum is $b_n = \frac{1-\sum_{i=1}^{n-1} b_i}{2}$. The condition $T_n \geq \frac{1}{2}$ guarantees that $\frac{1-\sum_{i=1}^{n-1} b_i}{2} \leq T_n$ and thus $b_n = max(0, \frac{1-\sum_{i=1}^{n-1} b_i}{2})$ is the optimum conditional on $a_n, b_n \geq 0$.

Now, we make the hypothesis that for player k, knowing that each player after him plays according to the optimal strategy, than player k should let $b_k = max(0, \frac{1-\sum_{i=1}^{k-1} b_i}{2})$. We will show this using backward induction. We know that the hypothesis is true for $k = n$. Assume it is true for all $n \geq k > j$, we show that it is true for $k = j$.

First, we consider the case when $\sum_{i=1}^{j} b_i \geq 1$, then by induction hypothesis, $b_k = 0$ for all $n \geq k > j$, and the cost for player j would be

$$T_j - b_j + b_j(\sum_{i=1}^{j-1} b_i + b_j).$$

To minimize the cost, we would choose $b_j = max(0, 1 - \sum_{i=1}^{j-1} b_i)$.

In the case when $\sum_{i=1}^{j} b_i \leq 1$, the induction hypothesis tells us that

$$b_{j+1} = \frac{1-\sum_{i=1}^{j} b_i}{2},$$
$$b_{j+2} = \frac{1-\sum_{i=1}^{j+1} b_i}{2} = \frac{1-\sum_{i=1}^{j} b_i}{4},$$

and so on. In general, we would have $b_k = \frac{1-\sum_{i=1}^{j} b_i}{2^{k-j}}$ for $n \geq k > j$ and the cost for player j would be

$$\begin{aligned}
T_j - b_j + b_j(\sum_{i=1}^{n} b_i) &= T_j - b_j + b_j(\sum_{i=1}^{j} b_i + \sum_{i=j+1}^{n} b_i) \\
&= T_j - b_j + b_j(\sum_{i=1}^{j} b_i + (1-\sum_{i=1}^{j} b_i) \sum_{k=j+1}^{n} \frac{1}{2^{k-j}}) \\
&= T_j - b_j + b_j(\sum_{i=1}^{j} b_i + (1-\frac{1}{2^{n-j}})(1-\sum_{i=1}^{j} b_i)) \\
&= T_j - b_j + b_j(1-\frac{1}{2^{n-j}} + \frac{1}{2^{n-j}}\sum_{i=1}^{j} b_i) \\
&= T_j - \frac{1-\sum_{i=1}^{j-1} b_i}{2^{n-j}} b_j + \frac{b_j^2}{2^{n-j}}.
\end{aligned}$$

To minimize the cost, we would thus take $b_j = max(0, \frac{1-\sum_{i=1}^{j-1} b_i}{2})$. Therefore, in general, we would have either $b_j = max(0, \frac{1-\sum_{i=1}^{j-1} b_i}{2})$ or $b_j = max(0, 1 - \sum_{i=1}^{j-1} b_i)$. Note that in the case when $\sum_{i=1}^{j-1} b_i \leq 1$, choosing $b_j = 1 - \sum_{i=1}^{j-1} b_i$ would make $\sum_{i=1}^{j} b_i \leq 1$, and thus by optionality condition, the cost would be smaller or equal when choosing $\frac{1-\sum_{i=1}^{j-1} b_i}{2}$ instead. So, we have $b_j =$

$max(0, \frac{1-\sum_{i=1}^{j-1} b_i}{2})$ as desired. Note that the condition $T_n \geq \frac{1}{2}$ guarantees $a_j \geq 0$. This proves our hypothesis. As a result, $b_1 = \frac{1}{2}$, $b_2 = \frac{1-b_1}{2} = \frac{1}{4}$, $b_3 = \frac{1-b_1-b_2}{2} = \frac{1}{8}$ and so on and in general we have $b_i = 2^{-i}$ and $a_i = T_i - 2^{-i}$ as desired. □

In particular, we see that the values of a_i and b_i do not depend on n. In addition, as $n \to \infty$, $\sum_{i=1}^{n} b_i \to 1$.

3 The two player case

In this section, we investigate the case where there are only two players

3.1. *Theorem* *In the case there are only two players, Player 1 should choose*

$$b_1 = \begin{cases} min(\frac{1-T_2}{2}, T_1), & T_2 \leq 1/4, \\ min(\frac{1-T_2}{2}, T_1), & 1/4 \leq T_2 \leq 1/3 \\ min(\frac{1}{2}, T_1), & Otherwise. \end{cases} \quad and \ \frac{(1-T_2)^2}{4} \geq \frac{min(\frac{1}{2},T_1)}{2} - \frac{min(\frac{1}{2},T_1)^2}{2}, \tag{1}$$

Proof. Similar as in the proof of Theorem 2.1, we first consider the decision process of player 2. He needs to minimize

$$a_2 + b_2(b_1 + b_2),$$

subject to $a_2 = T_2 - b_2$ and $a_2, b_2 \geq 0$. Substituting $a_2 = T_2 - b_2$, the problem can be transformed into minimizing

$$b_2^2 + b_2(b_1 - 1),$$

subject to $0 \leq b_2 \leq T_2$. We solve this optimization problem and obtain

$$b_2 = \begin{cases} 0, & \\ T_2, & \\ \frac{1-b_1}{2}, & \text{otherwise.} \end{cases} , \tag{2}$$

Now, we consider player 1's strategy. He will try to minimize his cost

$$a_1 + b_1(b_1 + b_2),$$

subject to $a_1, b_1 \geq 0$ and $a_1 + b_1 = T_1$, where b_2 can be thought as a function of b_1 described in Equation 4. Substitute $a_1 = T_1 - b_1$, denote $c(b_1)$ the cost of player 1 after choosing b_1, we have

IOP Conf. Series: Journal of Physics: Conf. Series **1039** (2018) 012006 doi:10.1088/1742-6596/1039/1/012006

$$c(b_1) = \begin{cases} T_1 - b_1 + b_1^2, & b_1 \geq 1, \\ T_1 + b_1(T_2 - 1) + b_1^2, & b_1 \leq 1 - 2T_2, \\ T_1 - \frac{b_1}{2} + \frac{b_1^2}{2}, & \text{otherwise} \end{cases} \quad (3)$$

We now investigate the optimal b_1 in each case. We note that $T_1 - b_1 + b_1^2$ is strictly increasing in b_1 when $b_1 \geq 1$ and $c(b_1)$ is a continuous for $b_1 \geq 0$. Therefore, player 1 will not choose $b_1 > 1$.

$T_1 + b_1(T_2 - 1) + b_1^2$ is a quadratic function in b_1 with minimum attained at $\frac{1-T_2}{2}$ and $T_1 - \frac{b_1}{2} + \frac{b_1^2}{2}$ is a quadratic function in b_1 with minimum attained at $\frac{1}{2}$. We now solve the constrained minimization problem in each case. First, we minimize

$$T_1 + b_1(T_2 - 1) + b_1^2,$$

subject to $0 \leq b_1 \leq min(1 - 2T_2, T_1)$. When $T_2 > \frac{1}{2}$, there is no feasible solution. When $T_2 < \frac{1}{2}$, the minimum is attained at $min(T_1, 1 - 2T_2, \frac{1-T_2}{2})$. Now, we minimize

$$T_1 - \frac{b_1}{2} + \frac{b_1^2}{2}$$

subject to $\max(0, 1 - 2T_2) \leq b_1 \leq T_1$. When $1 - 2T_2 > T_1$, there will be no feasible solution. Otherwise, the minimum attained at

$$b_1 = \begin{cases} T_1, & T_1 < \frac{1}{2} \\ 1 - 2T_2, & 1 - 2T_2 > \frac{1}{2} \\ \frac{1}{2}, & \text{otherwise} \end{cases} \quad (4)$$

In other words, the minimum is attained at $Median(T_1, 1 - 2T_2, \frac{1}{2})$.

We now investigate the minimum of $c(b_1)$ in the case when $T_2 \leq \frac{1}{2}$ and $1-2T_2 \leq T_1$. In this case, $min(T_1, 1 - 2T_2, \frac{1-T_2}{2}) = min(1 - 2T_2, \frac{1-T_2}{2})$. As c is continuous, if $T_2 \geq \frac{1}{3}$, then $1 - 2T_2 \leq \frac{1-T_2}{2}$, the minimum of $c(b_1)$ will be attained at $b_1 = Median(T_1, 1 - 2T_2, \frac{1}{2})$. Similarly, when $T_2 \leq \frac{1}{4}$, then $Median(T_1, 1 - 2T_2, \frac{1}{2}) = 1 - 2T_2$ and the minimum of $c(b_1)$ will be attained at $\frac{1-T_2}{2}$. Finally, when $\frac{1}{4} \leq T_2 \leq \frac{1}{3}$, we need to compare

$$c(\frac{1-T_2}{2}) = T_1 - \frac{(1-T_2)^2}{4}$$

and

$$c(min(\frac{1}{2}, T_1)) = T_1 - \frac{min(\frac{1}{2}, T_1)}{2} + \frac{min(\frac{1}{2}, T_1)^2}{2}$$

and choose b_1 to between $min(\frac{1}{2}, T_1)$ and $\frac{1-T_2}{2}$ whichever attains a smaller c.
Now, we summarize the result, we have player 1 should choose

$$b_1 = \begin{cases} min(\frac{1-T_2}{2}, T_1), & T_2 \leq 1/4, \\ min(\frac{1-T_2}{2}, T_1), & 1/4 \leq T_2 \leq 1/ \\ min(\frac{1}{2}, T_1), & Otherwise, \end{cases} \quad 3 \text{ and } \frac{(1-T_2)^2}{4} \geq \frac{min(\frac{1}{2},T_1)}{2} - \frac{min(\frac{1}{2},T_1)^2}{2}, \tag{5}$$

as desired. □

Note that, when e.g. $T_1 \geq \frac{1}{2}$, we would have player 1 should choose

$$b_1 = \begin{cases} \frac{1-T_2}{2}, & T_2 \leq 1 - \frac{\sqrt{2}}{2} \\ \frac{1}{2}, & Otherwise. \end{cases} \tag{6}$$

In particular, there is a discontinuity in the choice of b_1 as a function of T_2 at $T_2 = 1 - \frac{\sqrt{2}}{2}$. Moreover, when $T_2 = 0$, $b_1 = \frac{1}{2}$ and b_1 decreases as T_2 increases until $T_2 = 1 - \frac{\sqrt{2}}{2}$ and b_1 jumps back to $\frac{1}{2}$ afterwards.

4.Acknowledgement

We would like to express our gratitude to Dr. Li Guolong for his advice on game theory, induction hypothesis and techniques for calculating optimum.

References

[1] T. Roughgarden, E, Tardos *How bad is selfish routing*. Princeton University PressJournal of the ACM, Vol 49, Issue 2, 2002
[2] H.Z. Aashtiani, T.L. Magnati Equilibria on a congested transportation network. SIAM J. Alg. Disc. Meth. 2, 3, 1981
[3] D. Braess *Uber ein paradoxon der verkehrsplanung*. Unternehmensforschung 12, 1968
[4] J.E. Cohen, F.P. Kelly *A paradox of congestion in a queuing network*. J. Appl. Prob. 27, 1990
[5] S. Dafermos Traffic equilibrium and variational inequalities. Transport. Sci. 14, 1, 1980
[6] J.D. Murchland *Braess's paradox of traffic flow*. Transport. Res. 4, 1970

IOP Conf. Series: Journal of Physics: Conf. Series **1039** (2018) 012007 doi:10.1088/1742-6596/1039/1/012007

A NOTE ON THE DIOPHANTINE EQUATION qx + py = z2

LU LEI

Guangdong Country Garden School Foshan, Guangdong 528312, China.
leitonylu@hotmail.com

Abstract. In this paper, we investigate the non-negative integer solutions of the equation of the form $q^x + p^y = z^2$ for x, y and z, with p,q primes. In particular, we consider the equation $3^x + p^y = z^2$, with p a prime congruent to 5 modulo 12. We prove that (1,0,2) is the unique non-negative integer solution of this equation. Moreover, we prove that (1,0,2) is also the unique non-negative integer solution for the equation $3^x + b^y = z^2$ where b is a positive integer congruent to 1 modulo 4 and has a prime factor congruent to 5 modulo 12 or congruent to 7 modulo 12.

1. Introduction

The Diophantine equation problem is one of the major areas in number theory. Diophantine equations refer to the study of the solutions of polynomial equations in two or more variables in either integers or rational numbers, rather than in real or complex numbers. If a Diophantine equation has an additional variable or variables occurring as exponents, it is an exponential Diophantine equation, such as am + bn = ck. One of the most famous Diophantine equations is Fermat's equation an +bn = cn (in the year 1637), known as Fermat's Last Theorem, which states that there are no integer solutions (a,b,c) with abc 6= 0 and n ≥ 3. However, Fermat did not give a proof except the case n = 4. Until the year 1994, Andrew Wiles gave a complete successful proof.

Theorem 1.1. (A.Wiles)[3]. There are no solutions in integers x,y,z with xyz 6= 0 of xn + yn = zn when n ≥ 3.

In the present paper, we investigate the non-negative integer solutions of the equation of the form qx + py = z2 for x, y and z, with p,q primes. In particular, we consider the equation 3x + py = z2, with p a prime. This equation has been solved in several cases. For example, Sroysang [1][2] proved that the equation has only one non-negative integer solutions (1,0,2) when p = 5,17. In this paper, we prove the following theorem (Theorem3.1).

Theorem 1.2. When p is a prime congruent to 5 modulo 12, the equation 3x + py = z2 has only one non-negative integer solution (1,0,2).

2010 Mathematics Subject Classification. 11D61.
Keywords: Exponential Diophantine equation.

We remark here that in the above theorem the condition p ≡ 5 mod 12 is equivalent to p ≡ 1 mod 4 and p ≡ 2 mod 3 by the Chinese Remainder Theorem. Here are a list of primes which are all congruent to 5 modulo 12:
5,17,29,41,53,89,101,113,137,149,173,197,233,257,269,281,293,317,353,389,....
We know that there are infinitely many primes which are congruent to 5 modulo 12 by Dirichlet's theorem on arithmetic progressions, which asserts that there are infinitely many primes that are congruent to a modulo d, for any two positive coprime integers a and d.
Moreover, in the last section we will give a more general result for the equation 3x+by = z2. In particular, we prove the following theorem (Theorem 4.1).

Theorem 1.3. The equation 3x + by = z2 has only one non-negative integer solution (1,0,2), when b is a positive integer congruent to 1 *modulo* 4 *and has a prime factor congruent to* 5 *modulo* 12 *or a prime factor congruent to* 7 *modulo* 12.

For example, in above theorem we can take $b = 21,25,45,49,57,65,77,85,\ldots$. Obviously, we can take infinitely many such integers *b*, and *b* can be arbitrary large.

2. Preliminaries
In this section, we will quickly first recall the basic knowledge of quadratic residue and quadratic reciprocity law, and then prove two useful results for the main theorem. Let *p* be an odd prime and *a* be an integer coprime to *p*. We say that *a* is a quadratic residue modulo *p* if *a* is a square modulo *p*. Otherwise, it is a quadratic non-residue. Recall that the Legendre symbol is defined to be

$$\left(\frac{a}{p}\right)=\begin{cases}1 & \text{if a is quadratic residue modulo p}\\ -1 & \text{if a is quadratic non-residue modulo p}\end{cases}$$

for odd prime p, when (a,p) = 1. Moreover, it is known to all that we have the following quadratic reciprocity law

$$\left(\frac{q}{p}\right)\left(\frac{p}{q}\right)=(-1)\left(\frac{p-1}{2}\right)\left(\frac{q-1}{2}\right)=\begin{cases}1 & \text{if } p\equiv 1 \bmod 4 \text{ or } q\equiv 1 \bmod 4\\ -1 & \text{otherwise}\end{cases}$$

where p and q are both odd primes.
The following proposition is crucial in the proof of our main theorem. Although the following result has been proved in [1] for q = 3, but here we will give a more general and simpler proof.

Proposition 2.1. For the Diophantine equation qx +1 = z2 with q a prime, we have the following results.

(1) It has only one non-negative integer solution (3,3) for q = 2.
(2) It has only one non-negative integer solution (1,2) for q = 3.
(3) It has no non-negative integer solution for q ≥ 5.

Proof. For the equation qx+1 = z2, we have that qx = z2−1, therefore qx = (z+1)(z−1). So we have that q|(z +1)(z −1). As (z +1)−(z −1) = 2, so (z +1) and (z −1) can not be both divided by q ≥ 3.

(1) For q = 2, we must have z + 1 = 2x1 and z − 1 = 2x2 with x1 + x2 = x. Then we get (z + 1) − (z − 1) = 2x1 − 2x2 = 2, i.e. 2x1−1 − 2x2−1 = 1. Hence x1 = 2 and x2 = 1. So x = 3 and z = 3.
(2) For q = 3, we have 3|(z + 1), but 3 - (z − 1). So we must have 3x = z + 1 and 1 = z − 1. So x = 1 and z = 2.

IOP Conf. Series: Journal of Physics: Conf. Series **1039** (2018) 012007 doi:10.1088/1742-6596/1039/1/012007

(3) For q ≥ 5, we have q|(z + 1), but q - (z − 1). So we must have qx = z + 1 and 1 = z − 1. Hence z = 2 and qx = 3, which is obviously not soluble for x. So there is no non-negative integer solution in this case.

Lemma 2.2. For the equation 3x + ny = z2, if n ≡ 1 mod 4, then x must be odd.

Proof. As 3x ≡ 3,1 mod 4, and ny ≡ 1 mod 4, and z2 ≡ 0,1 mod 4, we must have 3x ≡ 3 mod 4, and z2 ≡ 0 mod 4, so z must be even and x must be odd.

3. Proof of the theorem

In this section, we are going to prove the following main theorem.

Theorem 3.1. When p is a prime congruent to 5 modulo 12, the equation 3x + py = z2 has only one non-negative integer solution (1,0,2).

Proof. When y = 0, (1,0,2) is the only solution of the equation by Proposition 2.1. When y 6= 0, we consider the equation 3x + py = z2 modulo p, we have that 3x ≡ z2 mod p.
The above congruence has a solution if and only if 3x is a quadratic residue of modulo p, that is, the Legendre symbol $\left(\frac{3^x}{p}\right)$= 1. We first have that

$$\left(\frac{3^x}{p}\right) = \left(\frac{3^{x-1}}{p}\right)\left(\frac{3}{p}\right).$$

Since x is an odd number by Lemma 2.2, we have that x − 1 is even, so $\left(\frac{3^{x-1}}{p}\right) = 1.$
Hence

$$\left(\frac{3^x}{p}\right) = \left(\frac{3}{p}\right).$$

Note that p ≡ 1 mod 4, we now have

$$\left(\frac{3}{p}\right) = \left(\frac{p}{3}\right)(-1)^{\frac{3-1}{2}\frac{p-1}{2}} = \left(\frac{p}{3}\right),$$

by the quadratic reciprocity law. Since when p ≡ 2 mod 3, we get

$$\left(\frac{p}{3}\right) = \left(\frac{2}{3}\right) = -1.$$

Therefore it is a contradiction, so the equation has no solution when y 6= 0. This completes the proof of the theorem.
By the above theorem, we can immediately get the following result.

Corollary 3.2. When p is a prime congruent to 5 modulo 12, the equation 3x +py = z2k has no non-negative integer solution for x,y,z, where k ≥ 2 is an integer.

Proof. According to Theorem 3.1, when p ≡ 5 mod 12, the equation 3x + py = z2k has a solution if and only if x = 1,y = 0,zk = 2, which is not soluble since z is an integer and k ≥ 2, so there is no non-negative integer solution for x,y,z.

4. Generalisation

In this section, we will generalise the main theorem to a more general equation $3^x+b^y = z^2$, where *b* is a positive integer congruent to 1 modulo 4. In particular, we prove the following theorem.

IOP Conf. Series: Journal of Physics: Conf. Series **1039** (2018) 012007 doi:10.1088/1742-6596/1039/1/012007

Theorem 4.1. The equation 3x + by = z2 has only one non-negative integer solution (1,0,2), when b is an positive integer congruent to 1 modulo 4 and has a prime factor congruent to 5 modulo 12 or a prime factor congruent to 7 modulo 12.

Proof. When y = 0, (1,0,2) is the only solution of the equation by Proposition 2.1. When y 6= 0, we consider the equation 3x +by = z2 modulo p, where p|b and p is a prime congruent to 5 modulo 12 or congruent to 7 modulo 12, so we have that 3x ≡ z2 mod p.

According to Lemma 2.2, we know that x is odd. By the same arguments of Theorem 3.1, we know that there is no non-negative integer solution in the case that p is congruent to 5 modulo 12. While if p ≡ 7 mod 12, i.e. p ≡ 3 mod 4 and p ≡ 1 mod 3, we still have that

$$\left(\frac{3^x}{p}\right) = \left(\frac{3^{x-1}}{p}\right)\left(\frac{3}{p}\right) = \left(\frac{3}{p}\right).$$

Note that p ≡ 3 mod 4, we now have

$$\left(\frac{3}{p}\right) = \left(\frac{p}{3}\right)(-1)^{\frac{3-1}{2}\frac{p-1}{2}} = -\left(\frac{p}{3}\right),$$

by the quadratic reciprocity law. Since when $p \equiv 1 \bmod 3$, we get

$$\left(\frac{3}{p}\right) = -\left(\frac{p}{3}\right) = -\left(\frac{1}{3}\right) = -1.$$

So we have that there is no non-negative integer solution in the case that p is congruent to 7 modulo 12. Hence (1,0,2) is the only non-negative integer solution for the equation $3^x + b^y = z^2$ where b is a positive integer congruent to 1 modulo 4 and has a prime factor congruent to 5 modulo 12 or congruent to 7 modulo 12.

Corollary 4.2. When b is an positive integer congruent to 1 modulo 4 and has a prime factor congruent to 5 modulo 12 or a prime factor congruent to 7 modulo 12, the equation 3x+py = z2k has no non-negative integer solution for x,y,z, where k is an integer greater than 1.

Proof. According to Theorem 4.1, under the above assumptions, the equation 3x+by = z2k has a solution if and only if x = 1,y = 0,zk = 2, which is not soluble since z is an integer, so there is no non-negative integer solution for x,y,z.

In conclusion, we remark that for the exponential equation of the type $q^x + p^y = z^2$ with p,q primes, we can try to find all non-negative integer solutions for x,y,z by similar methods.

References

[1] B. Sroysang, On the Diophantine equation 3x + 5y = z2, International Journal of Pure and Applied Mathematics, Volume 81, No.4, 2012, 605–608.
[2] B. Sroysang, On the Diophantine equation 3x + 17y = z2, International Journal of Pure and Applied Mathematics, Volume 89, No.1, 2013, 111–114.
[3] A. Wiles, Modular elliptic curves and Fermat's last theorem, Ann. of Math. **141** (1995), 443–551.

IOP Conf. Series: Journal of Physics: Conf. Series **1039** (2018) 012008 doi:10.1088/1742-6596/1039/1/012008

An implicit numerical method for the space-time variable-order fractional Bloch-Torrey equation[1]

Tao Xu, Shujuan Lü, Haonan Li

School of Mathematics and System Science, Beihang University, Beijing, 100191.

E-mail: lsj@buaa.edu.cn

Abstract. Recently, a new space-time fractional Bloch-Torrey equation (ST-FBTE) has been proposed to study anomalous diffusion in the human brain. In a paper, we consider the ST-FBTE with variable fractional order derivatives. The time and space derivatives are replaced by the variable-order Caputo and variable-order Riesz fractional derivatives, respectively. An implicit numerical method for the equation are proposed. Unconditional stability and a special case convergence of the proposed method are proved by the mathematical induction. Some numerical examples are given to support our theoretical analysis.

1. Introduction

In a paper, we study the space-time variable-order fractional Bloch-Torrey equation(ST-VO-FBTE) which is stated as follows

$$\begin{cases} {}_0^cD_t^{\alpha(x,t)}u(x,t) = \kappa(x,t)\dfrac{\partial^{\beta(x,t)}u(x,t)}{\partial \mid x \mid^{\beta(x,t)}} + f(x,t), & (x,t) \in \Omega = (a,b) \times (0,T], \\ u(x,0) = \phi(x), & a \le x \le b, \\ u(a,t) = 0, u(b,t) = 0, & 0 \le t \le T, \end{cases} \tag{1}$$

where $0 < \underline{\alpha} \le \alpha(x,t) \le \overline{\alpha} \le 1$ and $1 < \underline{\beta} \le \beta(x,t) \le \overline{\beta} \le 2$, $\kappa(x,t)$ stands for the generalized diffusion coefficient, ${}_0^cD_t^{\alpha(x,t)}u(x,t)$ denotes Caputo fractional derivative of variable-order $\alpha(x,t)$ to the function $u(x,t)$, the notation $\partial^{\beta(x,t)}u(x,t)/\partial \mid x \mid^{\beta(x,t)}$ presents the generalized Riesz fractional derivative of variable-order $\beta(x,t)$ to function $u(x,t)$, which has introduced in the literature of [1] and [2], but for readability, we restate it in the following

$$\frac{\partial^{\beta(x,t)}u(x,t)}{\partial \mid x \mid^{\beta(x,t)}} = -\frac{1}{\cos(\frac{\beta(x,t)\pi}{2})}(\rho_a D_x^{\beta(x,t)} + \sigma_x D_b^{\beta(x,t)})u(x,t),$$

in which ${}_aD_x^{\beta(x,t)}u(x,t)$ and ${}_xD_b^{\beta(x,t)}u(x,t)$ represent the left-side and right-side variable-order Riemann-Liouville fractional derivative to the function, respectively, defined by

$${}_aD_x^{\beta(x,t)}u(x,t) = [\frac{1}{\Gamma(2-\beta(x,t))}\frac{d^2}{d\theta^2}\int_a^\theta (\theta-\xi)^{1-\beta(x,t)}u(\xi,t)d\xi]_{\theta=x},$$

[1] This work is supported by National Nature Science Foundation under Grant 11272024, 11672011.

$$_xD_b^{\beta(x,t)}u(x,t) = [\frac{1}{\Gamma(2-\beta(x,t))}\frac{d^2}{d\theta^2}\int_\theta^b (\xi-\theta)^{1-\beta(x,t)}u(\xi,t)d\xi]_{\theta=x}.$$

Finally, in order to simplify the process in the sequel, we set $A(x,t) = \kappa(x,t) \times (-\frac{1}{\cos\beta(x,t)\pi/2})$ $(A(x,t) > 0)$ to conclude the section.

2. Implicit numerical method for the ST-VO-FBTE

In this section, we shall derive a numerical approximation of Eq.(1). To this end, we let $h = (b-a)/m$ and $\tau = T/n$ as the spatial step and temporal step, respectively, moreover, using u_i^k as the numerical approximation to $u(x_i, t_k)$, and similar for $\alpha_i^k = \alpha(x_i, t_k)$, $\beta_i^k = \beta(x_i, t_k)$ and $f_i^k = f(x_i, t_k)$. Thus, suppose $u \in C^{(2)}(\Omega)$, adopting the discrete scheme given in [4], we may discretize the variable order time fractional derivative by

$$^c_0D_t^{\alpha_i^{k+1}} u(x_i, t_{k+1}) = \frac{\tau^{-\alpha_i^{k+1}}}{\Gamma(2-\alpha_i^{k+1})}\sum_{j=0}^{k}((u(x_i, t_{k+1-j}) - u(x_i, t_k - j)))b_{i,j}^{k+1} + O(\tau). \quad (2)$$

In what follow, by letting $r_i^{k+1} = \tau^{\alpha_i^{k+1}}\Gamma(2-\alpha_i^{k+1})$, $A_i^k = A(x_i, t_k)$ and $b_{i,j}^k = (j+1)^{1-\alpha_i^k} - j^{1-\alpha_i^k}$. Combining (2) and Lemma 3.1 of [2], we can discrete (1) by the following implicit difference scheme :

$$[1 - r_i^1 A_i^1(\rho h^{-\beta_{i+1}^1} g_{\beta_{i+1}^1}^1 + \sigma h^{-\beta_{i-1}^1} g_{\beta_{i-1}^1}^1)]u_i^1 - r_i^1 A_i^1(\rho h^{-\beta_{i+1}^1}\sum_{j=0,j\neq 1}^{i+1} g_{\beta_{i+1}^1}^j u_{i+1-j}^1$$
$$+\sigma h^{-\beta_{i-1}^1}\sum_{j=0,j\neq 1}^{m-i+1} g_{\beta_{i-1}^1}^j u_{i-1+j}^1) = u_i^0 + r_i^1 f_i^1, \quad (3)$$

$$[1 - r_i^{k+1} A_i^{k+1}(\rho h^{-\beta_{i+1}^{k+1}} g_{\beta_{i+1}^{k+1}}^1 + \sigma h^{-\beta_{i-1}^{k+1}} g_{\beta_{i-1}^{k+1}}^1)]u_i^{k+1} - r_i^{k+1} A_i^{k+1}(\rho h^{-\beta_{i+1}^{k+1}}\sum_{j=0,j\neq 1}^{i+1} g_{\beta_{i+1}^{k+1}}^j u_{i+1-j}^{k+1}$$
$$+\sigma h^{-\beta_{i-1}^{k+1}}\sum_{j=0,j\neq 1}^{m-i+1} g_{\beta_{i-1}^{k+1}}^j u_{i-1+j}^{k+1}) = \sum_{j=0}^{k-1}(b_{i,j}^{k+1} - b_{i,j+1}^{k+1})u_i^{k-j} + b_{i,k}^{k+1}u_i^0 + r_i^{k+1}f_i^{k+1}, \quad (4)$$

in which $i = 1, 2, \ldots, m-1; k = 1, 2, \ldots, n-1$.

The initial and boundary conditions are discretized as

$$u_i^0 = \phi(ih),\ i = 0, 1, 2, \cdots, m, \quad u_0^k = 0, \quad \text{and} \quad u_m^k = 0,\ k = 0, 1, 2, \cdots, n. \quad (5)$$

Hence, the previous equation can be expressed by the next matrix form

$$\begin{cases} W^1U^1 = U^0 + r^1 f^1, \\ W^{K+1}U^{K+1} = \sum_{j=0}^{k-1}(b_j^{k+1} - b_{j+1}^{k+1})U^{k-j} + b_k^{k+1}U^0 + r^{k+1}f^{k+1}, \\ U^0 = \phi. \end{cases}$$

in which $U^k = (u_1^k, u_2^k, \ldots, u_{m-1}^k)^T, b_j^k = (b_{1,j}^k, b_{2,j}^k, \ldots, b_{m-1,j}^k)^T, r^k = (r_1^k, r_2^k, \ldots, r_{m-1}^k)^T, f^k =$

$(f_1^k, f_2^k, \ldots, f_{m-1}^k)^T$, $\phi = (\phi_1, \phi_2, \ldots, \phi_{m-1})^T$, and $W^k = (w_{ij}^k)$:

$$w_{ij}^k = \begin{cases} -r_i^k A_i^k \sigma h^{-\beta_{i-1}^k} g_{\beta_{i-1}^k}^{j-i+1}, & j \geq i+2 \\ -r_i^k A_i^k (\rho h^{-\beta_{i+1}^k} g_{\beta_{i+1}^k}^0 + \sigma h^{-\beta_{i-1}^k} g_{\beta_{i-1}^k}^2), & j = i+1 \\ 1 - r_i^k A_i^k (\rho h^{-\beta_{i+1}^k} g_{\beta_{i+1}^k}^1 + \sigma h^{-\beta_{i-1}^k} g_{\beta_{i-1}^k}^1), & j = i \\ -r_i^k A_i^k (\rho h^{-\beta_{i+1}^k} g_{\beta_{i+1}^k}^2 + \sigma h^{-\beta_{i-1}^k} g_{\beta_{i-1}^k}^0), & j = i-1 \\ -r_i^k A_i^k \rho h^{-\beta_{i+1}^k} g_{\beta_{i+1}^k}^{i-j+1}, & j \leq i-2, \end{cases} \quad (6)$$

for $i = 1, 2, \ldots, m-1, j = 1, 2, \ldots, m-1, k = 1, 2, \ldots, n$.

In accordance with Lemma 3 of [3], which associated with (6) to see that the matrix W is strictly diagonally dominant with positive diagonal terms and nonpositive off-diagonal terms. Hence, finally, the following theorem is established.

Theorem 2.1. *The discretization matrix W is invertible. Furthermore, the system* (3)-(5) *has a unique solution.*

3. Stability and Convergence of the Implicit numerical method

Theorem 3.1. *The implicit difference scheme* (3)-(5) *is unconditionally stable.*

Proof. Suppose that $\widetilde{u}_i^k$ is the approximate solution of u_i^k, which is the exact solution of the implicit scheme (3) and (4). The error $\varepsilon_i^k = \widetilde{u}_i^k - u_i^k$, satisfies

$$[1 - r_i^1 A_i^1 (\rho h^{-\beta_{i+1}^1} g_{\beta_{i+1}^1}^1 + \sigma h^{-\beta_{i-1}^1} g_{\beta_{i-1}^1}^1)]\varepsilon_i^1 - r_i^1 A_i^1 (\rho h^{-\beta_{i+1}^1} \sum_{j=0, j\neq 1}^{i+1} g_{\beta_{i+1}^1}^j \varepsilon_{i+1-j}^1$$
$$+ \sigma h^{-\beta_{i-1}^1} \sum_{j=0, j\neq 1}^{m-i+1} g_{\beta_{i-1}^1}^j \varepsilon_{i-1+j}^1) = \varepsilon_i^0,$$

$$[1 - r_i^{k+1} A_i^{k+1} (\rho h^{-\beta_{i+1}^{k+1}} g_{\beta_{i+1}^{k+1}}^1 + \sigma h^{-\beta_{i-1}^{k+1}} g_{\beta_{i-1}^{k+1}}^1)]\varepsilon_i^{k+1} - r_i^{k+1} A_i^{k+1} (\rho h^{-\beta_{i+1}^{k+1}} \sum_{j=0, j\neq 1}^{i+1} g_{\beta_{i+1}^{k+1}}^j \varepsilon_{i+1-j}^{k+1}$$
$$+ \sigma h^{-\beta_{i-1}^{k+1}} \sum_{j=0, j\neq 1}^{m-i+1} g_{\beta_{i-1}^{k+1}}^j \varepsilon_{i-1+j}^{k+1}) = \sum_{j=0}^{k-1} (b_{i,j}^{k+1} - b_{i,j+1}^{k+1}) \varepsilon_i^{k-j} + b_{i,k}^{k+1} \varepsilon_i^0, (k \geq 1).$$

Next, we let $\|X^1\|_\infty = |\varepsilon_l^1| = \max_{1 \leq i \leq m-1} | \varepsilon_i^1 |$ and notice that $\sum_{j=0}^{l} g_{\beta_i^k}^j < 0$, thus

$$\|X^1\|_\infty = |\varepsilon_l^1| \leq |\varepsilon_l^1| - r_l^1 A_l^1 (\rho h^{-\beta_{l+1}^1} \sum_{j=0}^{l+1} g_{\beta_{l+1}^1}^j | \varepsilon_l^1 | + \sigma h^{-\beta_{l-1}^1} \sum_{j=0}^{m-l+1} g_{\beta_{l-1}^1}^j | \varepsilon_l^1 |)$$
$$\leq | [1 - r_l^1 A_l^1 (\rho h^{-\beta_{l+1}^1} g_{\beta_{l+1}^1}^1 + \sigma h^{-\beta_{l-1}^1} g_{\beta_{l-1}^1}^1)]\varepsilon_l^1 - r_l^1 A_l^1 (\rho h^{-\beta_{l+1}^1} \sum_{j=0, j\neq 1}^{l+1} g_{\beta_{l+1}^1}^j \varepsilon_{l+1-j}^1$$
$$+ \sigma h^{-\beta_{l-1}^1} \sum_{j=0, j\neq 1}^{m-l+1} g_{\beta_{l-1}^1}^j \varepsilon_{l-1+j}^1) | = | \varepsilon_l^0 | \leq \| X^0 \|_\infty .$$

In what follow, in view of $\|X^{k+1}\|_\infty = |\varepsilon_l^{k+1}| = \max\limits_{1\le i\le m-1} \mid \varepsilon_i^{k+1} \mid$ and assume that $\|X^j\|_\infty \le \|X^0\|_\infty (j\le k)$, together with Lemma 3 of [3] and Lemma 1 of [4] to see that

$$\|X^{k+1}\|_\infty = |\varepsilon_l^{k+1}| \le |\varepsilon_l^{k+1}| - r_l^{k+1} A_l^{k+1}(\rho h^{-\beta_{l+1}^{k+1}} \sum_{j=0}^{l+1} g^j_{\beta_{l+1}^{k+1}} \mid \varepsilon_l^{k+1} \mid + \sigma h^{-\beta_{l-1}^{k+1}} \sum_{j=0}^{m-l+1} g^j_{\beta_{l-1}^{k+1}} \mid \varepsilon_l^{k+1} \mid)$$

$$\le \sum_{j=0}^{k-1} (b_{l,j}^{k+1} - b_{l,j+1}^{k+1}) \mid \varepsilon_l^{k-j} \mid + b_{l,k}^{k+1} \mid \varepsilon_l^0 \mid \le \sum_{j=0}^{k-1} (b_{l,j}^{k+1} - b_{l,j+1}^{k+1}) \parallel X^{k-j} \parallel_\infty + b_{l,k}^{k+1} \parallel X^0 \parallel_\infty$$

$$\le \sum_{j=0}^{k-1} (b_{l,j}^{k+1} - b_{l,j+1}^{k+1}) \parallel X^0 \parallel_\infty + b_{l,k}^{k+1} \parallel X^0 \parallel_\infty = \parallel X^0 \parallel_\infty .$$

Therefore, the implicit difference scheme defined by (3)-(5) is unconditionally stable.

In what follows, we study a special case of discrete scheme(3)-(5) for the time fractional derivatives is independent of time variable t.

Theorem 3.2. *Suppose that u_i^k $(i = 1, 2, \cdots, m-1,\ k = 1, 2, \cdots, n.)$ is the approximation solution of $u(x_i, t_k)$ computed by scheme (3)-(5), then there exist a positive constant C independent of τ and h, such that*

$$\parallel u(\cdot, t_k) - u^k \parallel_\infty \le C(\tau + h),$$

Proof. Firstly, denote $u(x_i, t_k)(i = 1, 2, \ldots, m-1; k = 1, 2, \ldots, n)$ by the exact solution of Eq.(1) at mesh point (x_i, t_k) and define that $\eta_i^k = u(x_i, t_k) - u_i^k$, $Y^k = (\eta_1^k, \eta_2^k, \ldots, \eta_{m-1}^k)^T$. Using $Y^0 = \mathbf{0}$, and $u_i^k = u(x_i, t_k) - \eta_i^k$, it follows from (3) and (4) to get that

$$[1 - r_i A_i^1(\rho h^{-\beta_{i+1}^1} g^1_{\beta_{i+1}^1} + \sigma h^{-\beta_{i-1}^1} g^1_{\beta_{i-1}^1})]\eta_i^1 - r_i A_i^1(\rho h^{-\beta_{i+1}^1} \sum_{j=0, j\ne 1}^{i+1} g^j_{\beta_{i+1}^1} \eta^1_{i+1-j}$$

$$+ \sigma h^{-\beta_{i-1}^1} \sum_{j=0, j\ne 1}^{m-i+1} g^j_{\beta_{i-1}^1} \eta^1_{i-1+j}) = R_i^1,$$

$$[1 - r_i A_i^{k+1}(\rho h^{-\beta_{i+1}^{k+1}} g^1_{\beta_{i+1}^{k+1}} + \sigma h^{-\beta_{i-1}^{k+1}} g^1_{\beta_{i-1}^{k+1}})]\eta_i^{k+1} - r_i A_i^{k+1}(\rho h^{-\beta_{i+1}^{k+1}} \sum_{j=0, j\ne 1}^{i+1} g^j_{\beta_{i+1}^{k+1}} \eta^{k+1}_{i+1-j}$$

$$+ \sigma h^{-\beta_{i-1}^{k+1}} \sum_{j=0, j\ne 1}^{m-i+1} g^j_{\beta_{i-1}^{k+1}} \eta^{k+1}_{i-1+j}) = \sum_{j=0}^{k-1} (b_{i,j}^{k+1} - b_{i,j+1}^{k+1}) \eta_i^{k-j} + R_i^{k+1},$$

in which $\alpha_i = \alpha(x_i), b_{i,j} = (j+1)^{1-\alpha_i} - j^{1-\alpha_i}$, and $r_i = \tau^{\alpha_i}\Gamma(2 - \alpha_i)$.

According to the formula (2) and Lemma 3.1 of [2], it is easy to obtain that the truncation error R_i^k satisfies $\mid R_i^k \mid \le C\tau^{\alpha_i}(\tau + h)$, $\quad i = 1, 2, \cdots, m-1,\ k = 1, 2, \cdots, n$.

Next, letting $\parallel Y^1 \parallel_\infty = \mid \eta_l^1 \mid = \max\limits_{1\le i\le m-1} \mid \eta_i^1 \mid$ and by calculating, we have

$$\|Y^1\|_\infty = |\eta_l^1| \le |\eta_l^1| - r_l A_l^1(\rho h^{-\beta_{l+1}^1} \sum_{j=0}^{l+1} g^j_{\beta_{l+1}^1} \mid \eta_l^1 \mid + \sigma h^{-\beta_{l-1}^1} \sum_{j=0}^{m-l+1} g^j_{\beta_{l-1}^1} \mid \eta_l^1 \mid)$$

$$\le \mid [1 - r_l A_l^1(\rho h^{-\beta_{l+1}^1} g^1_{\beta_{l+1}^1} + \sigma h^{-\beta_{l-1}^1} g^1_{\beta_{l-1}^1}) - r_l A_l^1(\rho h^{-\beta_{l+1}^1} \sum_{j=0, j\ne 1}^{l+1} g^j_{\beta_{l+1}^1} \eta^1_{l+1-j}$$

$$+\sigma h^{-\beta_{l-1}^{1}}\sum_{j=0,j\neq 1}^{m-l+1} g_{\beta_{l-1}^{1}}^{j}\eta_{l-1+j}^{1}) =\mid \eta_l^0 + R_l^1 \mid,$$

thus, by the facts of $Y^0 = 0$ and $\mid R_l^1 \mid\leq C\tau^{\alpha_l}(\tau+h)$,, we have $\|Y^1\|_\infty \leq C\tau^{\alpha_l}(\tau+h)$.

Furthermore, suppose $\|Y^j\|_\infty \leq C(b_{l,j-1})^{-1}\tau^{\alpha_l}(\tau+h)$, $j = 2,3,\cdots,k$, and $\mid \eta_l^{k+1}\mid = \max_{1\leq i\leq M-1}\mid \eta_i^{k+1}\mid$, which combined with $b_{l,j}^{-1}\leq b_{l,k}^{-1}, j=0,1,\ldots,k$, to get that

$$\|Y^{k+1}\|_\infty = |\eta_l^{k+1}| \leq |\eta_l^{k+1}| - r_l A_l^{k+1}(\rho h^{-\beta_{l+1}^{k+1}}\sum_{j=0}^{l+1} g_{\beta_{l+1}^{k+1}}^{j}\mid \eta_l^{k+1}\mid + \sigma h^{-\beta_{l-1}^{k+1}}\sum_{j=0}^{m-l+1} g_{\beta_{l-1}^{k+1}}^{j}\mid \eta_l^{k+1}\mid)$$

$$\leq\mid \sum_{j=0}^{k-1}(b_{l,j}-b_{l,j+1})\eta_l^{k-j} + R_l^{k+1}\mid\leq \sum_{j=0}^{k-1}(b_{l,j}-b_{l,j+1})\parallel Y^{k-j}\parallel_\infty + \mid R_l^{k+1}\mid.$$

Hence, by $b_{l,j}^{-1}\leq b_{l,k}^{-1}, j=0,1,\ldots,k$ and $\mid R_l^{k+1}\mid\leq C\tau^{\alpha_l}(\tau+h)$, we obtain the conclusion

$$\parallel Y^{k+1}\parallel_\infty\leq C(b_{l,j-1})^{-1}[\sum_{j=0}^{k-1}(b_{l,j}-b_{l,j+1}) + b_{l,k}]\tau^{\alpha_l}(\tau+h) \leq C(b_{l,k})^{-1}\tau^{\alpha_l}(\tau+h).$$

Since the fact of $\lim_{k\to\infty}\frac{(b_{l,k})^{-1}}{k^{\alpha_l}} = \frac{1}{1-\alpha_l}$, which implies that there exist a constant C such that $\parallel Y^{k+1}\parallel_\infty\leq C(\tau+h)$ while $k\tau\leq T$ is finite, i.e., the conclusion is valid.

Table 1: Example 1

x	Numerical	Exact	Error
0.80	0.1157	0.1152	0.0005
1.60	0.4096	0.4096	0.0000
2.40	0.8067	0.8064	0.0003
3.20	1.2300	1.2288	0.0012
4.00	1.6017	1.6000	0.0017
4.80	1.8439	1.8432	0.0007
5.60	1.8786	1.8816	0.0030
6.40	1.6293	1.6384	0.0091
7.20	1.0232	0.0368	0.0136

Table 2: Example 2

x	Numerical	Exact	Error
0.1	0.0171	0.0162	0.0009
0.2	0.0521	0.0512	0.0009
0.3	0.0885	0.0882	0.0003
0.4	0.1144	0.1152	0.0008
0.5	0.1228	0.1250	0.0022
0.6	0.1119	0.1152	0.0033
0.7	0.0846	0.0882	0.0036
0.8	0.0483	0.0512	0.0029
0.9	0.0149	0.0162	0.0013

4. Numerical examples

Example 1. *We take the various data, respectively, by*

$$\begin{cases} \alpha(x,t) = 0.5 + 0.01\sin(5xt), \\ \beta(x,t) = 1.5 + 0.01x^2t^2, \\ \kappa(x,t) = -0.5\cos(\beta(x,t)\pi/2), \\ \rho = 1,\ \sigma = 0,\ (x,t)\in\Omega = (0,8)\times(0,T]. \end{cases}$$

The corresponding forcing term is

$$f(x,t) = \frac{x^2(8-x)t^{1-\alpha(x,t)}}{80\Gamma(2-\alpha(x,t))} - (t+1)[\frac{\Gamma(3)x^{2-\beta(x,t)}}{20\Gamma(3-\beta(x,t))} - \frac{\Gamma(4)x^{3-\beta(x,t)}}{160\Gamma(4-\beta(x,t))}].$$

Then Eq.(1) *has the exact solution* $u(x,t) = \frac{(t+1)x^2(8-x)}{80}$.

IOP Conf. Series: Journal of Physics: Conf. Series **1039** (2018) 012008 doi:10.1088/1742-6596/1039/1/012008

Table 1 gives the numerical solution, exact solution and absolute error at $T = 1$ of Example 1. Fig. 1 shows the solution behavior of Example 1 at $T = 0, T = 0.5$ and $T = 1$, respectively.

Example 2. *In the example, we reselect the data, respectively, by*

$$\begin{cases} \alpha(x,t) = 1 - 0.8e^{-xt}, \\ \beta(x,t) = 1.7 + 0.1e^{-\frac{x^2}{1000}-\frac{t}{50}-1}, \\ \kappa(x,t) = 1,\ \rho = \sigma = \frac{1}{2},\ (x,t) \in \Omega = (0,1) \times (0,T]. \end{cases}$$

The corresponding forcing term is

$$f(x,t) = 2x^2(1-x)^2\left[\frac{t^{2-\alpha(x,t)}}{\Gamma(3-\alpha(x,t))} + \frac{t^{2-\alpha(x,t)/2}}{\Gamma(3-\alpha(x,t)/2)}\right] + \frac{\sec(\beta(x,t)\pi/2)}{2}(1+t^2)$$
$$\left[\frac{2(x^{2-\beta(x,t)} + (1-x)^{2-\beta(x,t)})}{\Gamma(3-\beta(x,t))} - \frac{12(x^{3-\beta(x,t)} + (1-x)^{3-\beta(x,t)})}{\Gamma(4-\beta(x,t))} + \frac{24(x^{4-\beta(x,t)} + (1-x)^{4-\beta(x,t)})}{\Gamma(5-\beta(x,t))}\right].$$

Then Eq.(1) *has the exact solution* $u(x,t) = (1+t^2)x^2(1-x)^2$.

Table 2 gives the numerical solution, exact solution and absolute error at $T = 1$ of Example 2. Fig. 2 shows the solution behavior of Example 2 at $T = 0.25, T = 0.5, T = 0.75$, respectively. It can be seen that the numerical solution is in good agreement with the exact solution.

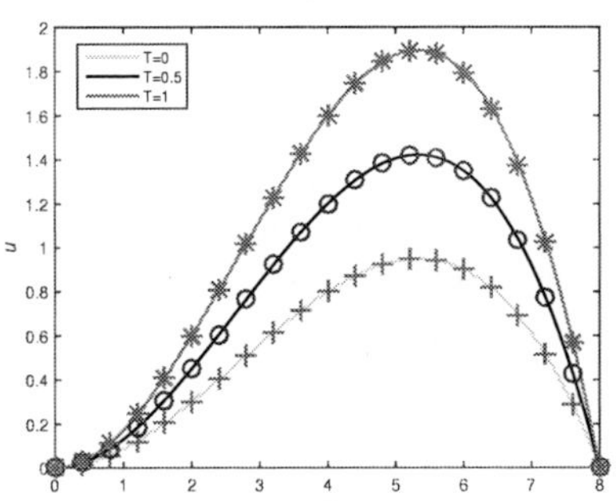

Figure 1: The solution behavior of Example 1 at $T = 0, T = 0.5, T = 1$.

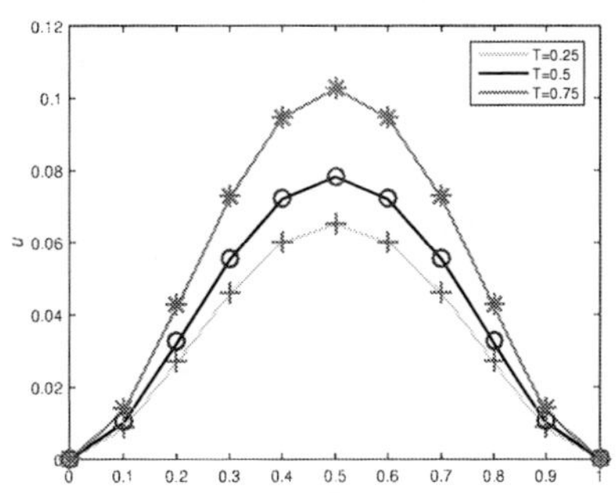

Figure 2: The solution behavior of Example 2 at $T = 0.25, T = 0.5, T = 0.75$.

References

[1] R. Lin, F. Liu, V. Anh, and I. Turner, Stability and convergence of a new explicit finitedifference approximation for the variable-order nonlinear fractional diffusion equation, Appl. Math. Comput. 212(2009) 435-445.

[2] P. Zhuang, F. Liu, V. Anh, I. Turner, Numerical methods for the variable-order fractional advection-diffusion equation with a nonlinear source term, SIAM J. Numer. Anal. 47 (3) (2009) 1760-1781.

[3] H. Zhang, F. Liu, P. Zhuang, I. Turner, V. Anh, Numerical analysis of a new space-time variable fractional order advection-dispersion equation, Appl. Math. Comput. 242 (2014)541-550.

[4] H. Zhang, F. Liu, M.S. Phanikumar, M.M. Meerschaert, A novel numerical method for the time variable fractional order mobile-immobile advectionC dispersion model, Comput. Math. Appl. 66 (2013) 693-701.

[5] G.D. Smith, Numerical Solution of Partial Differential Equations: Finite Difference Methods, Clarendon Press, Oxford, 1985.

An explicit solution to a random walk game

Guolong Li, Tony Tian

Address: Whitgift School, Haling Park, South Croydon, London, UK CR2 6YT

Email: tt36206856@163.com

Abstract. We study a two player zero sum game where a stone is initially placed on an integer point (i,j) with $0 < i < m$ and $0 < j < n$ on the Cartesian plane. At each step, player A makes the choice between +1 and −1 and player B makes the choice between x-axis and y-axis and they move the stone accordingly. Player A's objective is to let the stone first hit the x-axis or the line $x = m$ while player B's objective is to let the stone first hit the y-axis or the line $y = n$. We present an explicit formula for the Nash equilibrium of this game.

Key words. Game theory, Mixed-strategy, Trigonometry, Value.

1. Introduction

In this paper, we consider a game where a stone is placed on an integer point (i,j) with $0 < i < m$ and $0 < j < n$ on the Cartesian plane. At each turn, the stone will move to one of its neighbouring integer point. As illustrated in Figure 1, Player B chooses in which axis it moves and player A chooses whether it moves in the +1 direction or −1 direction. Player A wins if the stone hits the blue boundary lines $y = 0$ or $x = m$ before it hits the red boundary lines $x = 0$ or $y = n$ and Player B wins otherwise. Both players want to maximise their winning probability.

We assume the payoff for player A to be 1 if player A wins and -1 if player B wins. Let $V_{i,j}$ be the expected payoff for player A of the game starting at {i,j} provided both players play according to the Nash equilibrium solution. At each step, it can thus be viewed as a two player zero sum game with payoff matrix shown in Table 1. In [1], it has been shown that this type of two person

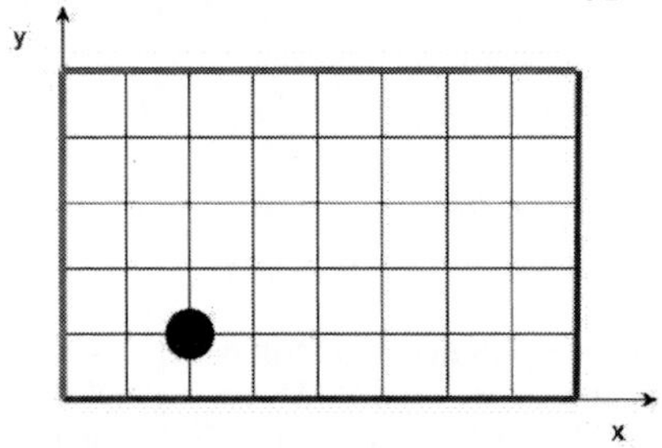

Figure 1: An illustration of the game: a stone is placed at {2,1}, it moves according to player A and player B's choices and player A wants it to hit the blue lines while player B wants it to hit the red lines.

IOP Conf. Series: Journal of Physics: Conf. Series **1039** (2018) 012009 doi:10.1088/1742-6596/1039/1/012009

	x-axis	y-axis
+1	$V_{i+1,j}$	$V_{i,j+1}$
-1	$V_{i-1,j}$	$V_{i,j-1}$

Table 1: Payoff matrix at each step of the game

zero-sum games has a Nash equilibrium point. Intuitively, we would expect Vi,j−1,Vi+1,j > Vi,j > Vi,j+1,Vi−1,j and therefore the players should play mixed strategies in the Nash equilibrium solution. In this paper, we investigate the Nash equilibrium solution of this game and obtain an explicit expression for the solution Vi,j.

2. Solution of the random walk game

In this section, we will present the Nash equilibrium solution of the game described in the Introduction. As discussed earlier, at each step of the game, we assume the two players to play the Nash equilibrium of the game shown in Table 1. Therefore, we can express Vi, j as a function of Vi+1,j, Vi,j+1, Vi−1,j and Vi,j−1. In addition, we know the boundary conditions Vi,0 = 1, Vi,n = −1, V0,j = −1,
Vm,j = 1 for 0 < i < m and 0 < j < n. We now present a explicit solution of this system of equations and show that it is indeed the Nash equilibrium solution of this game.

Lemma 2.1. For a game with payoff matrix $\begin{pmatrix} a & b \\ c & d \end{pmatrix}$, if min(a,d) > max(b,c), then the value of this game is $\frac{ad-bc}{a+d-b-c}$.

Proof. As min(a,d) > max(b,c), we know that both players must play mixed strategies in the Nash equilibrium solution. Let p be the probability that player A chooses the first row and q the probability that player A chooses the second row. At equilibrium, we must have

$$p + q = 1,$$

$$ap + cq = bp + dq.$$

Solving this, we deduce

$$p = \frac{d-c}{a+d-b-c}$$
$$q = \frac{a-b}{a+d-b-c},$$

And the value of this game is therefore

$$ap + cq = a\frac{d-c}{a+d-b-c} + c\frac{a-b}{a+d-b-c}$$
$$= \frac{ad-bc}{a+d-b-c}.$$

So, we now expect that the Nash equilibrium solution satisfies

$$V_{i,j} = \frac{V_{i+1,j}V_{i,j-1} - V_{i,j+1}V_{i-1,j}}{V_{i+1,j} + V_{i,j-1} - V_{i-1,j} - V_{i,j+1}}.$$

Solving this type of systems of equations in general is usually difficult. However, we have found a particular solution of this system of equations by making an educated guess.

IOP Conf. Series: Journal of Physics: Conf. Series **1039** (2018) 012009 doi:10.1088/1742-6596/1039/1/012009

Theorem 2.2. *Let z ∈ R be a constant and for 0 ≤ x, 0 ≤ y, x + y < π and x,y not both zero*

define the function

$$f_z(x,y) = \begin{cases} 1 & y = 0, \\ -1 & x = 0, \\ \frac{cot(y)-cot(x)+z}{cot(y)+cot(x)} & otherwise. \end{cases}$$

Now, for $0 < \theta < \frac{\pi}{2}$ *and θ ≤ x + y < π − θ, let*

$$a = f_z(x+\theta, y),$$
$$b = f_z(x, y+\theta),$$
$$c = f_z(x-\theta, y),$$
$$d = f_z(x, y-\theta),$$

then

$$\frac{ac - bd}{a + d - b - c} = f_z(x,y).$$

Proof. This can be shown using standard trigonometric identities. Instead, we have used Mathematica and verified these results. □

Theorem 2.3. Define fz as in Theorem 3.2 $\theta = \frac{\pi}{m+n}$ let and z = 2cot(mθ), then Vi,j = fz(iθ, jθ).

Proof. Let $\bar{V}_{i,j} = f_z(i\theta, j\theta)$ By Theorem 3.2, we have

$$\bar{V}_{i,j} = \frac{\bar{V}_{i+1,j}\bar{V}_{i,j-1} - \bar{V}_{i,j+1}\bar{V}_{i-1,j}}{\bar{V}_{i+1,j} + \bar{V}_{i,j-1} - \bar{V}_{i-1,j} - \bar{V}_{i,j+1}}.$$

Now, note that for 0 < x, y, x < mθ and x + y < π

$$\frac{\partial f_z(x,y)}{\partial x} = \frac{1}{\sin^2(x)} \times \frac{2\cot(m\theta) + 2\cot(y)}{(\cot(y) + \cot(x))^2} > 0.$$

So, fz is strictly increasing in x. Further, for 0 < i ≤ m, 0 < j < n, we have

IOP Conf. Series: Journal of Physics: Conf. Series **1039** (2018) 012009 doi:10.1088/1742-6596/1039/1/012009

$$\begin{aligned} f_z(i\theta, j\theta) &= \frac{\cot(j\theta) - \cot(i\theta) + z}{\cot(j\theta) + \cot(i\theta)} \\ &= -1 + \frac{2\cot(m\theta) + 2\cot(j\theta)}{\cot(i\theta) + \cot(j\theta)} \\ &> -1, \end{aligned}$$

as iθ+jθ < π and mθ+jθ < π. Therefore, we have fz(iθ, jθ) > fz((i−1)θ, jθ). Similarly, we have

$$\frac{\partial f_z(x,y)}{\partial y} = \frac{1}{\sin^2(y)} \times \frac{2\cot(m\theta) - 2\cot(x)}{(\cot(y) + \cot(x))^2} < 0.$$

And, for 0 < i < m, 0 < j ≤ n, we have

$$\begin{aligned} f_z(i\theta, j\theta) &= \frac{\cot(j\theta) - \cot(i\theta) + z}{\cot(j\theta) + \cot(i\theta)} \\ &= 1 + \frac{2\cot(m\theta) - 2\cot(i\theta)}{\cot(i\theta) + \cot(j\theta)} \\ &< 1. \end{aligned}$$

Hence, we have fz(iθ,jθ) < fz(iθ,(j − 1)θ). Recall $\theta = \frac{\pi}{m+n}$ when j = n, 0 < i < m, we have

$$\begin{aligned} f_z(i\theta, j\theta) &= \frac{\cot(n\theta) - \cot(i\theta) + 2\cot(m\theta)}{\cot(n\theta) + \cot(i\theta)} \\ &= \frac{\cot(n\theta) - \cot(i\theta) - 2\cot(n\theta)}{\cot(n\theta) + \cot(i\theta)} \\ &= \frac{-\cot(i\theta) - \cot(n\theta)}{\cot(n\theta) + \cot(i\theta)} \\ &= -1, \end{aligned}$$

and when *i* = *m*, 0 < *j* < *n*

$$\begin{aligned} f_z(i\theta, j\theta) &= \frac{\cot(j\theta) - \cot(m\theta) + 2\cot(m\theta)}{\cot(j\theta) + \cot(m\theta)} \\ &= \frac{\cot(j\theta) + \cot(m\theta)}{\cot(j\theta) + \cot(m\theta)} \\ &= 1. \end{aligned}$$

So, all the boundary conditions are satisfied by $\bar{V}_{i,j}$. Moreover, for 0 < i < m, 0 < j < n, min(V¯i+1,j,V¯i,j−1) > max(V¯i−1,j,V¯i,j+1). So, the Nash equilibrium of the game with payoff matrix $\begin{pmatrix} \bar{V}_{i+1,j} & \bar{V}_{i,j+1} \\ \bar{V}_{i-1,j} & \bar{V}_{i,j-1} \end{pmatrix}$ has mixed strategies and the value of this game is $\bar{V}_{i,j}$. Let (it,jt) be the position of the stone at step t of the game. Now, suppose player A plays the Nash equilibrium strategy of the payoff matrix $\begin{pmatrix} \bar{V}_{i_t+1,j_t} & \bar{V}_{i_t,j_t+1} \\ \bar{V}_{i_t-1,j_t} & \bar{V}_{i_t,j_t-1} \end{pmatrix}$ at each step, then no matter what player B does, $\bar{V}_{i_t,j_t}$ will be a martingale. So, suppose the stone starts at position i,j, By optional stopping theorem, player A can guarantee an expected payoff $\bar{V}_{i,j}$. Similarly, if player B plays the Nash equilibrium strategy of the payoff matrix $\begin{pmatrix} \bar{V}_{i_t+1,j_t} & \bar{V}_{i_t,j_t+1} \\ \bar{V}_{i_t-1,j_t} & \bar{V}_{i_t,j_t-1} \end{pmatrix}$, then according to [2], no matter what player A does, $\bar{V}_{i_t,j_t}$ will be a martingale and thus player B can guarantee that the expected payoff for player A in this game is $\bar{V}_{i,j}$. So, Vi,j = $\bar{V}_{i,j}$ is the expected payoff of player A in the Nash equilibrium solution. □

We now look at some properties of this solution. We now write $V_{i,j}^{m,n}$ be the solution of the expected payoff for player A if the stone starts at i,j on a m × n board.

Corollary 2.4. Let s > 0 be an integer, then $V_{is,js}^{ms,ns} = V_{i,j}^{m,n}$.

Proof. By Theorem 3.3,

$$\begin{aligned} V_{is,js}^{ms,ns} &= f_{2\cot(\frac{ms\pi}{ms+ns})}(\frac{is\pi}{ms+ns}, \frac{js\pi}{ms+ns}) \\ &= f_{2\cot(\frac{m\pi}{m+n})}(\frac{i\pi}{m+n}, \frac{j\pi}{m+n}) \\ &= V_{i,j}^{m,n}, \end{aligned}$$

as desired.

Corollary 2.5.

$$\lim_{s\to\infty} V_{i,j}^{ms,ns} \to \frac{i-j}{i+j}.$$

Proof. The result is trivial when i = 0 or j = 0. When i,j ≠ 0, we have

$$\begin{aligned} \lim_{s\to\infty} V_{i,j}^{ms,ns} &= \lim_{s\to\infty} \frac{\cot(\frac{j\pi}{ms+ns}) - \cot(\frac{i\pi}{ms+ns}) + 2\cot(\frac{m}{m+n})}{\cot(\frac{j\pi}{ms+ns}) + \cot(\frac{i\pi}{ms+ns})} \\ &= \lim_{s\to\infty} \frac{\cot(\frac{j\pi}{ms+ns}) - \cot(\frac{i\pi}{ms+ns})}{\cot(\frac{j\pi}{ms+ns}) + \cot(\frac{i\pi}{ms+ns})} \\ &= \lim_{s\to\infty} \frac{\cos(\frac{j\pi}{ms+ns})sin(\frac{i\pi}{ms+ns}) - cos(\frac{i\pi}{ms+ns})sin(\frac{j\pi}{ms+ns})}{\cos(\frac{j\pi}{ms+ns})sin(\frac{i\pi}{ms+ns}) + cos(\frac{i\pi}{ms+ns})sin(\frac{j\pi}{ms+ns})} \\ &= \lim_{s\to\infty} \frac{sin(\frac{i\pi}{ms+ns}) - sin(\frac{j\pi}{ms+ns})}{sin(\frac{i\pi}{ms+ns}) + sin(\frac{j\pi}{ms+ns})} \\ &= \frac{i-j}{i+j} \quad \text{(because } \sin(x) \sim x \text{ when x tends to zero.)} \end{aligned}$$

Now, we simplify the solution in the case of a square board.

Corollary 2.6. In the case when $m = n,\ V_{i,j}^{m,n} = \frac{\sin(\frac{i-j}{2m})}{\sin(\frac{i+j}{2m})}$.

Proof. Apply Theorem 3.3, z will equal to zero in this case and we have

$$\begin{aligned} V_{i,j}^{m,n} &= \frac{\cot(\frac{j\pi}{2m}) - \cot(\frac{i\pi}{2m})}{\cot(\frac{j\pi}{2m}) + \cot(\frac{i\pi}{2m})} \\ &= \frac{\cos(\frac{j\pi}{2m})\sin(\frac{i\pi}{2m}) - \sin(\frac{j\pi}{2m})\cos(\frac{i\pi}{2m})}{\cos(\frac{j\pi}{2m})\sin(\frac{i\pi}{2m}) + \sin(\frac{j\pi}{m+n})\cos(\frac{i\pi}{2m})} \\ &= \frac{\sin(\frac{i-j}{2m})}{\sin(\frac{i+j}{2m})} \end{aligned}$$

Intuitively, when the starting point is in the midpoint of the board, the expected payoff of the game will be zero. We show this result in the following corollary.

Corollary 2.7. *When m = 2i and n = 2j, then*

$$V_{i,j}^{m,n} = 0.$$

Proof. Using the notations in Theorem 3.3, denote. $x = \frac{i\theta}{m+n}$, $y = \frac{j\theta}{m+n}$ Recall $\theta = \frac{\pi}{m+n}$, $z =$ 2cot($m\theta$) we have $y + x = \frac{\pi}{2}$ and z = 2cot(2x). Hence,

$$\begin{aligned} V_{i,j}^{m,n} &= \frac{\cot(y) - \cot(x) + z}{\cot(x) + \cot(y)} \\ &= \frac{\tan(x) - \cot(x) + 2\cot(2x)}{\cot(x) + \tan(x)} \\ &= \frac{\tan(x) - \frac{1}{\tan(x)} + \frac{2}{\tan(2x)}}{\cot(x) + \tan(x)} \\ &= \frac{\tan(x) - \frac{1}{\tan(x)} + \frac{1-\tan^2(x)}{\tan(x)}}{\cot(x) + \tan(x)} \\ &= 0, \end{aligned}$$

as desired.

3. Conclusion

As a result,the given solution can be used to show the Nash equilibrium solution of the game. We have shown this with the examples as written above. We have also shown that the given solution satisfies the boundary conditions. Therefore, we have proven that the given solution is an explicit solution of the game.

4. Appendix

We have discussed that when $m = n$, $V_{i,j}^{m,n} = \frac{\sin(\frac{i-j}{2m})}{\sin(\frac{i+j}{2m})}$. In this section, we will present the proof for this statement.

Proof. Let $V_{i,j}^{m,n} = \frac{\sin(\frac{i-j}{2m})}{\sin(\frac{i+j}{2m})}$. By applying Thorem3.2, we have $\theta = \frac{\pi}{2m}$ and

$$\bar{V}_{i,j} = \frac{ad - bc}{a + d - b - c.}$$

Now, for $0 < \theta < \frac{\pi}{2}$ and $\theta \leq x + y < \pi - \theta$, let$((i - j)\theta)$ be A and $((i + j)\theta)$ be B, we then have

$$a = \frac{sin(A+\theta)}{sin(B+\theta)},$$
$$b = \frac{sin(A-\theta)}{sin(B+\theta)},$$
$$c = \frac{sin(A-\theta)}{sin(B-\theta)}$$
$$d = \frac{sin(A+\theta)}{sin(B-\theta)},$$

$$\begin{aligned} ad - bc &= \frac{sin^2(A+\theta)}{sin(B+\theta)sin(B-\theta)} - \frac{sin^2(A-\theta)}{sin(B+\theta)sin(B-\theta)} \\ &= \frac{sin^2(A+\theta) - sin^2(A-\theta)}{sin(B+\theta)sin(B-\theta)} \end{aligned}$$

$$\begin{aligned} a + d - b - c &= \frac{sin(A+\theta)sin(B-\theta) + sin(A+\theta)sin(B+\theta) - sin(A-\theta)sin(B+\theta) - sin(A-\theta)sin(B-\theta)}{sin(B+\theta)sin(B-\theta)} \\ &= \frac{sin(A+\theta)(sin(B-\theta) + sin(B+\theta)) - sin(A-\theta)(sin(B+\theta) + sin(B-\theta))}{sin(B+\theta)sin(B-\theta)} \end{aligned}$$

We substitute these results to

$$\bar{V}_{i,j} = \frac{ad - bc}{a + d - b - c}$$

and obtain

$$\begin{aligned} \bar{V}_{i,j} &= \frac{sin^2(A+\theta) - sin^2(A-\theta)}{sin(A+\theta)\big[sin(B-\theta) + sin(B+\theta) - sin(A-\theta)(sin(B+\theta) + sin(B-\theta))\big]} \\ &= \frac{sin(A+\theta) + sin(A-\theta)}{sin(B-\theta) + sin(B+\theta)} \\ &= \frac{2sin(A)cos(\theta)}{2sin(B)cos(\theta)} \\ &= \frac{sin(A)}{sin(B)} \\ &= \frac{sin((i-j)\theta)}{sin((i+j)\theta)} \end{aligned}$$

IOP Conf. Series: Journal of Physics: Conf. Series **1039** (2018) 012009 doi:10.1088/1742-6596/1039/1/012009

as desired.

5. References

[1] J. von Neumann and O. Morgenstern Theory of Games and Economic Behavior. Princeton University Press

[2] P.Hall/C.C Heyde *Martingale Limit Theorem and Its Application.* Academic Press

6. Acknowledgement

We would like to express our gratitude to Dr. Si Cheng for his advices on trigonometry formulae and techniques for calculating limits and differentials.

Cop and Robber with road building ability

Guolong Li, Yanchen Pu

Address: Whitgift School, Haling Park, South Croydon, London, UK CR2 6YT
Email: pyc.china@outlook.com

Abstract. We investigate the game of cop and robber, playing on a finite graph between one cop and one robber. The robber can spend one move to build a road to any other vertices on the graph or stay at his position. Several results are provided toward to find characteristics of graphs on which the robber can win.

1. Introduction

The game of Cop and Robber is played on a reflexive graph, that is a graph with loops at every vertex. The cop chooses a vertex to occupy then the Robber chooses one. The two sides then move alternatively, Cop first, where a move is to slide along an edge. The loops are technical devices which allow the cop and the robber to pass. There is perfect information, that is each side knows the location of the other. The Cop wins if he and the Robber are on the same vertex at the same time. Graphs in which the Cop can win are called cop-win graphs and are characterised in [1] and [2]. In [3], a corner ranking procedure was described and it had proved that a graph is cop-win if and only if it has a finite corner rank.

In our research, we consider the situation in which the Robber can build a road to any other vertices or just stay at his position using his move. To avoid confusion, we regard the graph that the winner is the robber as builder-win in such a game. We have made some contribution in characterising builder-win graphs.

We show that a graph is builder-win if for every vertex on the graph, if there exists another vertex that is 3 unit of distance from it. However, the converse is not true. We give an example of a cop-win graph, which has a vertex that its distance from any other vertices on the graph is at most 2, but the graph is also builder-win. In addition, we show that there exist graphs with arbitrarily high corner rank which are not builder-win.

2. Sufficient conditions for builder-win graphs

As the robber has the extra ability to build roads in our game compared to the traditional cop and robber game, any robber-win graph is builder-win. Therefore, we will focus on cop-win graphs in our research. On a cop-win graph, if the robber can find a vertex which is far from the cop, he may be able to have time to build roads before the cop comes and finally win the game. The
results are as follows:

2.1. Theorem 2.1. For a graph G, for every vertex v on the graph, if we can find another vertex u with distance at least 3 from v, then the graph is builder-win.

Proof. Consider a graph G satisfying the conditions in the theorem. We assume G is a connected graph as otherwise the graph is clearly builder-win. Let v be the vertex chosen by the cop, the robber can then choose a vertex u with distances 3 from the cop. Let v0, u0 be two vertices so that v,v0,u0,u forms a path.

We only need to consider the case in which the cop never stays at its own position because the robber can pass as well. Now, the cop moves away from *v*, the robber can build a road from *u* to *v*. As the distance from *u* and *v* is 3, *u* and *v* has no common neighbours and the cop can not catch the robber in the next step.

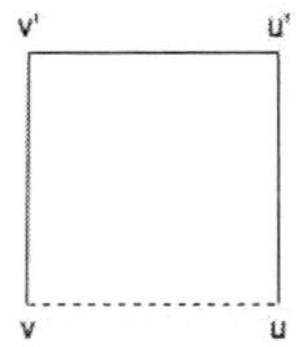

Figure 1: A cycle formed by the four vertices u, v, u0, v0

Now, we show that the robber can win by staying in the cycle shown on Figure 1. We consider the following strategy played by the robber:

- If the robber's position is at v (or u), stay there unless the cop is at one of the robber's neighbours. If the cop is next to the robber's position, but not at u (or v), the robber then can move to u (or v). If the cop is at u (or v), the robber can move to v0 (or u0). Since u and v has no common neighbours, the cop cannot catch the robber in the next step.
- So, the robber can only be chased away from vertices u and v if the cop is on the other vertex. Assume without loss of generality, the cop moves to u, chasing the robber to v0. If the cop moves to v, the robber moves to u0. Otherwise, if the cop moves to any other verteices, the robber moves to v.
- So, after the movement of robbers, the situations are
  -The robber is at u or v, the cop is not at robber's neighbours.
  -the robber is at u0 or v0 and the cop is at the diagonally opposite vertex of the robber as shown on Figure 1.

In both cases, the cop can not catch the robber.

One might expect that if, on a cop-win graph, the cop can choose a vertex which has distance at most 2 to any other vertices, then the cop can win. However, we testify that this is not the case through a counter example.

2.2. Theorem 2.2. There exist cop-win graphs, in which the distance between any two vertices is at most 2, are also builder-win.

Proof. We set the following graph with vertices set *(i,j)*, $1 \leq i \leq 3$, $1 \leq j \leq 4$ and there will be an edge between vertices *(i,j)* and *(k,l)*, if and only if at least one of the following two conditions holds.

- $i = k = 3$
- $|i - k| \leq 1$ and $j - l$ 6≡ 2 (mod 4)

First, we show that this graph is cop-win. We consider the following strategies of cops.

- Place at (3,1) initially.
- If the robber places at (i1,j1) as the beginning point, go to (3,j1).
- Now, the robber goes to (i2,j2), if i2 6= 1, cops either catch the robbers or go to (2,j2).
- Then, if the robber is not caught yet and goes to (i3,j3), the cop can always catch the robber.

Also , we note that every two vertices in the graph have distances at most 2. We now show that the graph can also be builder-win by conceiving the following strategies adopted by the robber.

- Suppose the cop chooses the position (i1,j1) initially, the robber then can move to (1,l1) where l1 ≡ 2 + j1 (mod 4).
- At step n, if the cop's position is at (in,jn), go to (1,ln) with ln ≡ 2 + jn (mod 4) whenever
- possible.

- The robber can always do this unless the cop moves to (3,jn) at step n with jn = ln−1. The robber then builds a road between (1,ln−1) and (3,jn−1).

Now we show that the robber can win after building the road between $(1, l_{n-1})$ and $(3, j_{n-1})$ while the cop is at $(3, l_{n-1})$ after both players made their n's turn. We can assume without loss of generality that $l_{n-1} = 1$. Now, consider the set of vertices $S=\{(1,1),(2,1),(2,2),(2,4),(3,1),(3,3)\}$. After the robber builds the road between (1,1) and (3,3), we observe that if the robber is at a vertex in S and the cop moves to a neighbour of the robber, the robber can always move to another vertex in S which is not the cop's neighbour. Therefore, the robber can always stay in S and make sure that the cop cannot catch him.
To conclude, we have constructed a graph which is cop-win, the distance between any two vertices on the graph is at most 2, and the graph is also builder-win.

3. Builder-win graphs and corner ranking

In this section, we will investigate the relationship between builder-win graphs and corner rank developed in [3].
First, we state the relevant definitions about corner rank which are necessary for this paper. For a reflexive graph G, V (G) refers to the vertices of G and E(G) refers to the edges of G. If G is a graph and X is a vertex or set of vertices in G, then we let G−X be the subgraph of G induced by V (G)\X. We say that a vertex u dominates a set of vertices X if u is adjacent to every vertex in X. Given a vertex v in a graph, by N(v), the closed neighbourhood of v, we mean the set of vertices adjacent to v. As the graph is reflexive, we always have v ∈ N(v). For distinct vertices v and w, if N(v) ⊆ N(w) then we say that v is a corner and w corners v; if N(v) ⊂ N(w), we say that v is a strict corner and w strictly corners v; if N(v) = N(w), we call v and w twins. We now define the corner ranking procedure as described in [3].

3.1. Definition 3.1. For any graph G, we define a corresponding corner rank function, cr, which maps each vertex of G to a positive integer or ∞. We also define a sequence of associated graphs $G[1],\ldots,G[\alpha]$.

- Initialize G[1] = G, and k = 1.
- If G[k] is a clique, then:
  -Let cr(x) = k for all x ∈ G[k].
  -Then stop.
- Else if G[k] is not a clique and has no strict corners, then:
  -Let cr(x) = ∞ for all x ∈ G[k].
  -Then stop.
- Else
  -Let V be the set of strict corners in G[k].
  -For all x ∈ V , let cr(x) = k.
  -Let G[k+1] = G[k] − V , increment k by 1 and return to step 1.

Furthermore, we denote cr(G) = maxx∈G cr(x).

Lemma 3.2. For a graph G and integer k, if cr(G) > k, then any vertex u ∈ G with cr(u) = k is strictly cornered by a vertex v ∈ G[k+1] in the graph G[k].

Proof. Let S be the set of vertices in G[k] which strictly corners u, as cr(u) = k, S cannot be empty. For a vertex x, denote d(x) the number of adjacent vertices of x in G[k]. Then, there exists a vertex v ∈ S such that d(v) ≥ d(x) for all x ∈ S. We show that this v is in G[k+1]. Suppose we have a vertex *w* which strictly corners *v* in $G^{[k]}$. As *v* strictly corners *u* in $G^{[k]}$, *w* will also strictly corner *u* in $G^{[k]}$ and thus $w \in S$. However, we would then have $d(v) \geq d(w)$, which means w cannot strictly corner v. Therefore, we have $v \in G^{[k+1]}$ as desired.

Now, we show some results on the relationship between builder-win graphs and corner rank.

Theorem 3.3. Consider a graph G, if there exists a vertex v $\in$ G[2] which is adjacent to all other vertices in G[2], then G cannot be a builder-win graph.

Proof. The cop can win with the following strategy:

- The cop chooses to be at v at the start.
- If the robber chooses to be in G[2], the cop can catch the robber immediately.
- If the robber chooses to be in a vertex w which is not in G[2], then there must be a vertex u in G[2] which strictly corners w. The cop can then move to u.
- As u strictly corners w, the cop can catch the robber in the next step no matter where the robber goes to.

Theorem 3.4. For any positive integer n, there exists a graph G with cr(G) = n such that G is not a builder-win graph.

Proof. We consider the graph G with vertices (i,j), $1 \leq i \leq 2$, $1 \leq j \leq n$. Vertices (i,j) and (k,l) are adjacent if any of the following conditions hold:

- i = k, • |j − l| = 1,
- l = j = n.

We now show that G has corner rank n by showing that cr(i,j) = j. We prove this statement using induction. We note that (1,1) is strictly cornered by (1,3) and (2,1) is strictly corned by (2,3) in G. None of the other vertices in G is strictly cornered as they all have the same degree. So, we have cr((1,1)) = cr((2,1)) = 1 and (i,j) $\in$ G[2] if and only if $j \geq 2$.
Now, assume for an integer $1 < k < n$, cr((i,j)) = j for all $1 \leq i \leq 2$ and $j < k$. So, (i,j) $\in$ G[k] if and only if $j \geq k$. Now, in G[k], if k 6= n − 1, (1,k) will be strictly cornered by (1,k + 2) and (2,k) will be strictly cornered by (2,k + 2), all other vertices will not be strictly cornered as they all have the same degree in G[k]. So, we have cr((1,k)) = cr((2,k)) = k and (i,j) $\in$ G[k+1] if and

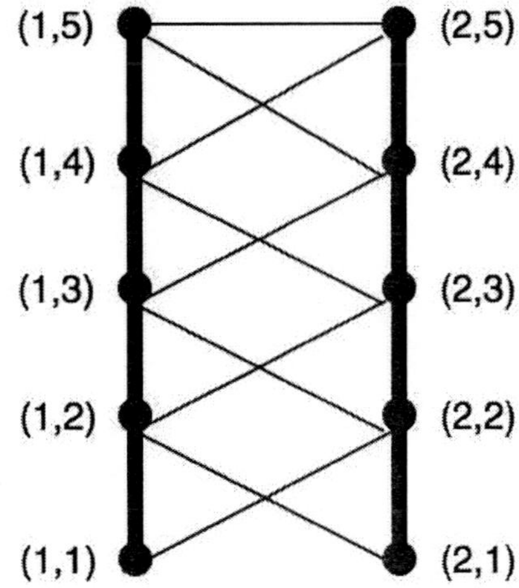

Figure 2: An illustration of the graph when n = 5. Vertices are adjacent if they are connected by a line segment or if they are on the same bold line.

only if $j \geq k + 1$ as desired. If $k = n - 1$, then $(1,k)$ and $(2,k)$ will both be strictly cornered by $(1,n)$ and $(2,n)$ and $G^{[n]}$ will just contain $(1,n)$ and $(2,n)$. This concludes the inductive proof.

Finally, we prove that the cop can win in such graph by adopting the following strategies:

- Choose (1,n) at the start.
- If the robber chooses (1,l1), the cop can catch him immediately. So, suppose the robber chooses (2,l1). Choose (2,j2) with j2 = n if n ≡ l1(mod 2) and j2 = n − 1 otherwise.
- After round m, suppose the robber is at (km,lm) and the cop is at (im,jm), if im = km, the cop can catch the robber directly, otherwise the cop goes to (im+1,jm+1) with im+1 = km and jm+1 = jm − 1.

We show that the cop can indeed win with this strategy. To avoid being caught, the robber will have to choose $k_2 = 1$, $k_3 = 2$, $k_4 = 1$ and so on. Therefore, we only need to consider the case when $k_m = i_{m+1} = 1$ if $m \equiv 0 (\mathrm{mod}\ 2)$ and $k_m = i_{m+1} = 2$ otherwise. Moreover, as long as $j_{m+1} > l_m$, $l_{m+1} = l_m + 1$ or $l_m - 1$. As $j_{m+2} = j_{m+1} - 1$, we would have $j_{m+2} - l_{m+1} = j_{m+1} - l_m$ or $j_{m+2} - l_{m+1} = (j_{m+1} - l_m) - 2$. As $j_2 \equiv l_1 (\mathrm{mod}\ 2)$ and $j_2 \geq l_1$, eventually, for some m we will have $j_{m+1} = l_m$ and thus the robber will be caught. □

4. References

[1] R. Nowakowski, P. Winkler Vertex to vertex pursuit in a graph, Discrete Mathematics 43 (1983), 235-239.
[2] A. Quillot, Th`ese d'Etat, Universit´e de Paris VI (1983), 235-239.
[3] D. Offner, K. Ojakian, Cop-Win Graphs: Optimal Strategies and Corner Rank submitted

IOP Conf. Series: Journal of Physics: Conf. Series **1039** (2018) 012011 doi:10.1088/1742-6596/1039/1/012011

Abdollahi's Conjectures and a Class of Prime Power Group

K Moradipour[1] and S Ilangovan[2]

[1]Department of Mathematics, Faculty of Khorramabad, Lorestan Branch, Technical and Vocational University (TVU), Tehran, Iran.
[2]Foundation of Science, Faculty of Science, University of Nottingham Malaysia Campus, Jalan Broga, 43500 Semenyih, Selangor, Malaysia.

Email: kayvanmrp@yahoo.com,

Abstract. In this paper, we give necessary and sufficient conditions for metacyclic p-groups H and K to have isomorphic non-commuting graphs.

1. Introduction

Let G be a non-abelian group and $Z(G)$ be its center. The non-commuting graph $\nabla(G)$ of G is a graph whose vertex set is $G - Z(G)$ and two vertices x and y are adjacent if $xy \neq yx$. In 1975 [1], Paul Erdos first considered the non-commuting graph of a group. Many researchers have studied the non-commuting graph (e.g [2],[3],[4]). In [5] Abdollahi *et al.* put forward the following two conjectures:

1. Let G and H be two finite non-abelian groups such that $\nabla(G) \cong \nabla(H)$. Then $|G| = |H|$
2. If G and H are two non-abelian finite groups and H is simple such that $\nabla(G) \cong \nabla(H)$, then $G \cong H$

The above conjectures were negative generally, but they hold for various classes of groups. Conjecture 1 was refuted by an example in [6], however it is valid whenever one of G or H is a non-abelian finite simple group [7]. Darafsheh [7] showed that if H or K is a finite non-abelian simple groups, then Conjecture 1 holds and if H or K is a finite non-abelian simple groups satisfying the Thompson's conjecture, then Conjecture 2 is true. Abdollahi and Shahverdi [8] proved that if H or K is an alternating group, then Conjecture 2 is true.

In this study, we use a family of non-abelian metacyclic p-groups of Beuerle's calasification [9] to investigate the above conjectures. If the Conjecture 2 holds, the groups in question need not be isomorphic. Indeed, the smallest scale are the dihedral group D_8 and the group Q_8 of quaternions. Thus, it is an interest to investigate the question of isomorphism for various restricted families of groups. In the present paper, we carry out this investigation for the family $\Im$ of non-abelian metacyclic groups of prime power order. That is, assuming $\nabla(G) \cong \nabla(H)$ for H, K $\in \Im$ determine necessary and sufficient conditions for the groups H and K to be isomorphic.

If G is a metacyclic p-group, then G is fallen in one of the following cases in which for all cases $\alpha, \beta \in \mathrm{N}$ and $\varepsilon, \gamma \geq 0$ are integers.

1. **Case A**:

$G = G(\alpha,\beta,\gamma,\varepsilon) = \langle a,b \mid a^{p^\alpha} = 1, b^{p^\beta} = a^{p^{\alpha-\varepsilon}}, a^b = a^{p^{\alpha-\gamma}+1} \rangle$, for some $\alpha,\beta,\gamma,\varepsilon$ where $\beta \geq \gamma \geq 1$ and p is an odd prime or $\alpha - \gamma \geq 2$.

2. **Case B**:
$G = G(\alpha,\beta,0,\varepsilon) = \langle a,b \mid a^{2^\alpha} = 1, b^{2^\beta} = a^{2^{\alpha-\lambda}}, a^b = a^{-1} \rangle$, where $\alpha \geq 2$.

3. **Case C**:
$G = G(\alpha,\beta,\gamma,\varepsilon) = \langle a,b \mid a^{2^\alpha} = 1, b^{2^\beta} = a^{2^{\alpha-\lambda}}, a^b = a^{2^{\alpha-\gamma-1}} \rangle$, where $\alpha - \gamma \geq 2$ and $\gamma > 0$.

In the following lemma some general properties of elements in the groups above are listed.

Lemma 1.1 Let G be a non-abelian metacyclic p-group (p is any prime) of Cases A, B or C. If $x, y \in G$ with $x = a^i b^j$ and $y = a^s b^t$, then the following hold in G.

(1) $b^j a^i = a^{ir^j} b^j$;

(2) $xy = a^{i+sr^j} b^{j+t}$;

(3) $x^y = a^{s(1-r^j)+ir^t} b^j$;

(4) $[x,y] = a^{i(1-r^t)+s(r^j-1)}$.

Proof. Since the proofs of all parts have similar way, we only give the proof of part (1). We first show $a^{b^j} = a^{r^j}$ by applying induction on j for each $j \geq 1$. In fact, if $a^b = a^r$ and $a^{b^j} = a^{r^j}$, then $a^{b^{j+1}} = \left(a^{b^j}\right)^b = \left(a^{r^j}\right)^b = \left(a^b\right)^{r^j} = \left(a^r\right)^{r^j} = a^{r^{j+1}}$. It follows that $a^{ib^j} = \left(a^{b^j}\right)^i = \left(a^{r^j}\right)^i$, that is $b^j a^i b^{-j} = a^{ir^j}$. Hence $b^j a^i = a^{ir^j} b^j$.

For future reference, we need the following two lemmas that give formulas for the order, center and the order of the center of metacyclic p-groups. For the proof, we refer to [9], Proposition 2.5.

Lemma 1.2 Let G be a metacyclic p-group of Case A. Then $|G| = p^{\alpha+\beta}$, $Z(G) = \langle a^{p^\gamma}, b^{p^\gamma} \rangle$ and $|Z(G)| = p^{\alpha+\beta-2\gamma}$.

Proof. For (i) and in the case that p is odd prime, we know that $\langle a \rangle \cap \langle b \rangle = \langle a^{p^{\alpha-\varepsilon}} \rangle$ has order p^ε and $G = \langle a \rangle \langle b \rangle$. An easy computation shows that the order of G is $p^{\alpha+\beta}$. For the case $p = 2$ we know that $G = \langle a \rangle \langle b \rangle$, $|a| = 2^\alpha$ and $|b| = 2^{\beta+\varepsilon}$. Also, the order of $\langle a \rangle \cap \langle b \rangle = \langle a^{2^{\alpha-\varepsilon}} \rangle$ is 2^ε, then the order of G is $2^{\alpha+\beta}$. Since $Z(G) = \langle a^{2^\gamma}, b^{2^\gamma} \rangle$. There are $2^{\alpha-\gamma}$ elements generated by a^{2^γ} and $2^{\beta-\gamma}$ elements generated by b^{2^γ}. Since $[b,a] = a^{2^{\alpha-\gamma}}$, thus $[b,a] \in \langle a^{2^{\alpha-\gamma}} \rangle$. Hence the order of $Z(G)$ is $2^{\alpha-\gamma} . 2^{\beta-\gamma} = 2^{\alpha+\beta-2\gamma}$.

Lemma 1.3 Let G be a metacyclic 2-group of Cases B and C. Then $Z(G) = \langle a^{2^{\alpha-1}}, b^{2^{\max\{1,\gamma\}}} \rangle$, $|Z(G)| = 2^{\beta-\max\{1,\gamma\}+1}$ and $|G| = 2^{\alpha+\beta}$.

These groups were studied through their centralizers in [10], which are summarized in the following three propositions.

Proposition 1.4 If $G = G(\alpha,\beta,\gamma,\varepsilon)$ be a group of case A and let $x = a^i b^j \in G$. Then $|C_G(x)| = p^{\alpha+\beta-\gamma+\min\{e_p(i),\, e_p(j)\}}$, where $e_p(i)$ denotes the largest exponent of p in i.

Proposition 1.5 If $G = G(\alpha,\beta,0,\varepsilon)$ be a group of case B and let $x = a^i b^j \in G$. Then

$$|C_G(x)| = \begin{cases} 2^{\alpha+\beta}, & a^i b^j \in Z(G) \\ 2^{\alpha+\beta-1}, & a^i b^j \notin Z(G), j \ even, \\ 2^{\beta+1}, & a^i b^j \notin Z(G), j \ odd. \end{cases}$$

Proposition 1.6 If $G = G(\alpha,\beta,\gamma,\varepsilon)$ is a group of case C, then for $x = a^i b^j \in G$,

$$|C_G(x)| = \begin{cases} 2^{\alpha+\beta-\gamma+e_2(i)}, & j \ even,\ e_2(j) \geq \gamma \ and \ e_2(i) < \gamma, \\ 2^{\alpha+\beta-\gamma+e_2(j)}, & j \ even,\ e_2(i) \geq \alpha-1, \ or \\ & e_2(i)-e_2(j) \geq \alpha-\gamma-1 \ and \ e_2(i) \geq \gamma, \\ 2^{\alpha+\beta-\gamma+e_2(j)-1}, & j \ even,\ e_2(i) \geq \gamma,\ e_2(i) < \alpha-1 \\ & and \ e_2(i)-e_2(j) < \alpha-\gamma-1, \\ 2^{\alpha+\beta-\gamma+e_2(j)}, & j \ even,\ e_2(j) \leq e_2(i) < \gamma \ and \\ & e_2(i)-e_2(j) \geq \alpha-\gamma-1, \\ 2^{\alpha+\beta-\gamma+e_2(j)-1}, & j \ even,\ e_2(j) \leq e_2(i) < \gamma \ and \\ & e_2(i)-e_2(j) < \alpha-\gamma-1, \\ 2^{\alpha+\beta-\gamma+e_2(j)}, & j \ even \ and \ e_2(i) < e_2(j) < \gamma, \\ 2^{\beta+1}, & j \ odd. \end{cases},$$

We show the case of a metacyclic p-group G by $\text{case}(G)$ and the set of all degrees of vertices of a graph $\nabla(G)$ by $D(\nabla(G))$.

2. Main Results

In this section, we give necessary and sufficient conditions for two prime power metacyclic groups to have isomorphic non-commuting graphs.

If G is a group, then the set of all conjugacy class sizes of G called the *conjugacy vector type* of G. The following lemma is used to prove the next theorems.

Lemma 2.1 Let H and K be two groups with the same orders. If the non-commuting graphs of these two groups are isomorphic, then the size of conjugacy classes of H and K are identical.

Proof. Let $\phi : V(\nabla(H)) \to V(\nabla(K))$ be a one-to-one correspondence of the vertices of the graphs $\nabla(H)$ and $\nabla(K)$. For $x \in V(\nabla(H))$, let x^H denote the conjugacy class of x in H. Suppose $\phi(x) = y$ such that $\deg_{\nabla(H)}(x) = \deg_{\nabla(K)}(y)$. Then $|C_H(x)| = |C_H(y)|$, since $|H| = K|$, we have

$|x^H| = |H : C_H(x)| = |H| \; / \; |C_H(x)| = |K| \; / \; |C_K(y)| = |K : C_K(y)| = |y^K|$. That is, H and K have the same number of conjugacy class sizes. □

In this section, we let $H = H(\alpha, \beta, \gamma, \varepsilon)$ be a non-abelian metacyclic $p-$group and $K = K(\alpha', \beta', \gamma', \varepsilon')$ be a non-abelian metacyclic q-group. The following theorem shows that Conjecture 1 holds for the family $\Im$ non-abelian metacyclic prime power groups.

Theorem 2.2. If $\nabla(H)$ is isomorphic to $\nabla(K)$, then H and K have the same orders and centers.

The following theorem gives necessary and sufficient conditions under which two non-abelian metacyclic prime power groups have isomorphic non-commuting graphs.

Theorem 2.3 $\nabla(H)$ is isomorphic to $\nabla(K)$ if and only if one of the following holds:

(1) case (*H*) = case (*K*) = *I*, $\alpha + \beta = \alpha' + \beta'$ *and* $\gamma = \gamma'$,
(2) case (*H*) = case (*K*) = *II*, $\alpha = \alpha'$ *and* $\beta = \beta'$,
(3) case (*H*) = case (*K*) = *III*, $\alpha = \alpha', \beta = \beta'$ *and* $\gamma = \gamma'$,
(4) case (*H*) = *I*, case (*K*) = *II*, $\gamma = 1, \alpha' = 2$ *and* $\beta' = \alpha + \beta - 2$,
(5) case (*H*) = *II*, case (*K*) = *III*, $\alpha = \alpha' \geq 3, \beta = \beta'$ *and* $\gamma' = 1$.

3. Conclusions

In this study, some basic properties of the non-commuting graphs of metacyclic p-groups were investigated. If the non-commuting graph Γ_H of the metacyclic p-group H and the non-commuting graph Γ_K of the 20 metacyclic q-group K are isomorphic, then H and K have the same orders and centers.

References

[1] Neumann BH 1976 *J. Aust. Math Soc. Ser. A* **21** pp 467-472
[2] Abdollahi A, Akbari S, Dorbidi H and Shahverdi H 2013 *Comm. Algebra* **41** pp 451-461
[3] Abdollahi A and Shahverdi H 2012 *J. Algebra* **37** pp 206-207
[4] Darafsheh MR and Yousefzadeh P 2013 *J. Group Theory* **2**(2) pp 47-72
[5] Abdollahi A, Akbari S and Maimani HR 2006 *J. Algebra* **298** pp 468-492
[6] Moghaddamfar AR 2005 *Siberian Math. Journal* **47**(5) pp 1112-1116
[7] Darafsheh MR 2009 *Discrete App. Math* **157** pp 833-837
[8] Abdollahi A and Shahverdi H 2014 *Comm. Algebra* **42** pp 3944-3949
[9] Beuerle JR 2005 *Algebra Colloq.* **12**(4) pp 553-562
[10] Moradipour K 2013 Size of conjugacy classes and centralizer of metacyclic p-groups *Thesis* UTM Malaysia

Acknowledgment

First author would like to acknowledge Technical and Vocational University, Tehran, Iran for their supports. Besides, the second author is also indebted to University of Nottingham Malaysia campus for the financial funding.

IOP Conf. Series: Journal of Physics: Conf. Series **1039** (2018) 012012 doi:10.1088/1742-6596/1039/1/012012

Solution of the general multi-objective De-Novo programming problem using compensatory operator under fuzzy environment

D Bhattacharya[1] **and S Chakraborty**[2]
1 Department of Mathematics, NIT Agartala, India
2 Department of Mathematics , NIT Agartala, India

Email: bhattacharyad_nita@yahoo.co.in

Abstract. This paper presents an alternative approach for the solution of the general multi-objective De-Novo Programming Problem under fuzzy environment in one step using Luhandjula's compensatory μ_θ -operator. Luhandjula used compensatory μ_θ -operator to solve vector maximization problem, whereas in this paper the use of compensatory μ_θ -operator has been extended to solve the general multi-objective De-Novo Programming Problem. Also the solution obtained by the proposed approach elicits an efficient solution of the problem considered under the assumption of the uniqueness of the solution. The method has been illustrated by a numerical example.

1. Introduction

One of the salient features of the traditional multi-objective optimization problem (MOOP) is trade-offs among the objectives due to the conflicting nature of the objectives. Thus under a given allocation of resources, it may not be always possible to find a single point in the feasible region, where all the objectives can be simultaneously maximized or minimized and thus trade-offs among the objectives become inevitable. De-Novo Programming Problem (DNPP) introduced by Zeleny [1] is popularly used to design an optimal system by extending existed resources (if necessary) instead of finding an optimum in a given system with fixed resources. Another interesting aspect of DNPP is that the decision-maker can obtain a trade-off free solution for its objectives. In his papers, Zeleny considered MOOPs (with maximizing type of objectives only)[1-12] where the number of objectives is equal to the number of unknowns.

But the cases where there are only minimizing type of objective functions or the general situation involving both maximizing and minimizing type of objective functions, were not addressed. Also there exists no algorithm for solving a general multi-objective De-Novo Programming Problem (MODNPP). For solving a general type of MODNPP, Li and Lee [13] introduced a two-step fuzzy approach using the concept of ideal and negative ideal points. Sasaki,et.al. [14] investigated the problem of the system reliability with fuzzy goals formulating it as DNPP and solved by Genetic Algorithm. Zhang,et.al. [15] presented an interval DNPP approach for designing optimal water resource management system mingling interval programming and DNPP. Pert FIALA [16] discussed some possible extensions of DNPP e.g. Fuzzy DNPP, Interval DNPP etc. Huang et.al. [15] used DNPP to solve Taiwan Freeway's maintenance work problem. Saeedi et.al. [17] determined the capacity of recovery facilities in the reverse flow system formulating it as a mixed integer non-linear programming problem with DNPP. Further works towards the enrichment of the concept of De-Novo Programming could be seen in the works of Chen [8] and Nurullah Umarusman [18]. Nurullah Umarusman [18] used Min-Max Goal Programming technique to obtain optimum solution of general DNPP under fuzzy environment to get the highest closeness to ideal solutions in two steps.

In the present paper a new method has been introduced to solve general MODNPP in one step only. The proposed approach is computationally easier for the solution of MODNPP in comparison with the existing ones and also it elicits efficient solution of the problem in one step only assuming uniqueness of such solution. Actually in this paper Luhandjula's technique [19] for the solution of vector maximization problem using compensatory min-bounded sum operator μ_θ has been extended to general MODNPP. Using μ_θ operator, a fuzzy approach for the solution of general MODNPP has been proposed. Here the optimistic expectations of the fuzzy objectives have been taken as usual the ideal values [1], whereas the pessimistic objective values or the threshold values have been computed using Luhandjula's comparison technique [19].The method has been illustrated with a numerical example. It is revealed that the same solution as that of Li and Lee could be obtained by the proposed technique but with greater computational ease. A theorem has also been stated here which guarantees an efficient solution of the problem considered. With this aim the paper has been organised in four sections. Section-1 contains introduction and a brief account of the relevant contemporary works; in section-2 we actually introduce the proposed method of solution and a theorem has been stated which establishes the existence of efficient solution of the problem; section-3 illustrates the new approach with numerical example and section-4 contains the conclusion.

2. General Multi-Objective De-Novo Programming Problem (MODNPP)

Let us first consider an MODNPP [20]

$$Max\ \boldsymbol{Z} = \boldsymbol{Cx};\ Subject\ to\ \boldsymbol{Ax} \leq \boldsymbol{b};\ \boldsymbol{pb} \leq B; \boldsymbol{x} \geq 0 \tag{1}$$

where $\boldsymbol{C} \in R^{q\times n}$ and $\boldsymbol{A} \in R^{m\times n}$ are the matrices of dimensions $q \times n$ and $m \times n$ respectively, $\boldsymbol{b}^T = [b_1, b_2, \dots, b_m]^T \in R^m$ be the m-dimensional vectors of resources, $\boldsymbol{x}^T = [x_1, x_2, \dots, x_n]^T \in R^n$ be the n-dimensional vectors of decision variables. Here B be the total available budget and $\boldsymbol{p} = (p_1, p_2, \dots \dots, p_m) \in R^m$ be the vector of unit prices of m resources. Here $Z_k(x) = \boldsymbol{C_k x} = \sum_{j=1}^{n} c_{kj}x_j$ is the k-th objective where $\boldsymbol{C_k} = (c_{k1}, c_{k2}, \dots, c_{kn})$ and $\boldsymbol{C} = [C_1, C_2, \dots, C_q]$. Throughout this paper row vectors and column vectors are respectively denoted by parenthesis '( )' and square brackets '[ ]'and the members of R^n are treated as row vectors.

The main feature of a De-Novo Program is that the resource vector $\boldsymbol{b}$ is unknown, one has to evaluate $\boldsymbol{b}$ in such problem. Under a given budgetary provision B, the resources have to be determined in such a way that either all the objectives are optimized with respect to the constraints simultaneously or an efficient/non-dominated solution can be obtained.

In the system (1) only maximizing type of objectives has been considered. But in practical situations, at the same time the decision-maker may have to consider both maximizing and minimizing type of objectives. Thus a general De-Novo set-up where both maximizing and minimizing type of objectives coexist, could be represented as [13]

$$\left.\begin{aligned} &\text{Max } Z_k = \textstyle\sum_{j=1}^{n} c_{kj}x_j,\ k = 1,2,\dots,l \\ &\text{Min } W_s = \textstyle\sum_{j=1}^{n} c_{sj}x_j,\ s = 1,2,\dots,r \\ &subject\ to\ \textstyle\sum_{j=1}^{n} a_{ij}x_j - b_i \leq 0,\ i = 1,2,\dots,m \\ &\textstyle\sum_{i=1}^{m} p_i b_i \leq B \\ &x_j \geq 0,\ j = 1,2,\dots,n \end{aligned}\right\} \tag{2}$$

In respect of system (2), the decision-maker maximizes each of the objectives $Z_1, Z_2, \dots, Z_l$ and minimizes each of the objectives $W_1, W_2, \dots, W_r$ independently subject to the given constraints (without considering the budgetary restriction) to find the ideal point for the present condition based on the present configuration and available resources. Thus the ideal point $\boldsymbol{\alpha}^*$ of the system for the present state is given by $\boldsymbol{\alpha}^* = (Z_1^*, Z_2^*, \dots, Z_l^*,\ W_1^*, W_2^*, \dots, W_r^*)$ where $Z_k^* = Max\ Z_k(\boldsymbol{x})$ and $W_s^* = Min\ W_s(\boldsymbol{x})$.

But there may not exist a feasible solution $\boldsymbol{x}^*$, determined by the present portfolio of the resources b_i which satisfies $Z_k^* = Z_k(\boldsymbol{x}^*)$ and $W_s^* = W_s(\boldsymbol{x}^*)$ simultaneously. In other words, the ideal point of the system-(2) for the given portfolio of the resource b_i may not correspond to a feasible solution. If such a solution (that is ideal solution) belongs to the feasible space, then the system is optimally designed. Otherwise by extending the available resources under the given budgetary provision B, our aim is to reach either to the ideal solution or to a non-dominated solution under the given budgetary provision. A feasible solution $\bar{x} \in X$ is said to be efficient or non-dominated for the MOOP (1) involving both maximizing and minimizing objective functions, if there is no $x \in X$ so that $Z_k(\overline{\boldsymbol{x}}) \le Z_k(\boldsymbol{x}),\ \forall\, \boldsymbol{x} \in X, k = 1,2,\dots,l$ and $W_s(\overline{\boldsymbol{x}}) \ge W_s(\boldsymbol{x}),\ \forall\, \boldsymbol{x} \in X, s = 1,2,\dots,r$ and for atleast one k, $Z_k(\overline{\boldsymbol{x}}) < Z_k(\boldsymbol{x})$ or for atleast one s, $W_s(\overline{\boldsymbol{x}}) > W_s(\boldsymbol{x})$ whereZ_k's and W_s's are respectively the maximizing and minimizing objectives. As the system configuration changes, the decision-maker has to search for a new ideal point and corresponding ideal solution. In this way continuous designing of the system could be accomplished via DNPP.

Here $\sum_{i=1}^{n} p_i a_{ij}$ represents per unit cost of the product j. Let us denote it by $\boldsymbol{v}_j$. Thus $\sum_{i=1}^{m} p_i a_{ij} = \boldsymbol{v}_j$, j=1,2,...,n. Using v_j, the constraints of system (2) is rewritten as $\sum_{i=1}^{n} p_i \sum_{j=1}^{n} a_{ij} x_j \le \sum_{i=1}^{m} p_i\, b_i$ or, $\sum_{j=1}^{n}(\sum_{i=1}^{m} p_i a_{ij}) x_j \le B$. That is $\sum_{j=1}^{n} v_j x_j \le B$.

Thus using $\mathrm{v_j}$, system (2) can be reframed as $\mathrm{Max}\ \mathrm{Z_k} = \sum_{j=1}^{n} \mathrm{c_{kj} x_j},\ \mathrm{k} = 1,2,\dots,\mathrm{l};$

$$\mathrm{Min}\ \mathrm{W_s} = \sum_{j=1}^{n} \mathrm{c_{sj} x_j},\ \ \mathrm{s} = 1,2,\dots,\mathrm{r};\ \text{Subject to } \sum_{j=1}^{n} \mathrm{v_j x_j} \le \mathrm{B};\ \ \mathrm{x_j} \ge 0,\ \ \mathrm{j} = 1,2,\dots,\mathrm{n} \tag{3}$$

The advantage of system (3) is that there is only one constraint involving n unknowns and hence the basic feasible solutions (at most n) could easily be determined. Further it is to be mentioned here that in the process of restructuring of system (2) to (3), the merit of De-Novo program is no way being distorted. This is because the budgetary restriction is reflected in the constraint of system (3) and every feasible solution of system (2) is also a feasible solution of system (3) and conversely [21]. Thus the ideal point of the system (3) is also an ideal point for the system (2). Hence considering the objective functions independently together with the single constraint $\sum_{j=1}^{n} \mathrm{v_j x_j} \le \mathrm{B}$, one can easily find the ideal point [5].

2.1. Proposed Approach for the Solution of General MODNPP under Fuzzy Environment Using Compensatory Operator

As mentioned earlier, there exists no general algorithm to solve the general MODNPP (2). However, in [13], Li and Lee solved a crisply defined general MODNPP in two steps in section 2 of [13] using fuzzy techniques. They actually fuzzified the crisp objectives between their absolute boundaries, the ideal values and the negative ideal values [13]. In the first step of their two-step approach, Li and Lee solved the MODNPP (3) under fuzzy environment using the non-compensatory 'min' operator λ [22]. The solution obtained in step 1, however, may not be an efficient one. In step 2, they actually derived an efficient solution of the problem (3) under fuzzy environment using fully compensatory operator.

In the present paper, it has been illustrated that only by a proper choice of the aspiration levels of the objectives and hence of their membership functions, an efficient solution of the problem (3) could be obtained in one step only. Here we solved a crisp general MODNPP in one step by fuzzifying the objectives between their ideal values and some practical threshold values obtained by using Luhandjula's comparison technique [19] and considering the compensatory min-bounded sum operator μ_θ [19]. To demonstrate the new approach we consider the general MODNPP (2) and restructured it as system (3).

Let $(Z_1^*, Z_2^*, \ldots, Z_l^*, W_1^*, W_2^*, \ldots, W_r^*)$ be the ideal point of the system (3). Further, let the pessimistic values of the objective functions obtained by Luhandjula's technique be respectively given by $\hat{Z}_1, \hat{Z}_2, \ldots, \hat{Z}_l, \widehat{W}_1, \widehat{W}_2, \ldots, \widehat{W}_r$ .The linear membership functions [22,23] for the objectives of (3) are constructed as follows:

$$\mu_{Z_k} = \begin{cases} 0, Z_k \le \hat{Z}_k \\ \frac{Z_k - \hat{Z}_k}{Z_k^* - \hat{Z}_k}, \hat{Z}_k < Z_k < Z_k^*, k = 1,2,..,l \\ 1, Z_k \ge Z_k^* \end{cases} \quad (4)$$

and $$\mu_{W_s} = \begin{cases} 0, W_s \ge \widehat{W}_s \\ \frac{\widehat{W}_s - W_s}{\widehat{W}_s - W_s^*}, W_s^* < W_s < \widehat{W}_s, s = 1,2,..,r \\ 1, W_s \le W_s^* \end{cases} \quad (5)$$

These membership functions indicate the degree of satisfaction of the decision-maker for the attainment of the aspiration levels of the corresponding objectives.

2.1.1. Solution of the General MODNPP Using Compensatory Min –Bounded Sum Operator Under Fuzzy Environment

Luhandjula in [19] defined an alternative operator μ_θ other than the usual Zimmermann's operator, called min-bounded sum operator, to solve vector maximization problem.

Let μ_i, i=1,2,…,q be the membership functions for the fuzzy objectives. Thus μ_θ is defined as follows: $\mu_\theta = \gamma \min_i \mu_i + (1-\gamma)\min(1, \sum_{i=1}^{q} \mu_i)$. $\gamma \in (0,1)$ is the grade of compensation between aggregated membership functions.

In this paper the use of this min-bounded sum operator μ_θ has been extended to the general MODNPP (3) and also it has been shown that an efficient solution could be obtained assuming uniqueness of such solution.

In this regard we first state some results whose validity could easily be checked following Luhandjula's proposition 2[19].

Theorem 2.1. $\bar{\boldsymbol{x}}$ will be the optimal solution for
$\max \mu_\theta$; Subject to $\mu_\theta = \gamma\, min(\mu_k, \mu_s) + (1-\gamma)\min(1, \sum_{k=1}^{l} \mu_k + \sum_{s=1}^{r} \mu_s)$;
$\boldsymbol{Vx} \le; \boldsymbol{x} \ge 0; \gamma \in (0,1); \mu_\theta \in [0,1]$ (6)
iff $(\bar{\boldsymbol{x}}, \bar{\lambda}, \bar{\mu})$ be the optimum solution of
$\max \mu_\theta = \gamma\lambda + (1-\gamma)\mu$; Subject to $\lambda \le min(\mu_k, \mu_s)$; $\mu \le 1$;
$\mu \le \sum_{k=1}^{l} \mu_k + \sum_{s=1}^{r} \mu_s$; $\boldsymbol{Vx} \le B$; $\boldsymbol{x} \ge 0$ (7)
$\lambda \in [0,1]$; $\mu \in [0,1]$; $\gamma \in (0,1)$; $\mu_k \in [0,1]$; $\mu_s \in [0,1]$; $\mu_\theta \in [0,1]$ where $\mu_{Z_k} \equiv \mu_k$ and $\mu_{W_s} \equiv \mu_s$ ((4) and (5)).

Lemma 2.1. $(\bar{\boldsymbol{x}}, \bar{\lambda}, \bar{\mu})$ is optimal for (7) then $\bar{\lambda} = \min(\mu_k(\bar{\boldsymbol{x}}), \mu_s(\bar{\boldsymbol{x}}))$ and $\bar{\mu} = \min(1, \sum_{k=1}^{l} \mu_k(\bar{\boldsymbol{x}}) + \sum_{s=1}^{r} \mu_s(\bar{\boldsymbol{x}}))$.

Note

In general, the optimum solution of (6) may not be efficient. However, the compensatory min-bounded sum operator μ_θ may also yield an efficient solution for the De-Novo Program (3) if $\bar{\boldsymbol{x}}$ is unique. This can be easily proved by the proposed theorem 2.2 with the help of theorem 2.1 and lemma 2.1.

But to prove theorem 2.2, an extension of the concept of domination of solutions to the following general type of multi-objective linear programming problem (8) where both maximizing and minimizing type of objectives are involved, is needed.

Let $\bar{\boldsymbol{x}}$ and $\boldsymbol{x}$ be two feasible solutions of the general multi-objective linear programming

$$\text{Max } Z_k = \sum_{j=1}^{n} c_{kj}x_j,\ k = 1,2,\dots,l;\ \text{Min } W_s = \sum_{j=1}^{n} c_{sj}x_j,\ s = 1,2,\dots,r;\ \text{Subject to } \boldsymbol{Ax} \le \boldsymbol{b};\quad \boldsymbol{x} \ge 0 \tag{8}$$

Definition 2.1. For the feasible solutions $\overline{\boldsymbol{x}}$ and $\boldsymbol{x}$ of (8), $\overline{\boldsymbol{x}}$ is said to dominate strictly $\boldsymbol{x}$ if for maximizing objectives $Z_k(\overline{\boldsymbol{x}}) \ge Z_k(\boldsymbol{x}), \forall k$ and for minimizing objectives $W_s(\overline{\boldsymbol{x}}) \le W_s(\boldsymbol{x}), \forall s$ and $Z_k(\overline{\boldsymbol{x}}) > Z_k(\boldsymbol{x})$ or $W_s(\overline{\boldsymbol{x}}) < W_s(\boldsymbol{x})$for at least one k or one s respectively.

*Theorem 2.2.*The optimum solution $\overline{\boldsymbol{x}}$ of the system (6) will be an efficient solution of (3) if $\overline{\boldsymbol{x}}$ is unique.

3. Numerical Example Illustrating the Proposed Approach

Let us consider the MODNPP [Example 2 of [13]

$$\max Z_1 = 2x_1 + 5x_2 + 7x_3 + x_4;\ \max Z_2 = 4x_1 + x_2 + 3x_3 + 11x_4;\ \max Z_3 = 9x_1 + 3x_2 + x_3 + 2x_4$$
$$\min W_1 = 1.5x_1 + 2x_2 + 0.3x_3 + 3x_4;\ \min W_2 = 0.5x_1 + x_2 + 0.73x_3 + 2x_4$$
$$subject\ to{:}\ 2x_1 + 2x_2 + 2x_3 + 8x_4 \le b_1;\ 4x_1 + 4x_2 + 2x_3 + 4x_4 \le b_2;\ 5x_2 + 5x_4 \le b_3$$
$$2.5x_1 + 2.5x_2 + 2.5x_4 \le b_4;\ 0.5b_1 + 0.25b_2 + 0.3b_3 + 0.4b_4 = 150;\ x_1, x_2, x_3, x_4 \ge 0 \tag{9}$$

To solve the MODNPP (9) of example 3.1, we first recast it in the following form using the procedure as mentioned in section 2.

$$\max Z_1 = 2x_1 + 5x_2 + 7x_3 + x_4;\ \max Z_2 = 4x_1 + x_2 + 3x_3 + 11x_4;\ \max Z_3 = 9x_1 + 3x_2 + x_3 + 2x_4;\ \min W_1 = 1.5x_1 + 2x_2 + 0.3x_3 + 3x_4;\ \min W_2 = 0.5x_1 + x_2 + 0.73x_3 + 2x_4$$
$$subject\ to{:}\ 3x_1 + 4.5x_2 + 1.5x_3 + 7.5x_4 \le 150;\ x_1, x_2, x_3, x_4 \ge 0 \tag{10}$$

For the solution of the problem (10), we first calculate the ideal points. For this, five LPPs are solved considering each objective independently subject to the single constraint. The solutions obtained with corresponding objective values are furnished below:

$$Z_1^* = 700;\ x_1 = 0, x_2 = 0, x_3 = 100, x_4 = 0$$
$$Z_2^* = 300;\ x_1 = 0, x_2 = 0, x_3 = 100, x_4 = 0$$
$$Z_3^* = 450;\ x_1 = 50, x_2 = 0, x_3 = 0, x_4 = 0$$
$$W_1^* = 30;\ x_1 = 0, x_2 = 0, x_3 = 100, x_4 = 0$$
$$W_2^* = 25;\ x_1 = 50, x_2 = 0, x_3 = 0, x_4 = 0$$

Thus the ideal point is $\alpha = (Z_1^*, Z_2^*, Z_3^*, W_1^*, W_2^*) = (700,300,450,30,25)$. To a decision-maker, attainment of objective values equal to those of ideal values is the most expected one. So in the construction of Zimmermann type of membership function for the fuzzy objectives, the ideal values are the natural choice of optimistic values or upper limits of the corresponding objective functions. Now to obtain a practical threshold value for the objectives, we apply Luhandjula's technique and the calculations are shown in Table 1.

Table 1. Calculation of Pessimistic Object Values.

Solutions				Maximizing Objective Values			Minimizing Objective Values	
x_1	x_2	x_3	x_4	Z_1	Z_2	Z_3	W_1	W_2
0	0	100	0	700	300	100	30	73
0	0	100	0	700	300	100	30	73
50	0	0	0	100	200	450	75	25
0	0	100	0	700	300	100	30	73
50	0	0	0	100	200	450	75	25
Corresponding Pessimistic Objective Values				*100* $(\hat{Z}_1)$	*200* $(\hat{Z}_2)$	*100* $(\hat{Z}_3)$	*75* $(\widehat{W}_1)$	*73* $(\widehat{W}_2)$

Now the membership functions for the maximizing and minimizing objective functions are constructed using the relations (4) and (5) of section 2.1 as follows:

For Maximizing Objectives: $\mu_{Z_1}(x) = \begin{cases} 0, & Z_1 < 100 \\ \frac{Z_1-100}{600}, & 100 \le Z_1 < 700; \\ 1, & Z_1 \ge 700 \end{cases}$

$\mu_{Z_2}(x) = \begin{cases} 0, & Z_2 < 200 \\ \frac{Z_2-200}{100}, & 200 \le Z_2 < 300 \\ 1, & Z_2 \ge 300 \end{cases}$ and $\mu_{Z_3}(x) = \begin{cases} 0, & Z_3 < 100 \\ \frac{Z_3-100}{350}, & 100 \le Z_3 < 450 \\ 1, & Z_3 \ge 450 \end{cases}$

For Minimizing Objectives

$\mu_{W_1}(x) = \begin{cases} 0, & W_1 \ge 75 \\ \frac{75-W_1}{45}, & 30 \le W_1 < 75 \\ 1, & W_1 < 30 \end{cases}$ and $\mu_{W_2}(x) = \begin{cases} 0, & W_2 \ge 73 \\ \frac{73-W_2}{48}, & 25 \le W_2 < 73 \\ 1, & W_2 < 25 \end{cases}$

Now introducing Luhandjula's μ_θ-operator, the system (10) under fuzzy environment becomes

$$\max \mu_\theta = \gamma\lambda + (1-\gamma)\mu$$

Subject to $\lambda \le \frac{1}{600}(2x_1 + 5x_2 + 7x_3 + x_4 - 100)$; $\lambda \le \frac{1}{100}(4x_1 + x_2 + 3x_3 + 11x_4 - 200)$;
$\lambda \le \frac{1}{350}(9x_1 + 3x_2 + x_3 + 2x_4 - 100)$; $\lambda \le \frac{1}{45}(75 - 1.5x_1 - 2x_2 - 0.3x_3 - 3x_4)$;
$\lambda \le \frac{1}{48}(73 - 0.5x_1 - x_2 - 0.73x_3 - 2x_4)$; $\mu \le 1$; $\mu \le \frac{1}{600}(2x_1 + 5x_2 + 7x_3 + x_4 - 100) + \frac{1}{100}(4x_1 + x_2 + 3x_3 + 11x_4 - 200) + \frac{1}{350}(9x_1 + 3x_2 + x_3 + 2x_4 - 100) + \frac{1}{45}(75 - 1.5x_1 - 2x_2 - 0.3x_3 - 3x_4)$

$+\frac{1}{48}(73 - 0.5x_1 - x_2 - 0.73x_3 - 2x_4)$; $3x_1 + 4.5x_2 + 1.5x_3 + 7.5x_4 \leq 150$; $\lambda \in [0,1]$;$\mu \in [0,1]$;
$\gamma \in (0,1)$; $\mu_k \in [0,1]$;$\mu_s \in [0,1]$;$\mu_\theta \in [0,1]$; $x_1, x_2, x_3, x_4 \geq 0$

Solving the above system again by using LINGO software, the following results are obtained:
$\mu_\theta = 0.5, \lambda = 0.5, \mu = 1, x_1 = 25, x_2 = 0, x_3 = 50, x_4 = 0$; $\max Z_1 = 400, maxZ_2 = 250, maxZ_3 = 275, \min W_1 = 52.5, \min W_2 = 49$
The results obtained by our proposed approach and that of Li and Lee's two step approach are identical.

4. Conclusion

In the present treatise a new approach for the solution of the general MODNPP under fuzzy environment has been introduced which yields efficient solution to the problems in one step only. The proposed technique could be used for the solution of a class of problems, like academic resource allocation problems, portfolio analysis, product – mix problems etc., which could be easily converted into multi-objective De-Novo pattern under a given budgetary provision. The proposed method is computationally easier than the existing ones for the solution of MODNPP. The construction of the proposed method is such that they could be successfully applied for the solution of MODNPP involving even large number of objectives, constraints as well as variables. Finally, apart from the computational simplicity, the proposed method requires less processing time in comparison with the existing ones because through the proposed method, the general MODNPP can be solved in one step only. To make the problem more flexible, instead of crisp coefficients, fuzzy, type2 fuzzy coefficients can be considered and solution procedure can be investigated.

References

[1] Zeleny M 1986 Optimal system design with multiple criteria:De-Novo Programming approach, *Engineering Costs and Production Economics*, **10** , pp-89-94.

[2] Zeleny M 1976 Multi-objective design of high-productivity systems. *In: Proc. Joint Automatic Control Conf., paper APPL94*, New York.

[3] Zeleny M 1981 On the squandering of resources and profits via linear programming. *Interfaces*, 11(S): 1Ol-107.

[4] Zeleny M 2009 On the essential multidimensionality of an Economic Problem: Towards tradeoffs-free Economics, *Acta Universitatis Carolinae Oeconomica*, **3**,pp-154-175.

[5] Zeleny M The Evolution of Optimality: De-Novo Programming *EMO, LNC 3410*, pp.1-13

[6] Zeleny M 2005 Optimization, Optimal design and De-Novo Programming: Discussion Notes, *Fordham University, New York, USA.*

[7] Zeleny M 1974 A concept of compromise solutions and the method of the displaced ideal", *Compute and Operation Research*, v.**1**, pp.479-496.

[8]Hessel M and Zeleny 1987 M Optimal system designs: Towards new interpretation of shadow prices in linear programming, *Computers and Operations Research*, v.**14(4)**, pp.265-271.

[9] Zeleny M 1982, Multiple Criteria Decision Making. *New York, McGraw-Hill.*

[10] Zeleny M 1998 Multiple Criteria Decision Making: Eight concepts of optimality, *Human Systems Management*, v.**17(2)**, pp.97-107.

[11] Zeleny M 1981a A Case Study in Multi-objective Design: De-Novo Programming, in Multiple Criteria Analysis: Operational Methods, edited by P. Nijkamp and J. Spronk, Lexington Books, Lexington, MA.

[12] Zeleny M 1981b "Multiple Criteria Decision-Making", *McGraw Hill, New York.*

[13] Li R J and Lee.E.S 1990 Fuzzy Approaches to Multicriteria De-Novo Programs, *Journal of Mathematical Analysis and Applications*, **153**, pp-97-111.

[14] Sasaki M, Gen M, Yamashiro. M (1995), "A method for solving Fuzy De-Novo Programming Problem by Genetic Algorithms", *Computers ind. Engng*, v.**29** (1-40), pp. 507-511.
[15] Y.M. Zhang, G.B. Huang, X.D. Zhang 2009) Inexact De-Novo Programming for water resources system planning *European Journal of operation Research*, v.**199**, pp. 531-541.
[16] FIALA. P, (2011), "Multi-objective De-Novo Linear Programming", *Acta Univ. Palacki. Olomuc., Fac. Rer. Nat Mathematics*, v.**5092**, pp.29-36.
[17] Saeedia. S, Mohammadia. M and Torabib. S.A, (2014), "A De-Novo Programming approach for a robust closed-looop supply chain design under uncertainty: an M/M/1 queuing model", *International Journal of Industrial Engineering Computations*, v.**6**, pp.211-228.
[18] Umarusman N 2013 Min-Max Goal Programming Approach For Solving Multi-Objective De Novo Programming Problems, *International Journal of Operations Research*, V. **10**, No. **2**, pp. 92-99.
[19] Luhandjula M K 1982 Compensatory operators in fuzzy linear programming with multiple objectives, *Fuzzy Sets and Systems*, **8**, pp- 24.
[20] Chen Y W and Hsieh H E 2006 Fuzzy multi-stage De-Novo programming problem, *Applied Mathematics & Computation*, **181**,pp-1139-1147.
[21] Zeleny, M. (1980)," Mathematical programming with multiple objectives (Special issue)", *Computer Ops. Res.*, v. **7**, pp. 101 – 107.
[22] Zimmermann H J 1978 Fuzzy programming and linear programming with several objective functions, *Fuzzy Sets and Systems*, **1**, pp-45-55.
[23] Chakraborty S and Bhattacharya D 2012 A New Approach of Solution of Multi-Stage and Multi-Objective Decision-Making Problem using De-Novo Programming, *European Journal of Scientific Research* , **79 (3)**, pp.393-417.

The Variance Formation of M9.3 Solar Flare Associated with Solar Radio Burst Type V Occurred at Eight (8) Different Sites of the CALLISTO System on 4th August 2011

N. N. M. Shariff[1], Z. S. Hamidi[2], S.N.U. Sabri[3]

[1]Academy of Contemporary Islamic Studies (ACIS), 40450 Shah Alam, Selangor, Malaysia
[2]Institute of Science (IOS), Universiti Teknologi MARA, 40450 Shah Alam, Selangor, Malaysia
[3]Faculty of Applied Sciences, Universiti Teknologi MARA, 40450 Shah Alam, Selangor, Malaysia

Email: nur.nafhatun.ms@gmail.com

Abstract. We report a detailed examination of the variance formation of M9.3 solar flare associated with solar radio burst Type V that occurred on 2011 August 4. Solar radio burst type V was identified occurred with present of Coronal Mass Ejection (CME) towards earth which moving estimated about 1,950 km/s which can gives major impact to our Earths in terms of satellites. The event was observed by using Compound Astronomical Low-cost Low-Frequency Instrument for Spectroscopy Transportable Observatory (CALLISTO). CALLISTO is a radio spectrometer that has been used 24 hour per day which designed for solar observation. The solar radio burst was occurred at 21.1 MHz at 3 minutes (0358 UTC till 0401 UTC). Noted that all the sites are less suffering from local RFI with solar wind speed and density reached 411.6 km/sec and 3.8 proton/cm^3. Solar radio burst Type V is known as smooth with short-lived continuum which sometime followed by a type III burst. We show radio flux density in eight (8) different sites. Finally, we discuss the X-ray flux density to determine classes of the solar flares.

1. Introduction

Significance formation of M9.3 solar flare associated with Type V solar burst almost exceed the X-territory. The strong and energetic particle can cause major disruption to human technology and health risk to astronaut. Solar Radio Burst divided into five (5) types (I-V) and it can differentiate by referring to the frequency and duration of the burst on that day. Solar Radio Burst Type I can be classified as solar storm which has short and narrow-bandwidth bursts. The Type II burst is known as Coronal Mass Ejection (CMEs) which cause by shock waves [1] has slow frequency drift bursts and accompanied by a stronger intensity and take about three to thirty minutes in duration for an event occurred. Next, Solar Radio Burst Type III is known as solar flares which also known as fast frequency drift burst and occurred in three situations which are singular, in groups or storms. Since mid-1996 the SOHO Large Angle Spectrometric Coronagraph (LASCO) instruments have observed numerous halo or partial-halo CMEs [2].

In this paper, we examine solar radio burst Type V known as the combination of Type I and Type III burst with slightly different in a range of frequency and the duration of the burst. Type V usually in 1-3 minutes which usually appears caused by electron streams pass through a region which was heated from the passage of the electron stream which associated with type III burst (Warwick, 1967). We

IOP Conf. Series: Journal of Physics: Conf. Series **1039** (2018) 012013 doi:10.1088/1742-6596/1039/1/012013

show radio flux density in eight (8) different sites. Finally, we discuss the X-ray flux density to determine classes of the solar flares.

2. Experimental Setup and Methodology

Compound Astronomical Low-cost Low-Frequency Instrument for Spectroscopy Transportable Observatory (CALLISTO) is used to interpret the radio bursts that have been detected by the antenna before the results appeared on the computer's screen [3, 4]. This system is one of the most outstanding project under ISWI with there are more than 80 instruments in more than 43 locations with users from more than 116 countries [4]. In order to study solar activities that relating to solar flare or Coronal Mass Ejection (CMEs) is usually related to solar burst and sometimes can affect the telecommunication system and radio navigation operating systems and even electric power supply [5]. The aim of this project is to monitor the solar activity in radio region for 24/7 in order to support develop country participate direct or indirect with the latest technologies of instruments [6].

Log Periodic Dipole Antenna (LPDA) has a twisted balanced transmission line that, fed by a coplanar linear array of equal and unequal spaced parallel linear dipoles [7]. The antenna has set up with range of frequency 45 MHz till 870 MHz. The LPDA that situated at National Space Centre (ANGKASA), Selangor at (3.0833333°N 101.53333333E) which has been connected will CALLISTO system that minimum Radio Frequency Interference (RFI) with average (85-100) dBm and has 45-870 MHz range of frequency [7, 8]. The other different sites used different pattern or types of radio telescopes so that can received good signal from sun radiation.

3. Result and Discussion

Data below showed that the radio flux density for 8 different sites where the antenna has been set up with the CALLISTO spectrometer. From this solar radio burst, we can conclude that it was from Type V Solar Radio Burst which has high intensity of burst from 20 MHz – 200 MHz (21.1 MHz) with the duration of burst about 3 minutes from 03:58 till 04:01 UTC. It has smooth and short lived continuum burst. There are several associated phenomena tend to occur which are eruptive prominences and magnetohydronamic shock waves and on 5th August 2011, a major geomagnetic storm occurred impact from the CMEs occurred on 4th August 2011. Although it has been set up at different sites with different frequency range of the antenna, the radio burst still can detect if the burst was in the range of frequency of the type of burst and the antenna has been set up with different polarization and connection with the presence of low in noise and RFI sources thus gave the clearer data.

Sometimes the data that have been collected are not showing the clearer image because of the different value gain of antenna that has been used and the different polarization of the antenna also affect the amount of radiation that the antenna could be received. The higher the value of the gain of an antenna, the clearer the image of data. Some of the data does not show the same pattern of burst because of the different amount of radiation of the signal that the antenna could be received due to the different RFI level of noise and the different in polarization of the antenna itself. This will affect the intensity of the solar radio burst detected thus give different pattern in solar radio flux density. At the conclusion, the best data solar radio burst collected by Radio Flux Density on 4th August 2011 was from Ooty, India.

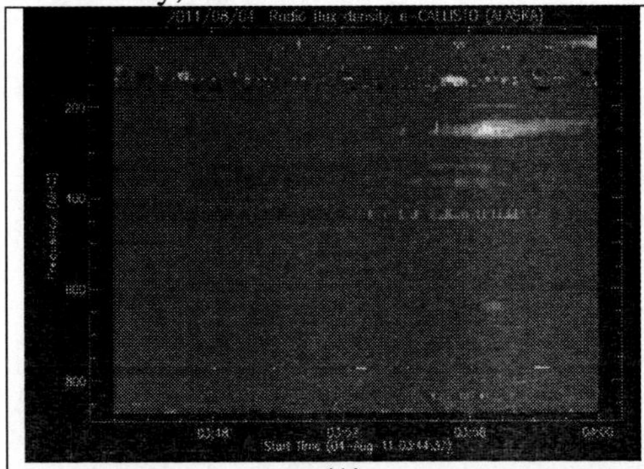

(A)

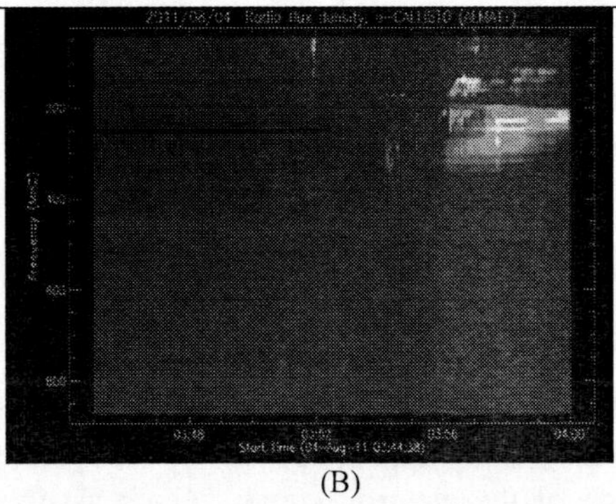

(B)

IOP Conf. Series: Journal of Physics: Conf. Series **1039** (2018) 012013 doi:10.1088/1742-6596/1039/1/012013

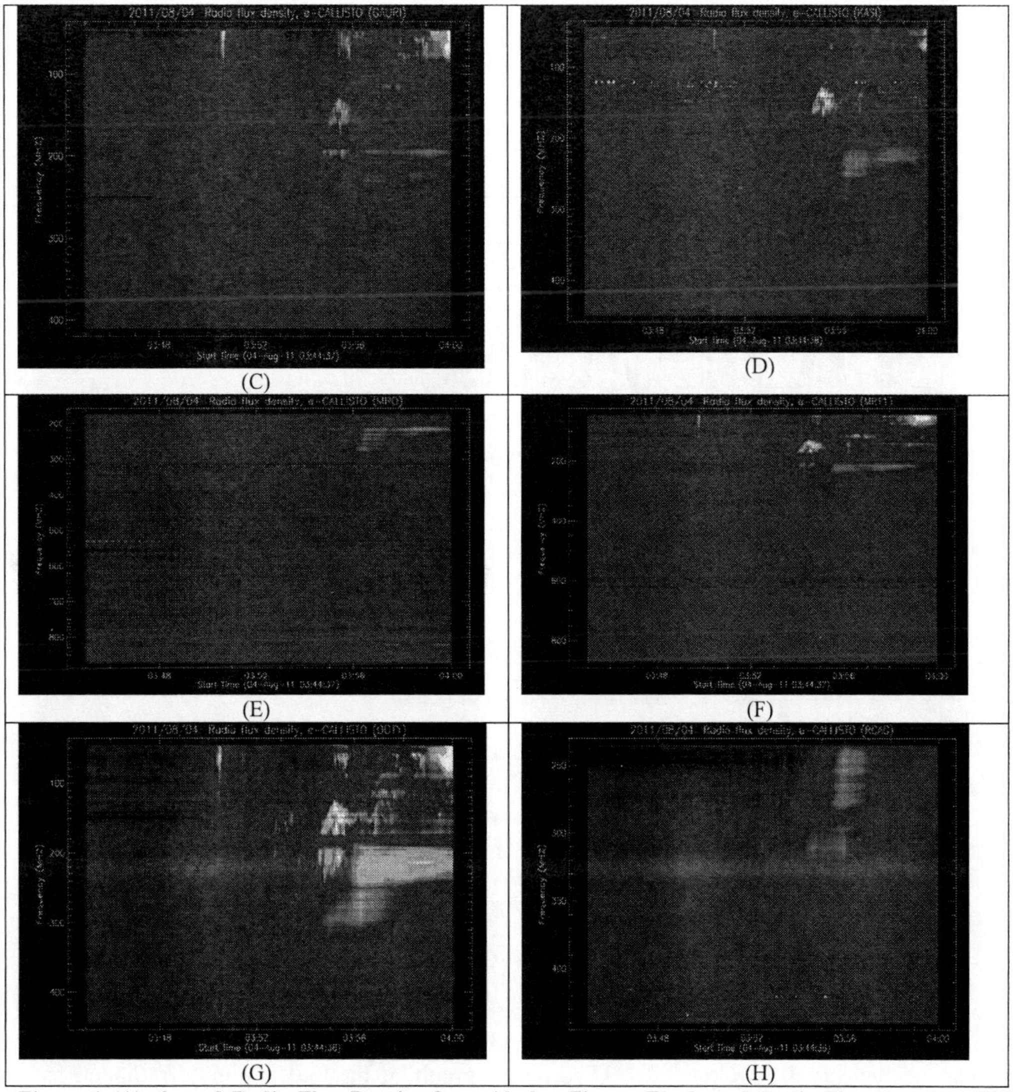

(C) (D) (E) (F) (G) (H)

Figure 1 (A) showed Radio Flux Density from Alaska, Figure (B) showed Radio Flux Density (ALMATY) from Kazakhstan Figure (C) showed Radio Flux Density, e-CALLISTO (GAURI) from India Figure (D) showed Radio Flux Density, e-CALLISTO (KASI) from Daejeon, South Korea, Figure (E) showed Radio Flux Density (MRO) from Metsähovi (Finland), Figure (F) showed Radio Flux Density (MRT1) from Mauritius, Figure (G) showed Radio Flux Density (Ooty) India and last but not least, Figure 8. Radio Flux Density (RCAG), Ulaan Baatar from Mongolia (Credited to: e-CALLISTO Website)

By referring GOES X-Ray flux below, it showed that the solar flares have the highest value at 0358UT which in M9.3 class of flares where it can cause brief radio blackouts that affect Earth's polar regions. Besides that, minor radiation storms sometimes follow M-class of flares. It quite highest rate in M- Class solar flares where it can exceed to X1-class of flares which can caused planet-wide radio blackouts and long lasting radiation storms.

IOP Conf. Series: Journal of Physics: Conf. Series **1039** (2018) 012013 doi:10.1088/1742-6596/1039/1/012013

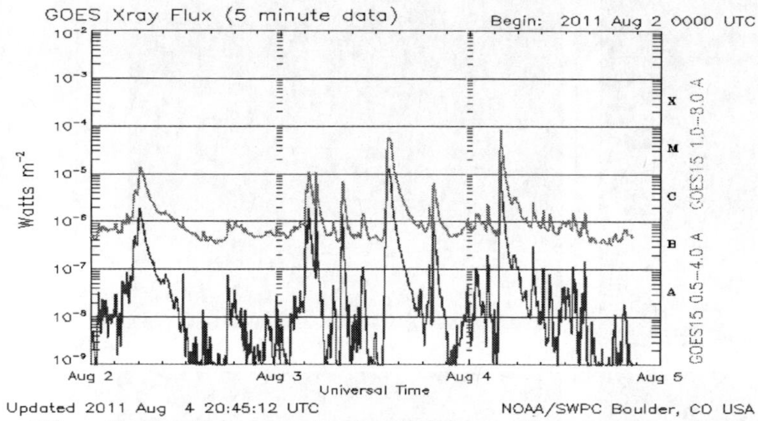

Figure 2 X-ray flux density to determine classes of the solar flares (credit to: www.solarmonitor.org)

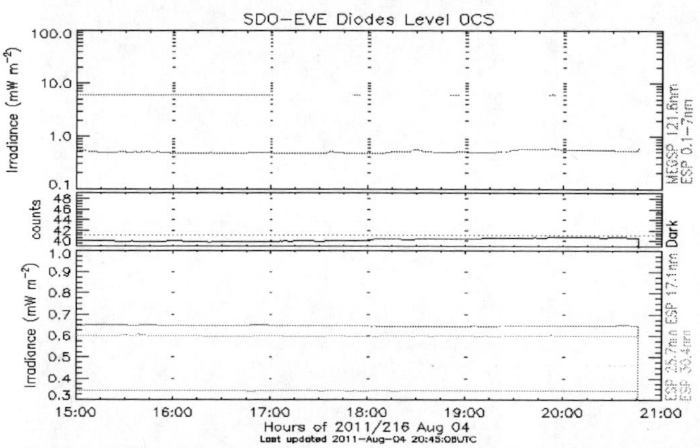

Figure 3 Solar Extreme Ultraviolet Irradiance on 4th August 2011 (credit to: www.solarmonitor.org)

From above graph, it showed that the graph was used to measure the solar extreme ultraviolet (EUV) irradiance with 0.1NM spectral resolution, which include with array detectors so all EUV wavelengths can be measured simultaneously. From this graph, it has 6.0 mWm^{-2} irradiance from 15:00 till 17:00 UTC where the solar radio burst type V occurred during that time. Solar irradiance is the output of the light energy from the entire disk of the sun, which measured at the Earth and it can be used to measure for any glowing object (stars, the moon) which can be seen in many wavelengths (from visible and infrared rays) through UV to EUV and X-ray.

4. Conclusion

The data that have been collected on 4th August, 2011 showed the same pattern of solar radio burst type V with different range of frequency that has been set up at the antenna at different sites and gave the same pattern of burst at the same time 03:58 till 04:01 UTC in 3 minutes. The difference of the intensity of radio burst received caused by the difference in polarization of the antenna and RFI level of noise which also will affect the signal that received. A major geomagnetic storms followed the following day caused by CMEs and the data showed by radio flux density influenced by the gain of the antenna, the maximum amount of radiation or signal that received from the sun and the lower noise level thus give the best and the clearest image of the burst. The antenna should not located near the RFI sources so that it does not affect the data received.

References

[1] N. Gopalswamy, Recent advances in the long-wavelength radio physics of the Sun, Planetary and Space Science 52(15) (2004).

[2] N.N.M. Shariff, Z.S. Hamidi, N. Zaino, Analysis of Slow Partial Halo CME Events with Velocity of 100-200 km/s: An Observation through CALLISTO System, International Conference on Industrial Engineering, Management Science and Application (ICIMSA), Seoul, 2017.
[3] N. Husien, N.H. Zainol, Z.S. Hamidi, S.N.U. Sabri, N.N.M. Shariff, M.S. Faid, M.O. Ali, C. Monstein, Solar Radio Bursts Detected by CALLISTO System and Their Related Events, International Conference on Industrial Engineering, Management Science and Application (ICIMSA), IEEE, Jeju Island, South Korea, 2016.
[4] Z.S. Hamidi, N.N.M. Shariff, C. Monstein, A.B.M.A. Ibrahim, M.I.M. Yusof, The International Cooperation of Solar Radio Burst Project Using e-Callisto System Network, 7th International Conference on Information Science and Applications, Springer, Ho Chi Minh City, 2016.
[5] N.H. Zainol, S.N.U. Sabri, Z.S. Hamidi, M.O. Ali, N.N.M. Shariff, N. Husien, C. Monstein, M.S. Faid, Effective Data Collection and Analysis of Solar Radio Burst Type II Event Using Automated CALLISTO Network System, International Conference on Industrial Engineering, Management Science and Application (ICIMSA), IEEE, Jeju Island, South Korea, 2016.
[6] Hamidi, Z. S., N.N.M. Shariff, Automatic Solar Radio Burst Type II and III Image Processing Tracking by Using CALLISTO System, International Conference on Industrial Engineering, Management Science and Application (ICIMSA), Seoul, 2017.
[7] S.U.M. Sabri, Z.S. Hamidi, N.N.M. Shariff, The Significance of e-Callisto System and Construction of Log Periodic Dipole Antenna (LPDA) and CALLISTO System for Solar Radio Burst Study, 7th International Conference on Information Science and Applications, Springer, Ho Chi Minh City, 2016.
[8] Y. Nishimura, T. Ono, F. Tsuchiya, M. Hiroaki, A. Kumamoto, Y. Katoh, S. Masuda, Y. Miyoshi, Narrowband frequency-drift structures in solar type IV bursts, Earth, Planets and Space 65(12) (2013) 1555–1562.

Acknowledgement
We are grateful to CALLISTO network; STEREO, LASCO, SDO/AIA,NOAA and SWPC make their data available online. This work was partially supported by Ministry of Higher Education (MOHE) (600-RMI/FRGS 5/3 (0077/2016)).

IOP Conf. Series: Journal of Physics: Conf. Series **1039** (2018) 012014 doi:10.1088/1742-6596/1039/1/012014

Logarithmic quantum wave equation with variable nonlinear coupling

Konstantin G. Zloshchastiev

Institute of Systems Science, Durban University of Technology, P.O. Box 1334, Durban 4000, South Africa

E-mail: `http://bit.do/kgz`

Abstract. We study a possibility of a dynamical behavior of the nonlinear coupling in the quantum wave equation of a logarithmic type. In order to construct a viable model, this coupling is associated with some kind of the quantum temperature conjugated to the Everett-Hirschman entropy. A combined quantum-mechanical and field-theoretical model is proposed, which leads to the logarithmic equation with a variable nonlinear coupling. It is expected that such models will shed light on a fundamental nature of the temperature – as a notion which can be derived from the basic quantum-mechanical principles preceding the classical notions of heat and entropy.

1. Introduction

Assuming three spatial dimensions (lower-dimensional cases can be studied by analogy), let us consider the logarithmic quantum wave equation, for historical reasons referred here as the logarithmic Schrödinger equation (LogSE):

$$\begin{aligned} i\hbar\partial_t\Psi &= \left(\hat{\mathbf{H}} - b\ln(a^3|\Psi|^2)\right)\Psi \\ &= \left[-\frac{\hbar^2}{2m}\Delta + V_{\text{ext}} - b\ln(a^3|\Psi|^2)\right]\Psi, \end{aligned} \tag{1}$$

where the coupling

$$b = b(\vec{r}, t) \tag{2}$$

measures the strength of a nonlinear interaction, a is a constant parameter of dimensionality length required to make the argument of the logarithm dimensionless, and m is a particle's mass (also assumed constant here), and $V_{\text{ext}} = V_{\text{ext}}(\vec{r}, t)$ is an external potential. Here $\Psi = \Psi(\vec{r}, t)$ is a wavefunction normalized to a number N:

$$\langle\Psi|\Psi\rangle \equiv \int_{\mathcal{V}} |\Psi|^2 d^3\vec{r} = N \geq 1, \tag{3}$$

and $\Delta = \vec{\nabla}\cdot\vec{\nabla}$ is the Laplacian (in the relativistic version the derivative part would be replaced by a d'Alembertian), and $\mathcal{V}$ is a volume occupied by our system.

The simplest case of Eq. (2),

$$b(\vec{r}, t) = b_0 = \text{const}, \tag{4}$$

was historically first to study [1, 2, 3, 4]. The corresponding models were proven to be very instrumental when dealing with extensions of quantum mechanics [3, 5], physics of quantum fields and particles [1, 6, 7, 8, 9, 10, 11], optics and transport or diffusion phenomena [12, 13, 14], nuclear physics [15, 16], theory of dissipative systems and quantum information [17, 18, 19, 20, 21, 22, 23], theory of quantum liquids and superfluidity [24, 25, 27, 28], and theory of physical vacuum and classical and quantum gravity [29, 30, 31].

Notwithstanding the success of the models with a constant coupling (4), few questions remained open. Firstly, what is a specific value of this coupling? In Refs. [3, 15, 16] it was shown that for a large class of conserved quantum systems the logarithmic term must be dominated by an external potential, therefore, the nonlinear coupling must be rather small. On the other hand, in theory of superfluid He-4 this term plays a crucial role [25]. Therefore, it seems that this coupling can take different values depending on physical conditions, i.e., it can vary from system to system. Essentially, it means that its value cannot be a fundamental constant but rather some dynamical function, whereas the relation (4) can be a leading-order approximation with a finite domain of applicability.

Secondly, if this coupling is a nontrivial notion then what would be its physical meaning? One possible idea is advocated in Refs. [3, 18, 24] to relate this coupling to some kind of a (non-thermal) temperature T_Ψ, which is conjugated to the Everett-Hirschman entropy [32, 33], the latter being defined as

$$S_\Psi = -\langle\Psi|\ln(a^3|\Psi|^2)|\Psi\rangle = -\int\limits_{\mathcal{V}} |\Psi|^2 \ln(a^3|\Psi|^2) d^3\vec{r}, \tag{5}$$

where the Boltzmann constant is assumed to be $k_B = 1$ from now on. Therefore, the logarithmic quantum wave equation can be viewed as a minimization condition not for the energy of a system

$$E = \langle\Psi|\hat{\mathbf{H}}|\Psi\rangle,$$

but for the quantum "internal energy", defined as

$$\mathcal{U}_\Psi = \langle\Psi|\hat{\mathbf{H}}|\Psi\rangle + T_\Psi S_\Psi, \tag{6}$$

where

$$T_\Psi = \left(\frac{\partial \mathcal{U}_\Psi}{\partial S_\Psi}\right)_{\mathcal{V}} = b, \tag{7}$$

some details can be found in Ref. [24]. In Eq. (6), the term $\langle\Psi|\hat{\mathbf{H}}|\Psi\rangle$ comes from the system's dynamics, while the term $T_\Psi S_\Psi$ is responsible for dissipative effects related to handling the information about a system

$$I_\Psi = -\log_2(a^3|\Psi|^2) = -\ln(a^3|\Psi|^2)/\ln 2. \tag{8}$$

These entropy considerations result in uncertainty relations which can complement the Heisenberg one [3, 34, 35]. Therefore, in what follows we shall focus on an interpretation of the nonlinear coupling and logarithmic term in terms of T_Ψ and Everett-Hirschman entropy.

Thirdly, while the Everett-Hirschman entropy's consideration seem to be an important step towards understanding the dynamical behavior of temperature T_Ψ, which we refer as the Everett-Hirschman's or simply *quantum temperature* from now on, remains an open question. If it is a dynamical quantity then what are the laws governing it?

These three questions will be a main subject of our study.

2. The model

Since we assume the nonlinear coupling to be a dynamical value, we can define it as a scalar function of a field $\sigma = \sigma(\vec{r}, t)$, *i.e.*, $b = b(\sigma)$ and $T_\Psi = T_\Psi(\sigma)$. Due to gauge invariance of a field, the corresponding wave equation must depend not on a field itself but on its derivative, such as the gradient. Since the latter is a vector, whereas the coupling must be a scalar function, we assume

$$b = b\left(\vec{n} \cdot \vec{\nabla}\sigma, \vec{\nabla}\sigma \cdot \vec{\nabla}\sigma, ...\right), \tag{9}$$

where $\vec{n} = \vec{r}/r = \vec{r}/\sqrt{\vec{r}\cdot\vec{r}}$ is a normal radius vector, and $r = \sqrt{\vec{r}\cdot\vec{r}}$ is an absolute value of the radius vector.

With these assumptions in hands, we can define the simplest ("minimal") variable-coupling model as

$$i\hbar\partial_t\Psi = \left[-\frac{\hbar}{2m}\Delta + V_{\text{ext}} - \vec{n}\cdot\vec{\nabla}\sigma \ln(a^3|\Psi|^2)\right]\Psi, \tag{10}$$

$$\Delta\sigma = 4\pi\rho_\sigma, \tag{11}$$

where $\rho_\sigma = \rho_\sigma(\vec{r}, t)$ is the energy density of the field σ. Furthermore, it is natural to assume that this field's distribution is correlated with our system, therefore, we can impose

$$\rho_\sigma = f(\rho), \tag{12}$$

where f is a function which is being chosen in a specific model, and

$$\rho = |\Psi|^2 \tag{13}$$

is a probability density of our system; in case of condensates ($N \gg 1$), ρ would be an actual particle density of a condensate.

One can see that the resulting model (10) - (13) is not only nonlinear but also coupled, therefore some robust simplifications must be made in order to extract an essential physical information.

From now on we shall be interested in a case of a trapless system,

$$V_{\text{ext}} = 0. \tag{14}$$

In this case, we can set

$$\sigma = \sigma(r), \tag{15}$$

and rewrite Eqs. (10) and (11) in the form:

$$i\hbar\partial_t\Psi + \frac{\hbar}{2m}\Delta\Psi + \frac{d\sigma}{dr}\ln(a^3|\Psi|^2)\Psi = 0, \tag{16}$$

$$\Delta\sigma - 4\pi f(|\Psi|^2) = 0. \tag{17}$$

Furthermore, we also assume our quantum-mechanical system to be localized

$$\lim_{r\to+\infty}\rho_\sigma = \lim_{r\to+\infty} r^2|\Psi|^2 = 0. \tag{18}$$

Under these conditions, Eq. (17) can be written in a series form, and Eqs. (16) and (17) can be decoupled from each other [36].

Our model thus suggests that the nonlinear coupling is not a constant in general. If for some system, the σ-field's charge Q_σ, is nonzero then at $r \to 0$ the coupling must diverge polynomially, whereas at large r the coupling tends to a constant, so that one recovers Eq. (4). It is interesting to notice that the constant part of the coupling, b_0, is actually induced by a regular localized part of the σ-field. This explains why the solutions of the conventional logarithmic equation (4) are so instrumental in describing regular extended objects, such as Q-balls and extended particles [6, 7, 8, 9, 10, 11], and superfluid droplets [24, 25]. The extended equation's properties give us some hope that this model will be also instrumental when dealing with singular or point-like objects.

3. Conclusion

We study a possibility of a dynamical behavior of the nonlinear coupling in the quantum wave equation of a logarithmic type. In order to construct a viable model, this coupling is associated with some kind of the quantum temperature conjugated to the Everett-Hirschman entropy. A combined quantum-mechanical and field-theoretical model is proposed, which leads to the logarithmic equation with a variable nonlinear coupling.

This coupling function approaches a constant at large distance from a system's center, where our equation becomes the conventional logarithmic Schrödinger equation. At short distances, the minimally-coupled model's behavior must differ from that of the conventional logarithmic Schrödinger equation. We also expect this model to be more applicable to the conservative quantum-mechanical systems.

In the full model, the coupling constant b_0 is no longer a built-in parameter of a theory, cf. Eq. (4), but one of the integration constants of evolution equations, such as Eqs. (16) and (17), therefore, its value can vary from system to system. Thus, the full model allows us to reduce a number of parameters of the theory and make it more self-consistent and self-sufficient.

Furthermore, in the models with variable nonlinear coupling, a nature of the quantum EH temperature (6) starts to become clearer. In this regard, it is also interesting to study a possible connection between this temperature and the conventional (thermal) one. The latter is still a notion which is well-defined at a classical level only (*e.g.*, as measured by bringing a system into a thermal equilibrium with some calibrated system), but its quantum analogue's nature is still obscure.

This problem manifests itself most strikingly when dealing with quantum gases and liquids such as Bose-Einstein condensates (BEC). Strictly speaking, one can not measure the temperature of a condensate without affecting it – for instance, the standard method consists of switching a trap off, using a laser upon the condensate's atoms and measuring the scattered light to deduce their 'temperature' from experimental profiles of density and momentum distributions, assuming that the energy spectrum is also known. Whether such "statistical" temperature is a genuine (thermal) temperature of a condensate remains an open question from a rigorous approach's viewpoint, especially bearing in mind that: (i) experimental condensates are energetically isolated, so no thermal equilibrium can be achieved with a calibrated system, (ii) the relations between temperature and the above-mentioned distributions and spectrum are derived from one or another theoretical model which must be thus presumed to be *a priori* valid for that particular condensate, and (iii) this detrapping measurement method is unlikely to be reliable for strongly-interacting quantum Bose liquids which tend to confine their atoms into a droplet in absence of trapping potentials [24, 25, 28].

Therefore, we can make a conjecture here that the origins of the conventional thermal temperature come from basic quantum-mechanical notions, such as those discussed in the introductory section above, whereas its current statistical or classical thermodynamical definitions are derivative concepts. In other words, the concept of temperature could be introduced without referring to classical concepts, such as the heat or average kinetic energy, but rather underlies them.

Finally, properties of this logarithmic equation with a variable nonlinear coupling and its solutions will also be a subject of our future work. It is expected that they gives us more insights regarding the nature of the Everett-Hirschman entropy and conjugate temperature.

Acknowledgments

This work is based on the research supported by the National Research Foundation of South Africa under Grants Nos. 95965 and 98892.

References

[1] G. Rosen 1968 *J. Math. Phys.* **9** 996
[2] G. Rosen 1969 *Phys. Rev.* **183** 1186
[3] I. Bialynicki-Birula and J. Mycielski 1976 *Ann. Phys. (N. Y.)* **100** 62
[4] I. Bialynicki-Birula and J. Mycielski 1975 *Commun. Math. Phys.* **44** 129
[5] I. Bialynicki-Birula and J. Mycielski 1979 *Phys. Scripta* **20** 539
[6] K. Enqvist and J. McDonald 1998 *Phys. Lett.* B **425** 309
[7] T. Hiramatsu, M. Kawasaki, and F. Takahashi 2010 *J. Cosmol. Astropart. Phys.* **2010** 008
[8] V. Dzhunushaliev and K. G. Zloshchastiev 2013 *Central Eur. J. Phys.* **11** 325
[9] I. E. Gulamov, E. Ya. Nugaev, and M. N. Smolyakov 2014 *Phys. Rev.* D **89** 085006
[10] I. E. Gulamov, E. Ya. Nugaev, A. G. Panin, and M. N. Smolyakov 2015 *Phys. Rev.* D **92** 045011
[11] V. Dzhunushaliev, A. Makhmudov, and K. G. Zloshchastiev 2016 *Phys. Rev.* D **94** 096012
[12] H. Buljan, A. Šiber, M. Soljačić, T. Schwartz, M. Segev, and D. N. Christodoulides 2003 *Phys. Rev.* E **68** 036607
[13] S. De Martino, M. Falanga, C. Godano and G. Lauro 2003 *Europhys. Lett.* **63** 472
[14] T. Hansson, D. Anderson, and M. Lisak 2009 *Phys. Rev.* A **80** 033819
[15] E. F. Hefter 1985 *Phys. Rev.* A **32** 1201
[16] V. G. Kartavenko, K. A. Gridnev and W. Greiner 1998 *Int. J. Mod. Phys.* E **7** 287
[17] K. Yasue 1978 *Ann. Phys. (N.Y.)* **114** 479
[18] J. D. Brasher 1991 *Int. J. Theor. Phys.* **30** 979
[19] D. Schuch 1997 *Phys. Rev.* A **55** 935
[20] M. P. Davidson 2001 *Nuov. Cim.* B **116** 1291
[21] J. L. López 2004 *Phys. Rev.* E **69** 026110
[22] J. L. López and J. Montejo-Gámez 2013 *Nanoscale Syst. Math. Model. Theory Appl.* **2** 49
[23] M. Znojil, F. Růžička, and K. G. Zloshchastiev 2017 *Symmetry* **9** 165
[24] A. V. Avdeenkov and K. G. Zloshchastiev 2011 *J. Phys.* B: *At. Mol. Opt. Phys.* **44** 195303
[25] K. G. Zloshchastiev 2012 *Eur. Phys. J.* B **85** 273
[26] D. A. Meyer and T. G. Wong 2014 *Phys. Rev.* A **89** 012312
[27] B. Bouharia 2015 *Mod. Phys. Lett.* B **29** 1450260
[28] K. G. Zloshchastiev 2017 *Z. Naturforsch.* A **72** 677
[29] K. G. Zloshchastiev 2010 Grav. Cosmol. **16** 288
[30] K. G. Zloshchastiev 2011 *Acta Phys. Polon.* B **42** 261
[31] K. G. Zloshchastiev 2011 *Phys. Lett.* A **375** 2305
[32] H. Everett III 1955 *Theory of the universal wave function* PhD thesis (Princeton) p 140
[33] I. I. Hirschman, Jr. 1957 *Am. J. Math.* **79** 152
[34] K. I. Babenko 1961 *Amer. Math. Soc. Transl.* **44** 115
[35] W. Beckner 1975 *Annals Math.* **102** 159
[36] K. G. Zloshchastiev, in preparation.

IOP Conf. Series: Journal of Physics: Conf. Series **1039** (2018) 012015 doi:10.1088/1742-6596/1039/1/012015

The impact of slip conditions on magnetohydrodynamics radiating fluid beyond an exponentially extended sheet

Siti Suzilliana Putri Mohamed Isa[1, 2*], Norihan Md. Arifin[1, 3], and Umer Farooq[4, 5]

[1]Institute for Mathematical Research, Universiti Putra Malaysia, 43400 UPM Serdang, Selangor Darul Ehsan, MALAYSIA.
[2]Centre of Foundation Studies for Agricultural Science, Universiti Putra Malaysia, 43400 UPM Serdang, Selangor Darul Ehsan, MALAYSIA.
[3]Department of Mathematics, Universiti Putra Malaysia, 43400 UPM Serdang, Selangor Darul Ehsan, MALAYSIA.
[4]Department of Mathematics, Faculty of Science, Jiangsu University, 212013 Zhenjiang, CHINA.
[5]Department of Mathematics, COMSATS Institute of Information Technology, Park Road, Tarlai Kalan Islamabad 44000, PAKISTAN.

Email: ctsuzilliana@upm.edu.my

Abstract. A mathematical model is performed to analyse the magnetohydrodynamics (MHD) fluid flow and transmission of heat in the case of an exponentially extended sheet. The system is supressed by thermal radiation, which is constituted in the energy equation. The boundary contains velocity, thermal and mass slips. Similarity transformation is applied, to convert the governing equations (continuity, momentum, energy and mass diffusion equations) into ordinary differential equations (ODE). Then, the purpose of numerical shooting method is to perform the solutions in the ODE part. Finally, results are presented in the forms of table and graphs (skin friction coefficient, heat transfer rate, concentration gradient at the plate, and the profiles of velocity, temperature, concentration and concentration gradient) for various values of controlling parameters. The consequences of the controlling parameters on the system are described in details.

1. Introduction

The theoretical reports of fluid flow and transmission of heat past a continuous extending surface contribute numerous applications in cooling of metallic plate in a bath, glass fibre and paper production, extrusion of polymer sheet from a die, and continuous casting [1]. This is due to the desired characteristics of final product in industrial manufacturing processes mostly depend on the rate of extending and heat transfer. In addition, the radiation effects and the study of magnetohydrodynamics (MHD) are also included in a variety of extending sheet problems. The thermal radiation impact is very important in gas turbines and devices for satellites, nuclear plants, space vehicles and aircraft [1]. Meanwhile, the study of MHD is applied in crystals growing, electromagnetic field pumps, plasma jets, MHD couples and bearing, and chemical synthesis [2]. Therefore, many researchers [3−16] described the radiating MHD fluid flow caused by a extending surface for non-slip boundary conditions.

In engineering industry, the requirement of selecting the condition regarding to the partial slip as the boundary conditions, instead of non-slip is due to the reason that partial slip boundary condition is more consistent with all physical features in real situations. The presence of boundary slip in the flow field has applications in a technology, namely polishing of artificial heart valves and internal cavities [17]. Slip for velocity aspect arises when the fluid is not sticking to a solid boundary. Beavers and Joseph [18] described the fluid flow due to the condition of slipped boundary over a permeable wall. Fluid flow bounded by a partial slip, in the case of an extending sheet were investigated by Andersson [19], Wang [20], Ariel [21] and Sajid et al. [22]. The transmission of heat in MHD fluid flow, and also has a radiative characteristics are reported by Ibrahim and Shankar [23], Mukhopadhyay [24], Loganathan and Vimala [25] and Abdul Hakeem et al. [26] with the exerted slip conditions at the boundary. Ibrahim and Shankar [23] studied the impact of partial slips at boundary conditions (thermal, velocity and solutal slip) on MHD radiating nanofluid flow. On the other hand, Mukhopadhyay [24] analysed the impact of slips (velocity and thermal) at boundary conditions on flow of MHD fluid and transmission of heat due to a extending sheet with thermal radiation. This sheet is assumed to extend with exponential variation in velocity. However, Loganathan, and Vimala [25] developed the model of unsteady state in MHD nanofluid with thermal radiation. They also add another effect in the system related to the exponentially extending sheet, which is heat generation/absorption. The internal heat generation or absorption, velocity slip impact and thermal radiation on MHD porous medium flow regarding to a linear velocity of the extending surface with wall mass transfer are analysed by Abdul Hakeem et al. [26]. In addition, many researchers [27−32] investigated the different flow problems in the case of extending sheet. The system developed by these investigators is bounded by the partial slip and suppressed by magnetic field field and thermal radiation. Furthermore, some of the researchers [16, 33−38] are interested to present the illustrations of concentration distribution, as an addition to the profiles of temperature and velocity. To produce the profile of fluid concentration, they need to incorporate mass diffusion equation, together with momentum and energy equations as governing equations.

Motivated by the work by previous investigator (Mukhopadhyay [24]) and others, we study the effect of partial slips boundary conditions (velocity, thermal and mass slips) on MHD radiating viscous fluid. The transmission of heat in this flow are caused by sheet which is extending in exponential distributions. We introduce similarity variables to convert momentum, energy, and mass diffusion equations into nonlinear ordinary differential equations. Shooting technique is used to solve the ODE section. Computed numerical results are performed through table and graphs. Characteristics of transferring heat and mass in fluid flow is reported, subjected to certain controlling parameters.

2. Mathematical formulation

Consider the steady boundary layer flow fluid over an isothermal extending surface, which illustrated in Figure 1. The fluid is assumed to be a viscous incompressible electrically conducting and radiating, and modelled in two-dimensional Cartesian coordinates. The direction of the extending surface is propagates into x-axis, and the y-axis is directed to this x-axis. The presence of an externally applied magnetic field field $B(x)$ is directed to the y-axis. The governing equations, to develop mathematical problem equations are written as

$$\frac{\partial u}{\partial x}+\frac{\partial v}{\partial y}=0 \tag{1}$$

$$u\frac{\partial u}{\partial x}+v\frac{\partial u}{\partial y}=\upsilon\frac{\partial^2 u}{\partial y^2}-\frac{\sigma B^2(x)}{\rho}u \tag{2}$$

$$u\frac{\partial T}{\partial x}+v\frac{\partial T}{\partial y}=\frac{\kappa}{\rho c_p}\frac{\partial^2 T}{\partial y^2}-\frac{1}{\rho c_p}\frac{\partial q_r}{\partial y} \tag{3}$$

$$u\frac{\partial C}{\partial x}+v\frac{\partial C}{\partial y}=D\frac{\partial^2 C}{\partial y^2} \tag{4}$$

where u and v are the x and y component of velocity in Cartesian coordinate, $\upsilon=\mu/\rho$ is the kinematic viscosity, μ is the viscosity, ρ is the fluid density, σ is the electrical conductivity, q_r is the radiative heat flux, c_p is the specific heat at constant temperature, κ is the thermal conductivity of the fluid, C is the concentration of the fluid, D is the coefficient of mass diffusivity and T is the temperature of the fluid. The induced magnetic field field in Equation (2) is disregarded since the magnetic field Reynolds number for the flow is assumed to be very small. In our design, the magnetic field field is formed as $B(x) = B_0 e^{(x/2L)}$, where B_0 being a constant.

Using Rosseland approximation for radiation [39], radiative heat flux q_r is written as

$$q_r=-\frac{4\sigma^*}{3k^*}\frac{\partial T^4}{\partial y} \tag{5}$$

where σ^* is the Stefan-Boltzman constant and k^* is the absorption coefficient. The higher orders in Equation (5) is neglected, and Equation (3) becomes

$$u\frac{\partial T}{\partial x}+v\frac{\partial T}{\partial y}=\frac{\kappa}{\rho c_p}\frac{\partial^2 T}{\partial y^2}+\frac{16\sigma^* T_\infty^3}{3\rho c_p k^*}\frac{\partial^2 T}{\partial y^2} \tag{6}$$

where T_∞ is the temperature of the fluid at infinity.

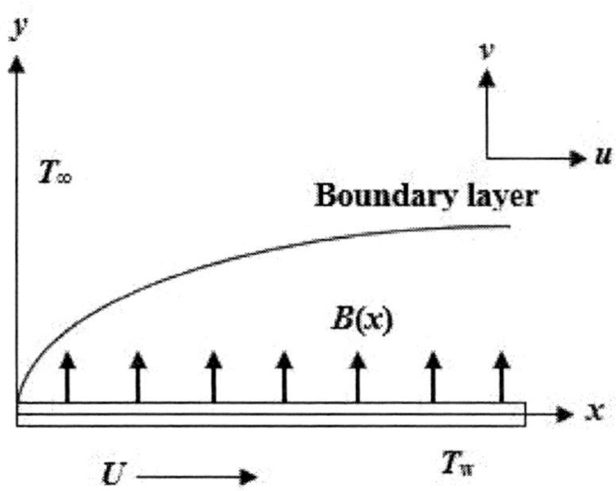

Figure 1. Sketch of physical problem.

The boundary conditions are

$$\begin{aligned} & u=U+N\frac{\partial u}{\partial y},\quad v=-V(x), \\ & T=T_w+D\frac{\partial T}{\partial y},\quad C=C_w+F\frac{\partial C}{\partial y} \qquad \text{at } y=0, \\ & u\to 0,\quad T\to T_\infty,\quad C\to C_\infty \qquad \text{as } y\to\infty \end{aligned} \tag{7}$$

where $U = U_0 e^{(x/L)}$ is the extending velocity, $T_w(x) = T_\infty + T_0 e^{(x/2L)}$ is temperature at the sheet, $C_w(x) = C_\infty + C_0 e^{(x/2L)}$ is the concentration of the plate, C_∞ is the concentration in the free stream, U_0, T_0 and C_0 are the reference velocity, temperature and concentration respectively, $N = N_1 e^{(-x/2L)}$ is the velocity slip factor, N_1 is the initial value of velocity slip factor, $D = D_1 e^{(-x/2L)}$ is the thermal slip factor, D_1 is the initial value of thermal slip factor, $F = F_1 e^{(-x/2L)}$ is the mass slip factor, and F_1 is the initial value of mass slip factor. The condition of the no-slip case is attained when $N = D = F = 0$. The velocity at the wall is defined as $V(x) = V_0 e^{(x/2L)}$, where V_0 is the initial strength of suction. The velocity of suction is $V(x) > 0$, and the velocity of blowing is denoted by $V(x) < 0$.

Introducing new similarity variables by Bala Anki Reddy [27]:

$$\eta = y\left(\frac{U_0}{2\upsilon L}\right)^{1/2} e^{(x/2L)}, \quad u = U_0 e^{(x/L)} f'(\eta), \quad v = -\left(\frac{\upsilon U_0}{2L}\right)^{1/2} e^{(x/2L)}\left[f(\eta) + \eta f'(\eta)\right], \quad T = T_\infty + T_0 e^{(x/2L)}\theta(\eta), \quad C = C_\infty + C_0 e^{(x/2L)}\phi(\eta) \tag{8}$$

where prime denotes differentiation with respect to η.

$$f''' + ff'' - 2(f')^2 - M^2 f' = 0 \tag{9}$$

$$\left(1 + \frac{4}{3}R\right)\theta'' + \Pr\left(f\theta' - f'\theta\right) = 0 \tag{10}$$

$$f'\phi - f\phi' = \frac{1}{Sc}\phi'' \tag{11}$$

and the boundary conditions reduce to

$$\begin{aligned} &f' = 1 + \lambda f'', \quad f = s, \quad \theta = 1 + \delta\theta', \quad \phi = 1 + \gamma\phi' \quad \text{at } \eta = 0, \\ &f' \to 0, \quad \theta \to 0, \quad \phi \to 0 \quad \text{as } \eta \to \infty \end{aligned} \tag{12}$$

where $M = \left(2\sigma B_0^2 L/\rho U_0\right)^{1/2}$ is the magnetic field parameter, $\lambda = N_1\left(U_0/2L\upsilon\right)^{1/2}$ is the velocity slip parameter, $\delta = D_1\left(U_0/2\upsilon L\right)^{1/2}$ is the thermal slip parameter, $\gamma = F_1\left(U_0/2\upsilon L\right)^{1/2}$ is the mass slip parameter, $R = 4\sigma^* T_\infty^3/\kappa k^*$ is the radiation parameter, $\Pr = \mu c_p/\kappa$ is the Prandtl number, $Sc = \upsilon/D$ is the Schmidt number and $s = V_0/\left(U_0 \nu/2L\right)^{1/2} > 0$ is the suction parameter.

The skin friction coefficient C_f, Nusselt number Nu_x and local Sherwood number Sh_x are stated as below:

$$C_f = \frac{\tau_w}{\rho u_w^2(x)}, \quad Nu_x = \frac{Lq_w}{k(T_w - T_\infty)}, \quad Sh_x = \frac{Lm_w}{D(C_w - C_\infty)} \tag{13}$$

where the shear stress at the surface τ_w, the heat transfer rate at the surface flux at the wall q_w and the mass transfer rate at the surface flux at the wall m_w are given by

$$\tau_w = \mu\left(\frac{du}{dy}\right)_{y=0}, \quad q_w = -k\left(\frac{dT}{dy}\right)_{y=0}, \quad m_w = -D\left(\frac{dC}{dy}\right)_{y=0} \tag{14}$$

Substituting Equation (8) into Equations (13)-(14), we get

$$C_f\left(2\mathrm{Re}_x\right)^{1/2} = f''(0), \quad Nu_x\left(\frac{2}{\mathrm{Re}_x}\right)^{1/2} = -\theta'(0), \quad Sh_x\left(\frac{2}{\mathrm{Re}_x}\right)^{1/2} = -\phi'(0) \tag{15}$$

where local Reynolds number is defined as $\mathrm{Re}_x = Lu_w(x)/\upsilon$.

2.1. Numerical method for solution

In this section, shooting method is performed in Maple programming to solve the boundary value problems in equations (9)–(12). These equations are transformed into an initial value problem. The programming has applied fourth-order Runge–Kutta integration scheme to solve the initial value problem. The following equations are fixed:

$$f' = fp, \qquad fp' = fpp, \qquad fpp' = -f\left(fpp\right) + 2\left(fp\right)^2 + M^2 fp \tag{16}$$

$$\theta' = \theta p, \qquad \theta p' = \frac{-\Pr\left(f\theta p\right) + \Pr\left(fp\theta\right)}{\left(1+\frac{4}{3}R\right)} \tag{17}$$

$$\phi' = \phi p, \qquad \phi p' = Sc\left(fp\phi - f\phi p\right) \tag{18}$$

with the boundary conditions

$$\begin{aligned} &fp = 1+\lambda fpp, \quad f = s, \quad \theta = 1+\delta\theta p, \quad \phi = 1+\gamma\phi p \quad \text{at } \eta = 0, \\ &fp \to 0, \quad \theta \to 0, \quad \phi \to 0 \quad \text{as } \eta \to \infty \end{aligned} \tag{19}$$

The first step in this shooting technique is to set predicted values for $fpp(0)$, $\theta p(0)$ and $\phi p(0)$ $\left(f''(0),\ \theta'(0)\ \text{and}\ \phi'(0)\right)$. A highest finite value for η is also predicted, denoted as η_∞. These predicted values $\left(\eta_\infty,\ f''(0),\ \theta'(0)\ \text{and}\ \phi'(0)\right)$ must satisfy the boundary conditions in Equation (19) at $\eta = 0$ and as $\eta \to \infty$. In addition, the predicted values are fixed when two consecutive values of $f''(0)$, $\theta'(0)$ and $\phi'(0)$ vary significantly by a specified value.

3. Results and discussion

The numerical calculations, namely as shooting method is used to solve equations (9)–(12). The graphs of skin friction coefficient $f''(0)$, local Nusselt number $-\theta'(0)$, local Sherwood number $-\phi'(0)$, velocity profile $f'(\eta)$, temperature profile $\theta(\eta)$, concentration profile $\phi(\eta)$ and concentration gradient profile $\phi'(\eta)$ are illustrated in this section. These illustrations are controlled by parameters of velocity slip λ, thermal slip δ, mass slip γ, magnetic field field M, radiation R and Schmidt number Sc. When the comparison of the presents findings with the previous results (Ishak [7] and Mukhopadhyay [24]) is performed, then the accuracy of current numerical method in this problem is achieved. This comparison is tabulated in Table 1. As a conclusion, the present values of $-\theta'(0)$ for some special cases are in good agreement with those reported by the previous investigators. Therefore, the good comparison proves that our numerical shooting method is applicable to present our theoretical study and numerical computation.

IOP Conf. Series: Journal of Physics: Conf. Series **1039** (2018) 012015 doi:10.1088/1742-6596/1039/1/012015

Table 1.
Values of $-\theta'(0)$ for several values of Prandtl number, radiation parameter and magnetic field field in the absence of suction, velocity, thermal and mass slips parameter and Schmidt number.

Pr	R	M	Ishak [7]	Mukhopadhyay [24]	Present
1	0	0	0.9548	0.9547	0.9548
2			1.4715	1.4714	1.4715
3			1.8691	1.8691	1.8691
5			2.5001	2.5001	2.5001
10			3.6604	3.6603	3.6604
1		1	0.8611	0.8610	0.8615
	0.5	0	-	-	0.6765
	1		0.5312	0.5311	0.5313
		1	0.4505	0.4503	0.4506
2	0.5	0	-	1.0734	1.0735
	1		-	0.8626	0.8629
3	0.5		-	1.3807	1.3807
	1		-	1.1213	1.1214

In Figure 2, the representation of velocity distribution is shown for different values of velocity slip parameter λ. In this figure, the velocity $f'(\eta)$ at a point diminishes at large distance from the sheet. This figure also displays that the fluid velocity declines with an enhancement in parameter λ, which happens because fluid experiences less drag with increase in velocity slip. The occurrence of slip will affect the value of the fluid velocity close to the sheet. As a result, the velocity of the fluid at this point is slightly differ to the velocity of the extending sheet. It is observed that the velocity of the fluid reduces due to the enhancement of velocity slip. This result due to the pulling of the extending sheet can be only transmitted partially to the fluid. Figure 3 represents that fluid velocity reduces with the enhancement in parameter *M*. This happens due to setting up of Lorentz force, caused the fluid motion is opposed by this force.

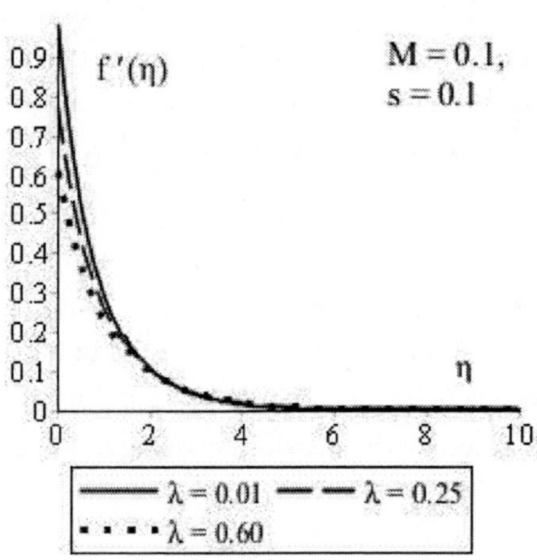

Figure 2. Variation of velocity for various λ.

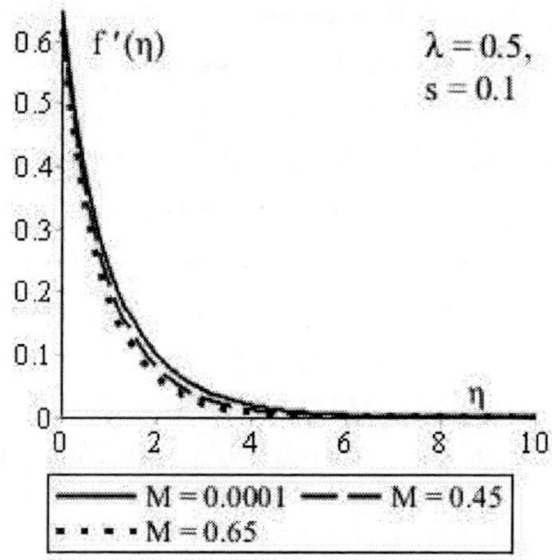

Figure 3. Variation of velocity for various *M*.

The temperature profiles $\theta(\eta)$ with the impact of various values of related parameters (λ, δ, R and M) has been showed in Figures 4–7, respectively. Figures 4, 6 and 7 show that the fluid temperature increase with an increment in the values of λ, R and M. Otherwise, the effect of thermal slip in Figure 5 is to minimize the fluid temperature, together with the thickness of thermal boundary layer.

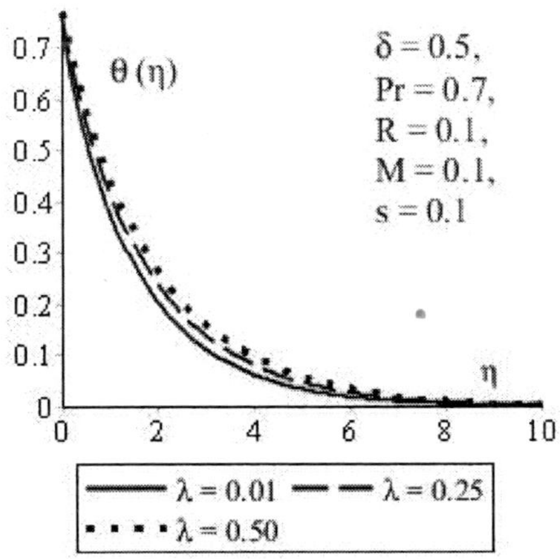

Figure 4. Variation of temperature for various λ.

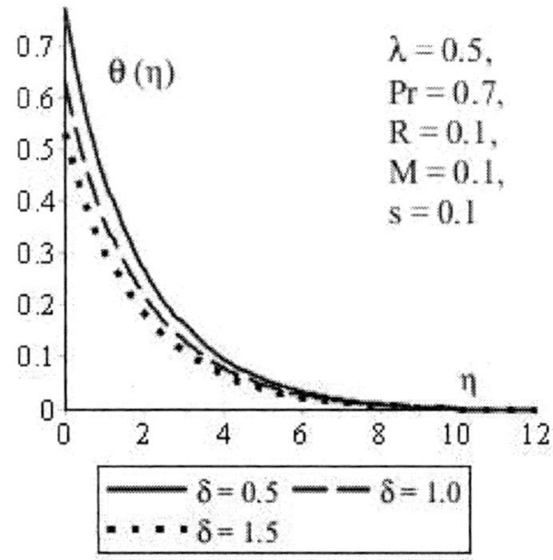

Figure 5. Variation of temperature for various δ.

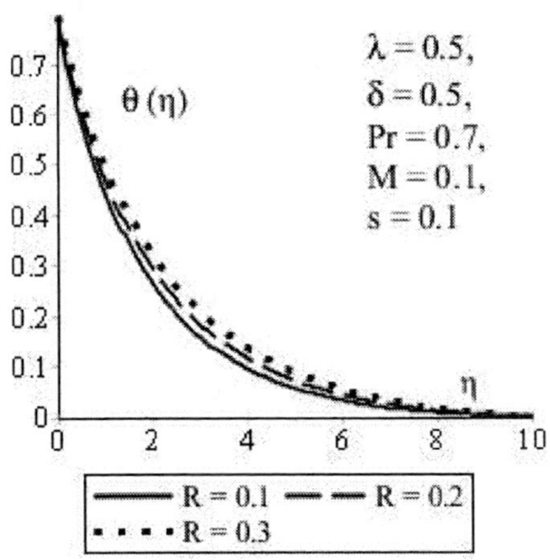

Figure 6. Variation of temperature for various R.

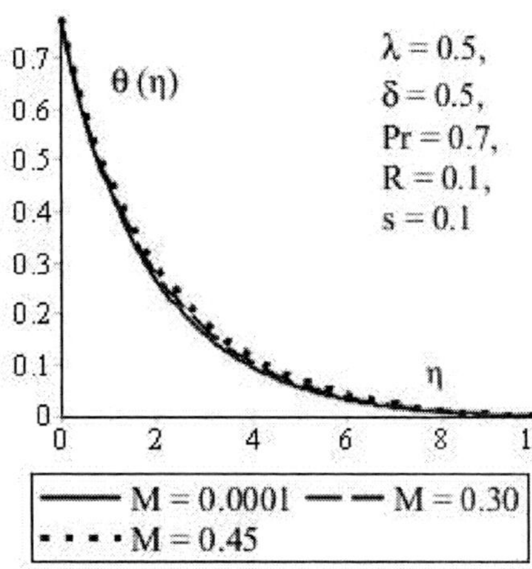

Figure 7. Variation of temperature for various M.

Figures 8−11 show the effect of related parameters (velocity slip λ, mass slip γ, magnetic field M and Schmidt number Sc) on the concentration profile $\phi(\eta)$. From these figures, the concentration is decreasing until it reaches zero at a point far from the extending sheet. Refer to Figures 8 and 10, the concentration enhance with the increase of λ and M. Figure 9 reveals that the values of concentration of the fluid reduces as we increase the values of mass slip parameter. The concentration is reduced when the mass transfer from the plate to the fluid reduced, due to the effect of mass slip. Refer to the Figure 11, the increment in the value of Sc will lead to the decrement in the rate of concentration and the thickness of concentration boundary layer. This is because due to an enhancement in Schmidt number, the diffusion coefficient diminishes. Consequently, the effect of Sc causes the decrement in the concentration boundary layer.

Figures 12−15 illustrate the concentration gradient profile $\phi'(\eta)$, with the various value of λ, γ, M and Sc. The increment of the concentration gradient profile at the point near to the wall can be seen by adding the effect of λ and M. These can be observed in Figures 12 and 14. From Figure 13, it is showed that the concentration gradient at a fixed point increase with increase of mass slip parameter γ. In Figure 15, the values of $\phi'(\eta)$ increase at the point far from the extending sheet with the impact of parameter Sc. However, this profile shows an opposite trend due to the various values of Sc at the point near to the extending sheet.

The illustrations of skin friction coefficient $f''(0)$ versus velocity slip parameter λ by including the effect of magnetic field are figured in Figure 16. From Figure 16, it is certain that the values of wall drag is minimized caused by the impact of magnetic field whereas it is enhanced when velocity slip is included. The variation of $-\theta'(0)$ with thermal slip parameter δ for various values of λ is displayed in Figure 17. The purpose of the velocity slip parameter is to lessen the transmission of heat, for a large value of thermal slip δ. However, the opposite nature can be observed for a small rate of thermal slip. The decrement on heat transfer rate is presented by adding the impact of magnetic field and radiation, which are illustrated in Figures 18 and 19. It is also noted that the values of $-\theta'(0)$ reduce when the thermal slip increases. The plots of local Sherwood number $-\phi'(0)$ against mass slip parameter γ for the various values of λ, M and Sc are showed in Figures 20−22. It is observed from Figures 20 and 21 that the magnitude of $-\phi'(0)$ decreases with larger value of λ and M, while Figure 22 shows that the values of $-\phi'(0)$ increase by increasing the values of Sc. Moreover, it can be seen that an increment in the values of γ lead to a decrement of the local Sherwood number $-\phi'(0)$.

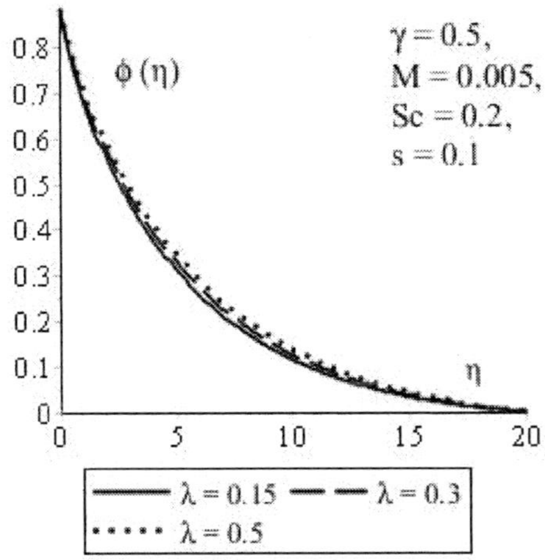

Figure 8. Variation of concentration for various λ.

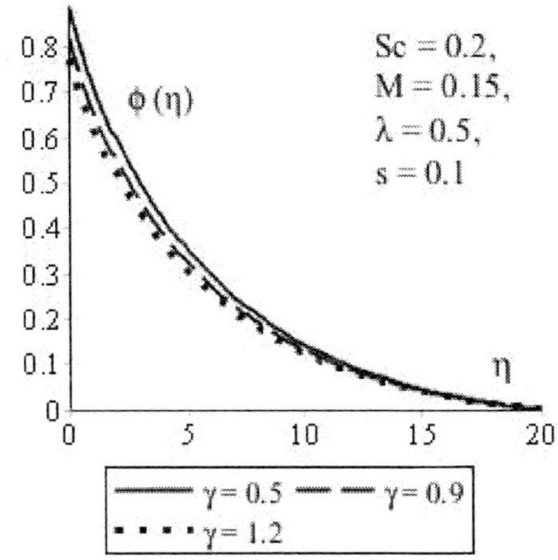

Figure 9. Variation of concentration for various γ.

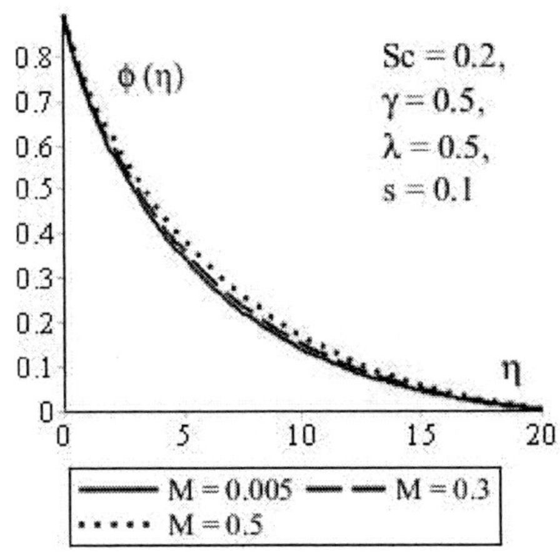

Figure 10. Variation of concentration for various *M*.

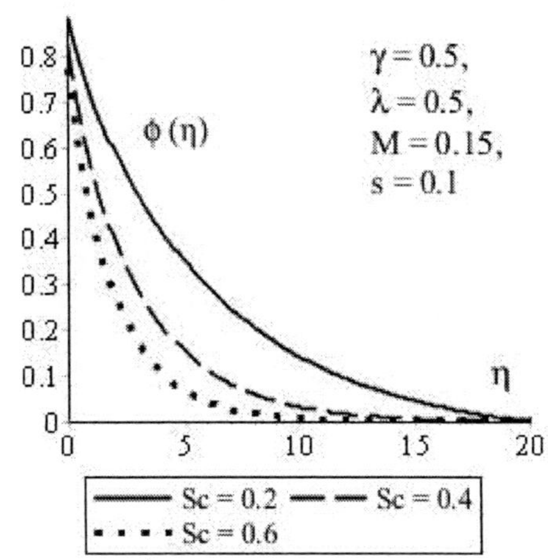

Figure 11. Variation of concentration for various *Sc*.

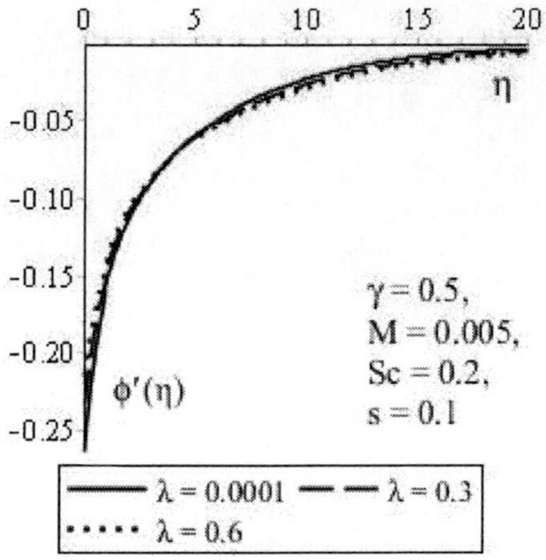

Figure 12. Variation of concentration gradient for various *λ*.

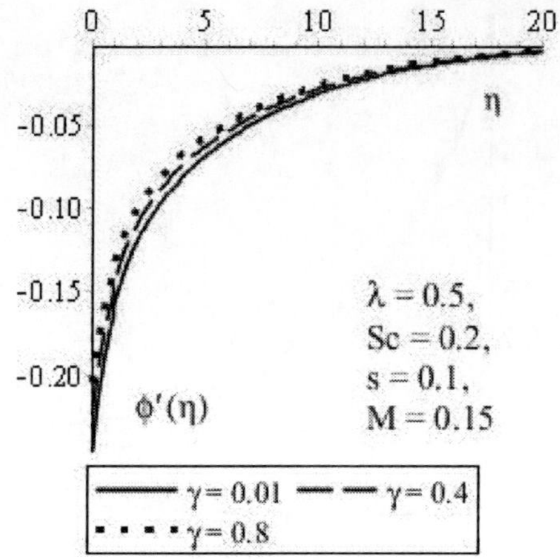

Figure 13. Variation of concentration gradient for various *γ*.

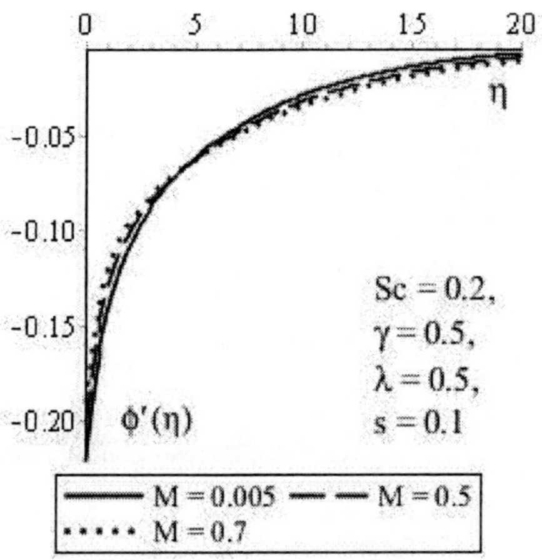

Figure 14. Variation of concentration gradient for various *M*.

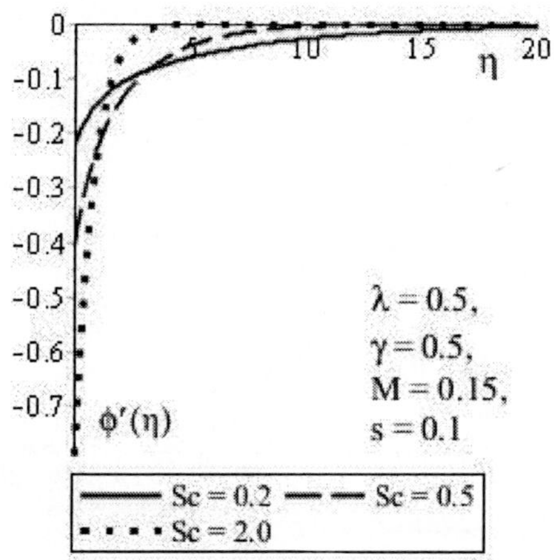

Figure 15. Variation of concentration gradient for various *Sc*.

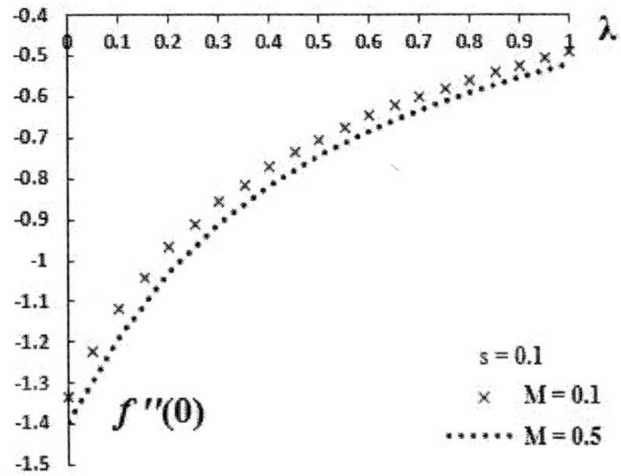

Figure 16. Variation of $f''(0)$ versus λ for various M.

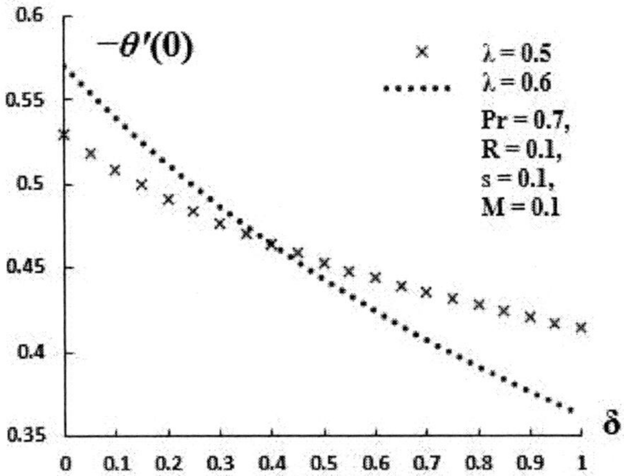

Figure 17. Variation of $-\theta'(0)$ versus δ for various λ.

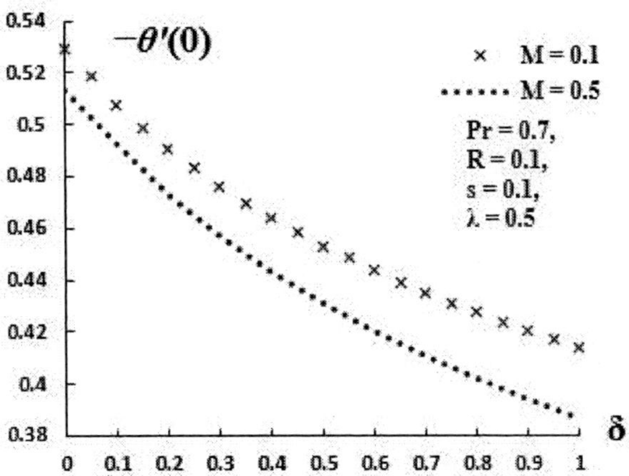

Figure 18. Variation of $-\theta'(0)$ versus δ for various M.

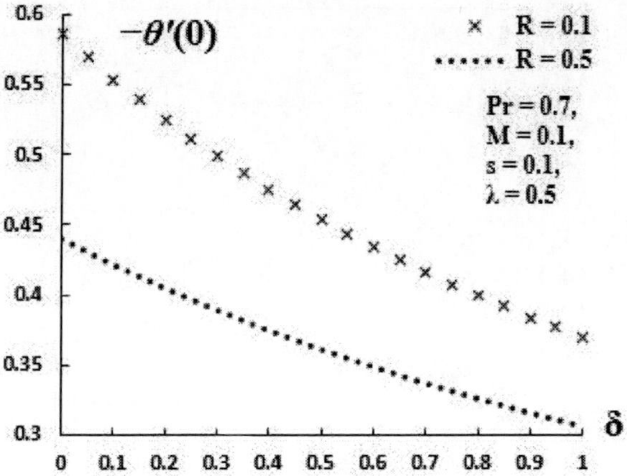

Figure 19. Variation of $-\theta'(0)$ versus δ for various R.

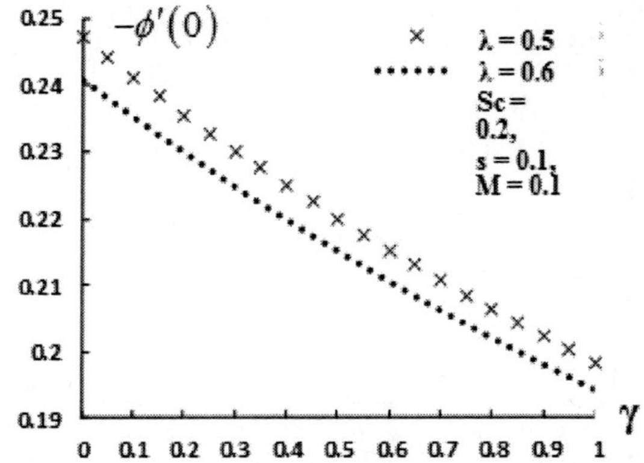

Figure 20. Variation of $-\phi'(0)$ versus γ for various λ.

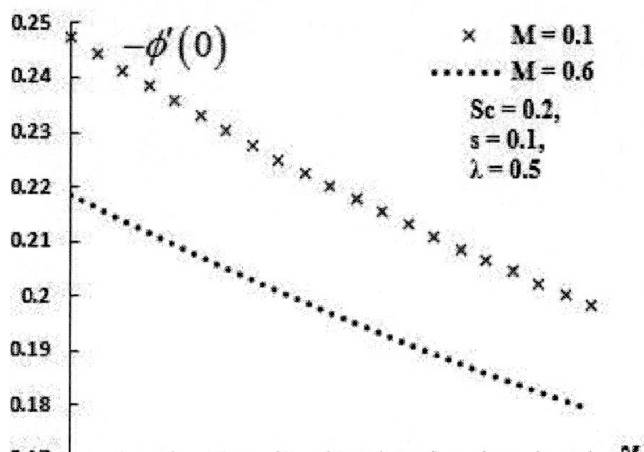

Figure 21. Variation of $-\phi'(0)$ versus γ for various M.

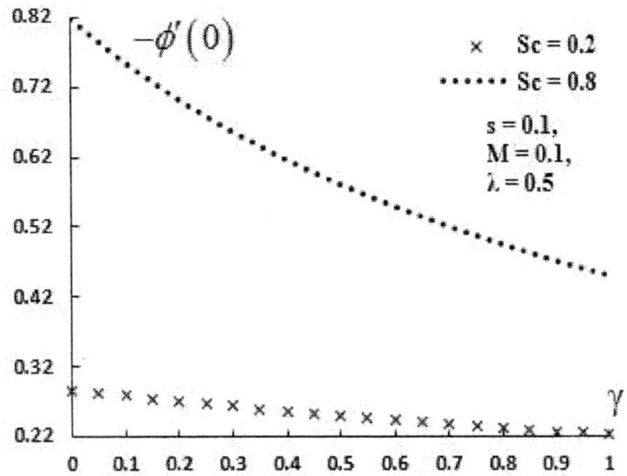

Figure 22. Variation of $-\phi'(0)$ versus γ for various *Sc*.

4. Conclusion

The numerical solution for the problem of MHD fluid flow towards a permeable exponentially extending sheet is developed. The system of the radiating fluid flow is bounded by slip condition. The following observations can be made from the present findings:

a) The drag experienced by the fluid reduces with the increment in magnetic parameter while it shows a reverse trend with the impact of velocity slip parameter.
b) The heat transfer rate is decreased when the parameters of velocity and thermal slip, magnetic and radiation increase.
c) The concentration gradient at the plate increase with increase of Schmidt number, whereas it drops when parameters of velocity slip, magnetic and mass slip enhance.
d) Parameter of magnetic field and velocity slip lessen the fluid velocity.
e) The fluid temperature rises with the addition in the parameters of velocity slip, radiation or magnetic field. On the other hand, the rate of the temperature reduces with the addition in thermal slip rate.
f) The raise of the value in mass slip parameter and Schmidt number lead to the decrement in the concentration of the fluid. However, by considering the impact of magnetic field and velocity slip, the temperature of the fluid shows an increment.
g) The concentration gradient profile enhances when the mass slip parameter occurs in the system. In addition, the impact of velocity slip and magnetic field lead to the decrement of concentration gradient of adjacent fluid. The increment of concentration gradient of adjacent fluid is due to the increasing Schmidt number.

References

[1] Hayat T, Shafiq A, Alsaedi, A 2014 Effect of Joule heating and thermal radiation in flow of third grade fluid over radiative surface. *Plos One* **9** e83153.
[2] Barik R N 2013 Radiation effect and MHD flow on moving vertical porous plate with variable temperature. *Bulletin of Engineering* **Fascicule 3** 45–48.
[3] Ouaf M E M 2005 Exact solution of thermal radiation on MHD flow over a stretching porous sheet. *Applied Mathematics and Computation* **170** 1117–1125

[4] Abel M S, Nandeppanavar M M 2007 Effects of thermal radiation and non-uniform heat source on MHD flow of viscoelastic fluid flow and heat transfer over a stretching sheet. *International Journal of Applied Mechanics and Engineering* **12** 903–918.

[5] Aliakbar V, Alizadeh-Pahlavan A, Sadeghy K 2009 The influence of thermal radiation on MHD flow of Maxwellian fluids above stretching sheets. *Communications in Nonlinear Science and Numerical Simulation* **14** 779–794.

[6] Chen C H, 2010 On the analytic solution of MHD flow and heat transfer for two types of viscoelastic fluid over a stretching sheet with energy dissipation, internal heat source and thermal radiation. *International Journal of Heat and Mass Transfer* **53** 4264–4273.

[7] Ishak A 2011 MHD boundary layer flow due to an exponentially stretching sheet with radiation effect. *Sains Malaysiana* **40** 391–395.

[8] Pal D, Mandal G, Vajravelu K 2013 MHD convection–dissipation heat transfer over a non-linear stretching and shrinking sheets in nanofluids with thermal radiation. *International Journal of Heat and Mass Transfer* **65** 481–490.

[9] Ahmad I, Sajid M, Awan W, Rafique M, Aziz W, Ahmed M, Abbasi A, Taj M 2014 MHD flow of a viscous fluid over an exponentially stretching sheet in a porous medium. *Journal of Applied Mathematics* **2014** 256761.

[10] Rashidi M, Rostami B, Freidoonimehr N, Abbasbandy S 2014 Free convective heat and mass transfer for MHD fluid flow over a permeable vertical stretching sheet in the presence of the radiation and buoyancy effects. *Ain Shams Engineering Journal* **5** 901–912

[11] Chaudhary S, Singh S, Chaudhary S 2015 Thermal radiation effects on MHD boundary layer flow over an exponentially stretching surface. *Applied Mathematics* **6** 295−303.

[12] Loganthan P, Vimala C 2015 MHD flow of nanofluids over an exponentially stretching sheet embedded in a stratified medium with suction and radiation effects. *Journal of Applied Fluid Mechanics* **8** 85−93.

[13] Mahanta G, Shaw S 2015 3D Casson fluid flow past a porous linearly stretching sheet with convective boundary condition. *Alexandria Engineering Journal* **54** 653–659

[14] Sharada K, Shankar B 2015 MHD mixed convection flow of a Casson fluid over an exponentially stretching surface with the effects of Soret, Dufour, thermal radiation and chemical reaction. *World Journal of Mechanics* **5** 165−177.

[15] Naramgari S, Sulochana C 2016 Dual solutions of radiative MHD nanofluid flow over an exponentially stretching sheet with heat generation/absorption. *Applied Nanoscience* **6** 131−139.

[16] Raju C S K, Sandeep N, Sugunamma V, Jayachandra Babu M, Ramana Reddy J V 2016 Heat and mass transfer in magnetohydrodynamic Casson fluid over an exponentially permeable stretching surface. *Engineering Science and Technology, an International Journal* **19** 45-52.

[17] Mahantesh M, Nandeppanavar K, Vajravelu K, Subhas Abel M, Siddalingappa M N 2013 MHD flow and heat transfer over a stretching surface with variable thermal conductivity and partial slip. *Meccanica* **48** 1451–1464.

[18] Beavers G S, Joseph D D 1967 Boundary conditions at a naturally permeable wall. *Journal of Fluid Mechanics* **30** 197–207.

[19] Andersson H I 2002 Slip flow past a stretching surface. *Acta Mechanica* **158** 121–125.

[20] Wang C Y 2002 Flow due to an stretching boundary with partial slip - an exact solution of the Navier–Stokes equation. *Chemical Engineering Science* **57** 3745–3747.

[21] Ariel P D 2007 Axisymmetric flow due to a stretching sheet with partial slip. *Computers & Mathematics with Applications* **54** 1169-1183.

[22] Sajid M, Ali N, Abbas Z, Javed T 2010 Stretching flows with general slip boundary condition. *International Journal of Modern Physics B* **24** 5939–5947.

[23] Ibrahim W, Shankar B 2013 Boundary layer flow and heat transfer of a nanofluid past a permeable stretching sheet with velocity, thermal and solutal slip boundary conditions. *Computers and Fluids* **75** 1–10.

[24] Mukhopadhyay S 2013 Slip effects on MHD boundary layer flow over an exponentially stretching sheet with suction/blowing and thermal radiation. *Ain Shams Engineering Journal* **4** 485–491.

[25] Loganathan P, and Vimala C 2013 Effects of radiation and magnetic field on unsteady boundary layer flow of a nanofluid past an exponentially stretching sheet in the presence of heat source/sink. *International Review of Mechanical Engineering (IREME)* **7** 1291-1298.

[26] Abdul Hakeem A K, Kalaivanan R, Vishnu Ganesh N, Ganga B 2014 Effect of partial slip on hydromagnetic flow over a porous stretching sheet with non-uniform heat source/sink, thermal radiation and wall mass transfer. *Ain Shams Engineering Journal* **5** 913–922.

[27] Bala Anki Reddy P 2016 Magnetohydrodynamic flow of a Casson fluid over an exponentially inclined permeable stretching surface with thermal radiation and chemical reaction. *Ain Shams Engineering Journal* **7** 593–602

[28] Yazdi M H, Abdullah S, Hashim I, Sopian K 2011 Slip MHD liquid flow and heat transfer over non-linear permeable stretching surface with chemical reaction. *International Journal of Heat and Mass Transfer* **54** 3214–3225.

[29] Turkyilmazoglu M 2011 Analytic heat and mass transfer of the mixed hydro-dynamic/thermal slip MHD viscous flow over a stretching sheet. *International Journal of Mechanical Sciences* **53** 886–896

[30] Mukhopadhyay S, Gorla R S R 2012 Effects of partial slip on boundary layer flow past a permeable exponential stretching sheet in presence of thermal radiation. *Heat Mass Transfer* **48** 1773–1781.

[31] Mukhopadhyay S, Arif M G, Ali M W 2013 Effects of partial slip on chemically reactive solute transfer in the boundary layer flow over an exponentially stretching sheet with suction/blowing. *Journal of Applied Mechanics and Technical Physics* **54** 928-936.

[32] Ibrahim W, Makinde O D 2015 Magnetohydrodynamics stagnation point flow and heat transfer of Casson nanofluid past a stretching sheet with slip and convective boundary condition. *Journal of Aerospace Engineering* **29** 04015037.

[33] Yazdi M H, Abdullah S, Hashim I, Sopian K 2011 Slip MHD liquid flow and heat transfer over non-linear permeable stretching surface with chemical reaction. *International Journal of Heat and Mass Transfer* **54** 3214–3225.

[34] Ibrahim W, Shankar B 2013 MHD boundary layer flow and heat transfer of a nanofluid past a permeable stretching sheet with velocity, thermal and solutal slip boundary conditions. *Computers and Fluids* **75** 1–10.

[35] Singh G, Makinde O D 2013 MHD slip flow of viscous fluid over an isothermal reactive stretching sheet. *Annals of Faculty Engineering Hunedoara- International Journal of Engineering* **Tome 11** 41–46.

[36] Goyal M, Bhargava R 2014 Boundary layer flow and heat transfer of viscoelastic nanofluids past a stretching sheet with partial slip conditions. *Applied Nanoscience* **4** 761–767.

[37] Besthapu P, Bandari S 2015 Mixed convection MHD flow of a Casson nanofluid over a nonlinear permeable stretching sheet with viscous dissipation. *Journal of Applied Mathematics and Physics* **3** 1580-1593.

[38] Mahanta G, Shaw S 2015 Mixed convection stagnation point flow of Casson fluid with convective boundary conditions. *International Journal of Innovative Research in Science, Engineering and Technology* **4** 370–379.

[39] Brewster M Q 1992 *Thermal Radiative Transfer and Properties* (New York: John Wiley and Sons).

The influence of plume angle and energy on the mass separation of plasma beam

Hailong Zhao[1], Zhongying Lin[1], Chunsheng Wang[2] and Hongchen Liu[3]

1 Taizhou University, Taizhou, 318000, China
2 Yanshan University, Yanshan, 066004, China
3 Harbin Institute of Technology, Harbin, 150001, China

Email: fenmiao@hit.edu.cn

Abstract. Plasma optical mass separation is an important innovation in the field of isotopes separation. The magnetic field in the separation region only plays the role of magnetizing plasma and guiding the beam flow direction. The ion trajectory in the separation region containing only radial electric field is the focus of the current research. The plume angle and energy of the plasma beam also determine the effect of beam mass separation. Based on the equation of motion of the ion in the electric field and the mechanism of obtaining the circumferential velocity in the azimuthator, the trajectories of the lithium ions at different plume angles and different energies are analysed, and the parameters of plume angle and energy with the best separation effect are obtained, which provides parameters for the engineering implementation of the multi-mass beam separation structure. The research results can be applied to the development of some devices, such as the mass separator, the mass spectrum analyser and material purification. At the same time, it has certain reference significance for the related field's research of the special configuration electromagnetic field controlling multi-mass beam flow.

1. Introduction

The isotopes have been applied in aerospace, medical, atomic and scientific research and other fields. With the development of science and technology, isotopes production has become an important part of economic activities. Internationally existing plasma separation device is directly from the working fluid of the plasma state to extract the target ions, known as the industrial plasma separator, which is low the single-stage separation coefficient and high energy consumption [1]. The essence of many new plasma separation methods is the mass focus of quasi-neutral steady-state plasma beams under electromagnetic field control [2-5], the most detailed description of this method is the plasma optical mass separator (POMS) mentioned in [6-8]. The principle of POMS proposed in [6] is shown in figure 1, the separator consists of four parts: the quasi-neutral axisymmetric plasma beam (1); the azimuthator for giving the ion circumferential velocity (2); the separation region for beam mass separation (3); the ring collectors for collecting target ions (4).

Compared with the traditional electromagnetic separation method, this method uses a widely used steady-state plasma thruster (SPT) as an ion source. The azimuthator containing only the radial magnetic field is an important part of the mass separator. The plasma containing the axial and radial (depending on the size of the plume) component velocity is accelerated by the azimuthator to obtain

IOP Conf. Series: Journal of Physics: Conf. Series **1039** (2018) 012016 doi:10.1088/1742-6596/1039/1/012016

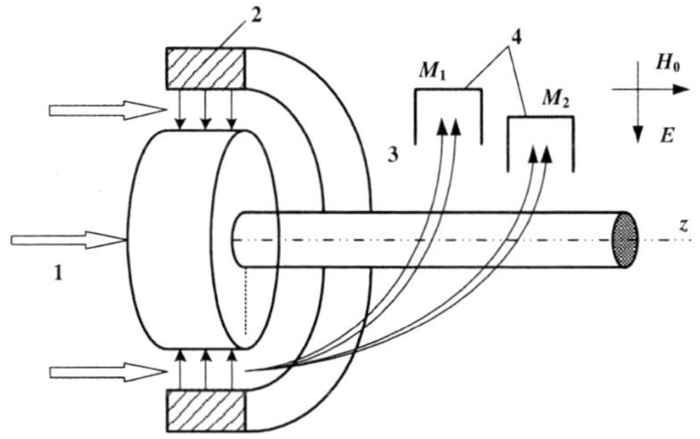

Figure 1. Schematic of POMS.
1-ion source; 2-azimuthator;
3-separating region; 4-collectors

the circumferential velocity, and the beam enters the separation region at a synthetic velocity containing three directional components. The separation region has a weak axial magnetic field and a radial electric field. Under the combined action of the electromagnetic field, the beam is focused at different positions by mass, that is, the mass separation is achieved. The ions through the azimuthator and separation region are ultimately collected by the collector. The ion current in POMS is not limited by its own electric field and the current density of the beam is large, which greatly improves the productivity, so POMS is the best choice to replace the traditional electromagnetic separation method [9-12].

For POMS, the study of the plasma source is more mature, the final realization of the azimuthator and the separation region is the focus of the isotopes separation. In [13], the cathode structure is introduced at the exit of the azimuthator. By comparing the key parameter of the beam, it is revealed that the introduction of the cathode structure is beneficial to improve the separation efficiency. The magnetic field in the separation region only plays the role of magnetizing plasma and guiding the beam flow direction, and is generally small, so the majority of the literature neglects its influence on ion motion. In the theoretical study, the separation region generally contains only an electric field. The SPT jet beam contains a certain plume angle, which result in the overlap phenomenon of different mass beams focusing, and cannot achieve mass separation. The existing literature also does not give the qualified conditions for effective separation.

In this paper, based on the ion trajectory model in the electric field of the separation region, the influence of the initial plume angle and energy on the trajectory of the lithium ion beam is analysed, and the plume angle and the incident energy range are obtained, which lays the theoretical foundation for the efficient separation of plasma beam.

2. The movement mechanism of beam in the azimuthator and the separation region

SPT discharge beam plume divergence angle is generally small, so the beam only contains axial velocity when plume divergence angle is ignored, and ions fly out of the SPT into the azimuthator which is 2 in figure 1.

In the azimuthator, there is only radial magnetic field, and the ion produces the circumferential velocity under the effect of the radial magnetic field in the azimuthator with the average radius r = R. Assuming that the radial magnetic field is uniform and constant, that is, B_r=B_R=const, and the magnetic field boundary effect can is ignored.

In cylindrical coordinate system (r,θ,z), the ions with energy ε, mass m, charge q haven't the circumferential component velocity ($\dot{\theta}=0$), and the circumferential component of the magnetic vector potential A_θ=0, before they enter the radial magnetic field along the axial direction. According to the Busch theorem, the generalized canonical angular momentum conservation equation of the ion in the radial magnetic field is followed:

$$D \equiv mr^2\dot{\theta} + qrA_\theta = 0 \tag{1}$$

The plasma toroidal flow containing different masses of ions passes through the azimuthator gap along the z-axis direction. Assuming the motion radius of all ions is R and constant. At the same time, the ions obtain the circumferential velocity $v_\theta = R\dot{\theta}$ in the azimuthator. whence the generalized canonical angular momentum can be expressed as follows:

IOP Conf. Series: Journal of Physics: Conf. Series **1039** (2018) 012016 doi:10.1088/1742-6596/1039/1/012016

$$D \equiv mR^2\dot{\theta} + qRA_\theta = mRv_\theta + qRA_\theta = 0 \tag{2}$$

From the equation (2) to obtain the circumferential velocity $v_\theta = -qA_\theta/m$. In fact, the width of the gap of the azimuthator is very small (about 0.1m), and the speed of the ions is very large, so the time of the ions stay in the azimuthator is very short. Since the radial magnetic field in the azimuthator is an axisymmetric static magnetic field, $\vec{A}$ only contains the circumferential component A_θ, so $B_r = -\partial A_\theta/\partial z$ (since $\vec{B} = \nabla \times \vec{A}$) . Defining $A_{\theta A}$ as the magnetic vector potential when the ion leaves the azimuthator (the "A" in the subscript means the moment of fling out the azimuthator), so:

$$A_{\theta A} = -\int_0^\Delta B_r dz$$

Where Δ is the average width of the azimuthator. In summary:

$$A_{\theta A} = -\int_0^\Delta B_r dz = -\int_0^\Delta B_R dz = -B_R\Delta \tag{3}$$

Equation (3) is put into formula $v_\theta = -qA_\theta/m$, to obtain the circumferential velocity when the ion mass m is flying out of the azimuthator:

$$v_{\theta A} = -\frac{qA_{\theta A}}{m} = \frac{qB_R\Delta}{m} \tag{4}$$

Assuming all kinds of mass ion energy are the same, and only kinetic energy expressed as $\varepsilon = mv_0^2/2 = M_0V_0^2/2$, that is $v_0 = \sqrt{2\varepsilon/m}$.(M_0 is defined as a central mass ion ,V_0 is the initial velocity of the M_0 mass ion before entering the azimuthator)

The azimuthator containing only magnetic field, does not affect the kinetic energy (ε is unchanged), so the ions contain the circumferential velocity $v_{\theta A}$ and the axial velocity v_{zA} when they fly out of the azimuthator, and $\varepsilon = \frac{1}{2}mv_{\theta A}^2 + \frac{1}{2}mv_{zA}^2$, whence:

$$v_{zA} = \sqrt{2\varepsilon/m - v_{\theta A}^2} \tag{5}$$

According to the momentum conservation $mv_{\theta A} = M_0 v_{\theta A0}$, $v_{\theta A} = M_0 v_{\theta A0}/m$ and $\varepsilon = M_0V_0^2/2$ are obtained, put them into equation (5) there:

$$v_{zA} = \sqrt{\frac{M_0V_0^2}{m} - \frac{M_0^2}{m^2}v_{\theta A0}^2} \tag{6}$$

According to $v_{\theta A0}^2 + v_{zA0}^2 = V_0^2$ (v_{zA0} is the axial velocity of the M_0 mass ion when it flies out of the azimuthator, different from V_0), the axial velocity v_{zA} of different mass ions near M_0 does not change much, and is approximately equal. Whence:

$$V_0^2 = 2v_{\theta A0}^2 = 2v_{zA0}^2 \tag{7}$$

Equation (7) is the basis for the choice of circumferential velocity, also is a basis for analyzing the ion trajectories in the separation region. At this time the axial speed is basically equal when the different quality of ions fly out of the azimuthator, The position of the ions focusing in the z-axis will be substantially the same when the ions enter the separation region (axial without action), that is, the focal plane is almost the same. According to the above characteristics, the size and position of the collector can be determined.

The magnetic field in the separation region is weaker, and it is generally assumed that there is no magnetic field in the theoretical analysis. It is known from [6] that the electric field set in the separation region is $\vec{E} = -\vec{e}_r E_0 R/r$, and according to particle motion equation, the trajectory of the ion with mass m is followed at B=0:

$$\frac{d^2r}{dt^2} = \frac{D^2}{mm_0R^2r}\left(\frac{m_0}{m}\frac{R^2}{r^2} - 1\right) \tag{8}$$

If the plume divergence angle is α, the initial radial velocity is $v_{r0} \approx \alpha v_{zA}$, then $\frac{v_{r0}}{v_{zA}} = \alpha$. From [14] we can see that the trajectories of ions in the (r,z) plane at different plume divergence angles are

IOP Conf. Series: Journal of Physics: Conf. Series **1039** (2018) 012016 doi:10.1088/1742-6596/1039/1/012016

followed:

$$(\frac{dr}{dz})^2 = \frac{M_0}{2m - M_0}(1 - \frac{R^2}{r^2}) + \frac{2m}{2m - M_0}\ln(\frac{R}{r}) + \alpha^2 \tag{9}$$

It can be seen that the mass separation effect is determined by the initial state of the plasma beam, and the influence of the plasma beam plume angle and energy on the separation effect is the most prominent. In order to obtain the maximum separation efficiency, it is necessary to depth study the influencing process of the initial state of the beam on its trajectory.

3. The influence of initial state on beam trajectory

From [2] we can see that plume angle and energy changes determine whether the focus rings overlap. Take ^6Li, ^7Li ions as an example, the center mass M_0=6. During the analysis of the beam mass separation, a radial voltage of 300V is applied in the cylindrical separator. Since the separation region is a symmetrical structure, the axial length of the separation region is set to 2 m and the plasma beam injection position is set at a radius R = 1.5 m. According to the above section set the beam plume angle in the range of 5° ~ 10°, the mass separation effect of the beam trajectory is analyzed by the energy size. Based on the component velocity value determined by the equation(4) ~ (9) and plume angle α value, the trajectory of the ions in the separation region is analyzed to determine the design parameters of the separation region.

3.1. influence of plume angle on ion trajectory

When the plume angle α of the SPT jet beam respectively is ±5°, ±7.5° and ±10°, it is assumed that the axial velocity is 100km/s and the radial velocity respectively is ±8.7489km/s, ±13.1652km/s, ±17.6326km/s, and the circumferential velocity is 30km/s. The ion trajectories obtained in these cases are shown in figure 2.

a)

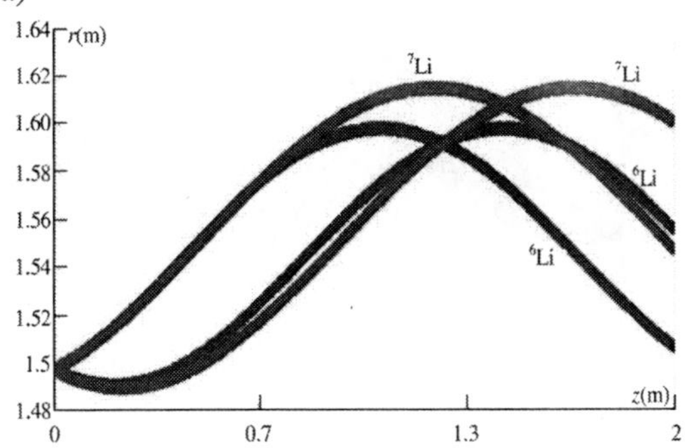

b)

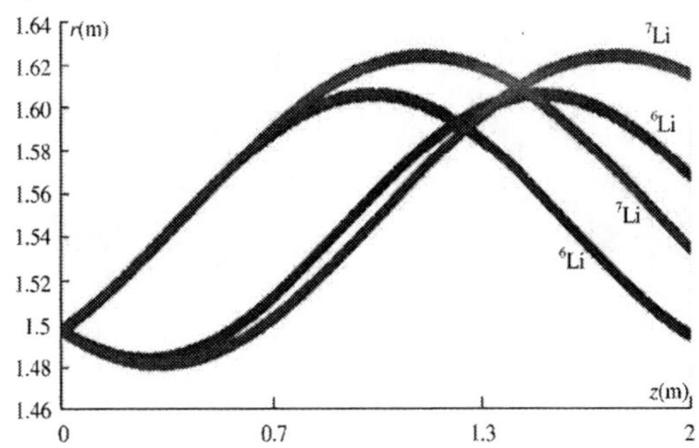

c)

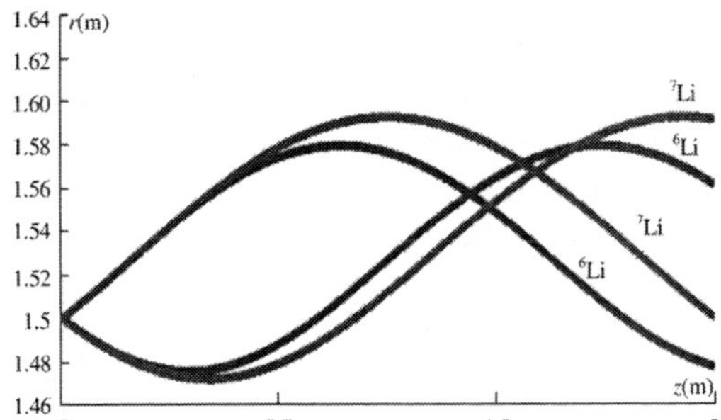

a) Li ions trajectories in the separation region at α=±5°
b) Li ions trajectories in the separation region at α=±7.5°
c) Li ions trajectories in the separation region at α=±10°

Figure 2. Li ions trajectory in the separation region.

It can be seen from figure a, when the ions enter the separation region with ±5° plume angle and the ^6Li ions meet the collection conditions, the position of the focus point is at the edge of the trajectory of the ^7Li ions, and the collected ion purity of the different isotopes is high, so icons can be effectively separated in this case.

From figure b of the trajectories of the ions in the separation region, when the ions enter the separation region with ±7.5° plume angle, the ions movement area gradually widened. When the ^6Li

ions meet the collection conditions, the position of the focus point is already within the ^{7}Li ion boundary line. If the required separation purity is not high, we can take such a situation of plume angle for ion separation

As can be seen from figure c, when the plume angle of the plasma beam is ± 10°, the collection area of the ^{6}Li ions is completely contained within the trajectory of the ^{7}Li ions, that is, the separation cannot be achieved at $\alpha > \pm 10°$

It can be seen that with the plume angle gradually increased, the mass separation of ion will be more and more difficult, and the collected ion purity will be reduced. The separation effect is obvious when the ions enter the separation region with ±5° plume angle, this conclusion is consistent with the analytical results of the divergent angle condition $\alpha < (\frac{9}{11\pi})^{1/3}(\frac{\delta m}{m_0})^{1/3}$ in [2], and it is consistent with the results of Morozov's research [6].

3.2. The influence of energy on ion trajectory

All ions are incident at a plume angle of ±5°. During the analysis, the ion trajectories in the separation region are obtained by changing the velocity component. Based on the difference in the focus point of each simulation process, the effect of mass separation is determined.

The ^{6}Li and ^{7}Li ions are injected into the separation region at an angle of $\alpha = \pm 5°$ with the horizontal direction. The beam is injected into the obliquely upward direction at an angle of ±5° with the axial direction. According to the equation (4)~(8), the axial velocity is 90km/s, the radial velocity is ±3.9295km/s, and the circumferential velocity is 27km/s. The ion trajectories obtained are shown in figure 3.

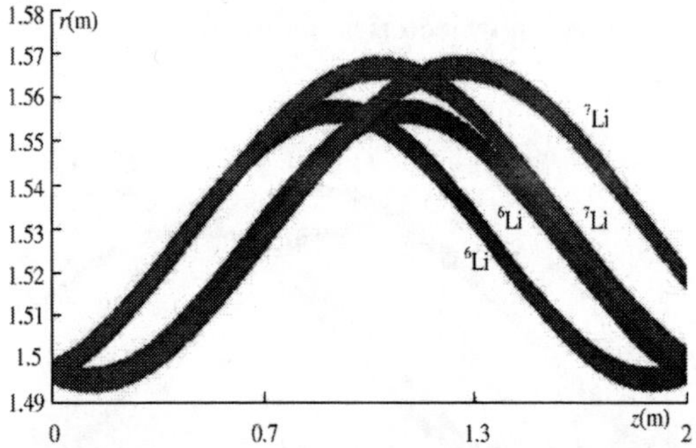

Figure 3. Li ions trajectories at axial speed of 90km/s.

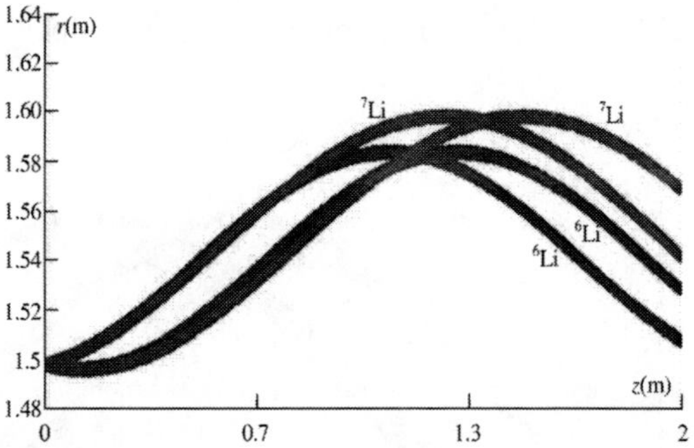

Figure 4. Li ions trajectories at axial speed of 95km/s.

In figure 3, the ions will be focused in advance within the same spatial range, the focus point of the ^{6}Li ions is in the range of ^{7}Li ions, and the position of the two focuses is also close, which brings some difficulty to the collection and leads to the decrease of the ion purity of the collected isotope.

When the energy is increased, that is, the axial velocity is increased to 95km/s, the radial velocity is ±4.1478km/s, and the circumferential velocity is 28.5km/s. The ion trajectories obtained are shown in figure 4. The figure 4 shows that the focus point of the two ions becomes closer. Although the collection position of ^{6}Li ions shifted backwards, but the overlap range of ^{6}Li ions focus position and ^{7}Li ions movement area slightly increased, which is conducive to mass separation. This means that the increase of incident energy is beneficial to the realization of mass separation.

The beam trajectories are shown in figure 5 when the axial velocity is 105km/s, the radial velocity is ±4.5844km/s, and the circumferential velocity is 31.5km/s.

The figure 5 shows that after increasing the ions incident energy which cannot be improved too much by the SPT jet beam limit, the ions will move the focus position backwards within the same spatial range, the overlap of the two ions motions increases little, ^{6}Li ions collection point will gradually move out of the ^{7}Li ions movement area and the change in the axial coordinate distance between the two collectors increases.

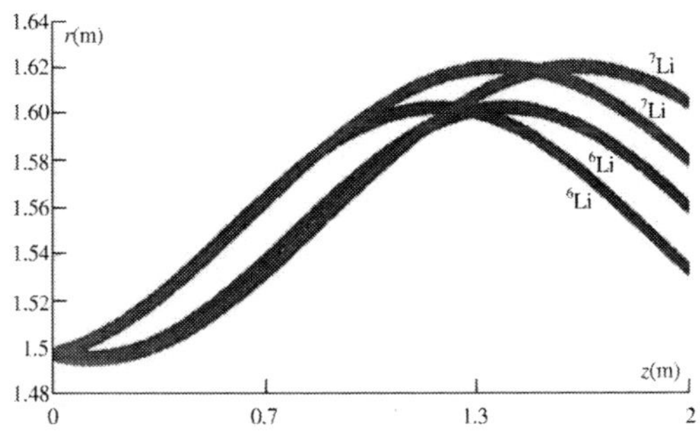

Figure 5. Li ions trajectories at axial speed of 105km/s.

By increasing the initial incident energy of ions to study the ions focusing and collection, it can be seen that the ions separation effect will get better and better when the incident energy of the plasma beam increases within a certain range, which is beneficial to the separation and collection of ions. This conclusion is consistent with the results of the equation (11) in [2] and is consistent with the law in [6]. However, the axial and radial dimensions of the separator increase with the increase of energy, which increases the input of the device. Therefore, the design of mass separators requires a combination of beam energy constraints and device size requirements.

4. Conclusions

In this paper, the influence of the initial plume angle and energy of the plasma beam on the separation process of the beam in the electric field of the separation region is analyzed in depth. The results show that as the plume angle increases, the mass separation of ions will be more and more difficult, and the purity of ions will be reduced. At the same time, when the incident energy of the plasma beam increases within a certain range, the ions separation effect will get better and better, which is beneficial to the mass separation and collection of ions. The conclusion of this paper will lay a theoretical foundation for the future realization of the azimuthator and the separation region.

References:

[1] Ю.В.Ковтун, Е.И. Скибенко and В.Б. Юферов. 2008 *Вопросы атомной науки и техники* **2** 149-154

[2] Bardakov V M, Kichigin G N and Strokin N A 2010 *Technical Physics Letters* **36(2)** 185-188

[3] Zhao hailong, Yan yuming, Wang chunsheng and Jiang binhao 2013 *Int. Conf. on Security Science and Technology* (Chengdu: ASME Press) pp372-378.

[4] Морозов А И and Семашко Н Н 2002 *Письма в ЖТФ* **28(24)** 63-66

[5] Zhao hailong, Jiang binhao, Wang chunsheng, Liu hongchen and Zhang zhonglin 2015 *Int. Conf. on Energy Environment and Green Building Materials*, (Guilin: Taylor & Francis Group) pp 379-384.

[6] А.И. Морозов and В.В. Савельев. 2005 *Физика плазмы.* **31(5)** 458-465

[7] E.I. Skibenko, Yu.V. Kovtun and A.M. Yegorov 2009 *Problems of Atomic Science and Technology* **4(2)** 286-292.

[8] А.И. Морозов 2008 *Введение в плазмо- динамику* (Физматлит) p 616.

[9] А.И.Бугрова, А.В.Десятсков and Ю.В.Коробкин, 2010 *Письма в ЖТФ* **36(19)** 31-36

[10] А.И. Бугрова, А.В.Десятсков and А.И.Морозов 2009 *Письма в ЖТФ* **35(22)** 89-94

[11] Zhao Hailong, Wang Chunsheng, Liu Hongchen, Zhang Le and Jiang Binhao 2014 *High power laser and particle beams* **26** 114001

[12] Zhao Hailong, Liu Hongchen, Wang Chunsheng, Zhang Le and Jiang Binhao 2015 *High power laser and particle beams* **27** 054001

[13] Zhao Hailong, Wang Chunsheng, Yan Yuming, Liu Hongchen and Jiang Binhao 2015 *Plasma Science and Technology* **17(1)**: 71-74

[14] Zhao Hailong, Jia Baisen, Liu Hongchen, Wang chunsheng and Jiang Binhao 2014 *Atomic Energy Science and Technology* **48** 28-32

The effect of numerical noises on statistics of a chaotic dynamical system

Xiaoming Li[1] **Xiaochen Li**[2] **and Shijun Liao**[1,3]

[1] School of Naval Architecture, Ocean and Civil Engineering, Shanghai Jiao Tong University, Shanghai 200240, China

[2] School of Civil Engineering and Transportation, South China University of Technology, Guangzhou 510641, China

[3] MoE Key Laboratory in Scientific and Engineering Computing, Shanghai 200240, China

E-mail: xiaochenli@scut.edu.cn

Abstract. It is well known that chaotic dynamical systems have the sensitive dependence on initial conditions (SDIC). Unfortunately, numerical noises (such as truncation error and round-off error) always exist in practice. Thus, due to the SDIC, long-term accurate prediction of chaotic dynamical systems is practically impossible. In this paper, using the so-called "clean numerical simulation" (CNS) whose numerical noises might be much smaller even than micro-level physical uncertainty and thus are negligible, we gain accurate prediction of a chaotic dynamical system in a long enough interval of time. Then, based on these reliable simulations, the influence of numerical noises on statistic computations is investigated. It is found that the influence of numerical noises is negligible when statistic results are time-independent. Unfortunately, when the statistics of a chaotic dynamical system are time-dependent, numerical noises have a great influence even on statistic computations.

1. Introduction

It is well known that chaotic dynamical systems have the sensitive dependance on initial conditions (SDIC) [1], which is called the "butterfly-effect". Unfortunately, numerical noises, i.e. truncation error and round-off error, always exist in practice. Thus, due to the SDIC, long-term accurate prediction of chaotic dynamical systems is practically impossible [1] and therefore numerical simulations of chaos are only mixtures of "true" solution with physical meanings and numerical noises without physical meanings. It is widely believed that turbulence [2, 3, 4, 5] is chaotic. Therefore, statistics are often used in the investigation of turbulence, and the direct numerical simulation (DNS) [6] has played an important role in statistical computation of turbulence. However, due to the SDIC, tiny numerical noises, which grow exponentially in time, may lead to spurious results. Thus, DNS results are in essence the mixtures of "true" solution with physical meaning and numerical noises without physical meanings. Even so, it is widely believed that numerical noises do not affect the statistics of turbulence by means of such kind of simulations. Unfortunately, Yee et al. [7] reported in 1999 that the DNS can produce a spurious solution that is completely different from the physical solution of their considered equations. In addition, Wang et al. [8] demonstrated a spurious evolution of turbulence originated from round-off error in DNS. Therefore, it is necessary to verify the reliability of statistic results for chaotic dynamical systems.

Recently, a numerical method with extremely small numerical noises, namely "clean numerical simulation" (CNS) [9], was proposed to gain reliable simulations of chaotic dynamical systems in a finite but large enough interval of time. The CNS is based on an arbitrary Taylor series method (TSM) [10, 11] and an arbitrary multiple precision [12] for every data, together with a kind of solution verification. By means of the CNS, the numerical noises can be so greatly reduced to be much smaller than the "true" solution that the numerical noises are negligible in a given interval of time even for chaotic dynamic systems. The CNS has been successfully applied in some chaotic dynamical systems, such as the famous Lorenz equation [9, 13, 14, 15] and the three-body problems [16, 17, 18]. Note that, using the traditional Runge-Kutta method and data in double precision, one can gain convergent chaotic results of Lorenz equation only in a few dozens of time interval. However, using the CNS, Liao and Wang [14] successfully obtained a convergent, reliable chaotic solution of the Lorenz equation in an interval [0,10000]. This shows the validity of the CNS for reliable simulations of chaotic dynamical systems.

2. Mathematical equations

It is well known that the famous Lorenz equation [1] is a very simplified model of the Rayleigh-Bénard (RB) flow of viscous fluid. There are exact Navier-Stokes equations for the Rayleigh-Bénard flow. Saltzman [19] deduced a family of highly truncated dynamical systems with different degrees of freedom, and the famous Lorenz equation [1] is only the simplest one among them. For example, in the case of the Prandtl number $\sigma = 10$, the highly truncated dynamical system with seven degrees of freedom (7-DOF) reads

$$\begin{cases} \dot{A} = 23.521BC - 1.500D - 148.046A, \\ \dot{B} = -22.030AC - 1.589E - 186.429B, \\ \dot{C} = 1.561AB - 0.185F - 400.276C, \\ \dot{D} = -16.284CE - 16.284BF - 13.958AG \\ \qquad\qquad - 1460.631\lambda A - 14.805D, \\ \dot{E} = 16.284CD - 16.284AF - 18.610BG \\ \qquad\qquad - 1947.508\lambda B - 18.643E, \\ \dot{F} = 16.284AE + 16.284BD - 486.877\lambda C \\ \qquad\qquad - 40.028F, \\ \dot{G} = 27.916AD + 37.220BE - 39.479G, \end{cases} \tag{1}$$

where $\lambda = R/R_c$ is dimensionless Rayleigh number, R_c is the critical Rayleigh number, A and D represent the cellular streamline and thermal fields for the Rayleigh critical mode, and G denotes the departure of the vertical variation, respectively. For details, please refer to Saltzman [19].

These are deterministic equations with chaotic solutions, and are greatly simplified models for the two dimensional Rayleigh-Bénard flow. In physics, the Rayleigh-Bénard flow with a large enough Rayleigh number R is an evolutionary process from an initial equilibrium state to turbulence after a long enough time, mainly because the flow is unstable and besides the micro-level physical uncertainty (such as thermal fluctuation) always exists. Such kind of initial micro-level physical uncertainty due to thermal fluctuation can be expressed by Gaussian random data, as illustrated by Wang et al. [3].

Without loss of generality, let us consider the deterministic 7-DOF equations (1) with the random initial conditions in normal distribution in case of $\lambda = 28$, corresponding to a turbulent flow. Considering the thermal fluctuation, we study here such kind of random initial condition in normal distribution with the mean

$$\langle A(0)\rangle = 1, \quad \langle B(0)\rangle = \langle C(0)\rangle = \langle D(0)\rangle = \langle E(0)\rangle = \langle F(0)\rangle = \langle G(0)\rangle = 10^{-3}$$

and the standard deviation $\sigma_0 = 10^{-30}$. By means of the CNS, we can gain reliable propagation of the micro-level physical uncertainty of a large numbers of random initial conditions. We compare the sample mean and unbiased estimate of standard deviation of $A(t)$.

3. Statistical Results

Two thousand samples of reliable numerical simulations of the 7-DOF system (1) are obtained in the time interval $[0, 10]$ by means of the CNS using the 80th-order Taylor series ($M = 80$), the 90 decimal-digit precision ($K = 90$) for every data, and the time-step $\delta t = 10^{-3}$. It is found that the numerical errors can be decreased to be much smaller than the micro-level physical uncertainty in the time interval [0,10] under consideration. These numerical simulations are so accurate that we can consider them as the "true" solutions of the chaotic dynamical system (1), which can be used to investigate the influence of numerical noises on statistic computations of chaotic systems. In this way, we successfully distinguish the "true" chaotic solutions with physical meanings from the numerical noises without physical meanings.

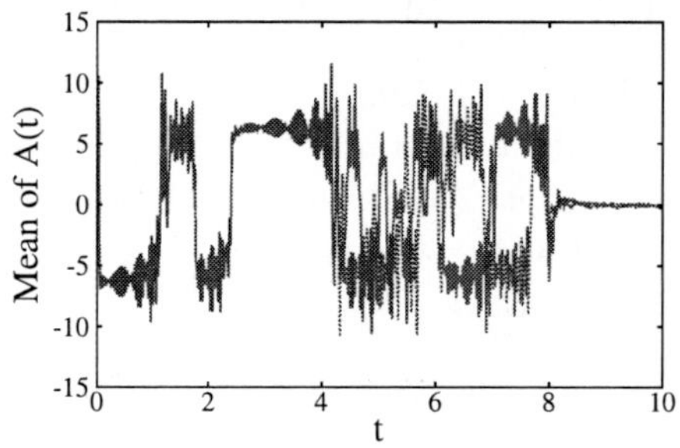

Figure 1. The influence of the truncation error on the mean of $A(t)$ of the 7-DOF chaotic systems obtained using the time-step $\delta t = 10^{-3}$ and the 90 decimal-digit multiple precision for every data with the negligible round-off error. Solid line in red: the reliable results given by the CNS using the 80th-order Taylor series expansion for t (i.e. $M = 80$); dashed line in blue: the unreliable result given by the 10th-order Taylor series expansion (i.e. $M = 10$).

Obviously, the larger the order M of the Taylor series in the frame of the CNS, the smaller the truncation error. For example, the traditional Runge-Kutta method corresponds to $M = 4$, the 4th-order Taylor series expansion for t. Thus, to investigate the influence of the truncation error, we use here the 10th-order Taylor series, i.e. $M = 10$, but retain the 90 decimal-digit multiple precision for every data. Note that our reliable CNS results are gained by means of $M = 80$, i.e. the 80th-order Taylor series, whose truncation errors are negligible in the considered interval of time $t \in [0, 10]$. However, when $M = 10$, the truncation error is not negligible in [0,10]. In this way, the round-off error is negligible in the considered interval of time $t \in [0, 10]$ so that the influence of the truncation error can be investigated independently.

The reliable mean (the red line, given by $M = 80$) of $A(t)$ of the 7-DOF chaotic dynamical system (1) is shown in Figure 1, compared with the unreliable result (the blue dashed line, given by $M = 10$). Note that the reliable mean of $A(t)$ given by the CNS becomes stable when $t > 8.2$, but is time-dependent when $t \leq 8.2$. However, as shown in Figure 1, the unreliable mean of $A(t)$ given by $M = 10$ has a noticeable difference from the reliable CNS result in the time interval $4.2 < t < 8.2$. It suggests that the truncation error has a great influence on the computation of unsteady statistical quantities. Note that the mean of $A(t)$ given by $M = 10$ agrees well with the reliable CNS result when $t > 8.2$. Thus, the truncation error has no influence on the time-independent statistics of chaotic dynamical systems in an equilibrium state.

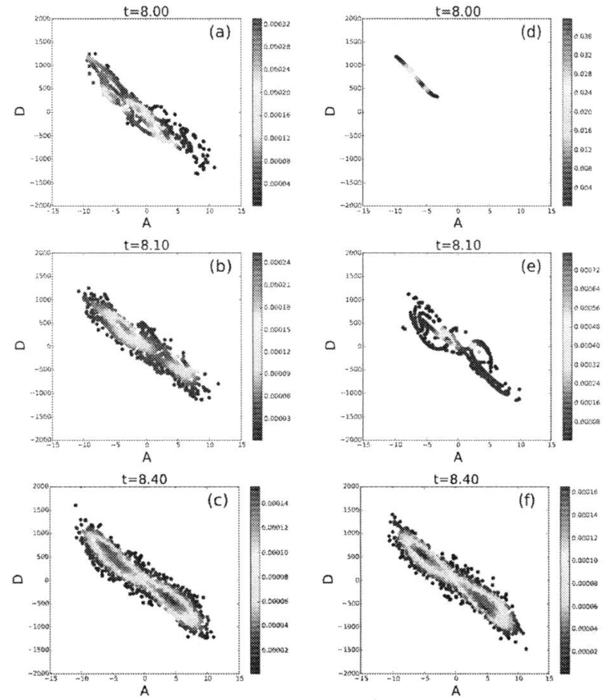

Figure 2. (color online). The scatter diagrams and probability distribution function in the A-D plane of the 7-DOF system at different time by means of the time-step $\delta t = 0.001$, 90 decimal-digit precision and different order of Taylor series. (a)-(c): $M = 10$; (d-f): $M = 80$.

The scatter diagrams of the probability distribution function (PDF) in the $A - D$ plane of the 7-DOF chaotic dynamical system (1) given by the two numerical schemes (i.e. $M = 10$ and $M = 80$) are as shown in Figure 2, where the PDF is obtained by using the Gaussian kernel density estimator. Note that the PDFs given by the two numerical approaches are almost the same at $t = 8.40$. However, at $t = 8.00$ and $t = 8.10$, the PDFs given by $M = 10$ are quite different from those given by the CNS using $M = 80$. This supports our previous conclusion that the truncation error has a significant influence on the computation of unsteady statistics, but has no influence on steady ones.

Note that the double precision is widely employed in numerical simulations, which leads to round-off errors at every time-step that increase exponentially for a chaotic dynamical system. To simulate the round-off error, we add a random data at each time-step with zero mean and the standard deviation 10^{-16}, while the 80th-order Taylor expansion ($M = 80$) is still used and all data are expressed in 90 decimal-digit so as to guarantee the negligible truncation error in the considered interval of time $t \in [0, 10]$.

Figure 3 shows the comparison between the reliable CNS statistics (using the 90 decimal-digit multiple precision for every data) and the unreliable results using the double precision. Note that the round-off error has a great influence on the standard deviation of $A(t)$ from the very beginning. At about $t \approx 5$ when the round-off error enlarges exponentially to reach the level of the "true" solution, the unsteady mean of $A(t)$ becomes unreliable. Note that the mean of $A(t)$ given by the double precision becomes time-independent more early, at $t \approx 5$, which is however wrong in physics. As statics of the system truly becomes steady when $t > 8.2$ (that is determined by the reliable CNS result), the round-off error has no influence on the computation of statistics. All of these suggest that the round-off error has a great influence on the unsteady

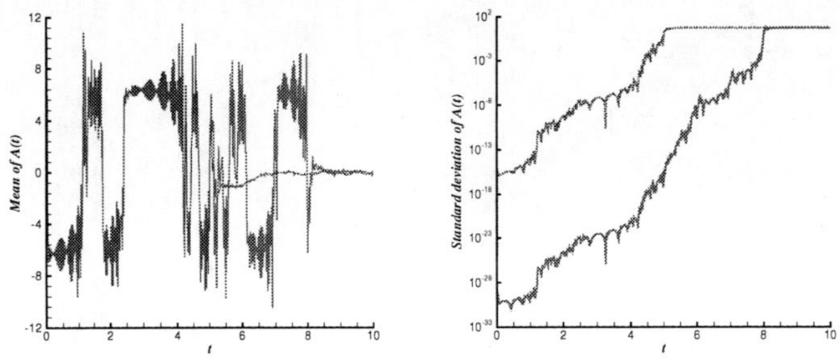

Figure 3. The influence of the round-off error on the mean (left) and the standard deviation (right) of $A(t)$ of the 7-DOF chaotic system (1) using the time-step $\delta t = 10^{-3}$ and the 80th-order Taylor expansion ($M = 80$) with the negligible truncation error. Solid line in red: the reliable CNS results given by the 90 decimal-digit multiple precision for every data; dashed line in blue: the unreliable results given by means of double precision.

statistics of chaotic dynamical system, but has a little impact on steady ones.

4. Conclusions

In this paper, we compare the statistic results of a chaotic dynamical model gained by two different approaches, one is the traditional numerical approach, the other is the so-called Clean Numerical Simulation (CNS) with extremely small numerical noises. It is found that the numerical noises have a great influence on the computation of unsteady statistics of chaotic dynamical systems, but have no influence on time-independent statistics for systems.

5. Reference

[1] Lorenz E N 1963 *Journal of the Atmospheric Sciences* **20** 130–141
[2] Lee M and Moser R D 2015 *Journal of Fluid Mechanics* **774** 395–415 ISSN 1469-7645
[3] Wang J, Li Q and Weinan E 2015 *Proceedings of the National Academy of Sciences* **112** 9518–9523
[4] Avila K, Moxey D, de Lozar A, Avila M, Barkley D and Hof B 2011 *Science* **333** 192–196
[5] Deike L, Fuster D, Berhanu M and Falcon E 2014 *Phys. Rev. Lett.* **112**(23) 234501
[6] Kim J, Moin P and Moser R 1987 *Journal of Fluid Mechanics* **177** 133–166 ISSN 1469-7645
[7] Yee H, Torczynski J, Morton S, Visbal M and Sweby P 1999 *International Journal for Numerical Methods in Fluids* **30** 675–711 ISSN 1097-0363
[8] Wang L P and Rosa B 2009 *Computers & Fluids* **38** 1943 – 1949 ISSN 0045-7930
[9] Liao S 2009 *Tellus A* **61** 550–564
[10] Corliss G and Chang Y 1982 *ACM Trans. Math. Software* **8** 114–144
[11] Barrio R, Blesa F and Lara M 2005 *Computers & Mathematics with Applications* **50** 93–111
[12] Oyanarte P 1990 *Computer Physics Communications* **59** 345–358
[13] Wang P, Li J and Li Q 2012 *Numer Algorithms* **59** 147 – 159
[14] Liao S and Wang P 2014 *Science China Physics, Mechanics & Astronomy* **57** 330–335
[15] Liao S 2014 *Int. J. Bifurcation and Chaos* **24** 1450119
[16] Liao S 2014 *Communications in Nonlinear Science and Numerical Simulation* **19** 601–616
[17] Li X and Liao S 2014 *Science China Physics, Mechanics & Astronomy* **57** 2121–2126
[18] Liao S and Li X 2015 *International Journal of Bifurcation and Chaos* **25** 1530023
[19] Saltzman B 1962 *Journal of the Atmospheric Sciences* **19** 329–341

Acknowledgment

This work is partly supported by National Natural Science Foundation of China (Approval No. 11432009).

IOP Conf. Series: Journal of Physics: Conf. Series **1039** (2018) 012018 doi:10.1088/1742-6596/1039/1/012018

The Non-thermal of a Large Solar Flares Associated with High Energy of Solar Burst Type III on 4th September 2017

***Z.S.Hamidi[1,3] , N.N.M.Shariff[2]**
[1] Institute of Science, Universiti Teknologi MARA, Shah Alam, Selangor, Malaysia
[2] Academy of Islamic and Contemporary Studies, Universiti Teknologi MARA, Shah Alam, Selangor, Malaysia
[3] Faculty of Applied Sciences, Universiti Teknologi MARA, Shah Alam, Selangor, Malaysia

Email: zetysh@salam.uitm.edu.my

Abstract. On 4th September 2017, the largest solar flare ever recorded this year saturated the GOES satellite X-ray detectors, making an assessment of its size difficult. We report a large solar flare that produces a hydrodynamic blast wave, moving out through interplanetary space with velocities of 573.6 km/sec and densities 18.1 protons/cm^3 of protons. The type III burst with split band and herring bones has been recorded from the Kangarlussuaq, Greenland site. It shows a signature of propagating beams of nonthermal electrons with a large and complex structure. During that time, R1-R2 (Minor-Moderate) radio blackouts have occurred on 4 September, 2017; with sunspot number and radio flux is 96 and 120 respectively. The active region,RGN 2673 is a magnetically complex and compact sunspot group that has produced numerous C-class flares and occasional M-class flares on 4 September; the largest thus far was an M5 (R2-Moderate) event at 2033 UTC (1633 ET). Meanwhile, the active region, RGN 2674 is not as magnetically complex and has been relatively inactive. However, due to the development of mixed polarities within RGN 2673, its continued growth, and activity trends, the forecast now calls for expected R1-R2 events the remainder of 4th September on into 5 September.

1. Introduction

Classification of solar flare can be divided into two regions depending on whether the electron energy input goes into radiation or explosive heating [1]. There are 5 classes of solar flare (A, B, C, M and X). Class A flares have purely thermal, compact sources while Class B flares are impulsive bursts which show double footpoints in hard X-rays. Class C flares have gradually varying hard X-ray and microwave fluxes from high altitudes and show hardening of the X-ray spectrum through the peak and on the decay [2]. During solar flares, the X-ray flux received at the Earth increases dramatically, often within a few minutes, and then decays again in times ranging from a few tens of minutes to several hours [3]. The X-rays radiation from solar flares ionize the neutral atmosphere at D region heights greatly increasing the electron densities there and thus markedly lowering the effective VLF reflection height [4]. Previous studies shown that the energy input to the radiative quasi-equilibrium region agrees with the observed flare energy output in optical, UV, and EUV radiation. The energy of solar flare dependence of the decrease may go inversely with particle rigidity, or it may be flatter, depending on the details of the magnetic configuration in and around the blast wave [5, 6]. Detailed analysis on the temperature of the underlying loops increased toward higher altitudes, while the temperature of the coronal source increased toward lower altitudes [7]. The previous results of the analysis indicate that (1) most of the flare plasma was at temperatures between 3 and 10 million degrees; (2) the peak temperature decreased with time from about 8 x 10^6 K to 5 x 10^6 K over a period of 3.5 hours; (3) the differential emission measure steadily decreased with time at nearly all temperatures; (4) both radiation and conduction were important cooling mechanisms for the plasma at $T>10^5$ K; and (5) a substantial amount of energy, of the order of 3 x 10^{31} ergs [8]. A time-intensity profiles for solar proton events are analyzed with respect to the acceleration of energetic particles from the ambient solar wind by an interplanetary shock [9]. The uncertainty in the determination of the level of the primary electron bremsstrahlung as well as the lack of measurements on the γ -ray emission above 100 MeV combine to allow rather a wide range of energy distribution parameters [10]. Some of these variations were irreversible, occurred in the vicinity of magnetic neutral lines, and likely were related to magnetic energy release in the flare [11]. It is shown that Sweet's mechanism is much more effective in a highly compressible medium if the merging magnetic fields are exactly antiparallel [12].

Meanwhile, the type III solar burst is a common signature of near-relativistic electrons streaming through the background plasma of the solar corona and interplanetary space, offering a means to remotely trace these electrons [13]. This burst shows the coalescence of electron plasma waves from a nonthermal distribution with electron plasma waves from the distribution of thermal charge fluctuations [14, 15]. It is estimated that the total number of electrons $\gtrsim$ 22 keV required to produce a type III burst is $\lesssim 10^{34}$ [16, 17]. This burst can occur singularly, in groups, or storm. This fast drift burst is the most common of the meter wavelength bursts [18].

2. Experimental

We use the Compound Astronomical Low cost, Low frequency Instrument for Spectroscopy and Transportable Observatory (CALLISTO) as a main system of solar radio bursts and RFI-monitoring for astronomical science and education [19-21]. The e-CALLISTO is the worldwide network connected via internet [22, 23]. This system can cover 24 hour data of solar burst, which operates identical spectrograph in different locations around the world. It is more than 133 instruments in more than 67 locations with users from more than 132 countries [24, 25].

However, it solar burst can be detected only after they have travelled far enough from the Sun [15, 26]. In Malaysia, we used 5.5 meters Log Periodic Dipole Antenna with a gain of 6 dB. The timing of the CALLISTO is controlled by a GPS clock in which the relative timing is accurate to within less than one millisecond, whereas the absolute time is uncertain to within a few milliseconds due to internal delays. This system has an individual channel has 300 kHz bandwidth and can be tuned by the controlling software in steps of 62.5 kHz [27].

3. Results and Discussion

Figure 1 shows the type III burst with split band and herring bones from the Kangarlussuaq, Greenland site. It shows a signature of propagating beams of nonthermal electrons in the solar atmosphere. This fast drift burst has a very different physical origin that produces such different properties. This event produces a type III burst with a large and complex structure. Figure 2 shows the GOES X-ray Flux data starting from 2nd to 4th September 2017. The Type III radio burst is used as the indicator of the starting point of the magnetic reconnection. Strong evidence for the escape of energetic electrons would be presence of radio burst Type III. The subject of nonlinear wave-wave interaction which involving interaction of electrostatic electron plasma that called as Langmuir waves.

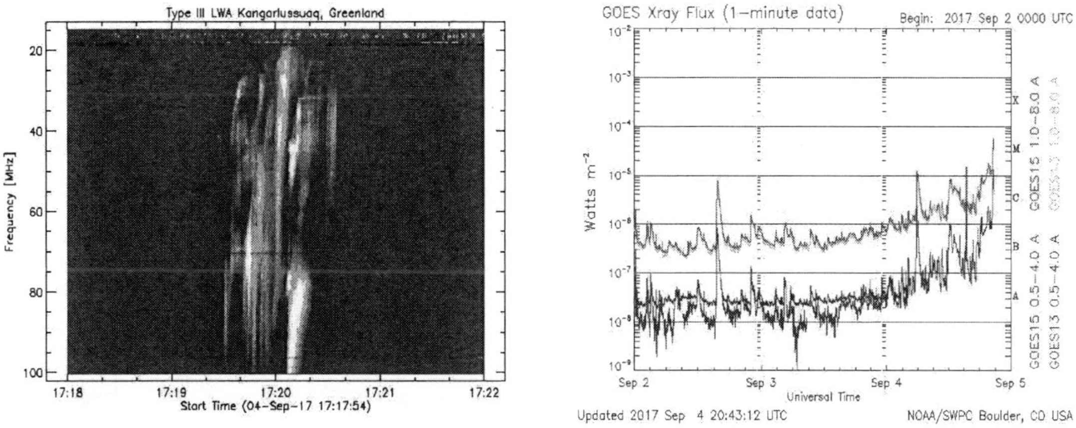

Figure 1. Strange group of type III solar radio bursts observed by Callisto in Kangarlussuaq, Greenland (credited to: e-CALLISTO data) and GOES X-ray Flux data starting on 2nd to 4th September 2017 (credited to: Solar monitor)

Table 1 shows the solar parameters during the explosion of solar flare. The active region AR2673 expanded more than 10-fold in a single day. Huge sunspot AR2673, which materialized with shocking speed over the few days period, is seething with activity. We can observe clearly the plasma currents surging inside the sunspot's magnetic canopy-apparently on the verge of an explosion in optical observation. It is then suddenly becomes one of the largest sunspots of the year. It has a 'beta-gamma' magnetic field that harbors energy for strong M-class solar flares. Any such explosion today would be geoeffective as the active region is directly facing Earth. Possible outcomes include moderately-strong shortwave radio blackouts, Earth-directed Coronal Mass Ejections (CMEs) and geomagnetic storms later this week.

Table 1. Solar Parameter during the explosion of solar flare

Parameter	**Value**
Solar wind speed	573.6 km/sec
Density Proton	18.1 protons/cm^3
Sunspot number	96
10.7 cm flux	120
Interplanetary Mag. Field, B_{total}	6.1 nT
Interplanetary Mag. Field, B_z	0.5 nT north

During that time, R1-R2 (Minor-Moderate) radio blackouts have occurred on 4 September, 2017. The active region, RGN 2673 is a magnetically complex and compact sunspot group that has produced numerous C-class flares and occasional M-class flares on 4 September; the largest thus far was an M5 (R2-Moderate) event at 2033 UTC (1633 ET). Meanwhile, the active region RGN 2674 is not as magnetically complex and has been relatively inactive. However, due to the development of mixed polarities within RGN 2673, its continued growth, and activity trends, the forecast now calls for expected R1-R2 events the remainder of 4th September on into 5th September.

IOP Conf. Series: Journal of Physics: Conf. Series **1039** (2018) 012018 doi:10.1088/1742-6596/1039/1/012018

4. Conclusion

The eruption mechanism of solar flares and radio burst Type III are currently an extremely active area research, especially during the solar cycle is towards maximum. It is normally found at the pre-flare stage that could be a signature of electron acceleration. Solar radio burst type III potentially reveals reveal about the acceleration site of the particles from the Sun to the Earth during large solar flare. In this case, the solar flare is one of the largest solar flare that occurred in 2017 due to the active region RGN 2673.

References

[1] R. Lin, H. Hudson, Non-thermal processes in large solar flares, Solar Physics, 50 (1976) 153-178.
[2] B.R. Dennis, Solar hard X-ray bursts, Solar Physics, 100 (1985) 465-490.
[3] Z.S. Hamid, N. Shariff, Z. Ibrahim, N. Ramli, Investigation of the statistical properties of solar radio burst type II and III, Space Science and Communication (IconSpace), 2015 International Conference on, IEEE, 2015, pp. 335-338.
[4] N.R. Thomson, C.J. Rodger, M.A. Clilverd, Large solar flares and their ionospheric D region enhancements, Journal of Geophysical Research: Space Physics, 110 (2005).
[5] E. Parker, Sudden Expansion of the Corona Following a Large Solar Flare and the Attendant Magnetic Field and Cosmic-Ray Effects, The Astrophysical Journal, 133 (1961) 1014.
[6] Z. Hamidi, N. Shariff, Z. Ibrahim, C. Monstein, W.W. Zulkifli, M. Ibrahim, N. Arifin, N. Amran, Magnetic Reconnection of Solar Flare Detected by Solar Radio Burst Type III, Journal of Physics: Conference Series, IOP Publishing, 2014, pp. 012006.
[7] L. Sui, G.D. Holman, Evidence for the Formation of a Large-Scale Current Sheet in a Solar Flare.
[8] G. Withbroe, The thermal phase of a large solar flare, Astrophysical Journal, 225 (1978) 641-649.
[9] D. Reames, Acceleration of energetic particles by shock waves from large solar flares, Astrophysical Journal, 358 (1990).
[10] N. Vilmer, A. MacKinnon, G. Trottet, C. Barat, High energy particles accelerated during the large solar flare of 1990 May 24: X/$\gamma $-ray observations, Astronomy & Astrophysics, 412 (2003) 865-874.
[11] A. Kosovichev, V. Zharkova, Magnetic energy release and transients in the solar flare of 2000 July 14, The Astrophysical Journal Letters, 550 (2001) L105.
[12] E. Parker, The Solar-Flare Phenomenon And The Theory Of Reconnection And Annihilation Of Magnetic Fields, Astrophysical Journal, Supplement Series (US), 8 (1963).
[13] H.A.S. Reid, H. Ratcliffe, A review of solar type III radio bursts, Research in Astronomy and Astrophysics, 14 (2014) 773.
[14] D. Melrose, On the theory of type II and type III solar radio bursts. I. The impossibility of nonthermal emission due to combination scattering off thermal fluctuations, Australian Journal of Physics, 23 (1970) 871-884.
[15] Z. Hamidi, N. Shariff, Automatic Solar Radio Burst Type II and III Image Processing Tracking by Using CALLISTO System, Industrial Engineering, Management Science and Application (ICIMSA), 2017 International Conference on, IEEE, 2017, pp. 1-4.
[16] S. Kane, Evidence for a common origin of the electrons responsible for the impulsive X-ray and type III radio bursts, Solar Physics, 27 (1972) 174-181.
[17] Z. Hamidi, N. Anim, N. Shariff, Z. Abidin, Z. Ibrahim, C. Monstein, Dynamical structure of solar radio burst type III as evidence of energy of solar flares, AIP Conference Proceedings, AIP, 2013, pp. 11-15.
[18] M. Ali, S. Sabri, Z. Hamidi, N. Husien, N. Shariff, N. Zainol, M. Faid, C. Monstein, e-CALLISTO Network System and the Observation of Structure of Solar Radio Burst Type III, Industrial Engineering, Management Science and Application (ICIMSA), 2016 International Conference on, IEEE, 2016, pp. 1-5.
[19] Z. Hamidi, Z. Abidin, Z. Ibrahim, N. Shariff, Indication of radio frequency interference (RFI) sources for solar burst monitoring in Malaysia, ICPAP 2011, AIP Conference Proceedings-American Institute of Physics, 1454 (2012).

[20] Roslan Umar, Zamri Zainal Abidin, Zainol Abidin Ibrahim, Mohd Saiful Rizal Hassan, Zulfazli Rosli, Z.S.Hamidi, Population density effect on radio frequencies interference (RFI) in radio astronomy, ICPAP 2012, AIP Conference Proceedings, Bandung Indonesia, 2012, pp. 4.
[21] Z.Z. Abidin, N.M. Anim, Z.S. Hamidi, C. Monstein, Z.A. Ibrahim, R. Umar, N.N.M. Shariff, N. Ramli, N.A.I. Aziz, I. Sukma, Radio frequency interference in solar monitoring using CALLISTO, New Astronomy Reviews, 67 (2015) 18-33.
[22] S. Sabri, N. Zainol, M. Ali, N. Shariff, N. Hussien, M. Faid, Z. Hamidi, C. Monstein, The Dependence of Log Periodic Dipole Antenna (LPDA) and e-CALLISTO Software to Determine the Type of Solar Radio Burst (I-V), Industrial Engineering, Management Science and Application (ICIMSA), 2016 International Conference on, IEEE, 2016, pp. 1-5.
[23] Z.H. NM Anim, ZZ Abidin, C Monstein, NS Rohizat, Radio frequency interference affecting type III solar burst observations, in: R.S. et.al (Ed.) PERFIK 2012, American Institute of Physics, Malaysia, 2013, pp. 5.
[24] Z. Hamidi, N. Zainol, M. Ali, S. Sabri, N. Shariff, M. Faid, N. Husien, C. Monstein, Signal Detection of the Solar Radio Burst Type III Based on the CALLISTO System Project Management, Industrial Engineering, Management Science and Application (ICIMSA), 2016 International Conference on, IEEE, 2016, pp. 1-4.
[25] N. Zainol, S. Sabri, Z. Hamidi, M. Ali, N. Shariff, N. Husien, C. Monstein, M. Faid, Effective Data Collection and Analysis of Solar Radio Burst Type II Event Using Automated CALLISTO Network System, Industrial Engineering, Management Science and Application (ICIMSA), 2016 International Conference on, IEEE, 2016, pp. 1-5.
[26] N. Husien, N. Zainol, Z. Hamidi, S.N. Sabri, N.N. Shariff, M. Faid, M. Ali, C. Monstein, Solar radio bursts detected by CALLISTO system and their related events, 3rd International Conference on Industrial Engineering, Management Science and Applications, ICIMSA 2016, Institute of Electrical and Electronics Engineers Inc., 2016.
[27] Z. Hamidi, N. Shariff, An Evaluation Performance of Log Periodic Dipole Antenna Based on the Parameter of Flux Density of the Solar Radio Burst Event, Industrial Engineering, Management Science and Applications 2015, Springer Berlin Heidelberg2015, pp. 685-692.

Acknowledgments

We are grateful to CALLISTO network, STEREO, LASCO, SDO/AIA, NOAA and SWPC make their data available online. This work was partially supported by the UITM internal BESTARI grant, 600-IRMI/DANA 5/3/BESTARI (067/2017) from the Kementerian Pengajian Tinggi Malaysia. Special thanks to the National Space Agency and the National Space Centre for giving us a site to set up this project and support this project. Solar burst monitoring is a project of cooperation between the Institute of Astronomy, ETH Zurich, and FHNW Windisch, Switzerland, Universiti Teknologi MARA and University of Malaya. This paper also used the NOAA Space Weather Prediction Centre (SWPC) for the sunspot, radio flux and solar flare data for comparison purpose. The research has made use of the National Space Centre Facility and a part of an initiative of the International Space Weather Initiative (ISWI) program.

IOP Conf. Series: Journal of Physics: Conf. Series **1039** (2018) 012019 doi:10.1088/1742-6596/1039/1/012019

Wave Induced Oscillation in an Irregular Domain by using Hybrid Finite Element Model

Prashant Kumar[1], Rajni[2], Rupali[3]

[1,3]Department of Applied Sciences, National Institute of Technology Delhi, India
[2]O.P. Jindal Global University, Sonepat, India

Email: prashantkumar@nitdelhi.ac.in

Abstract. Harbors are designed to provide the safe loading, unloading and sheltering for the moored vessels. In seasonal weather conditions, semi-enclosed harbors experienced high resonance, which amplifies the surface wave amplitude in the interior of the harbor due to combined effect of wave refraction, diffraction and partial reflection from the solid harbor walls. An accurate description boundary of the harbor is required to analyze the impact of resonant frequency waves over surface ocean wave height. The fluid domain is divided into two regions as bounded and open sea region. The bounded region consists of harbor boundary and interior of the harbor and open sea region consists of ocean area outside the bounded region. Firstly, the mild slope equation (MSE) is derived for both the regions in terms of a potential function using the energy conservation principle. The total wave energy in the bounded region is estimated by using a mathematical model based on Hybrid Finite Element Method (HFEM) is used to formulate the mild slope equation. In HFEM model, the finite element method is coupled with the boundary element method to solve the mild slope equation in both the region. Further, the present HFEM model is validated with existing studies lead by Lee (1971) and Ippen and Goda (1963). The current numerical model is implemented on realistic Pohang New Harbor (PNH), which is situated on the southeast coast of South Korea. The present numerical model can be used as an efficient engineering tool for planning and designing of the artificial industrial harbor and predict the incident wave response under the resonance conditions.

1. Introduction

Most of the existing harbors connecting with the open sea are experienced the harbor oscillations due to the incident waves propagating from the open sea into the harbor. The small amplitude waves are diffracted, refracted and partially or fully reflected by the exterior and interior boundaries of the harbor walls and a small portion of them are re-radiated through the entrance to the open sea again. The Pohang New harbor (PNH) is built to support the POSCO steel industry in South Korea, which is one of the most powerful steel production plants since the 1970s. Occasionally, the incoming waves to PNH have large amplitudes while they are generated by the seasonal typhoons, winds or currents and swells. Such phenomena are typically observed in Korea coastal region.

Several researchers have carried out significant investigations regarding the prediction of the wave field in a harbor due incident waves of the short period by adopting various numerical schemes. Initially, the mild-slope wave equation was solved by Berkhoff [1] using a finite element technique. Chen and Mei [2] introduced a different version of hybrid element method to solve the shallow water equation. Researchers have developed a different numerical model based on Helmholtz equation and mild slope equation (MSE) [3-8]. These models are very useful to modify or redesign the existing harbors. Lee and Williams [9] developed a boundary element numerical model to predict the wave field for the multidirectional random waves in a harbor due to diffraction and refraction. Researchers have designed the time-dependent numerical model with MSE and amplitude dispersion [10]. They transformed extended MSE to the simple time-dependent hyperbolic equation. Tsai and his colleagues [11] have proposed a wave model with both dispersive and nonlinearity for wave field without water depth restriction. They derived linear and nonlinear effects to the second order of MSE for wave steepness and bed slope for narrow-banded sea waves.

In this paper, MSE is utilized to determine the wave response in PNH. Further, the MSE is solved by using the Hybrid Finite Element method for an irregular domain. The kinematic and dynamics free surface boundary condition is utilized to determine the amplification factor in an irregular domain. The consideration of partial wave reflection, diffraction, dissipation losses due to absorption and wave absorption due to breakwaters is incorporated in the present model. Further incident waves with different wave frequencies and directions are also analyzed in PNH ocean surface wave field is estimated for interior and exterior harbor region under the resonance conditions. The safe location in the harbor can be identified under the resonance conditions during the seasonal weather for the safety of moored vessels and ships.

2. Mathematical Formulation

The inner bounded region Ω_b includes the domain D with partially reflecting wall ∂D with outward normal vector n_D and the boundary of the inner bounded region is denoted as $\partial\Omega_b$ shown in Fig. 1. The exterior region outside from bounded region (Ω_b) is defined as an unbounded region (Ω_u). The incident wave propagating from unbounded region towards inner bounded region are partially reflecting and diffracting from the solid boundary ∂D. Thus, partial reflection boundary is given by Isaacson and Qu [12]

$$\frac{\partial \phi_b}{\partial n_D} + ik\alpha_p \phi_b = 0 \tag{1}$$

where n_D is outward normal vector and $\alpha_p = \alpha_p^1 + i\alpha_p^2$ is transmission coefficient. It is defined in terms of reflection coefficient K_r, which can be written as the ratio of a reflected wave height to an incident wave height. Here, the transmission coefficient is considered as $\alpha_p^1 = 0$ and $\alpha_p^2 = 1 - K_r / 1 + K_r$. The velocity potential satisfied the mild slope equation in the inner bounded region Ω_b defined as

$$\nabla \cdot (CC_g \nabla \phi_b) + \omega^2 \frac{C_g}{C} \phi_b = 0 \tag{2}$$

where $\phi_b = \phi_b(x, y)$ is the potential function in Ω_b, ω is angular frequency, $C(x, y)$ is phase velocity, $C_g(x, y)$ is the group velocity, i.e., $C_g(x, y) = nC$ such that $n = (1 + 2kh / \sinh kh) / 2$, where k is the wave number related to water depth through dispersion relation as $\omega^2 = gk \tanh kh$. The region exterior to the boundary $\partial\Omega_b$ is denoted as an unbounded region Ω_u. Incident wave denoted as an arrow toward to bounded region. The Somerfield radiation boundary condition is given on fictitious boundary $\partial\Omega_b$. The matching boundary conditions is given over the boundary $\partial\Omega_b$, the velocity potential for unbounded region ϕ_u and bounded region ϕ_b are equal, i.e.,

$$\phi_b = \phi_u \quad \text{and} \quad \frac{\partial \phi_b}{\partial n_b} = \frac{\partial \phi_u}{\partial n_u} \tag{3}$$

In unbounded region, the velocity potential can be defined as

$$\phi_u = \phi_{inc} + \phi_{scr} \tag{4}$$

where ϕ_{inc} represent the incident wave potential and ϕ_{scr} is scattering wave potential. In the unbounded region, radiated waves propagate to far field from the center of Ω_u. Thus, the Somerfield radiation boundary condition must satisfy, i.e., $lim_{r\to\infty}\sqrt{r}\left(\partial/\partial r-ik\right)\phi_{scr}=0$, where r is the radius from the center of Ω_b to the fictitious far field boundary $\partial\Omega_b$ in unbounded region.

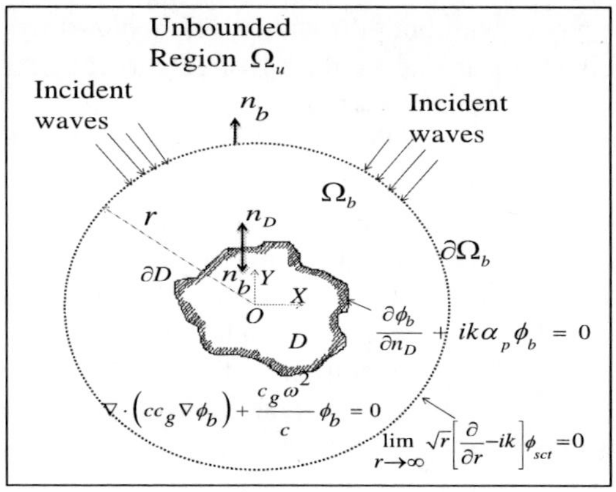

Fig. 1. In the bounded region Ω_b includes the domain D with curved solid boundary ∂D and outward normal vector n_D. A closed circular arc $\partial\Omega_b$ with outward normal vector n_b is also given.

Energy flux in the bounded region can be defined as

$$E_{f\,\Omega_b}=\frac{\rho}{g}\left[\left\{\frac{1}{s}\iint_{\Omega_b}\left(CC_g\left(\bar{\nabla}\phi_b\right)^2-\frac{C_g}{C}\phi_b\right)dxdy\right\}-\left\{\int_{\partial\Omega_b}\frac{1}{2}CC_g\phi_b\frac{\partial\phi_b}{\partial n_b}ds-\int_{\partial D}\frac{1}{2}i\alpha_p\omega C_g\phi_b^2ds\right\}\right]e^{-2i\omega t} \quad (5)$$

Here bar represents the complex conjugate. Moreover, the Energy flux acting in the unbounded region and along the boundary $\partial\Omega_b$ is given as

$$E_{f\,\Omega_u}=\frac{\rho}{g}\left[\frac{1}{2}\left\{\int_{\partial\Omega_b}CC_g\phi_u\frac{\partial\phi_u}{\partial n_u}ds-\int_{\partial\Omega_b}CC_g\phi_{scr}\frac{\partial\phi_{scr}}{\partial n_u}ds\right\}-\left\{\int_{\partial\Omega_b}\frac{1}{2}CC_g\phi_{inc}\frac{\partial\phi_{inc}}{\partial n_u}ds-\int_{\partial\Omega_b}CC_g\phi_{scr}\frac{\partial\phi_{inc}}{\partial n_u}ds\right\}\right]e^{-2i\omega t} \quad (6)$$

From conservation of energy, i.e., $\partial E_{\Omega_b}/\partial t=0$ and $\partial E_{\Omega_u}/\partial t=0$, therefore E_{Ω_b} and E_{Ω_u} must be constant. The matching boundary conditions Eq. (3) and Eq. (4) are utilized to formulate Eq. (5) and Eq. (6).

3. Numerical validation and discretization

In this experiment, the rectangular harbor is used as the same size which is used by Ippen and Goda (1963) in their experiment. The triangulation mesh is generated and the rectangular domain is discretized into 12564 elements.

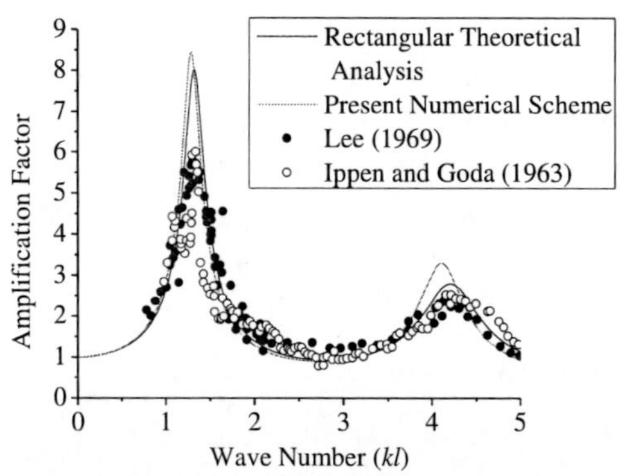

Fig. 2. Comparison of the present scheme (red line) with the experimental results (by Lee and rectngular harbor theory

As shown in Fig. 2, the hollow circle shows the experimental result of Ippen and Goda [13], a solid circle denotes the experimental result of Lee [14], the red line shows numerical data of present scheme, the black line shows the data of rectangular theory. The present numerical simulation results of HFEM model have validated with the rectangular harbor theoretical results and existing experimental results. Therefore, HFEM model can be applied to realistic harbor domain such as Pohang New Harbor (PNH).

4. Implementation on PNH

The present HFEM model is applied to realistic Pohang New Harbor (PNH). Recorder stations R3 to R8 are chosen as the interior of PNH boundary especially at locations of moored ships, and R1 and R2 are chosen as reference locations exterior of PNH (see Fig. 3a). After confirmation of validation of present numerical scheme, the HFEM model is applied to PNH for estimating the amplification factor at various recorder stations from R1 to R8. For each recorder station, the amplification factor varied in amplitude as shown with the different color line (see Fig. 3b).

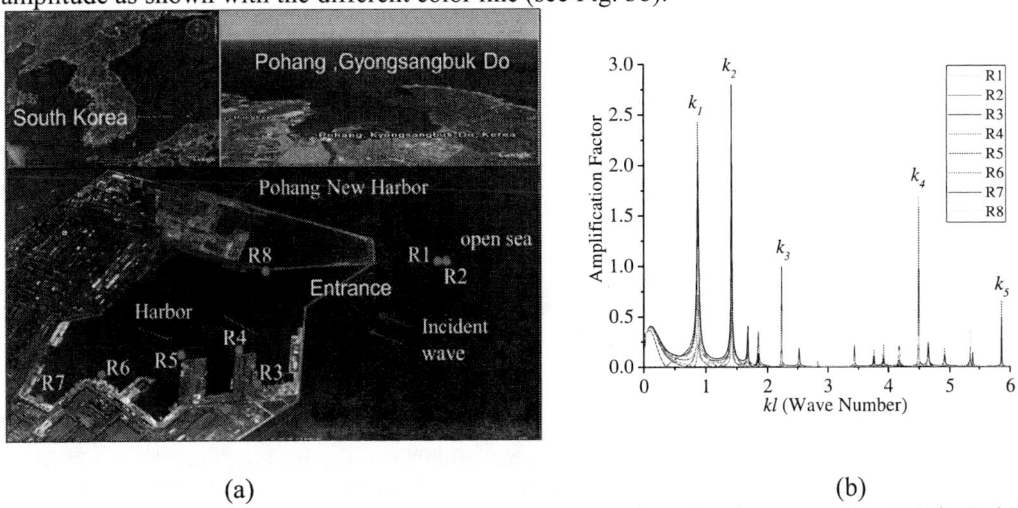

Fig. 3. The topographic view of PNH with recorder stations is given as R1 to R8 in Pohang New Harbor, including open sea region and incident wave (denoted by a red arrow). The amplification factor at different recorder stations from R1 to R8 is compared with respect to wave number (*kl*), where *l* is the length of the entrance of PNH.

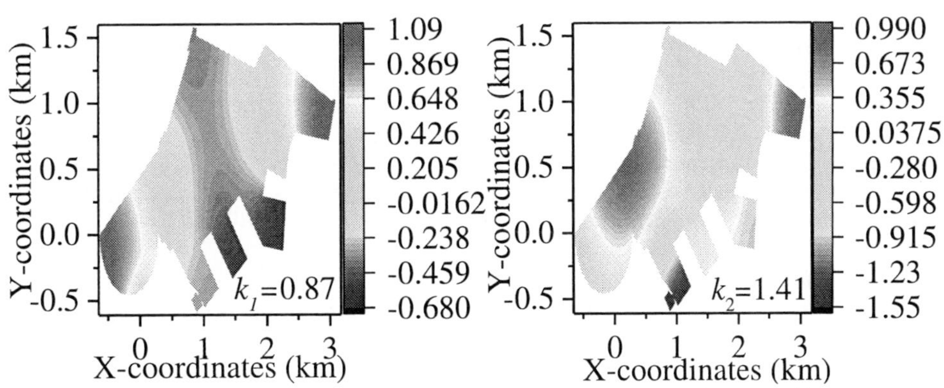

Fig. 4. Wave field analysis for the PNH at first and second resonance mode K_1=0.87, K_2=1.41.

The wave field analysis for the PNH is also conducted for the incident waves propagating at various incident angles. In Fig. 4, the ocean surface wave field is given for the first and second resonance modes with an incident angle $\alpha = 3\pi/4$ at the entrance. There are two major peaks appears in the PNH for first resonance mode and three major peaks for second resonance mode. Similarly, we can obtain the surface wave elevation in the PNH for different resonance modes and incident angles. On the basis of wave field information , the safe location for the moored ship can be identified inside the PNH during the seasonal weather conditions.

5. Discussion and conclusion

In this paper, the mild slope equation (MSE) is utilized to determine the wave potential function and MSE is solved using HFEM. In this formulation, partial reflection boundary at the harbor and Somerfield radiation boundary condition are applied for far field. The comparison of the amplification factor in the rectangular harbor for a present model with analytical approximation and experimental results of existing studies are given. Physical models are very expensive and time-consuming; however numerical model proposed here is effective and sufficiently accurate to analyze the resonance phenomenon in the harbor complex geometry. Therefore, we can utilize such models as a reliable engineering tool to predict the wave frequencies and amplitude in the harbor. Further, the moored ship can be protected from the high amplitude waves during seasonal weather conditions.

References

[1] Berkhoff J C W 1972 Computation of combined refraction-diffraction *Proc. 13th Coastal Engineering Conference, Vancouver, Canada, ASCE* **1** 471-490.

[2] Chen H S and Mei C C 1974 Oscillations and wave forces in an offshore harbor: applications of hybrid finite element method to water-wave scattering *Ralph M. Parsons Lab. for water resources and Hydrodyn., Massachusetts Institute of Technology* Rep. 190 215.

[3] Booij N 1983 A note on the accuracy of the mild slope equation *Journal of Coastal Engineering* **7(3)** 191-203.

[4]Panchang V 1991 Solution of the mild-slope wave problem by iteration *Journal of Applied Ocean Research* **13(4)** 187-199.

[5] Chou C R and Han W Y 1995 Oscillations induced by irregular waves in harbours *in Coastal Engineering Conference American Society of Civil Engineers* **3** 2987-2987.

[6] Bellotti G 2007 Transient response of harbors to long waves under resonance conditions *Coastal engineering* **55** 680-693.

[7] Kumar P, Zhang H, Kwang K I, and Yuen D A 2013 wave field analysis in as harbor with irregular geometry through boundary integral of Helmholtz equation with corner contribution *Computers & Fluids* **88** 287-297.

[8] Davis J R, Sheremet A, Tian M and Saxena S 2014 A numerical implementation of a nonlinear mild slope model for shoaling directional waves *Journal of Marine Science and Engineering* **2** 140-158.

[9] Lee H S and Williams A N 2002 Boundary element modeling of multidirectional random waves in a harbor with partially reflecting boundaries *Journal of Ocean Engineering* **29** 39–58.

[10] Zhao H, Song Z, Xu F and Li R 2010 An extended time-dependent numerical model of the mild-slope equation with weakly nonlinear amplitude dispersion *Acta Oceanol. Sin.* **29(2)** 5-13.

[11]Tsai C P, Chen H B and Hsu J R C 2014 Second-order time-dependent mild-slope equation for wave transformation *Mathematical Problems in Engineering* Article ID 341385.

[12] Isaacson M and Qu S Q 1990 Waves in a harbor with partially reflecting boundaries *Coastal Engineering* **14** 193-214.

[13]Ippen A T and Goda Y 1963 Waves induced oscillation in harbors, the solution for a rectangular harbor connected to the open sea *Massachusetts institute of technology, Cambridge, Hydrodynamics laboratory. Hydrodynamics Laboratory* Report no.59.

[14] Lee J J 1969 Wave-induced oscillations in harbors of arbitrary shape *W. M. Keck Laboratory of Hydraulics and Water Resources, California Institute of Technology, Pasadena*, Rep. KH-R-20, 266.

Acknowledgement

This research work is supported by CPDA grant, Department of Applied Sciences, National Institute of Technology Delhi, India.

IOP Conf. Series: Journal of Physics: Conf. Series **1039** (2018) 012020 doi:10.1088/1742-6596/1039/1/012020

Probabilistic sensitivity analysis for multivariate model outputs with applications to Li-ion batteries

V. Triantafyllidiis[1], W.W. Xing[1], P.K. Leung[2], A. Rodchanarowan[3], and A.A. Shah[1]

[1]School of Engineering, University of Warwick, Coventry CV4 7AL, UK
[2]Department of Materials, University of Oxford, Oxford, OX1 3PH, UK
[3]Department of Materials Engineering, Faculty of Engineering, Kasetsart University, 50 Ngamwongwan Rd., Ladyao, Chatuchak, Bangkok, 10900, Thailand

Abstract. Full battery models are highly complex, which limits their application to tasks such as optimization and uncertainty quantification. To lower the computational burden, sensitivity analysis (SA) can be used as a precursor to identify the most important parameters in the model, but SA itself relies on a high number of full model evaluations, which has motivated the use of emulators. For high-dimensional output problems, emulators are challenging to construct. In this paper we develop a probabilistic framework for SA of high-dimensional output models using a Gaussian process emulator based on dimensionality reduction. This allows us to perform SA under uncertainty for multi-ouput problems, providing error bounds for the emulator predictions of sensitivity measures. We show how this can be achieved using Monte Carlo sampling or possibly by using semi-analytical expressions with highly efficient sampling. Moreover, we can perform SA for multivariate outputs by ranking the sensitivity measures related to (uncorrelated) coefficients in a basis for the output space.

1. Introduction

The large number of parameters appearing in mathematical models and numerical codes for batteries complicates modelling efforts. Lowering the time cost of simulations by identifying the most influential parameters and studying their effects is an effective precursor to tasks such as design optimization and uncertainty quantification (UQ). This process is referred to as *sensitivity analysis (SA)* [1]. SA methods can be categorized in different ways. In *quantitative* SA the influence of a parameter (usually referred to as "factor" of an *input*) is assigned (reproducibly) a number called a *sensitivity index* or *importance measure*. In *local* SA, the output variability is studied by perturbing an input around a nominal (base) value, while methods that attempt to measure the output variability across the entire input space are termed *global*. For small variations in the inputs local methods may be more computationally efficient. In many cases involving complex nonlinear models, however, local SA methods are inadequate.

In all but the simplest of cases, sensitivity indices must be approximated numerically based on a sample of the inputs (sampling-based SA) together with the corresponding outputs. Output variations are measured by varying one input factor at a time (as in all local methods), or by varying all input factors simultaneously (as in most global methods) using factorial or fractional factorial methods, which takes into account the joint influences of factors due to correlations, but comes at the cost of a higher number of input samples in order to achieve accurate results.

In the context of battery models, very little attention has been paid to SA. In the majority of cases, formal SA methods are not used; the model is simply run multiple times by varying factors individually and inspecting the outputs, using *ad-hoc* measures or by employing visualization tools [2]. In a small number of studies more rigorous approaches have been used, but almost invariably with highly simplified models. Applying formal SA methods to complex battery and fuel cells models is computationally burdensome and often not feasible, particularly with brute force Monte Carlo (MC) approaches. In order to overcome this issue an emulator or meta-model can be used. The emulator itself can be difficult to construct when the input and/or output space is high-dimensional (e.g., input and/or output fields). If the quantity of interest (QoI), which is derived from the output, is a scalar, an alternative is to use an emulator directly between the inputs and QoI. It may be the case, however, that there are multiple QoIs, in which case it would be ideal to emulate the output, especially when other tasks (e.g. UQ) involving different quantities, including perhaps the original output, are to be performed subsequently.

To address these issues, we develop an approach for SA of a nonlinear Li-ion battery model by employing a Gaussian process emulator based on dimensionality reduction to approximate entire charge-discharge curves. QoIs are extracted from the curves in order to perform a SA. We show how it is possible to perform an efficient probabilistic SA in the context of a variance-based approach [4], extending previous results for the scalar case considered by Oakley and O'Hagan [5] to linear functional QoIs derived from the multi-dimensional output. Lastly, using the emulation method we are able to perform a SA of multiple outputs (including in high dimensional spaces) by ranking coefficients in a low-dimensional subspace approximation of the output space.

2. Problem setup and sensitivity analysis

Suppose that the model output is $y = \eta(\mathbf{x};\boldsymbol{\xi}) \in \mathcal{F}$ for some function space $\mathcal{F}$, where $\mathbf{x}$ represents, e.g., space, time or space-time and $\boldsymbol{\xi}^T = (\xi_1, \xi_2, ..., \xi_k) \in \mathcal{X} \subset \mathbb{R}^k$ is a vector of input factors; that is, a spatial, temporal or spatio-temporal field, *parameterized* by inputs $\boldsymbol{\xi}$. The computer model (simulator), on the other hand, provides a finite-dimensional approximation of $\eta(\mathbf{x};\boldsymbol{\xi})$, e.g., at a finite number of points in a spatial and/or temporal grid or in terms of a finite basis. We may write the simulator output as a vector $\mathbf{y} \in \mathbb{R}^d$, in which d is the number of degrees of freedom, e.g., the d components of $\mathbf{y}$ represent values of $\eta(\mathbf{x};\boldsymbol{\xi})$ at d points $\mathbf{x}$. We can therefore consider the simulator as a mapping $\boldsymbol{\eta} : \mathcal{X} \to \mathcal{Y} \subset \mathbb{R}^d$ between a feasible input space $\mathcal{X} \subset \mathbb{R}^k$ and an output space $\mathcal{Y}$, i.e., $\mathbf{y} = \boldsymbol{\eta}(\boldsymbol{\xi})$.

Let Q be a scalar QoI that is derived from y *via* a *linear* functional $G : \mathcal{F} \to \mathcal{Q}' \subset \mathbb{R}$. We can instead consider Q as a mapping $F = (G \circ \eta)(\boldsymbol{\xi}) : \boldsymbol{\xi} \mapsto Q$ directly between $\mathcal{X}$ and $\mathcal{Q}$, i.e., $Q = F(\boldsymbol{\xi}) = G(\eta(\mathbf{x};\boldsymbol{\xi}))$. In reality, we have an approximation $q = f(\boldsymbol{\xi})$ of Q, where the mapping $f : \mathcal{X} \to \mathcal{Q}$ is derived from a linear functional $g : \mathbb{R}^d \to \mathcal{Q} \subset \mathbb{R}$ that acts on the simulator outputs $\mathbf{y}$, that is $f(\boldsymbol{\xi}) = (g \circ \boldsymbol{\eta})(\boldsymbol{\xi}) = g(\boldsymbol{\eta}(\boldsymbol{\xi}))$. We develop a SA framework in which the outputs $\mathbf{y}$ are estimated by an emulator. We first describe the SA methods employed and the construction of the emulator. We then show how it is possible to perform SA on QoIs derived from the emulator output, before presenting a probabilistic SA analysis and a method for multi-output SA.

One approach to SA measures the sensitivity of the QoI by estimating the derivative $\partial_{\xi_i} f$ around a nominal (base) point $\overline{\boldsymbol{\xi}}$ using finite differences [6]. The main weakness of such local methods is that they provide no information on how the sensitivity to a given ξ_i depends on the values of the other factors. In order to extend this method to a global analysis we may use multiple base points $\{\overline{\boldsymbol{\xi}}_j\}_{j=1}^r$, which leads to the class of *elementary effect tests* (EETs). Suppose that $\mathcal{X}$ is the unit hypercube, and each direction ξ_i is discretized into p levels (points). The *elementary effect* of ξ_i at $\overline{\boldsymbol{\xi}}_j = (\overline{\xi}_{j,1}, \ldots, \overline{\xi}_{j,k})^T$ is: '

$$EE_i^{(j)} \equiv \left[f(\overline{\xi}_{j,1}, \ldots, \overline{\xi}_{j,i} + \delta_i, \ldots, \overline{\xi}_{j,k}) - f(\overline{\xi}_{j,1}, \ldots, \overline{\xi}_{j,i}, \ldots, \overline{\xi}_{j,k})\right] / \Delta_i, \tag{1}$$

where $\Delta_i \in \{1/(p-1), \ldots, 1-1/(p-1)\}$. Typically, $\Delta_i = \Delta$, $\forall i$. Morris [3] proposed two sensitivities measures for each ξ_i, namely the mean and the standard deviation of the finite distribution F_i over $\{EE_i^{(j)}\}_{j=1}^r$ (or $|EE_i^{(j)}|$). The mean μ_i of EE_i measures the influence of ξ_i, while the standard deviation σ_i measures the degree of interaction with the other factors.

To calculate the statistics for each elementary effect, i.e., μ_i and σ_i, we can randomly choose M base points $\{\bar{\boldsymbol{\xi}}_j\}_{j=1}^M$, then construct so-called *trajectories* in $\mathcal{X}$ of $k+1$ points $\bar{\boldsymbol{\xi}}_j \cup \{\boldsymbol{\xi}_{j,n}\}_{n=1}^k$ for each j. Setting $\boldsymbol{\xi}_{j,0} = \bar{\boldsymbol{\xi}}_j$, the trajectory point $\boldsymbol{\xi}_{j,n}$ is obtained by perturbing a randomly chosen factor of $\boldsymbol{\xi}_{j,n-1}$ by $\pm\Delta$ until all factors have been perturbed, but with the property that $\boldsymbol{\xi}_{j,n}$ and $\boldsymbol{\xi}_{j,n-1}$ differ in only one factor. The model is run at every point in each of the trajectories (a total of $M(k+1)$) to obtain $EE_{i,j}$, $i = 1, \ldots k$, $j = 1, \ldots M$, yielding estimates of μ_i and σ_i.

A more sophisticated approach to SA, embedded in probability, involves treating the inputs as stochastic variables, which leads to a distribution over the QoI [4, 6]. A variance based first-order effect of each input factor is given by $\mathbb{V}\mathrm{ar}_{\xi_i}(\mathbb{E}_{\boldsymbol{\xi}_{\sim i}}[q|\xi_i])$, where $\mathbb{E}[\cdot]$ and $\mathbb{V}\mathrm{ar}(\cdot)$ denote expectation and variance operators with respect to the distribution over a subscripted random variable; or with respect to the distribution $p(\boldsymbol{\xi})$ over $\boldsymbol{\xi}$ if no subscript is present, i.e. $\mathbb{E}[\cdot] \equiv \mathbb{E}_{\boldsymbol{\xi}}[\cdot]$. The quantity $\boldsymbol{\xi}_{\sim i}$ is the vector of all inputs factors *excluding* ξ_i (and similarly for multiple indices). The *first order sensitivity index* (or main effect index) for the input ξ_i is defined as $S_i \equiv \mathbb{V}\mathrm{ar}_{\xi_i}(\mathbb{E}_{\boldsymbol{\xi}_{\sim i}}[q|\xi_i])/\mathbb{V}\mathrm{ar}(q)$, which measures the contribution of the main effect of ξ_i to the total QoI variance. Another measure of sensitivity, defined below, is the *total effect index*, which incorporates interactions between the factors ξ_i:

$$S_{T_i} \equiv \mathbb{E}_{\boldsymbol{\xi}_{\sim i}}[\mathbb{V}\mathrm{ar}_{\xi_i}(q|\boldsymbol{\xi}_{\sim i})]/\mathbb{V}\mathrm{ar}(q) = 1 - \mathbb{V}\mathrm{ar}_{\boldsymbol{\xi}_{\sim i}}(\mathbb{E}_{\xi_i}[q|\boldsymbol{\xi}_{\sim i}])/\mathbb{V}\mathrm{ar}(q). \tag{2}$$

The variance-based SA framework can be couched in terms of the decomposition of the variance of q. Suppose $q = f(\boldsymbol{\xi}) \in L^2(\mathcal{X})$ (square integrable functions defined on $\mathcal{X}$) and $\mathcal{X}$ is (without loss of generality) a unit hypercube $\mathcal{X} = \{\boldsymbol{\xi}|0 \le \xi_i \le 1; i = 1, \ldots, k\}$. We also assume that the factors are independently and uniformly distributed within $\mathcal{X}$, so that the probability density functions satisfy $p(\xi_{i_1}, \ldots, \xi_{i_l}) = \mathbf{1}_{[0,1]^l}$ for $\{i_1, \ldots, i_l\} \subset \{1, \ldots, k\}$, where $\mathbf{1}_A$ is the indicator function on a set A. The expectation operators $\mathbb{E}_{\boldsymbol{\xi}_{\sim i_1 \ldots i_l}}[\cdot]$ are then unweighted integrals over $\xi_{i_1}, \ldots, \xi_{i_l}$. The function $f(\boldsymbol{\xi})$ can be decomposed in the following way (Hoeffding [7]):

$$f(\boldsymbol{\xi}) = f_0 + \sum_{i=1}^{k} f_i(\xi_i) + \sum_{i=1}^{k} \sum_{j=i+1}^{k} f_{ij}(\xi_i, \xi_j) + \ldots + f_{1\ldots k}(\xi_1, \ldots, \xi_k), \tag{3}$$

where f_0 is a constant, $f_i(\xi_i)$ (the main effect of ξ_i) is a function only of ξ_i, $f_{ij}(\xi_i, \xi_j)$ (the interaction) is a function only of ξ_i and ξ_j, and so on. The condition $\int_0^1 f_{i_1 i_2 \ldots i_s}(\xi_{i_1}, \xi_{i_2}, \ldots, \xi_{i_s}) d\xi_{i_w} = 0$ is imposed for $1 \le i_1 < i_2 < \ldots < i_s \le k$ and $i_w \in \{i_1, i_2, \ldots, i_s\}$ [7]. Thus, the summands are orthogonal, in the sense that:

$$\int_{\mathcal{X}} f_{i_1 i_2 \ldots i_m}(\xi_{i_1}, \xi_{i_2}, \ldots, \xi_{i_m}) f_{i'_1 i'_2 \ldots i'_n}(\xi_{i'_1}, \xi_{i'_2}, \ldots, \xi_{i'_n}) d\boldsymbol{\xi} = 0, \tag{4}$$

for $\{i_1, i_2, \ldots, i_m\} \neq \{i'_1, i'_2, \ldots, i'_n\}$. Moreover, $f_0 = \mathbb{E}[q] = \int_{\mathcal{X}} f(\boldsymbol{\xi}) d\boldsymbol{\xi}$, $f_i = \mathbb{E}_{\boldsymbol{\xi}_{\sim i}}[q|\xi_i] - f_0$, $f_{ij} = \mathbb{E}_{\boldsymbol{\xi}_{\sim ij}}[q|\xi_i, \xi_j] - f_i - f_j - f_0$, etc.. By squaring and integrating Eq. (3) and using the orthogonality property, we obtain a *decomposition of the total variance* $V = \mathbb{V}\mathrm{ar}(q) = \int_{\mathcal{X}} f^2(\boldsymbol{\xi}) d\boldsymbol{\xi} - f_0^2$, that is, $V = \sum_{i=1}^k V_i + \sum_{i=1}^k \sum_{j=i+1}^k V_{ij} + \ldots + V_{1\ldots k}$, in which $V_i = \mathbb{V}\mathrm{ar}_{\xi_i}(f_i(\xi_i)) = \mathbb{V}\mathrm{ar}_{\xi_i}(\mathbb{E}_{\boldsymbol{\xi}_{\sim i}}[q|\xi_i])$, $V_{ij} = \mathbb{V}\mathrm{ar}_{\xi_i \xi_j}(f_{ij}(\xi_i, \xi_j)) = \mathbb{V}\mathrm{ar}_{\xi_i \xi_j}(\mathbb{E}_{\boldsymbol{\xi}_{\sim ij}}[q|\xi_i, \xi_j]) - \mathbb{V}\mathrm{ar}_{\xi_i}(\mathbb{E}_{\boldsymbol{\xi}_{\sim i}}[q|\xi_i]) - \mathbb{V}\mathrm{ar}_{\xi_j}(\mathbb{E}_{\boldsymbol{\xi}_{\sim j}}[q|\xi_j])$, and so on. The terms $V_{i_1 \ldots i_l}$, $l \le k$ are called partial variances and it is obvious that the main effect indices S_i are simply the partial variances normalized by the total variance. We can also define higher order sensitivity indices by normalizing the $V_{i_1 \dots i_l}$, e.g. the second-order index $S_{ij} = V_{ij}/V$, which measures the effect of interactions between ξ_i and ξ_j on q.

The main and total indices can be computed using a quasi MC method [1] by first generating a matrix $\mathbf{X} = [\xi_{i,j}]$, $i = 1, \dots, 2k$, $j = 1, \dots, N$, of N points in the $2k$ hypercube, using a low-discrepancy sequence such as a Latin hypercube. This is done according to the distribution $p(\boldsymbol{\xi})$ over the factors, e.g., $p(\boldsymbol{\xi}) = \mathbf{1}_{[0,1]^k}$ for independent $\xi_i \sim \mathcal{U}[0,1]$. $\mathbf{X}$ is then partitioned into a matrix $\mathbf{A} \in \mathbb{R}^{N \times k}$ consisting of the first k columns and a matrix $\mathbf{B} \in \mathbb{R}^{N \times k}$ consisting of the remaining k columns. This provides two independent sets of N samples in the k hypercube. A third matrix $\mathbf{C}_i$ consists of the columns of matrix $\mathbf{B}$ except the i-th column, which is set to the i-th column of $\mathbf{A}$. The next step is to compute the QoI q by running the model at the selected inputs contained in the sample matrices $\mathbf{A}$, $\mathbf{B}$, and $\mathbf{C}_i$ to yields vectors $\mathbf{q_A} = f(\mathbf{A})$, $\mathbf{q_B} = f(\mathbf{B})$ and $\mathbf{q}_{\mathbf{C}_i} = f(\mathbf{C}_i)$ ($f(\mathbf{A})$ is used to denote vectorized q values from the set consisting of the rows of $\mathbf{A}$). The indices S_i and S_{T_i} are then estimated as follows:

$$S_i = \frac{(1/N)\sum_{j=1}^{N} q_{\mathbf{A},j} q_{\mathbf{C}_i,j} - f_0^2}{(1/N)\sum_{j=1}^{N} q_{\mathbf{A},j}^2 - f_0^2}, \quad S_{T_i} = \frac{(1/N)\sum_{j=1}^{N} q_{\mathbf{B},j} q_{\mathbf{C}_i,j} - f_0^2}{(1/N)\sum_{j=1}^{N} q_{\mathbf{A},j}^2 - f_0^2}, \tag{5}$$

where $q_{\mathbf{A},j}$ is the j-th coordinate of $\mathbf{q_A}$ (etc.) and $f_0 = (1/N)\sum_{j=1}^{N} q_{\mathbf{A},j}$ is the sample mean. This procedure is repeated for each $i = 1, \dots, k$. The first of Eqs. (5) follows from the basic definition $\mathbb{V}\mathrm{ar}_{\xi_i}(\mathbb{E}_{\boldsymbol{\xi}_{\sim i}}[q|\xi_i]) = \int_{[0,1]} \mathbb{E}^2_{\boldsymbol{\xi}_{\sim i}}[q|\xi_i] d\xi_i - (\int_{[0,1]} \mathbb{E}_{\boldsymbol{\xi}_{\sim i}}[q|\xi_i] d\xi_i)^2$. The last term is $\mathbb{E}^2[q] = f_0^2$, while the first term can be written as: $\int_{[0,1]} \mathbb{E}^2_{\boldsymbol{\xi}_{\sim i}}[q|\xi_i] d\xi_i = \int_{[0,1]^k} \int_{[0,1]^{k-1}} f(\xi_1, \dots, \xi_k) \times f(x_1', \dots, x_k') d\boldsymbol{\xi} d\boldsymbol{\xi}_{\sim i}'$, i.e., the expectation over $\boldsymbol{\xi}$ and $\boldsymbol{\xi}_{\sim i}'$ of $f(\xi_1, \dots, \xi_k) \times f(x_1', \dots, x_k')$, which explains the MC estimate in Eq. (5). A similar explanation can be given for the S_{T_i} estimate.

The cost of this procedure is $2N$ runs of the model to generate the matrices $\mathbf{A}$ and $\mathbf{B}$, and an additional Nk runs to obtain the QoIs corresponding to $\mathbf{C}_i$. This give a total of $N(k+2)$, which is much lower than the cost of brute-force MC estimates of $\mathbb{V}\mathrm{ar}_{\xi_i}(\mathbb{E}_{\boldsymbol{\xi}_{\sim i}}[q|\xi_i])$ and $\mathbb{V}\mathrm{ar}_{\boldsymbol{\xi}_{\sim i}}(\mathbb{E}_{\xi_i}[q|\boldsymbol{\xi}_{\sim i}])$. The former, e.g., would require $O(N)$ runs ($N \gg k$) to estimate the inner expectation for a fixed ξ_i, which we would be repeated $O(N)$ times to estimate the outer variance, leading to $O(N^2)$ runs for each i.

3. Gaussian process emulation of the model outputs

Suppose we are given *training points* $\{\mathbf{y}_j\}_{j=1}^m \subset \mathcal{Y}$, which are values of $\mathbf{y} = \boldsymbol{\eta}(\boldsymbol{\xi})$ at the *design points* $\{\boldsymbol{\xi}_j\}_{j=1}^m$. Without loss of generality we mean centre the training points: $\mathbf{y}_j \mapsto \mathbf{y}_j - \overline{\mathbf{y}}$, where $\overline{\mathbf{y}} = \sum_{k=1}^m \mathbf{y}_k$. Assume that $\mathcal{Y}$ is a low-dimensional linear subspace of $\mathbb{R}^d$. We derive an approximate basis for $\mathcal{Y}$ using principal component analysis (PCA) [8], i.e., we find a linear transformation $\mathbf{w}(\boldsymbol{\xi}) = \mathbf{V}^T\mathbf{y}$, in which $\mathbf{V} \in \mathbb{R}^{d \times d}$ has orthogonal columns $\mathbf{v}_i$ (a basis for $\mathbb{R}^d$) and the uncorrelated components $w_i(\boldsymbol{\xi})$ of $\mathbf{w}(\boldsymbol{\xi})$ have decreasing variance with i.

Let $\boldsymbol{\Sigma} = \mathbb{E}[\mathbf{y}\mathbf{y}^T]$ be the symmetric and positive definite variance-covariance matrix. The eigenvalue problem $\boldsymbol{\Sigma}\mathbf{v} = \lambda\mathbf{v}$ yields the $\mathbf{v}_i$ and corresponding positive eigenvalues $\lambda_1 > \cdots > \lambda_d$. The components of a point in this basis satisfy $\mathbb{V}\mathrm{ar}[w_i(\boldsymbol{\xi})] = \lambda_i$ and $\mathbb{E}[w_i(\boldsymbol{\xi})w_j(\boldsymbol{\xi})] = 0$ for $i \neq j$. Any point $\mathbf{y} \in \mathcal{Y}$ can be written in the form $\mathbf{y} = \mathbf{V}\mathbf{w}(\boldsymbol{\xi}) = \sum_{i=1}^d w_i(\boldsymbol{\xi})\mathbf{v}_i = \sum_{i=1}^d (\mathbf{v}_i^T\mathbf{y})\mathbf{v}_i$ and an r-dimensional approximation $\mathbf{y}_r \in \mathcal{Y}_r = \mathrm{span}(\mathbf{v}_1, \dots, \mathbf{v}_r)$ of $\mathbf{y}$ is given by $\mathbf{y}_r = \mathbf{V}_r\mathbf{w}_r(\boldsymbol{\xi}) = \sum_{i=1}^r w_i(\boldsymbol{\xi})\mathbf{v}_i$, where $\mathbf{V}_r = [\mathbf{v}_1 \dots \mathbf{v}_r]$ and $\mathbf{w}_r(\boldsymbol{\xi}) = (w_1(\boldsymbol{\xi}), \dots, w_r(\boldsymbol{\xi}))^T$. It can be demonstrated [8] that $\mathbb{E}\left[\|\mathbf{y} - \mathbf{y}_r\|^2\right] = \sum_{i=r+1}^d \lambda_i$, from which a value of r can be selected based on a chosen tolerance. $\boldsymbol{\Sigma}$ is approximated by the sample covariance matrix $\boldsymbol{\Sigma}_S = (1/m)\mathbf{Y}\mathbf{Y}^T$, where $\mathbf{Y} = [\mathbf{y}_1 \dots \mathbf{y}_m]$. The coefficients $w_i(\boldsymbol{\xi}) = \mathbf{v}_i^T\mathbf{y}$ of a point $\mathbf{y} = \boldsymbol{\eta}(\boldsymbol{\xi})$ are assumed to be realizations of uncorrelated (therefore mutually independent) GPs. We emulate these coefficients for test inputs $\boldsymbol{\xi}$ to find low-dimensional approximations $\mathbf{y}_r = \boldsymbol{\eta}_r(\boldsymbol{\xi}) \equiv \mathbf{V}_r\mathbf{w}_r(\boldsymbol{\xi}) \in \mathcal{Y}_r$ [9].

Let us focus on $w_i(\boldsymbol{\xi})$ for some $i \in \{1, \ldots, r\}$. We wish to approximate $w_i(\boldsymbol{\xi}) : \mathcal{X} \to \mathbb{R}$ given values of this function (from a PCA) at design points $\{\boldsymbol{\xi}_j\}_{j=1}^m$. In GPR, we place a GP *prior distribution* indexed by $\boldsymbol{\xi} \in \mathcal{X}$ over $w_i(\boldsymbol{\xi})$. The distribution $p(w_i(\boldsymbol{\xi}_1), \ldots, w_i(\boldsymbol{\xi}_m))$ for an arbitrary finite collection of indices $\{\boldsymbol{\xi}_1, \ldots, \boldsymbol{\xi}_m\}$ is Gaussian. The GP prior is $w_i(\boldsymbol{\xi})|\boldsymbol{\theta}_i \sim \mathcal{GP}(0, c(\boldsymbol{\xi}, \boldsymbol{\xi}'; \boldsymbol{\theta}_i))$, which has a zero mean function (we set $w_i(\boldsymbol{\xi}_j) \mapsto w_i(\boldsymbol{\xi}_j) - \overline{w}_i$, where $\overline{w}_i = (1/m)\sum_j w_i(\boldsymbol{\xi}_j)$) and a covariance function $c(\boldsymbol{\xi}, \boldsymbol{\xi}'; \boldsymbol{\theta}_i)$, dependent upon *hyperparameters* $\boldsymbol{\theta}_i$. We employ a square-exponential function for all i: $c(\boldsymbol{\xi}, \boldsymbol{\xi}'; \boldsymbol{\theta}_i) = \theta_0 \exp\{-(\boldsymbol{\xi} - \boldsymbol{\xi}')^T \mathrm{diag}(\theta_{i,1}, \ldots, \theta_{i,k})(\boldsymbol{\xi} - \boldsymbol{\xi}')\}$, where $\boldsymbol{\theta}_i = (\theta_{i,0}, \ldots, \theta_{i,k})^T$, in which $\theta_{i,1}, \ldots, \theta_{i,k}$ are the inverse square correlation lengths.

The distribution of $\mathbf{d}_i \equiv (w_i(\boldsymbol{\xi}_1), \ldots, w_i(\boldsymbol{\xi}_m))^T$ given $\boldsymbol{\theta}_i$ (i.e., the likelihood) is $p(\mathbf{d}_i|\boldsymbol{\theta}_i) = \mathcal{N}(\mathbf{0}, \mathbf{C}_i)$, with covariance matrix $\mathbf{C}_i = [c(\boldsymbol{\xi}_i, \boldsymbol{\xi}_j; \boldsymbol{\theta}_i)]_{i,j=1}^m$. The predictive distribution at new inputs $\boldsymbol{\xi} \in \mathcal{X}$ is obtained from the joint distribution $p(w_i(\boldsymbol{\xi}), \mathbf{d}_i|\boldsymbol{\theta}_i)$ by conditioning on $\mathbf{d}_i$ [10]:

$$
\begin{gathered}
w_i(\boldsymbol{\xi})|\mathbf{d}_i, \boldsymbol{\theta}_i \sim \mathcal{GP}\left(m_i'(\boldsymbol{\xi}), c_i'(\boldsymbol{\xi}, \boldsymbol{\xi}')\right), \\
m_i'(\boldsymbol{\xi}) = \mathbf{c}_i(\boldsymbol{\xi})^T \mathbf{C}_i^{-1} \mathbf{d}_i + \overline{w}_i, \quad c_i'(\boldsymbol{\xi}, \boldsymbol{\xi}') = c(\boldsymbol{\xi}, \boldsymbol{\xi}'; \boldsymbol{\theta}_i) - \mathbf{c}_i(\boldsymbol{\xi})^T \mathbf{C}_i^{-1} \mathbf{c}(\boldsymbol{\xi}'),
\end{gathered} \tag{6}
$$

in which $\mathbf{c}_i(\boldsymbol{\xi}) = (c(\boldsymbol{\xi}_1, \boldsymbol{\xi}; \boldsymbol{\theta}_i), \ldots, c(\boldsymbol{\xi}_m, \boldsymbol{\xi}; \boldsymbol{\theta}_i))^T$. The hyperparameters can be specified by point estimates [10] such as the maximum log likelihood estimate (MLE): $\boldsymbol{\theta}_{i,MLE} = \arg\max_{\boldsymbol{\theta}_i}(-\ln|\mathbf{C}_i|/2 - \mathbf{d}_i^T \mathbf{C}_i^{-1} \mathbf{d}_i/2)$. This procedure is repeated for each $i = 1, \ldots, r$ to obtain $\mathbf{w}_r(\boldsymbol{\xi}) = (w_1(\boldsymbol{\xi}), \ldots, w_r(\boldsymbol{\xi}))^T$. Using MLE estimates, we obtain $\mathbb{E}[\mathbf{w}_r(\boldsymbol{\xi})] = \mathbf{m}'(\boldsymbol{\xi}) \equiv (m_1'(\boldsymbol{\xi}), \ldots, m_r'(\boldsymbol{\xi}))^T$. The predicted variance of each coefficient is $\mathbb{V}\mathrm{ar}(w_i(\boldsymbol{\xi})) = c_i'(\boldsymbol{\xi}, \boldsymbol{\xi})$. The model outputs are therefore distributed as follows (noting that $\mathbb{C}\mathrm{ov}(w_i(\boldsymbol{\xi}), w_j(\boldsymbol{\xi})) = 0$ for $i \neq j$):

$$
\begin{aligned}
\mathbf{y}_r - \overline{\mathbf{y}} = \boldsymbol{\eta}_r(\boldsymbol{\xi}) = \mathbf{V}_r \mathbf{w}_r(\boldsymbol{\xi}) &\sim \mathcal{GP}\left(\mathbf{m}_{\mathbf{y}_r}, \mathbf{c}_{\mathbf{y}_r}(\boldsymbol{\xi}, \boldsymbol{\xi}')\right) \\
\mathbf{m}_{\mathbf{y}_r}(\boldsymbol{\xi}) = \mathbb{E}[\boldsymbol{\eta}_r(\boldsymbol{\xi})] &= \mathbf{V}_r \mathbf{m}'(\boldsymbol{\xi}), \\
\mathbf{c}_{\mathbf{y}_r}(\boldsymbol{\xi}, \boldsymbol{\xi}') = \mathbb{C}\mathrm{ov}\left(\boldsymbol{\eta}_r(\boldsymbol{\xi}), \boldsymbol{\eta}_r(\boldsymbol{\xi}')\right) &= \mathbf{V}_r \mathrm{diag}(c_1'(\boldsymbol{\xi}, \boldsymbol{\xi}'), \ldots, c_r'(\boldsymbol{\xi}, \boldsymbol{\xi}'))\mathbf{V}_r^T,
\end{aligned} \tag{7}
$$

3.1. Probabilistic and multivariate sensitivity analysis

The emulation method described above extends Bayesian GP modelling to multiple outputs in a probabilistic manner, furnishing an explicit distribution over the output (Eqs. (7)). We can always extract estimates of the statistics of sensitivity measures using full MC sampling. Take for example the main effect index $S_i = \mathbb{V}\mathrm{ar}_{\xi_i}(\mathbb{E}_{\boldsymbol{\xi}_{\sim i}}[q|\xi_i])/\mathbb{V}\mathrm{ar}(q)$. The expected value and variance of a quantity with respect to the distribution over $\boldsymbol{\eta}_r$ are denoted $\mathbb{E}_{\boldsymbol{\eta}_r}[\cdot]$ and $\mathbb{V}\mathrm{ar}_{\boldsymbol{\eta}_r}(\cdot)$, respectively. Since $q = f(\boldsymbol{\xi}) = g(\boldsymbol{\eta}(\boldsymbol{\xi}))$, a MC estimate of $\mathbb{E}_{\boldsymbol{\eta}_r}[S_i]$ is given by:

$$
\begin{aligned}
\mathbb{E}_{\boldsymbol{\eta}_r}[S_i] &= \mathbb{E}_{\boldsymbol{\eta}_r}\left[\frac{\mathbb{V}\mathrm{ar}_{\xi_i}(\mathbb{E}_{\boldsymbol{\xi}_{\sim i}}[g(\boldsymbol{\eta}(\boldsymbol{\xi}))|\xi_i])}{\mathbb{V}\mathrm{ar}(g(\boldsymbol{\eta}(\boldsymbol{\xi})))}\right] = \mathbb{E}_{\boldsymbol{\eta}_r}\left[\frac{\mathbb{E}_{\xi_i}\left[\mathbb{E}^2_{\boldsymbol{\xi}_{\sim i}}[g(\boldsymbol{\eta}(\boldsymbol{\xi}))|\xi_i])\right] - \mathbb{E}^2\left[g(\boldsymbol{\eta}(\boldsymbol{\xi}))\right]}{\mathbb{E}\left[g(\boldsymbol{\eta}(\boldsymbol{\xi}))^2\right] - \mathbb{E}^2\left[g(\boldsymbol{\eta}(\boldsymbol{\xi}))\right]}\right] \\
&\approx \frac{1}{J}\sum_{j=1}^J \frac{N^{-3}\sum_{l=1}^N\left(\sum_{n=1}^N g(\boldsymbol{\eta}^{(j)}(\boldsymbol{\xi}_{\sim i}^{(n)}, \xi_i^{(l)}))\right)^2 - N^{-2}\left(\sum_{n=1}^N g(\boldsymbol{\eta}^{(j)}(\boldsymbol{\xi}^{(n)})\right)^2}{N^{-1}\sum_{n=1}^N g(\boldsymbol{\eta}^{(j)}(\boldsymbol{\xi}^{(n)})^2 - N^{-2}\left(\sum_{n=1}^N g(\boldsymbol{\eta}^{(j)}(\boldsymbol{\xi}^{(n)})\right)^2}
\end{aligned} \tag{8}
$$

where $\boldsymbol{\eta}_r^{(j)}$ is drawn from $p(\boldsymbol{\eta}_r) = \mathcal{GP}(\mathbf{m}_{\mathbf{y}_r}(\boldsymbol{\xi}), \mathbf{c}_{\mathbf{y}_r}(\boldsymbol{\xi}, \boldsymbol{\xi}'))$ and the $\xi_i \sim \mathcal{U}[0,1]$ are independent. The notation $\boldsymbol{\eta}^{(j)}(\boldsymbol{\xi}_{\sim i}^{(n)}, \xi_i^{(l)})$ means that $\boldsymbol{\eta}^{(j)}(\boldsymbol{\xi})$ is evaluated at $\xi_i = \xi_i^{(l)}$, $\boldsymbol{\xi}_{\sim i} = \boldsymbol{\xi}_{\sim i}^{(n)}$ for some $i \in \{1, \ldots, k\}$. The generic number N in Eq. (8) need not be the same for all MC estimates.

In this MC procedure we interpret $S_i \equiv S_i(\boldsymbol{\eta})$ as a random function of the random vector $\boldsymbol{\eta}$ and obtain samples $\boldsymbol{\eta}^{(j)}(\boldsymbol{\xi})$ (deterministic functions of $\boldsymbol{\xi}$) from $p(\boldsymbol{\eta}_r)$ to approximate the outer integral. We then sample from $\boldsymbol{\xi}_{\sim i} \sim \mathcal{U}[0,1]^{k-1}$ and $\xi_i \sim \mathcal{U}[0,1]$ to approximate the inner integrals of $g(\boldsymbol{\eta}(\boldsymbol{\xi})) = f(\boldsymbol{\xi})$, which is a random function of $\boldsymbol{\xi}$. In practice, we generate the samples of $\boldsymbol{\xi}_{\sim i}$ and ξ_i, and then sample from the distributions over $w_i(\boldsymbol{\xi})$, $i = 1, \ldots, r$, to obtain partial realizations (at the sampled values of $\boldsymbol{\xi}$) of $\mathbf{w}_r(\boldsymbol{\xi})$, from which we can obtain (partial) realizations of $\boldsymbol{\eta}_r(\boldsymbol{\xi}) = \mathbf{V}_r\mathbf{w}_r(\boldsymbol{\xi})$. In fact, the last step is not necessary since we can work directly with $\mathbf{w}_r(\boldsymbol{\xi})$ to obtain realizations of the QoI $q = g(\boldsymbol{\eta}(\boldsymbol{\xi}))$, i.e., $g(\boldsymbol{\eta}(\boldsymbol{\xi})) = g(\mathbf{V}_r\mathbf{w}_r(\boldsymbol{\xi}))$. $\mathbb{V}\mathrm{ar}_{\boldsymbol{\eta}_r}(S_i)$ is estimated in the same way and the same procedure can be used for S_{T_i} or any other measure.

In the scalar output case (the output being the QoI), Oakley and O'Hagan derived semi-analytical expressions for estimating the expectations $\mathbb{E}_{\boldsymbol{\eta}_r}[\cdot]$ and possibly variances $\mathbb{V}\mathrm{ar}_{\boldsymbol{\eta}_r}(\cdot)$ of several sensitivity measures using only a very small number of MC runs (e.g. $O(1)$ *vs.* $O(N)$ for $\mathbb{E}_{\boldsymbol{\eta}_r}[S_i]$ as required in full MC to estimate $\mathbb{V}\mathrm{ar}_{\xi_i}(\mathbb{E}_{\boldsymbol{\xi}_{\sim i}}[g(\boldsymbol{\eta}(\boldsymbol{\xi}))|\xi_i]))$ [5]. Equivalent semi-analytical expressions can be established for certain types of scalar QoIs derived from the multivariate output emulator used in this paper, namely QoIs arising from a linear functional of the output. These expressions are derived in the Appendix. Another feature of this method is that we could investigate the *sensitivity of multivariate outputs under uncertainty* to the inputs by *separately* (due to their independence) ranking the sensitivity indices for the coefficients $w_i(\boldsymbol{\xi})$, $i = 1, \ldots, r$, using the procedures described above, i.e., setting $q = w_i(\boldsymbol{\xi})$. Since the $w_i(\boldsymbol{\xi})$ are scalar GPs, the probabilistic analysis of Oakley and O'Hagan [5] is directly applicable.

4. Li-on battery model

We consider a Li-ion battery comprised of a $LiMn_2O_4$ positive electrode and a graphite Li_xC_6 porous negative electrode. The electrolyte consists of a non-aqueous carbonate solvent mixture and a lithium salt $LiPF_6$ in a 1:1 mixture of ethylene carbonate and diethyl carbonate dispersed in an inert polymer matrix. The domain is 1-d (direction x) and the positive and negative current collectors are located at $x = 0$ and $x = L$, respectively. Intercalation of Li is described by a mass balance with diffusion in a pseudo dimension R (into spherical particles) [11]. The solid Li concentration in the positive (negative) electrode c_s^p (c_s^n) is given by:

$$\partial_t c_j^s = R^{-2}\partial_R(R^2 D_j^s \partial_R c_j^s), \tag{9}$$

where D_j^s is the diffusion coefficient of Li in the active material. Here and below, $j = p$ for the positive electrode, $j = n$ for the negative electrode and $j = s$ for the separator. The boundary conditions are $\partial_R c_j^s|_{R=0} = 0$ and $-D_j^s\partial_R c_j^s|_{R=R_p} = (1/aF)\partial_x i_2$, where R_p is the particle radius, a is the specific active surface area and i_2 is the current density in the electrolyte:

$$i_2 = -\kappa_2\partial_x\phi_2 + \kappa_2 R_U T F^{-1}(1 - t_+^0)(1 + \partial_{\ln c}\ln f_A)\partial_x \ln c, \tag{10}$$

where κ_2 is the effective ionic conductivity, T is the temperature, F is Faraday's constant, R_U is the universal gas constant, ϕ_2 is the electrolyte potential, f_A is the mean molar activity coefficient of the electrolyte, c is the lithium ion (Li^+) concentration and t_+^0 is the transference number of Li^+. The solid phase current density i_1 is governed by Ohm's law: $i_1 = -\kappa_1\partial_x\phi_1$, where κ_1 is the effective conductivity of the solid and ϕ_1 is the solid-phase potential. Charge conservation demands that $i_1 + i_2 = I$, for a total current density I. The boundary conditions for the potentials (galvanostatic) are $-\kappa_1\partial_x\phi_1 = I$ at $x = 0, L$ and $-\kappa_2\partial_x\phi_2 = 0$ at $x = 0, L$, while the electronic charge fluxes are zero and the ionic charge fluxes are continuous at the separator interfaces. The mass balance for Li^+ is:

$$\epsilon_j\partial_t c = \partial_x(\epsilon_j D_j \partial_x c) - (1 - t_+^0)(\nu_+ F)^{-1}\partial_x i_2, \tag{11}$$

where ϵ_j is the volume fraction of electrolyte, D_j is the effective diffusion coefficient of the Li^+ through the electrolyte, and ν_+ is the number of cations into which a mole of electrolyte dissociates. The flux at both ends of the cell ($-\epsilon_j D_j\partial_x c$ by virtue of the zero ionic charge flux) is set to zero. The current density is given by the Butler-Volmer equation:

$$\partial_x i_j = -aFk_j(c)^{\alpha_{a,j}}(c_t - c_j^s)^{\alpha_{a,j}}(c_j^s)^{\alpha_{c,j}}\left(e^{-\alpha_a F\eta_j/(R_U T)} - e^{\alpha_a F\eta_j/(R_U T)}\right), \tag{12}$$

where α_a and α_c are the charge transfer coefficients, c_t is the total concentration of lithium, k_j is the rate constant for the relevant reaction and $\eta_j = \phi_1 - \phi_2 - U_j$ is the overpotential at the relevant electrode, in which U_j is the corresponding equilibrium potential.

5. Results and discussion

The Li-ion battery model was implemented in COMSOL Multiphysics [12]. A total of 500 simulations were performed by varying the initial state of charge SOC_{in} (initial c_n^s divided by c_t), the particle diameter in the positive electrode R_p and the positive electrode porosity ϵ_p. We set $\boldsymbol{\xi} = (\mathrm{SOC}_{in}, R_p, \epsilon_p)^T \in \mathcal{X} \subset \mathbb{R}^k$ as the input, with the $k = 3$ factors given by the components. The inputs for the 500 simulations were selected using a Sobol sequence. A current pulse $i(t)$ consisting of 12 s of 120 A discharge, followed by 12 s of relaxation (0 A load), and then 12 s of 120 A charge was simulated (galvanostatic operation). The output was taken to be the cell voltage $E_{cell}(t)$[V] at 0.5 s intervals, yielding a total of 73 values at $t = 0, 0.5, 1, \ldots, 35.5, 36$ s. These values were vectorized to form the outputs: $\mathbf{y}^T = \boldsymbol{\eta}(\boldsymbol{\xi}) = (E_{cell}(0), E_{cell}(0.5), \ldots, E_{cell}(35.5), E_{cell}(36)) \in \mathcal{Y} \subset \mathbb{R}^d$, where $\boldsymbol{\eta}$ as before represents the simulator and $d = 73$. The first 100 outputs were reserved for training the emulator and the remaining $m_t = 400$ were used for testing the emulator. The training data set is denoted $\{(\boldsymbol{\xi}_j, \mathbf{y}_j)\}_{j=1}^m$, as before, and the test data set is denoted $\{(\boldsymbol{\xi}_j^*, \mathbf{y}_j^*)\}_{j=1}^{m_t}$.

We consider two QoIs: (i) the energy efficiency $q = f(\boldsymbol{\xi}) = \eta_E = \int_{[d]} i(t)E_{cell}(t)dt \Big/ \int_{[c]} i(t)E_{cell}(t)dt \in \mathcal{Q} = [0,1]$, in which $[d]$ ($[c]$) is the discharge (charge) time interval; .and (ii) the voltage drop during discharge, $q = \Delta V_c \in \mathcal{Q} = \mathbb{R}_+$

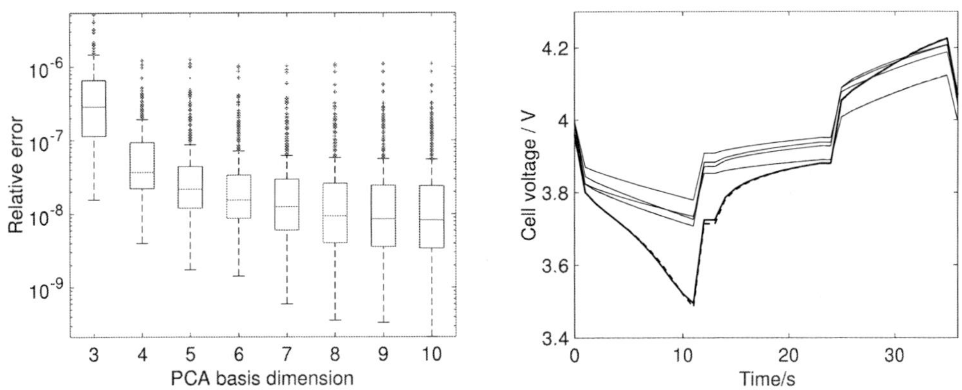

Figure 1. (Left) Boxplots of the emulator error on the test set $\{(\boldsymbol{\xi}_j^*, \mathbf{y}_j^*)\}_{j=1}^{m_t}$ using the training set $\{(\boldsymbol{\xi}_j, \mathbf{y}_j)\}_{j=1}^m$ with $m = 100$. (Right) Example predictions $\overline{\mathbf{y}}_{r,j}^*$ (dashed lines) of E_{cell} during the discharge-charge cycle $\mathbf{y}_j^*$ (solid lines) using $r = 10$. The worst case predictions (highest ϵ^*) are the thick lines and 4 further examples are shown.

Figure 1 shows Tukey boxplots of the relative errors on the test set $\{(\boldsymbol{\xi}_j^*, \mathbf{y}_j^*)\}_{j=1}^{m_t}$ using the training set $\{(\boldsymbol{\xi}_j, \mathbf{y}_j)\}_{j=1}^m$ with $m = 100$ for an increasing r. The errors were defined as $\epsilon^* = ||\overline{\mathbf{y}}_{r,j}^* - \mathbf{y}_j^*||/||\mathbf{y}_j^*||$, in which $\overline{\mathbf{y}}_{r,j}^*$ is the mean GP prediction of $\mathbf{y}_j^*$ using Eq. (7). Example predictions of the discharge-charge cycle are shown in Figure 1, for $r = 10$. The worst case predictions (highest ϵ^*) are shown, alongside 4 further examples. For the SA below we used $m = 100$ and $r = 10$.

IOP Conf. Series: Journal of Physics: Conf. Series **1039** (2018) 012020 doi:10.1088/1742-6596/1039/1/012020

5.1. Sensitivity analysis

The SA was performed using the SAFE package developed by Pianosi *et al.* [13]. For the variance-based method we placed a uniform distribution on the factors and sampled points $\mathbf{X} = [\xi_{i,j}]$, $i = 1, \ldots, 2k$, $j = 1, \ldots, N$ ($N = 5000$) in the $2k$ hypercube using a Latin hypercube design. The physical ranges were $0.1 \leq \epsilon_p \leq 0.4$, $0.5 \leq R_p[\mu\text{m}] \leq 2$ and $0.4 \leq \text{SOC}_{in} \leq 0.6$, and the factors were scaled to obtain $\mathcal{X} = [0, 1]^3$. The sampled inputs were used to produce the three input matrices $\mathbf{A} \in \mathbb{R}^{N \times k}$, $\mathbf{B} \in \mathbb{R}^{N \times k}$ and $\mathbf{C}_i \in \mathbb{R}^{N \times k}$, $i = 1, \ldots, k$, from which the QoI values $\mathbf{q_A} = f(\mathbf{A})$, $\mathbf{q_B} = f(\mathbf{B})$ and $\mathbf{q}_{\mathbf{C}_i} = f(\mathbf{C}_i)$ were extracted.

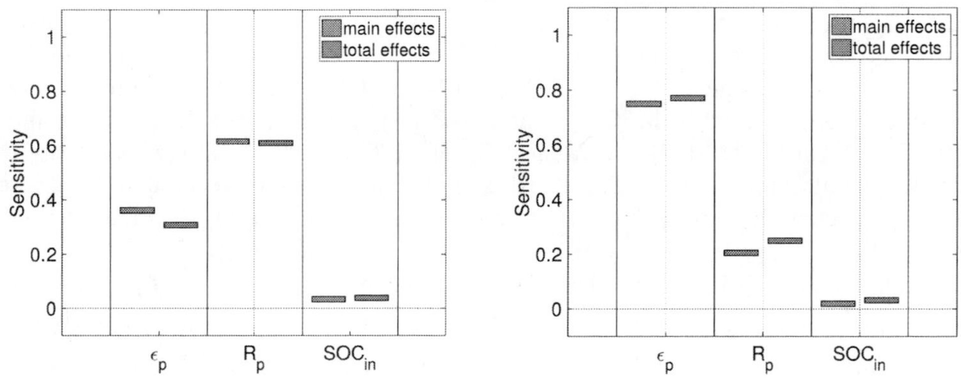

Figure 2. Main and total effects for η_E (left) and ΔV_c (right).

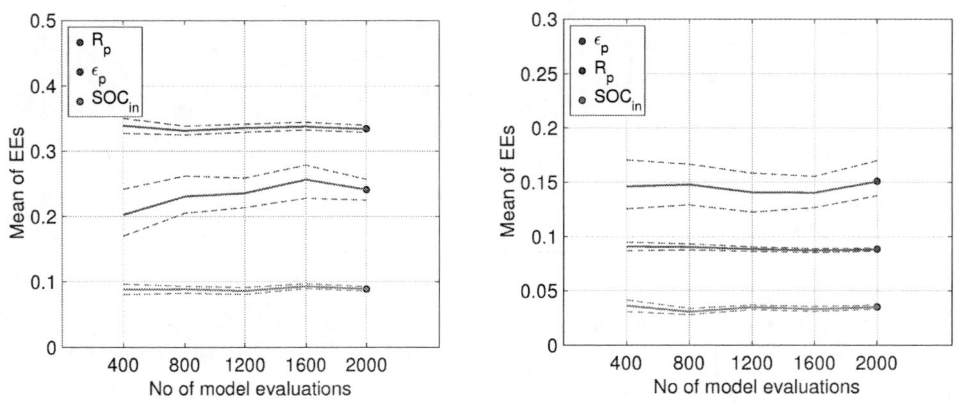

Figure 3. Convergence of the EEs for different numbers of model evaluations with 95% confidence intervals in the case of η_E (left) and ΔV_c (right).

Figure 2 presents both the main and total effects for the three factors. As expected, the particle size and porosity are the most influential, while the initial SOC mainly affects the open-circuit potential so has relatively little influence on q. The porosity determines the effective ionic conductivity (the volume fraction of electrolyte is ϵ_p) and since the ohmic loss is predominantly suffered in the ionic phase, ϵ_p has a major influence on the internal resistance. Moreover, the

IOP Conf. Series: Journal of Physics: Conf. Series **1039** (2018) 012020 doi:10.1088/1742-6596/1039/1/012020

reaction rate depends upon the concentration (per unit volume of the electrode) of Li^+ according to the Butler-Volmer law (12), so a restricted supply of Li^+ in the positive electrode will lead to a large concentration overpotential for a fixed current (the overpotential in (12) must increase as c decreases in order to maintain a fixed left hand side, i.e, applied current density). The particle radius determines the level of mass transport resistance for the solid Li (which has to diffuse through the particle to react at $R = R_p$) as well as the specific surface area for reaction (smaller particles lead to higher specific areas). Thus, increasing the particle radius will lead to a higher concentration overpotential and, therefore, a deterioration in performance. For the voltage drop during discharge (ΔV_c), ϵ_p has the greatest influence, followed by R_p and lastly SOC_{in}, as seen in Figure 2. The combination of an increased Ohmic drop and a higher concentration overpotential on the total polarization caused by a lower ϵ_p outweighs the effect of an increased concentration overpotential caused by lowering R_p.

For an EET, uniform distributions were selected for the three factors, which were again scaled to yield $\mathcal{X} = [0,1]^3$. A major difference between the variance-based method and the EET is the sampling strategy. The EET is highly efficient, requiring only $M(k+1)$ model evaluations *vs.* $N(k+2)$ to calculate the main effect indices; in the results above, $N(k+2) = 5000 \times (3+2) = 25000$, which is much higher than typical values of $M(k+1)$. The trends in the means of the μ_i with confidence intervals (CIs) are depicted in Figure 3 for an increasing number of model evaluations ($M(k+1)$). The CIs were established using bootstrapping [14], which consists of re-sampling the base points with replacement to produce P copies of the trajectories and for each of the P copies to use the EET to estimate μ_i and σ_i. This provides empirical distributions over μ_i and σ_i from which means and confidence bounds can be estimated.

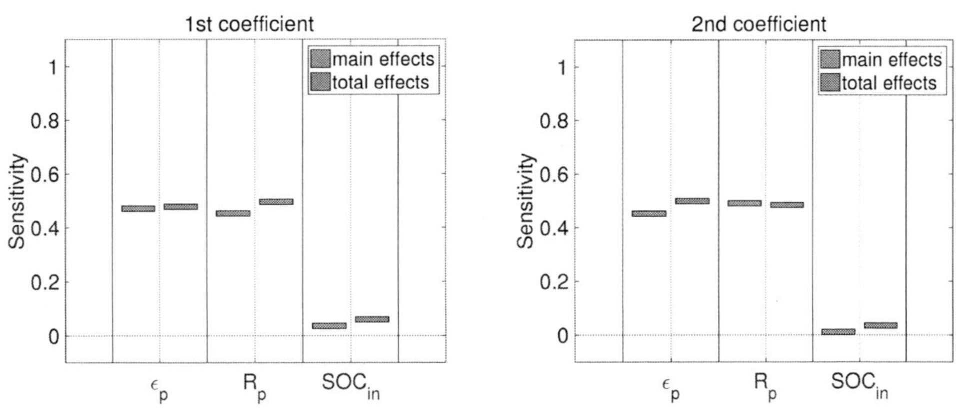

Figure 4. Main and total effects for the first two PCA coefficients (for the cell voltage curve).

The ranking of the inputs is the same as in the variance-based method. The means stabilize at around $M = 500$ (2000 model runs). There are small but noticeable fluctuations in the mean for ϵ_p around the value 0.25 even at much higher values of M but this behaviour is stable. The cause is the broad range of ϵ_p in comparison with the other factors and, therefore, the relatively small number of samples. Moving to a higher value of M the confidence intervals shrink, suggesting greater accuracy in the predictions. The ranking, however, is accurate even for very low numbers of M, which shows that the EET is more efficient than the variance based method. Although time cost is not an issue for the emulator, which provides extremely rapid predictions (on the order of a few seconds for 2000 predictions), in cases where a full simulator is used the much lower number of model runs for the EET represents an enormous advantage.

To investigate the *sensitivity of the charge-discharge curve*, we examine the main and total effects of the PCA coefficients $w_i(\boldsymbol{\xi})$ using the same Latin hypercube design and $N = 5000$ (we replace, e.g., η_E with $w_i(\boldsymbol{\xi})$, $i \in \{1, \ldots, r\}$). The results are depicted in Figure 4 for w_1 and w_2. Figure 5 shows an example (from the test set) of the contributions from the PCA eigenvectors ($w_i\mathbf{v}_i$, up to $i = 4$) towards the final (mean centred) voltage profile. The first two contributions can be seen to have by far the most influence. The sensitivities of the coefficients w_1 and w_2 are highest for ϵ_p and R_p, with roughly equal contributions from each, while higher-order coefficients (w_3 and w_4) were more heavily influenced by SOC_{in}.

6. Summary and conclusions

SA is often unfeasible with complex computer models. In such cases emulators, of varying degrees of sophistication, can be employed. Quantifying the uncertainty in the emulator predictions is desirable, but this is only achievable for certain approaches. For multivariate outputs (especially in high dimensional spaces), SA under uncertainty is especially challenging, even when the QoI is a scalar. In this paper we propose a GP emulator approach for performing SA under uncertainty when the model output is multivariate (possibly in a high-dimensional space). We present an example for a Li-ion battery, demonstrating that the method can be efficient and accurate. We are able to perform a probabilistic SA on scalar QoIs derived from the output (*via* a linear functional) and also on the output itself by focusing individually on each random principal coefficient. This can be achieved with either full MC sampling or by using semi-analytical expressions that are extensions of those derived by Oakley and O'Hagan [5] for scalar QoIs.

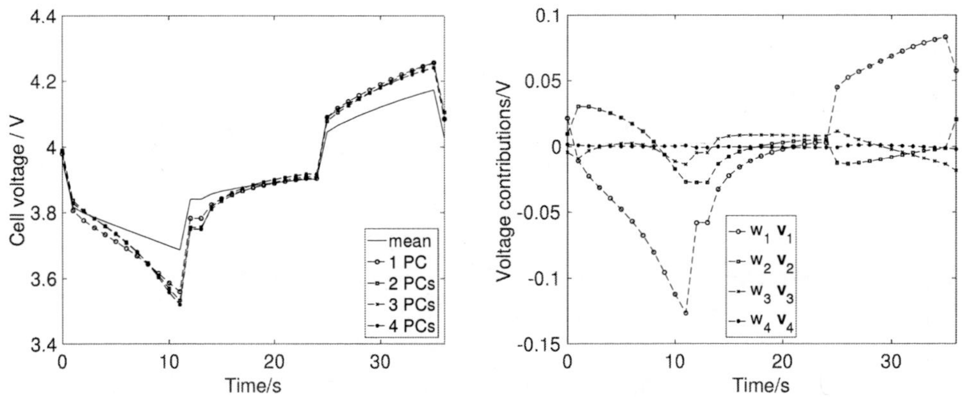

Figure 5. An example of the contributions from the PCA eigenvectors ($w_i\mathbf{v}_i$) towards the final (mean centred) voltage profile. In the left-hand figure, $w_i\mathbf{v}_i$ is successively added to the mean.

References

[1] A. Saltelli, M. Ratto, T. Andres, F. Campolongo, J. Cariboni, D. Gatelli, M. Saisana, and S. Tarantola, *Global sensitivity analysis: the primer.* John Wiley & Sons, 2008.

[2] D. Zhou, K. Zhang, A. Ravey, F. Gao, and A. Miraoui, Parameter sensitivity analysis for fractional-order modeling of lithium-ion batteries, *Energies*, 9, 123, 2016.

[3] M. D. Morris, Factorial sampling plans for preliminary computational experiments, *Technometrics*, 33, 161–174, 1991.

[4] I. M. Sobol, Sensitivity estimates for nonlinear mathematical models, *MMCE*, 1, 407–414, 1993.

[5] J. E. Oakley and A. O'Hagan, Probabilistic sensitivity analysis of complex models: a Bayesian approach, *R. Stat. Soc. Ser. B Stat. Methodol.*, 66, 751–769, 2004.

[6] F. Pianosi, K. Beven, J. Freer, J. W. Hall, J. Rougier, D. B. Stephenson, and T. Wagener, Sensitivity analysis of environmental models: A systematic review with practical workflow, *Environ. Model. Softw.*, 79, 214–232, 2016.

[7] A. Saltelli and I. M. Sobol', Sensitivity analysis for nonlinear mathematical models: numerical experience, *Matematicheskoe Modelirovanie*, 7, 16–28, 1995.
[8] I. Jolliffe, *Principal Component Analysis.* Springer Series in Statistics, Springer, 2002.
[9] D. Higdon, J. Gattiker, B. Williamsa, and M. Rightley, Computer model calibration using high-dimensional output, *J Am. Stat. Assoc.*, 103, 570–583, 2008.
[10] C. Rasmussen and C. Williams, *Gaussian Processes for Machine Learning.* MIT Press, Cambridge MA, USA, 2006.
[11] M. Doyle, J. Newman, A. S. Gozdz, C. N. Schmutz, and J.-M. Tarascon, Comparison of modeling predictions with experimental data from plastic lithium ion cells, *J. Electrochem. Soc.*, 143, 1890–1903, 1996.
[12] `https://www.comsol.com/model/1d-lithium-ion-battery-model-for-internal-resistance-and-voltage-loss-determinat-19131`, Last accessed 20 November 2017.
[13] F. Pianosi, F. Sarrazin, and T. Wagener, A matlab toolbox for global sensitivity analysis, *Environ. Model. Softw.*, 70, 80–85, 2015.
[14] B. Efron, Bootstrap methods: Another look at the jackknife, *Ann. Statist.*, 7, 1–26, 01 1979.

Appendix A. Efficient sampling for sensitivity analysis under uncertainty

We consider a scalar linear functional QoI $Q = F(\boldsymbol{\xi}) = (G \circ \eta)(\boldsymbol{\xi})$ as defined in section 2. For example, we may consider $G(y) = \int_{\mathcal{R}} \eta(\mathbf{x};\boldsymbol{\xi})w(\mathbf{x})d\mu(\mathbf{x})$, for some measure μ on a compact subset $\mathcal{R}$ of $\mathbb{R}^L$ representing space or time. To keep matters simple, we set $w(\mathbf{x}) \equiv 1/\mu(\mathcal{R})$, use Lebesgue measure and assume that $\eta(\mathbf{x};\boldsymbol{\xi})$ is continuous; then the Riemann and Lebesgue integrals coincide and approximations to $G(y)$ by Riemann sums or Gauss quadratures converge. In reality of course we have a discrete output $\mathbf{y}_r = (y_1, \ldots, y_d)^T = \boldsymbol{\eta}_r(\boldsymbol{\xi})$ that approximates $\eta(\mathbf{x};\boldsymbol{\xi})$ at, say, points $\{\mathbf{x}_l\}_{l=1}^d \subset \mathcal{R}$. Correspondingly, we have a discrete approximation $g(\mathbf{y})$ of $G(y)$ defined by a quadrature $g(\mathbf{y}_r) = \mu(\mathcal{R})^{-1}\sum_{j=1}^{d'} b_j y_{l_j}$, where $\{y_{l_j}\}_{j=1}^{d'} \subset \{y_l\}_{l=1}^d$ (approximating $\eta(\mathbf{x}_{l_j};\boldsymbol{\xi})$, $j = 1, \ldots, d'$) is a subset of the coefficients of $\mathbf{y}_r$ and b_j are quadrature weights. If we are using a Gauss quadrature, the points $\mathbf{x}_{l_j}$ are specified and must be included in the design $\{\mathbf{x}_l\}_{l=1}^d$. For ease of presentation, and without loss of generality, we use a mid-point Riemann sum, so that $f(\boldsymbol{\xi}) = g(\boldsymbol{\eta}_r(\boldsymbol{\xi})) = g(\mathbf{y}_r) = d^{-1}\sum_{l=1}^d y_l$.

Rather than a point estimate of $\mathbf{y}_r$ we have a distribution over functions (7), which leads to distributions over $q = f(\boldsymbol{\xi})$ and therefore over the sensitivity measures, as a consequence of the emulator uncertainty. The typical sensitivity measures employed are $S_i = V_i/V = \mathbb{V}\mathrm{ar}_{\xi_i}(\mathbb{E}_{\boldsymbol{\xi}_{\sim i}}[q|\xi_i])/\mathbb{V}\mathrm{ar}(q)$ and $S_{T_i} = 1 - \mathbb{V}\mathrm{ar}_{\boldsymbol{\xi}_{\sim i}}(\mathbb{E}_{\xi_i}[q|\boldsymbol{\xi}_{\sim i}])/\mathbb{V}\mathrm{ar}(q)$. Oakley and O'Hagan [5] also propose the main effects $f_i = \mathbb{E}_{\boldsymbol{\xi}_{\sim i}}[q|\xi_i] - f_0$ as useful graphical summaries of the influences of each variable. We derive approximate estimates of the means and variances of these various quantities, extending the analysis in [5] to multiple output problems. Recalling relationship (7), namely, $\mathbf{y}_r = \sum_{i=1}^r w_i(\boldsymbol{\xi})\mathbf{v}_i$, and denoting the l-th component of $\mathbf{v}_j$ by v_j^l, we obtain:

$$\begin{aligned}
\mathbb{E}_{\boldsymbol{\eta}_r}\left[\mathbb{E}_{\boldsymbol{\xi}_{\sim i}}[q|\xi_i]\right] &= \mathbb{E}_{\boldsymbol{\eta}_r}\left[\mathbb{E}_{\boldsymbol{\xi}_{\sim i}}\left[\frac{1}{d}\sum_{l=1}^d y_l\bigg|\xi_i\right]\right] = \frac{1}{d}\mathbb{E}_{\boldsymbol{\eta}_r}\left[\mathbb{E}_{\boldsymbol{\xi}_{\sim i}}\left[\sum_{l=1}^d\sum_{j=1}^r w_j(\boldsymbol{\xi})v_j^l\bigg|\xi_i\right]\right] \\
&= \frac{1}{d}\mathbb{E}_{\boldsymbol{\eta}_r}\left[\mathbb{E}_{\boldsymbol{\xi}_{\sim i}}\left[\sum_{j=1}^r b_j w_j(\boldsymbol{\xi})\bigg|\xi_i\right]\right] = \frac{1}{d}\mathbb{E}_{\boldsymbol{\xi}_{\sim i}}\left[\sum_{j=1}^r b_j\mathbb{E}_{\boldsymbol{\eta}_r}[w_j(\boldsymbol{\xi})]\bigg|\xi_i\right] \qquad \text{(A.1)} \\
&= \frac{1}{d}\mathbb{E}_{\boldsymbol{\xi}_{\sim i}}\left[\sum_{j=1}^r b_j m_j'(\boldsymbol{\xi})\bigg|\xi_i\right] = \frac{1}{d}\sum_{j=1}^r b_j\int_{[0,1]^{k-1}} m_j'(\boldsymbol{\xi})d\boldsymbol{\xi}_{\sim i} = \frac{1}{d}\sum_{j=1}^r b_j T_j(\xi_i)
\end{aligned}$$

where $b_j = \sum_{l=1}^d v_j^l$ and the functions $T_j(\xi_i)$ are defined by the integrals in the last line. Similarly:

$$\mathbb{E}_{\boldsymbol{\eta}_r}\left[\mathbb{E}[q]\right] = (1/d)\sum_{j=1}^r b_j\int_{[0,1]^k} m_j'(\boldsymbol{\xi})d\boldsymbol{\xi} = (1/d)\sum_{j=1}^r b_j T_j', \qquad \text{(A.2)}$$

where T_j' are now constants since the integration is over all input factors. Thus $\mathbb{E}_{\boldsymbol{\eta}_r}[f_i] = \mathbb{E}_{\boldsymbol{\eta}_r}[\mathbb{E}_{\boldsymbol{\xi}_{\sim i}}[q|\xi_i]] - \mathbb{E}_{\boldsymbol{\eta}_r}[\mathbb{E}[q]] = (1/d)\sum_{j=1}^{r} b_j(T_j(\xi_i) - T_j')$. The integrals defining T_j and T_j' can be approximated numerically at a very low computational cost.

We denote by $\boldsymbol{\xi}_{\sim\mathcal{I}}$ the vector of factors excluding those corresponding to the index set $\mathcal{I} \subset \{1,\ldots,k\}$ and denote by $\boldsymbol{\xi}_{\mathcal{I}}$ the subset of factors corresponding to $\mathcal{I}$. From the definition of covariance we have:

$$\begin{aligned}
&\mathbb{C}\text{ov}_{\boldsymbol{\eta}_r}\left(\mathbb{E}_{\boldsymbol{\xi}_{\sim\mathcal{I}}}[q|\boldsymbol{\xi}_{\mathcal{I}}], \mathbb{E}_{\boldsymbol{\xi}_{\sim\mathcal{J}}}[q|\boldsymbol{\xi}_{\mathcal{J}}]\right)\\
&= \mathbb{E}_{\boldsymbol{\eta}_r}\left[\mathbb{E}_{\boldsymbol{\xi}_{\sim\mathcal{I}}}[q|\boldsymbol{\xi}_{\mathcal{I}}]\,\mathbb{E}_{\boldsymbol{\xi}_{\sim\mathcal{J}}}[q|\boldsymbol{\xi}_{\mathcal{J}}]\right] - \mathbb{E}_{\boldsymbol{\eta}_r}\left[\mathbb{E}_{\boldsymbol{\xi}_{\sim i}}[q|\boldsymbol{\xi}_{\mathcal{I}}]\right]\,\mathbb{E}_{\boldsymbol{\eta}_r}\left[\mathbb{E}_{\boldsymbol{\xi}_{\sim\mathcal{J}}}[q|\boldsymbol{\xi}_{\mathcal{J}}]\right]\\
&= \int_{\boldsymbol{\xi}'_{\sim\mathcal{J}}}\int_{\boldsymbol{\xi}_{\sim\mathcal{I}}} \mathbb{E}_{\boldsymbol{\eta}_r}\left[f(\boldsymbol{\xi})f(\boldsymbol{\xi}')\right] d\boldsymbol{\xi}_{\sim\mathcal{I}}d\boldsymbol{\xi}'_{\sim\mathcal{J}} - \mathbb{E}_{\boldsymbol{\eta}_r}\left[\mathbb{E}_{\boldsymbol{\xi}_{\sim\mathcal{I}}}[q|\boldsymbol{\xi}_{\mathcal{I}}]\right]\,\mathbb{E}_{\boldsymbol{\eta}_r}\left[\mathbb{E}_{\boldsymbol{\xi}_{\sim\mathcal{J}}}[q|\boldsymbol{\xi}_{\mathcal{J}}]\right]\\
&= \int_{\boldsymbol{\xi}'_{\sim\mathcal{J}}}\int_{\boldsymbol{\xi}_{\sim\mathcal{I}}} \left\{\mathbb{C}\text{ov}_{\boldsymbol{\eta}_r}\left(f(\boldsymbol{\xi}), f(\boldsymbol{\xi}')\right) - \mathbb{E}_{\boldsymbol{\eta}_r}\left[f(\boldsymbol{\xi})\right]\mathbb{E}_{\boldsymbol{\eta}_r}\left[f(\boldsymbol{\xi}')\right]\right\} d\boldsymbol{\xi}_{\sim\mathcal{I}}d\boldsymbol{\xi}'_{\sim\mathcal{J}}\\
&\qquad\qquad - \mathbb{E}_{\boldsymbol{\eta}_r}\left[\mathbb{E}_{\boldsymbol{\xi}_{\sim\mathcal{I}}}[q|\boldsymbol{\xi}_{\mathcal{I}}]\right]\,\mathbb{E}_{\boldsymbol{\eta}_r}\left[\mathbb{E}_{\boldsymbol{\xi}_{\sim\mathcal{J}}}[q|\boldsymbol{\xi}_{\mathcal{J}}]\right] \qquad \text{(A.3)}\\
&= \int_{\boldsymbol{\xi}'_{\sim\mathcal{J}}}\int_{\boldsymbol{\xi}_{\sim\mathcal{I}}} \mathbb{C}\text{ov}_{\boldsymbol{\eta}_r}\left(f(\boldsymbol{\xi}), f(\boldsymbol{\xi}')\right) d\boldsymbol{\xi}_{\sim\mathcal{I}}d\boldsymbol{\xi}'_{\sim\mathcal{J}}\\
&= \int_{\boldsymbol{\xi}'_{\sim\mathcal{J}}}\int_{\boldsymbol{\xi}_{\sim\mathcal{I}}} \sum_{j=1}^{r}\sum_{p=1}^{r} b_j b_p \mathbb{C}\text{ov}_{\boldsymbol{\eta}_r}\left(w_j(\boldsymbol{\xi}), w_p(\boldsymbol{\xi}')\right) d\boldsymbol{\xi}_{\sim\mathcal{I}}d\boldsymbol{\xi}'_{\sim\mathcal{J}}\\
&= \frac{1}{d^2}\int_{\boldsymbol{\xi}'_{\sim\mathcal{J}}}\int_{\boldsymbol{\xi}_{\sim\mathcal{I}}} \sum_{j=1}^{r} b_j^2 c_j(\boldsymbol{\xi}, \boldsymbol{\xi}') d\boldsymbol{\xi}_{\sim\mathcal{I}}d\boldsymbol{\xi}'_{\sim\mathcal{J}} = U(\boldsymbol{\xi}_{\mathcal{I}}, \boldsymbol{\xi}_{\mathcal{J}})
\end{aligned}$$

in which the last step follows from the mutual independence of the w_i, and the posterior covariances $c_j(\boldsymbol{\xi}, \boldsymbol{\xi}')$ are given in Eqs. (6). Now, $\mathbb{V}\text{ar}_{\boldsymbol{\eta}_r}(f_i) = \mathbb{E}_{\boldsymbol{\eta}_r}[\mathbb{E}^2_{\boldsymbol{\xi}_{\sim i}}[q|\xi_i]] - 2\mathbb{E}_{\boldsymbol{\eta}_r}[\mathbb{E}_{\boldsymbol{\xi}_{\sim i}}[q|\xi_i]\mathbb{E}[q]] - \mathbb{E}_{\boldsymbol{\eta}_r}[\mathbb{E}^2[q]] - \mathbb{E}^2_{\boldsymbol{\eta}_r}[f_i]$ in which the last term on the right hand side is already known. The first to third terms are calculated by using the definition of covariance and Eq. (A.3) with $(\mathcal{I},\mathcal{J}) = (\{i\},\{i\})$, $(\{i\},\emptyset)$ and $(\emptyset,\emptyset)$, together with the previous expressions for $\mathbb{E}_{\boldsymbol{\eta}_r}\left[\mathbb{E}_{\boldsymbol{\xi}_{\sim i}}[q|\xi_i]\right]$ and $\mathbb{E}_{\boldsymbol{\eta}_r}[\mathbb{E}[q]]$. The integrals in Eq. (A.3) are again cheap to evaluate. The means of all the f_i, evaluated separately for selected values of ξ_i in $[0,1]$, can be combined on a single plot together with standard deviations to measure the influences of the factors [5] .

We next consider the partial variances V_i. We first note that $\mathbb{E}_{\boldsymbol{\eta}_r}[V_i] = \mathbb{E}_{\boldsymbol{\eta}_r}[\mathbb{V}\text{ar}_{\xi_i}(\mathbb{E}_{\boldsymbol{\xi}_{\sim i}}[q|\xi_i])] = \mathbb{E}_{\boldsymbol{\eta}_r}[\mathbb{E}_{\xi_i}[\mathbb{E}^2_{\boldsymbol{\xi}_{\sim i}}[q|\xi_i]] - \mathbb{E}^2_{\xi_i}[\mathbb{E}_{\boldsymbol{\xi}_{\sim i}}[q|\xi_i]]] = \mathbb{E}_{\boldsymbol{\eta}_r}[\mathbb{E}_{\xi_i}[\mathbb{E}^2_{\boldsymbol{\xi}_{\sim i}}[q|\xi_i]]] - \mathbb{E}_{\boldsymbol{\eta}_r}[\mathbb{E}^2[q]]$, the second term of which is already known. The first term is evaluated as follows:

$$\begin{aligned}
&\mathbb{E}_{\boldsymbol{\eta}_r}\left[\mathbb{E}_{\xi_i}[\mathbb{E}^2_{\boldsymbol{\xi}_{\sim i}}[q|\xi_i]]\right] = \mathbb{E}_{\boldsymbol{\eta}_r}\left[\int_{[0,1]}\left(\int_{[0,1]^{k-1}} f(\boldsymbol{\xi}^*)d\boldsymbol{\xi}^*_{\sim i}\int_{[0,1]^{k-1}} f(\boldsymbol{\xi})d\boldsymbol{\xi}_{\sim i}\right)d\xi_i\right]\\
&= \int_{[0,1]}\int_{[0,1]^{k-1}}\int_{[0,1]^{k-1}} \mathbb{E}_{\boldsymbol{\eta}_r}\left[f(\boldsymbol{\xi})f(\boldsymbol{\xi}^*)\right] d\boldsymbol{\xi}^*_{\sim i}d\boldsymbol{\xi}_{\sim i}d\xi_i\\
&= \int_{[0,1]}\int_{[0,1]^{k-1}}\int_{[0,1]^{k-1}} \left\{\mathbb{C}\text{ov}_{\boldsymbol{\eta}_r}\left(f(\boldsymbol{\xi}), f(\boldsymbol{\xi}^*)\right) - \mathbb{E}_{\boldsymbol{\eta}_r}[f(\boldsymbol{\xi})]\mathbb{E}_{\boldsymbol{\eta}_r}[f(\boldsymbol{\xi}^*)]\right\} d\boldsymbol{\xi}^*_{\sim i}d\boldsymbol{\xi}_{\sim i}d\xi_i \qquad \text{(A.4)}\\
&= \frac{1}{d^2}\sum_{j=1}^{r}\int_{[0,1]}\int_{[0,1]^{k-1}}\int_{[0,1]^{k-1}} b_j\left\{b_j c_j(\boldsymbol{\xi}, \boldsymbol{\xi}^*) - \sum_{p=1}^{r} b_p w_j(\boldsymbol{\xi}) w_p(\boldsymbol{\xi}^*)\right\} d\boldsymbol{\xi}^*_{\sim i}d\boldsymbol{\xi}_{\sim i}d\xi_i
\end{aligned}$$

in which $\boldsymbol{\xi}^* = (\xi_1^*,\ldots,\xi_i,\ldots,\xi_k^*)^T$. These integrals are readily and cheaply approximated numerically. A first order Taylor expansion yields $\mathbb{E}_{\boldsymbol{\eta}_r}[S_i] = \mathbb{E}_{\boldsymbol{\eta}_r}[V_i/V] = \mathbb{E}_{\boldsymbol{\eta}_r}[V_i]/\mathbb{E}_{\boldsymbol{\eta}_r}[V]$, from which the main effect indices can be approximated.

Robust portfolio selection based on Gaussian rank correlation estimator

Mei Yan* , Xin Zhao

School of Mathematical Science, Univresity of Electronic Science and Technology of China, Sichuan, China

Email: myanlxy920@hotmail.com

Abstract. Estimating the covariance matrix of stock returns is a fundamental question in empirical finance with implications for portfolio selection. In the presence of outliers, classical estimates of covariance and correlation matrices are not reliable. For this reason, many robust estimator methods are given. However, high dimensional data poses great challenges to the traditional estimation, the merit of robust estimator cannot retain and improved robust estimator is required. In this paper, we proposed a robust pairwise covarinace matrix with Gaussian rank correlation, which is acting an input to Markowitz portfolio model. There are two main advantages of the proposed way. The first is that the final covariance estimator is positive semidefinite, we need not to transform the estimator. The second is that the covariance estimator can be computed fast because of the pairwise covariance and the Gaussian rank crrelation estimation. Moreover, our numerical results on empirical data show that the proposed estimation is much better than the traditional method. Through empirical studies, we also found that estimation error in sample mean should be considerd in Mean-variance portfolios.

1. Introduction

Markowitz's mean-variance analysis sets the basis for moder portfolio optimization theory. In the Markowitz portfolio selection model, the "return" on a portfolio is measured by the expected value of the random portfolio return, and the associated "risk" is quantified by the variance of the portfolio return[1]. This model showed that, given either an upper bound on the risk that the investor is willing to take or a lower bound on the return the investor is willing to accept, the optimal portfolio can be obtained by solving a convex quadratic problem. To implement these portfolios in practice, one has to estimate the mean and the covariance matrix of asset returns from historical data.

However, the classical sample covariance estimator is highly sensitive to outlying observation. For the aberrant observation, robust estimators of covariance matrices are desired. Robust estimation of covariance matrices have received much attention by statistician in the past. Classical robust covariance matrix estimators include M-estimators[2,3], minimum volume ellipsoid(MVE)[4] and minimum covariance determinant(MCD) estimators[5-7], S-estimators[8], and the Stahel-Donoho estimate (SDE) based on projection [9]. These estimators are almost specifically designed for data with very low dimensions and large sample sizes. Simultaneously, to overcome the sensitivity to outliers, robust correlation estimator is investigated widely by researchers. Such as the Spearman correlation[10], Kendall correlation[11], quadrant correlation[12], Gaussian rank correlation (GRCor)[13] and so on. Compared with the Pearson correlation, these proposed correlation have a lot of advantages. It is worth mentioning that correlation matrices constructed from the Spearman or Kendall correlation or GRCor are robust and computing very fast. Nevertheless, to obtain consistently, the first two need to transformate and are not guaranteed to be positive semidefinite anymore[14].

In recent years, both economic and financial applications have encountered very large data sets which high-dimensional variables[15-17]. For example, the World Bank has data for about two hundred countries over forty years; in portfolio allocation, the number of stocks can be in thousands and be large or of the same order of the sample size[18]. Analysis of high-dimensional data pose many challenges, for instance covariance estimation will face with burdensome computaion. Clearly, all of these estimates mentioned above are facing a serious computational challenge. Realizing the limitation of the classical robust estimators in high-dimensional setting, some faster estimators with high breakdown points are researched and given in the past years[19,20].

Intuitively, it is simply to calculate a robust location estimate to each individual variable for multivariate location . In the case of multivariate scatter, we can similarly apply a robust covariance or correlation estimate to each pair of variables. Estimates of this way are called "coordinatewise" and "pairwise"[21-23].The simplest pairwise methods are based on pairwise robust correlation or covariance estimates, for example the Spearman's ρ, kendall's τ, the quadrant correlation and so on.

Unfortunately, those pairwise matrix estimates are not guaranteed to be positive definite. Recently, [24]proposed a new way to obtain positive-definite estimator based on pairwise robust method. In this article we will use the similar technique to solve the trouble of portfolio selection.

The remainder of this paper is structured as follows. Section 2 provides a brief review of the mean-variance portfolio selection model. The pairwise covariance matrix estimation and Gaussin rank correlationestimators are shown in section 3. In section 4 the results of simulation are given. In section 5, we compare the proposed estimator to the Classical and FMCD estimators under empirical asset returns. Finally, a brief discussion is given to conclude this work.

2. The mean-variance portfolio selection model

The objective of seminal work by Markowitz[1] is to seek a optimal portfolio that ensure the minimal variance with a given expected return. In this section, we review the man-variance optimization problem. Given p stocks whose returns are distributed with the expected returns vector μ and covariance matrix Σ. We consider the mean-variance model defined as follows:

$$\begin{aligned} \min_{\omega} \quad & \omega'\Sigma\omega \\ st. \quad & \omega' e = 1, \\ & \omega'\mu = q, \end{aligned} \tag{1}$$

where, q is the expected rate of return which is required on the portfolio, $1=(1,1,1,\ldots,1)'$ denotes a conformable vector of ones.

Solving Eq.(1) , the well-known solution is :

$$\omega = \frac{C-qN}{MC-B^2}\hat{\Sigma}^{-1}1+\frac{qM-N}{MC-N^2}\hat{\Sigma}^{-1}\hat{\mu}, \tag{2}$$

where $\hat{\Sigma}$ is the estimated covariance matrix of return, $\hat{\mu}$ is the estimated of expection vector. $M=1'\hat{\Sigma}^{-1}1$, $N=1'\hat{\Sigma}^{-1}\hat{\mu}$ and $C=\mu'\hat{\Sigma}^{-1}\hat{\mu}$. From the Eq.(2) , it is clearly that the ideal portfolio weights depend on the inverse of the estimated covariance matrix and vector of expected returns, which are estimated from historical data available in the empirical implementation. In other words, if the covariance matrix estimator is not invertible or if it is numerically ill-conditioned, the model did not work at all. For this reason, the resulting estimator matrix $\hat{\Sigma}$ requires be positive semidefinite.

3. Robust covariance matrix estimation

Gnanadesikan and Kettenring[25] use a simple method for turning scale estimators into covariance estimators to obtain a robust, pairwise covariance estimate. A modified version of the Gnanadesikan-Kettenring robust covariance estimate was introduced by Ma and Zamar[20] , which is to prominence in the context of robust estimation with shorter computer times. The idea is based on the identity,

$$cov(X,Y)=(4\alpha\beta)^{-1}[Var(\alpha X+\beta Y)-Var(\alpha X-\beta Y)], \tag{3}$$

where X and Y are random variable. In general, X and Y may have different scales, hence it is standard to let $\alpha=(var(X))^{-1/2}$ and $\beta=(var(Y))^{-1/2}$. A robust convariance estimator is found by replacing the variance in Eq. (3) with (squared) robust scale estimator. With identity Eq. (3), a symmetric matrix full of pairwise covariances can be constructed simply, however the changed estimator is no necessarily positive semidefinite.

Following the similar spirit of Ma and Zamar[20], Öllerer and Croux[24] studied a modification that using the Gaussian rank correlation to estimate the pairwise covariance matrix. This approach has two virtue. On the one hand, it compute fast and easily. On the other hand, the resulting covariance matrix is positive definiteness, thus we do not need to require a transformation to ensure the result be positive semidefinite. To achieve this goal, tow elements-correlation and scale matrix estimsator-are considered in the following.

It is well known that the classical Pearson correlation estimator is sensitive to the outliers, even only a small amount of atypical observations may seriously affect it. In view of the outliers, many proposals for the robust estimation correlation of variables were given. For example, the Spearman and Kendall correlation estimators[10,11] which are based on ranks; the quadrant correlation estimator based on winsorizating data. All of those estimators are robust and efficient to compute, even in high dimensions. Unfortunately, the resulting correlation matrix is no longer guaranteed to be positive semidefinite. To avoid the problem, Boudt et al. [13] proposed the Gaussian rank correlation $GR(\cdot)$, which is defined as the sample correlation estimator from the Van Der Waerden scores (or normal scores):

$$GRcor = \sum_{i=1}^{n}\Phi^{-1}(\frac{R(x_i)}{n+1})\sum_{i=1}^{n}\Phi^{-1}(\frac{R(y_i)}{n+1})(\sum_{i=1}^{n}\Phi^{-1}(\frac{R(x_i)}{n+1})^2\sum_{i=1}^{n}\Phi^{-1}(\frac{R(y_i)}{n+1})^2)^{-\frac{1}{2}} \tag{4}$$

where $R(x_i)$ and $R(y_i)$ denote the rank of x_i and y_i respectively, for $1 \le i \le n$. Specifically, the corresponding Gaussian (or normal) scores are obtained by plugging these ranks in the quantile function Φ^{-1} of the standard normal distribution.

Consequently, we use the $GR(\cdot)$ as a valuable robust correlation estimatior to estimate the pairwise covariance matrix $(s_{jk}) \in \Re^{p\times p}$. Concretely, we use the following equation:

$$\sigma_{jk} = scale(X^j)scale(X^k)r((X^j),(X^k)),\ j,k = 1,...,p \tag{5}$$

from the data $X = (X^1,...,X^p) \in \Re^{n\times p}$, where scale $(\cdot)$ is replaced by a robust scale estimation, such as the Q_n estimator or the median absolute deviation. Both the two mentioned scale estimators are equally robust with a breakdown point of 50%. To simplify, we opt for the median absolute deviation as scale estimate.

It is worth mentioning the pairwise covariance estimator matrix $\hat{\Sigma}$ using Gaussian rank correlation is positive semidefinite. We study estimating a robust and positive semidefinite covariance matrix are performed by one step. Therefore, we do not need to apply Nearest positive definite matrix procedure(NPD) or Orthogonalised Gnanadesikan Kettenriing procedure(OGK) to obtain a positive semidefinite covariance estimate. In the other words, the proposed method not only saves computation time, but also simplifies the final precision matrix estimator.

4. Numerical simulation

4.1. The simulated data set

In this section we run a simulation to comparing the FMCD and the pairwise robust covariance (PRC) estimates. To evaluate these statistical performance, we need the situations in which the "true values" are known. Considering the heavy-tailed of the returns, the simulated data generated by a multivariate Student-T distribution. Concretely, the simulated data following a distribution is a multivariate Student-T distribution with the degrees freedom 3, and we let Σ be the covariance matrix process, where $\Sigma = a^{|i-j|}$ characterizes the entry of in the row i and column j , a =0.7 in this simulatioin. For each sample i, the covariance is estimated by $\hat{\theta}_i$, and the Mean Squared error(MSE) is computed as $MSE = {}^{1}\!/_{n}\sum_{i=1}^{n}(\hat{\theta}_i - \theta_i)^2$.

4.2. Discussion of the simulation

In our experiments, to study the effect of dimension on the performance of various estimators, we consided different cases n=200 ,600,1000 with p taking the values 20,60, 100, 150,300 respectively. The MSE results are reported for different estimation methods in Table 1. In small p case, for p=20 and p=60, we find that the classical sample covariance(CSM) method substantially outperforms the PRC method, and markedly surpass the FMCD method. That means in relatively low dimension, the CSM method presents a promising alternative, the PRC is the second one. It can be found that the PRC estimator performs significantly better than the classical method when the value of p is large, see Table 1, for p=300, p=600. That means the PRC estimator is more robust than the CSM method in high dimension case. Note that the number of stocks are often large in practice, the high dimension ofen happen with the Age of Big Data. Focussing on the robust approach FMCD and PRC, we observed that FMCD performbetter than the PRC in the case of large p and small n. However, the procedure of FMCD is required huge values of subsample, this method cannot run when n is not large enough (see Table 1). Obviously, when n is not large we cannot get the estimator of FMCD. Moreover, from the simulation, we also find that the FMCD estimator is not ensure be positive definite in higher dimensions.

We also compared the running time of the two robust estimators in this simulation. Table 2 gives the simulation time in seconds. When n =200, we let p=20,40, 200 and run separately. The time of FMCD are 11 and 25 times for the time of PRC. The proposed method PRC is noticeably faster than the FMCD.

Table 1. Simulation results

	Estimate	p=20	p=60	p=100	p=150	p=300	p=600
n=200	FMCD	0.024	0.029	**0.018**	---	---	---
	PRC	0.012	0.027	0.039	**0.052**	**0.086**	**0.156**
	Classical	0.010	0.020	0.037	0.054	0.107	0.214
n=600	FMCD	0.007	0.018	0.014	---	---	---
	PRC	0.003	0.008	0.011	0.014	0.022	**0.036**
	Classical	0.002	0.004	0.006	0.010	0.020	0.041
n=1000	FMCD	0.004	0.020	0.025	---	---	---
	PRC	0.002	0.004	0.006	0.008	0.012	**0.019**
	Classical	0.001	0.002	0.003	0.005	0.009	**0.019**

Simulated MSE(the Mean Squared error) ,---represent simulation result NAN .

Table 2. Times for simulation data

Estimate	p=20	p=40	p=100
FMCD	14.5	25.6	109
PRC	0.13	0.7	4.4

Simulated time with n=200.

5. Real data

5.1. The real data set

In this section we use the real data to show the superior performance of the proposed robust portfolio selection approach in this paper. This datasets containing 20 assets were extracted from Yahoo Web-Finace-Stocks dataly database composing the S&P500 index. The data span from the sixth of Decmeber 2013 to the sixth of December 2016. Deleting the missing data in the sample, all same trading days are 163. This is that n=163, p=20 in the following.

5.2. Result

In order to detect whether the returns are norm distribution or not, we computed the Skewness and Kurtosis of each stock. For simplicity, we denoted the twenty stocks Si(i=1,2,...,20) respectively. To simplify discussion, we selected the first 10 to study their distributions. It is clearly from Table 1 that each stock Si's (i=1,2,...,10) Skewness and Kurtosis have different extent compared with standard normal distribution. Moreover, the Skewness and Kurtosis are differ considerably(see Table 3).

Another method to identify the difference between the return's distribution and the norm ones is frequency chart. Figure1 displays frequency chart of S6. Finally ,we also presented boxplots of some stock return (see figure 2). Similar to Table3, Figure 1 provide that the distributions of returns of risky assets have fat tail nature and are noticeably leptokurtic. For the mean- variance model, the properties of normal distribution are important hypothesis. In the tradeoff between income and risk, the variance and the mean of the securities return are best statistics if the stock return data is normally distributed. Many empirical researches have shown that if the return of risky assets no longer accord with normal distribution, variance is not the best risk. Furthermore, the influence of outliers are shown in Figure 2. We selected six stocks from the twenty stocks stochastically, then given the boxplots of the six stocks. Apparently, the outliers have impact on each stock more or less in Figure 2.

To check the estimation error from sample mean estimator, we came up with the adjust pairwise robust covariance way, which is replace the sample mean with robust mean estimator from the Fast-MCD(FMCD) and used the robust covariance estimator from the pairwise covariance estimator based on Gaussian rank correlation. Next, we used the classical, FMCD, pairwise robust covariance (PRC) and adjust pairwise robust covariance (APRC) estimator as input into portfolio selection model (1).

Figure 4 plots the Mean-Variance Efficient Frontier of the four estimators, the horizontal axis and vertical axis represents the standard deviation and expected return of portfolio selection separately. It is clearly that the FMCD, the PRC, the APRC are better than the classical estimator. With the same level of expected, the risk with robust estimation way is lower than the traditional way. Overall, both the adjust pairwise robust covariance way and the FMCD provide competitive performance relative to the pairwise robust covariance way. However, the FMCD has three drawbacks can not be neglect: (i) the FMCD requires substantial running time for large p, it means that if we need compute a few stocks, the FMCD is ineffective; (ii) it no longer retains its robust nature better anymore when n is large, it means that no high breakdown point; (iii) the result covariance estimator cannot ensure positive semidefinite when n is large, which is important in this paper. Note that the APRC estimation is better than the PRC, it verifies that sample mean estimator in mean-variance portfolios should be not overlooked. With the same robust covariance estimator, the robust mean estimator show a better quality than sample mean.

To compare the computing times of different estimates, the running times in seconds is given in Table 4. Clearly, pairwise robust covariance compute fast than the others.

Table 3. Skewness and kurtosis of the stock return

Name	Skewness	Kurtosis	Name	Skewness	Kurtosis
S1	-0.44	4.33	S6	-0.45	3.99
S2	0.34	4.13	S7	0.33	3.35
S3	0.29	5.45	S8	0.95	4.69
S4	0.05	4.58	S9	0.17	3.36
S5	0.22	8.50	S10	0.08	4.31

Table 4. Time(seconds) for real data

Estimate	Times
FMCD	11.7
PRC	0.6
APRC	10.2

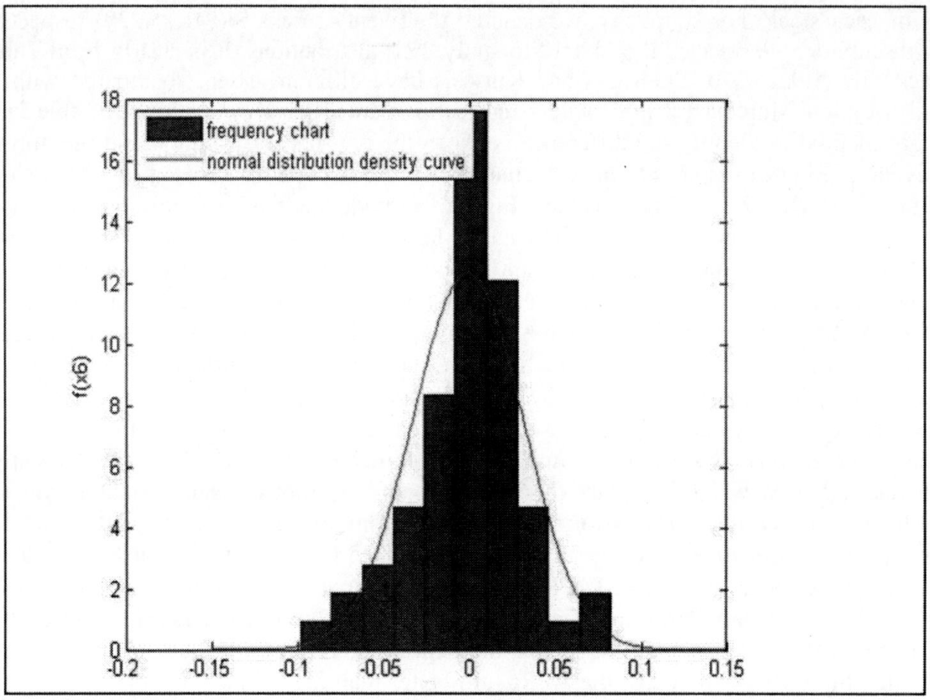

Figure 1. Frequency chart of stock S6. Performance comparison normal distribution density cure.

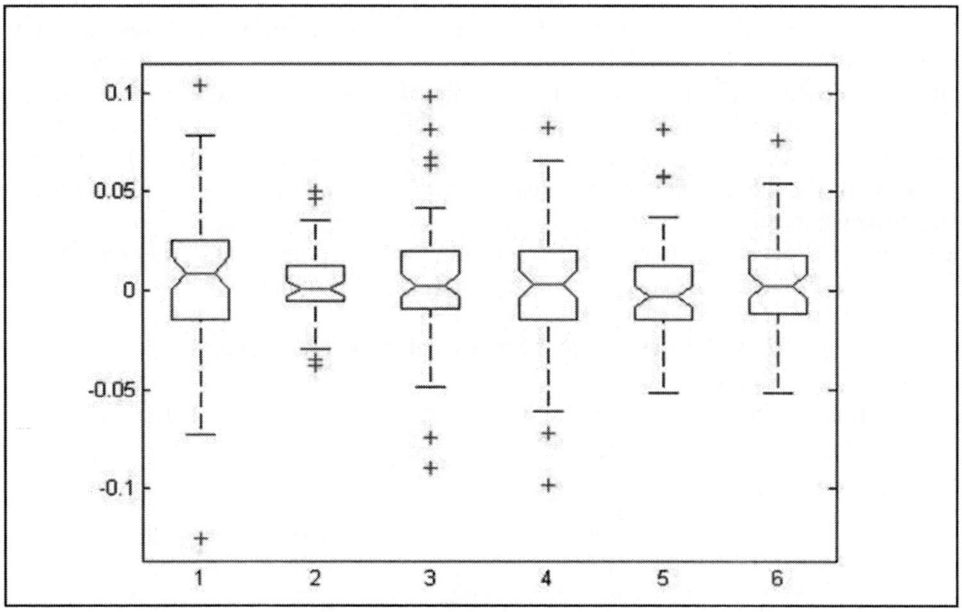

Figure 2. The boxplots of the stocks for S2, S4, S6, S7 S9, S10.

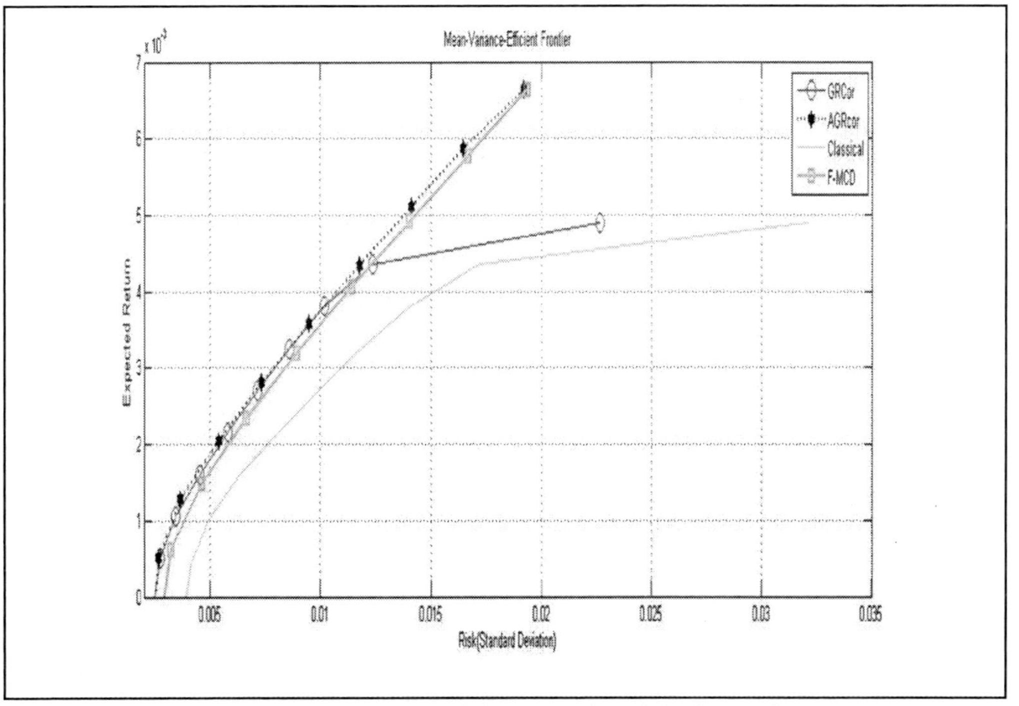

Figure 3. The Efficient-Frontier of the four estimators.

6. Conclusion

Markowitz portfolio selectioin model has had a profound impact on the economic modeling of financial markets and the pricing of assets. It makes the optimal portfolio can be obtained by solving a convex quadratic programming problem. Consequently, we can study the optimal portfolio quantitatively and qualitatively. The curse of dimensionality and the influence of perturbations in practice make the classical estimation perform poorly. Various aspects of those phenomenon have been disussed extensively, generally can be divided into re-established model, renewed the estimation method.

In this paper we developed a robust estimation into a large dimensional. Motivated by the well-known solution Eq.(2), we consider a pairwise approach to covariance estimation, which is based on Gaussian rank correlation estimator. The proposed way have main advantage focus on the following:

- The estimator is robust. As we all know, the stock returns is non-normal distribution, and is fat-tail. The returns are sensitive to outliers, robust estimator can ease the disturbance from outliers.
- The final covariance estimator is straightly positive semidefinite without transforming. From Eq.(2), the solution rely on the inverse of covariance matrix. The estimator we studied above-pairwise covariance and Gaussian rank correlation-are all positive semidefinite, therefore it is invertible. Furthermore, we need not special way to change the final estimator positively, which improves the efficiency in some degree.
- The estimation can be computed much faster than others. On account of the pairwise method and the rank correlation, the complexity of calculation is more simply, especially in the high dimension.

We compared the performance of the pairwise robust approach with the FMCD and classicial non-robust method for the covariance estimation of stock returns. Overall , the PRA method performs better than the others from the simulation and real data set. The numerical results also show that the improved PRA estimation perform well in the mean-variance model. Additional , we verified that errors of mean sample estimation also manipulate the result of the portfolio selection in our method.

References

[1] Markowitz H 1952 Portfolio selection *The journal of finance* vol 7 pp 77-91
[2] Cressie N and Hawkins M 1980 Robust estimation of the variogram: I *J. Int. Assoc. Math. Geol.* vol 7 pp 115-125
[3] Hubert M, Rousseeuw P J and Van Aelst S 2008 High-breakdown robust multivariate methods *Statist. Sci.* pp 92-119
[4] Hubert M and Debruyne M 2010 Minimum covariance determinant *Wiley Interdiscip. Rev. Comput. Stat.* vol 2 pp 36-43
[5] Rousseeuw P J and Driessen K V 1999 A fast algorithm for the minimum covariance determinant estimator *Technometrics* vol 41 pp 212-223
[6] Hardin J and Rocke D M 2004 Outlier detection in the multiple cluster setting using the minimum covariance determinant estimator *Comput. Statist. Data Anal.* vol *44* pp 625-638
[7] Maronna R A and Yohai V J 2017 Robust and efficient estimation of multivariate scatter and location *Comput. Statist. Data Anal.* vol *109* pp 64-75
[8] Rocke D M 1996 Robustness properties of S-estimators of multivariate location and shape in high dimension *Ann. Statist.* pp 1327-1345
[9] Zuo Y, Cui H and He X 2004 On the Stahel-Donoho estimator and depth-weighted means of multivariate data *Ann. Statist.* pp 167-188
[10] Zar J H 1972 Significance testing of the Spearman rank correlation coefficient *J. Amer. Statist. Assoc.* vol *67* pp 578-580
[11] Kendall M G 1938 A new measure of rank correlation *Biometrika* vol *30* pp 81-93
[12] Devlin S J, Gnanadesikan R and Kettenring J R 1975 Robust estimation and outlier detection with correlation coefficients *Biometrika* vol *62* pp 531-545
[13] Boudt K, Cornelissen J and Croux C 2012 The Gaussian rank correlation estimator: robustness properties *Stat. Comput.* vol *22* pp 471-483
[14] Croux C and Dehon C 2010 Influence functions of the Spearman and Kendall correlation measures *Stat. Methods Appl.* vol *19* pp 497-515
[15] Oh D H and Patton A J 2017 Modeling dependence in high dimensions with factor copulas J. *Bus. Econom. Statist.* vol 35 pp 139-154
[16] Ledoit O and Wolf M 2004 A well-conditioned estimator for large-dimensional covariance matrices *J. Multivariate Anal.* vol *88* pp 365-411
[17] Dissmann J, Brechmann E C, Czado C and Kurowicka D 2013 Selecting and estimating regular vine copulae and application to financial returns *Comput. Statist. Data Anal.* vol *59* pp 52-69
[18] Fan J, Liao Y and Mincheva M 2011 High dimensional covariance matrix estimation in approximate factor models *Ann. Statist.* vol *39* pp 3320
[19] Fritz H, García-Escudero L A and Mayo-Iscar A 2013 A fast algorithm for robust constrained clustering *Comput. Statist. Data Anal.* vol *61* pp 124-136
[20] Maronna R A and Zamar R H 2002 Robust estimates of location and dispersion for high-dimensional datasets *Technometrics* vol *44* pp 307-317
[21] Maronna R 2005 Principal components and orthogonal regression based on robust scales *Technometrics* vol *47* pp 264-273
[22] Maronna R A and Yohai V J 2017 Robust and efficient estimation of multivariate scatter and location *Comput. Statist. Data Anal* vol *109* pp 64-75
[23] Tarr G, Müller S and Weber N C 2016 Robust estimation of precision matrices under cellwise contamination *Comput. Statist. Data Anal.* vol *93* pp 404-420
[24] Öllerer V and Croux C 2015 Robust high-dimensional precision matrix estimation (*Springer International Publishing*) pp 325-350
[25] Gnanadesikan R and Kettenring J R 1972 Robust estimates, residuals, and outlier detection with multiresponse data *Biometrics* pp 81-124

Acknowledgments
This work was supported by a grant from the National Natural Science Foundation of China （No. G0511701069 ） and the Fundamental Research Funds for Central Universities. (No. ZYGX2015J100)

IOP Conf. Series: Journal of Physics: Conf. Series **1039** (2018) 012022 doi:10.1088/1742-6596/1039/1/012022

Agent-based models in financial market studies

L Wang[1], K Ahn[2], C Kim[3] and C Ha[4]

[1] Department of Economics, University of Pittsburgh, 4901 Wesley W. Posvar Hall, 230 South Bouquet Street, Pittsburgh, PA 15260, U.S.

[2] Graduate School of Future Strategy, Korea Advanced Institute of Science and Technology, 291 Daehak-ro, Yuseong-gu, Daejeon 34141, Republic of Korea

[3] Computational Economics Laboratory, Computational Science Center, Korean Institute of Science and Technology, Hwarangro-14 gil, Seongbuk-gu, Seoul 02792, Republic of Korea.

[4] HSBC Business School, Peking University, University Town, Nanshan District, Shenzhen 518055, P.R. China

E-mail: tree@econ.kist.re.kr, cyha@phbs.pku.edu.cn

Abstract. In this manuscript, we summarize prior research on the agent-based modeling of financial markets. While extensive research related to agent-based modeling has been done in various economic disciplines, we focus mainly on the evolution of the models and their applications to financial markets. A large number of studies have adopted agent-based modeling methodologies to explain various empirical findings in financial markets. Our summary shows the benefits of using such modeling to account for various financial market phenomena. We confirm that small changes in initial parameter values can lead to relatively large fluctuations through the financial markets that can be viewed as complex or chaotic systems. This also means that financial markets become volatile due to small unexpected changes in the parameters of the models that describe the market.

1. Introduction

Serving probably the best example of chaotic dynamics, the global financial system is among the most complex systems ever created by humanity. It consists of multiple interacting autonomous agents (elements) and emergent rules of engagement: the agents normally have heterogeneous goals, interaction rules, and decision-making strategies. The interactions among different agents generate global system dynamics toward an equilibrium or disequilibrium in the prices of various assets.

The majority of analytical tools used to explore financial markets apply general equilibrium theory in the representative agent framework. Within this framework, each agent is assumed to optimize his/her objective function with full rationality and knowledge. A major underlying tenet of this modeling philosophy is that studying every single element is sufficient to understand the system as a whole.[5] While this attempt is able to derive analytical solutions for subsequent analysis, it frequently fails to account successfully for stylized facts, especially in financial markets. This is largely because various models in this framework heavily rely on many unrealistic assumptions, such as market clearing, market convergence to equilibrium prices, perfect information, and rationality, while ignoring the emergent characteristics of agents' interactions and their diverse strategies. Failure to take into account heterogeneity and interaction among agents in the market suggests that this field still lacks a sound micro-level foundation, as well as the unimaginable results of their interactions.

Since it was first developed in the late 1940s, the bottom-up approach known as agent-based modeling (ABM)[6] has provided a more convincing theoretical framework than the prior top-down approaches used in general equilibrium analysis. In an agent-based model (ABM), agents' behaviors are both autonomous and heterogeneous: each agent is able to follow his/her own rules, and interacts with all others autonomously within a designated space that provides the unique information required to analyze the system.[7]

Due to the incorporation of these properties, as observed in real financial markets, ABMs successfully link the micro-level rules of investors' behavior with the macrobehavior of asset prices in real markets [1], [2].

ABM provides many benefits in exploring financial market dynamics.[8] First, ABM shows that agents' behavior is artificially intelligent, thereby avoiding the inaccuracy of mathematical equations derived under the general equilibrium theory. Second, the use of ABM makes it easy to manipulate the behavior of an agent as well as environmental parameters. Consequently, one basic model can spin off variant models through this manipulation. Third, observers can not only capture emergent systemic traits and phenomena but also identify the threshold of the model in the ABM setting, because the interaction of agents inevitably generates a unique and chaotic system as time passes.

Such benefits of simplicity, flexibility, and emergence help ABM explore new venues of economic research, especially in dynamic markets with asymmetric information, learning, and uncertainty. If combined, however, they pose particularly daunting technical challenges in terms of computational feasibility and practicality. Remarkable advances in computing technology and accumulation of the real-market data since the

5 Reductionism is the basis of modern scientific analysis; it implies that reducing one element to another can help us better understand the behavior of a whole system.

6 We use 'ABM' to refer to agent-based modeling or an agent-based model, depending on the context.

7 ABMs recognize each agent as an autonomous interacting unit in the system. As such, a clear definition of the designated space for each individual is essential for the workings of the model because it provides the information about the agent's location, neighborhood of mobility, and breadth of interaction with other agents. As demonstrated by the El Farol problem for the stock market, the location of an agent when its action is implemented provides pivotal information that can be used to select other agents with whom it will interact.

8 See, among many others, Schelling, who investigates relations among micromotives and macrobehavior [3], and El Farol's bar model [4], [5].

Agent-based models in financial market studies

1990s has increasingly facilitated the design, computation, and verification of ABMs that aim to assess the effects of the actions and interactions of autonomous agents on the system as a whole.

There exist challenges as well. ABM still remains complicated in most cases, making it difficult to find significant parameters governing agents' unique behaviors. Furthermore, the inherent complexity of this method might lead to a lack of analytical solutions. ABM also brings about new and untested algorithms, relatively weak fundamental assumptions, and likely omissions of important parameters. Therefore, careful design of an ABM is critical for overcoming potential weaknesses.

In this article, we summarize the papers that apply ABMs to financial markets in various settings, and account for market construction, agents' action rules and their dynamic evolution derived from the respective models.

2. Agent-based stock market

The theory of Rational Expectations (RE), initially proposed by Muth (Econometrica, 1961) and later popularized by Lucas, is one of the most influential theoretical frameworks used to solve most macroeconomic problems. However, not only does it unrealistically postulate that every agent behaves in the same optimal way, but its fundamental assumptions, such as complete information, perfect rationality, and common expectations, are too strong in most cases. It is, in large part, due to the complexity of the reality, say, in the stock market, that the theory tackles to explain. A more realistic analytical setting with bounded rationality, heterogeneity, and adaptiveness at the agent level can be provided only after relaxing the underlying assumptions of the RE and designing a platform for a number of single interacting agents, as is done in Schelling's works [3], [6], [7].

Palmer et al. [8] provide an ABM of the stock market where independent adaptive agents buy and sell shares of stocks. Those agents select or generate investment rules after observing the outcomes of the mental models adopted in each previous time step. Their model is designed to have each agent endowed with a set of condition-action rules that consists of three parts: a condition under which the agent actions are triggered; the action of either buying or selling; and the strength of the optimizing action to increase an agent's wealth. Each agent can adapt to a new environment, such as a changed price or dividend stream, by choosing a share of each stock. Genetic algorithm (GA) infuses evolutionary rules such as mutation, crossover and selection into the designated actions. This model successfully generates dynamic and non-equilibrium features, such as bubbles, crashes, wide wealth distribution, and higher volatility, which are rarely seen in a typical RE model.

This framework postulates that heterogeneous agents adopt inductive reasoning and form their expectations based on their guesses about others' expectations. Based on this framework, Arthur et al. [1] propose a theory of asset pricing with heterogeneous agents whose expectations are continually revised to adapt to the market created by their aggregate expectations, and explore how such a market would work and how it would react to price assets. They also provide carefully-controlled numerical experiments to investigate the equilibrium implications of their model. Specifically, the market resulting from their experiments confirms both the efficient market academic view and the traders' view each is valid under different regimes. That is, efficient market views and traders' views based on "market psychology" can both be verified in a market with inductively rational traders.

The key ingredients mentioned above, such as bounded rationality, heterogeneity, and adaptiveness (even though they seem simple) are still widely used as a basis on which to build ABMs to describe an economic system and to reveal available market making structure [9], [10], [11]. LeBaron [9], instead of focusing on learning speed, emphasizes length of memory as the major dimension of heterogeneity. Using a more standard, empirically calibrated economic structure with fewer parameters, LeBaron shows that short-memory traders could significantly contribute to the magnification of volatility and thus are essential to characterize market dynamics. The results in his paper are consistent with previous studies of the relationship between short-memory traders and market volatility.

Pastore [10] proposes an information-based artificial stock market with heterogeneous agents. Stock prices are formed by propagation of information among the interacting agents, who are subject to budget restraints and market feedback. The artificial stock market in his paper successfully replicates the stylized facts of real financial markets under various conditions, highlighting the role of information. Ponta et al. [11] propose a multi-asset, agent-based financial market which consists of zero-intelligence agents (ZIA) with finite financial resources.[9] In this model, agents follow a random allocation strategy with finite financial resources. The stock market in this framework is shown to produce stylized facts such as volatility clustering, and fat-tailed distribution and mean reversion of returns.

3. Sentiment effect

How do investors react to one another's sentiments (other traders' moods)? Weidlich proposes several, now-well known, ABMs [13], [14] to describe opinion-formation among agents. In his model, agents have the choice of voicing either one of the two opinions: optimistic or pessimistic. They change their beliefs continuously according to a Poisson process that formalizes switches between the optimistic and pessimistic groups from one moment to another. The corresponding transition rates, $\omega_\uparrow$ and $\omega_\downarrow$, are assumed to have exponential functions:

$$\begin{aligned} \omega_\uparrow &= \nu \exp(U) \\ \omega_\downarrow &= \nu \exp(-U), \end{aligned}$$

9 According to [12] they have no intelligence, do not seek or maximize profits, nor do they observe, remember, or learn.

where ν is a scale parameter that determines the transition frequency. U works as a threshold to change agents' belief.

Lux [15] investigates the influence of sentiment on asset returns. In his model, sentiment is classified into two different regimes: short-term and medium-term. This makes it possible to construct a stochastic ABM of opinion-formation with explicit social interactions. Using the model which incorporates returns as a representative environment parameter, he shows that there is a significant influence of sentiment on returns of the DAX. Results in the paper confirm the effect on market returns of strong social interaction in short-run sentiment. Even though there is an abrupt short-run sentiment change, medium-term sentiment moves more slowly.

Lux's ABM [15] has the following ingredients: 1) agents' actions and attributes describe short- and medium-term sentiment; and 2) the environment has prices and returns for the stock market, i.e., Deutscher Aktienindex (DAX). There exists no explicit equilibrium in the model, where fluctuations in returns and sentiment are observed during the whole sample period. Numerical maximum likelihood approaches taken from Lux's previous work [16] are used to estimate the parameters of the model.

While a series of ABM simulations finds evidence of strong social interactions in short-term sentiment, social influences on medium-run sentiment tend to become moderated. Previous research using vector autoregressive (VAR) models is consistent with Lux's ABM. This implies that simple macroscopic equation modeling can be replaced effectively with a full microscopic Markov process that is based only on agents' actions.

4. How do agents herd?

Herd behavior was first observed in ant colonies by entomologists. Kirman [17] develops a simple, but mathematically sophisticated, ABM to analyze and adapt ants' "recruiting" behavior to explain agents' herding behavior in financial markets. Ants have two identical food sources: 1) they first exploit one food source more intensively than the other; and then 2) they switch to the other source from time to time. As opposed to the scientists' prediction that the feeding level would settle on an equilibrium point where the ants would consume equal amounts from each source, ants' herding behavior leads to a situation in which the remaining amounts of food at the two sources stay severely lopsided in a ratio of around 80:20 over time. Kirman [17] demonstrates that this result arises from interactions between identical individuals.[10]

The similarity between ant societies and financial markets, in terms of herding, allowed Kirman [17] to apply a multi-agent concept to the dynamics in financial markets. This concept defines the probabilities that one agent will follow another agent's opinion ('herding') and that agents will change their decisions on their own ('self-conversion'). These processes of herding and conversion are intrinsically dynamic: 1) they involve

10 'Identical' in this context does not mean that they are exactly the same, but that each agent shows the same behavior.

agents making different choices; and 2) agents recruit other agents to their choice by several means. Kirman [17] made an excellent contribution to the foundations for subsequent studies on this subject by illuminating the ways in which economic agents interact with one another actively.

Herding is known to be a key source of endogenous fluctuations in asset prices. For the past several decades, finance academics have considered herding behavior in ABMs, trying to fit ABMs to stylized statistical features salient in empirical data [18], [19], [20]. Kirman [21] presents microeconomic behavioral models with interacting agents, in which the forecasts and trades of agents are influenced by other agents in the market. These models provide a theoretical foundation for the long-memory characteristics of agents' market perspectives. A series of studies by Kirman propose a novel explanation for bubbles in asset prices; bubble-like features could result from herding behavior and switching between opinions.

Alfarano et al. [22] investigate the effects of agents' behaviors on the stylized facts of asset returns in DAX and gold markets. Traditionally, the efficient market hypothesis (EMH) has explained the herding effects on the asset returns through the news arrival process. Alfarano et al. develop an asymmetric herding model with asymmetric transition probabilities that correspond to biased herding in a speculative market. Many behavioral finance studies have seen these effects as a universal pattern of interactions in the markets. That is to say, stylized facts, such as fat-tails and volatility clustering, emerge from traders' interactions.

Alfarano et al. [22] also suggest that the tail shape of the financial returns is determined by structural variables for agents' interactions, i.e., herding behavior and autonomous switching propensity. In a system with two different types of agents, namely, 'fundamentalists' and noise' traders,[11] the probability of finding a specific agent type becomes stabilized as opinion forms. Among the environmental parameters linking the agents and the market, the price is the key to identifying speculative market behavior. Alfarano's work designs agent's actions based on Kirman's work [17], assuming that the market dynamics follow a Markov jump process in continuous time.

5. Evolution as learning

According to LeBaron [23], GA can model learning behavior in many agent-based financial markets. From the modeling technique perspective, it serves as an improvement to some traditional learning frameworks such as a Bayesian learning model. Agents interact constantly to learn new information and update their strategies. In spite of its conceptual appeal, however, a GA is quite complex to build. Using the concept of a simple stock market in Section 2, LeBaron et al. [24] replicate salient time series features in real markets, such as fundamental and technical predictability, volatility persistence, and leptokurtosis. Agents' beliefs tend to be consistent with what the environmental

11 Here, noise traders are seen as mass traders; the fundamentalists focus on company-specific events to determine a strategy.

parameters suggest; these essentially represent aggregate level data. The environment is the world they directly construct, so LeBaron et al. use the environmental information that appears significant in the time series data in formulating the forecasting rules. When the learning speed slows down, agents learn homogeneously, which manifests the importance of forecast horizons.

Time horizon is incorporated into ABM to model learning in financial markets in [25]. Lux's model then investigates the evolutionary features of a market populated with both long- and short-horizon investors. The results uncover a plausible mechanism for real-market dynamics and can be summarized as follows: the selection pressure for long-horizon agents to overcome the high-frequency fluctuations appears to be weak. Short-horizon agents act as volatility generators, preventing the longer-horizon agents from getting a foothold in the market. The market containing different types of traders, who generate the various time series features observed in real markets.

Other ABM studies in this area have tried to combine two forms of learning (active and passive) in a single framework [26], [27]. LeBaron [28] compares the relative strengths and weaknesses of these two forms of learning. 'Active' agents are defined as those who choose strategies based on their well-defined objective functions, so that they move their wealth into the strategies that performed well in the past. Active learning moves at reasonable speeds and may better match observed behavior. However, active learning is hard to model and generates noise in financial time series. Passive agents keep their strategies fixed no matter how poorly these are doing. Passive learning is easy to model, but does not necessarily converge to utility maximization.

The temporal evolution of wealth is considered in [29]. This utilizes the former (established) concept of learning, which consists of passive and active stances. There are three features of agents that survive in the market. A 'buy and hold' strategy controls a large fraction of wealth. High-gain learners mainly adopt adaptive and fundamental strategies. All the strategies put a heavy weight on the recent past in terms of volatility estimates, which may lead to market instabilities.

6. ABMs and stochastic models

Lux and Marchesi [30] develop a stochastic multi-agent model of a financial market with the aim of understanding the origin of scaling laws in finance. There has been considerable effort made to introduce stochastic components into ABMs to account for empirical facts and explore interaction mechanisms in financial markets. Their model has three categories of agents: fundamentalists, optimistic noise traders, and pessimistic noise traders. The interactions among agents are then incorporated into the model: noise traders exchange opinions via their interactions. There are also switches between noise traders and fundamentalists, and the exponential function is used to describe the transition probabilities for switches.

Price changes are modeled by imbalances between demand and supply. Relative changes in fundamental values (input) are assumed to be independent Gaussian random variables, but the time series of returns (output) shows volatility clustering instead of normal distribution. The scaling behavior[12] of financial prices confirmed in the complex systems research results from the output [31]. More specifically, it arises from the interaction among market participants, rather than similar scaling in the input signal of fundamental values implied by the EMH.

In Feng et al.'s work [32], ABM is combined with stochastic approaches to capture statistical features of price dynamics. They construct a behavioral ABM in which price changes bring about a convergence in technical traders' opinions. The model also introduces different investment horizons for technical strategies and a technical threshold based on the returns over different horizons. Surprisingly, the model does not suffer from finite-size effects. The results demonstrate that technical traders' collective behavior contributes to fat-tails in return distributions. Feng et al. conjecture that long-term memory originates from the heterogeneous (investment) horizons of technical traders.

The bottom-up framework of ABMs is applied to look into deterministic and stochastic features. While all traders, in a stochastic buy model, 'buy' a random stock a fraction of the time with a deterministic 'buy' decision, they sell a random stock a fraction of the time with a deterministic 'sell' decision in a stochastic sell model. Comparing phase diagrams of three different markets (a dead market (phase I), a booming market (phase II), and a jammed market (phase III) as shown in Figure 1), Lye et al. [33] find that the primary impact of stochasticity is to eliminate the dead market phase as the region occupied by the dead market phase in the phase diagram decreases with the level of stochasticity.

The concept of the stochastic model is also linked to understanding returns in financial markets. It is well described in the work of Gontis and Kononovicius [34], which relates microscopic ABM to macroscopic phenomenological modeling, combining exogenous noise with the stochastic dynamics of agents by defining joint endogenous and exogenous volatilities. They uncover the origin of power-law (i.e., scaling behavior) in both developed and emerging markets. There also exists an interesting extension: an interacting epidemic system (a so-called S-I-R model[13]) is incorporated into an ABM of a stock market [36]. Its objective is to explore fluctuating prices in the financial system. Simulation results well match the statistical properties in real stock markets, verifying the suitability of ABM for financial market modeling.

7. Financial fragility

To check for financial fragility, an ABM of heterogeneous interacting firms and banks is considered; this naturally gives rise to complex dynamics [37], [38]. Heterogeneity is represented by different balance sheet conditions and firm sizes. The simulation results

12 This is an important index to measure the magnitude of collective/herding behavior in the market.
13 The model labels has three compartments: S, number of susceptible; I, number of infectious; and R, number recovered [35].

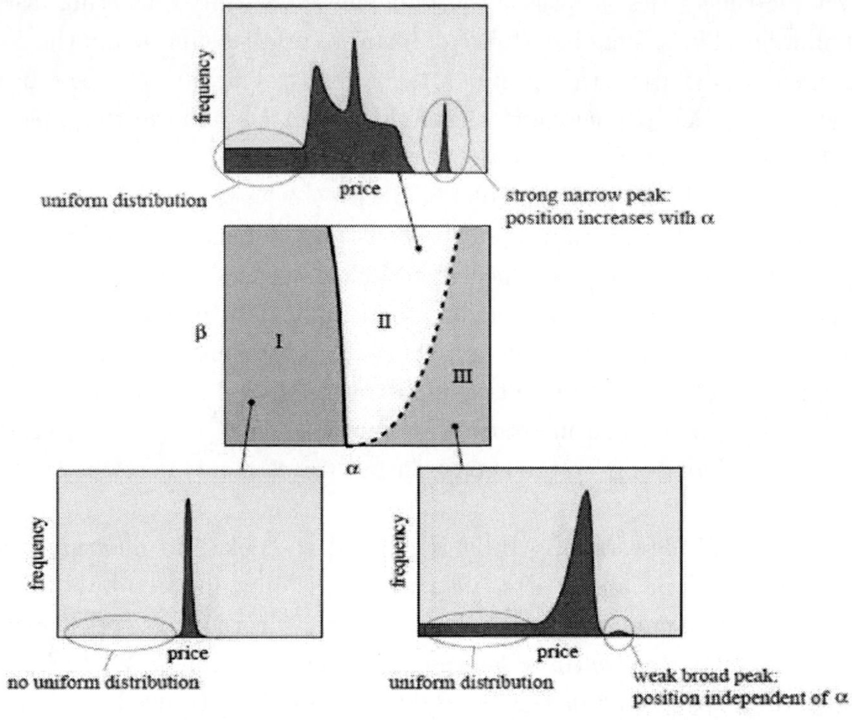

Figure 1. Schematic phase diagram of the deterministic model for phases I, II, and III, each of which characterizes their steady-state price distributions from Lye et al. [33]

fit surprisingly well a range of empirical facts related to complex industrial dynamics, as observed in U.S. firms.

In the goods market, the model in [37], [38] assumes a supply-driven economy, where products are demanded at a stochastic price. In the credit market, one monopolistic bank allocates credit to each firm according to its loan and collateral sizes. The interaction between firms and bank equity may generate a domino effect, making it possible to explain how large aggregate fluctuations can result from small idiosyncratic shocks.

Labor-saving technological innovation and the firm size-wage relationship are introduced to establish and account for a link between power-law shifts and business cycle theory [39]. That the technological level and productivity improvement depend not only on R&D investment but also on imitation provides a clear incentive to imitate other firms. They also show that power-law drifts are a consequence of changes in firms' capital accumulation behavior, which originates from technological progress and the wage-firm size relationship.

An ABM model is also used to reveal the power-law distribution of firm size [40]. The power-law distribution is a key to generating the Laplace distribution of growth rates. Large fluctuations in average output are observed even in response to small aggregate shocks, which are due to the power-law distribution of firms' sizes. This is also applied to other features, such as business fluctuations.

Future studies may well extend the model by incorporating other factors, e.g., aggregate demand, various degrees of market power, policy variables and learning processes. Market here may refer to goods or credit markets.

8. Concluding remarks

We summarize prior studies on ABM of financial markets from the perspective of model applicability and historical evolution. In particular, we highlight several key features of ABMs, how target markets are constructed and how their dynamic evolutions are characterized in their models.

First, we see that small changes in initial parameters in agents could lead to relatively large perturbations throughout the financial markets, i.e., 'complexity emergence.' This remains true even though the definitions of agents' actions are varied across the models. It is, indeed, now known that the financial markets could become volatile due to any unexpected small changes. Second, our survey confirms various benefits of using ABMs to account for financial market phenomena. In particular, we corroborate from their results that ABMs allow us to better understand how the market dynamically evolves to the equilibrium or drifts away from it and thus how asset bubbles are created.

Overall, ABM clearly has a number of benefits a simulation tool to show emergent behaviors (so-called 'macrobehavior from micromotives'), evolution dynamics, and consistent fit with real-market data. Our review summarizes those aspects as explored in the studies discussed here.

Acknowledgments

The authors of this research declare that there is no conflict of interest regarding the publication of this paper. We are grateful to the Korea Institute of Science and Technology (Economic and Social Science Research Initiative (ESRI) 2E28082 and Future Convergence 2E26940) (C.K.). One of us is supported by Bridge Trust Asset Management Research Fund (K.A.). In addition, we would like to thank Zhuang Chu for his excellent research assistant work. All remaining errors are solely ours.

References

[1] Arthur W B, Holland J H, LeBaron B, Palmer R G and Tayler P 1997 *The Economy as an Evolving Complex System II* ed B Arthur *et al* (MA: Addison-Wesley Reading) pp 15-44

[2] Wolfram S 1994 *Cellular Automata and Complexity: Collected Papers* vol 1 (MA: Addison-Wesley Reading)

[3] Schelling T C 1969 Models of segregation *Am. Econ. Rev.* **59** 488-93

[4] Arthur W B 1994 Inductive reasoning and bounded rationality (The El Farol Problem) *Am. Econ. Rev.: Papers and Proc.* **84** 406-11

Agent-based models in financial market studies

[5] Chen S-H and Gostoli U 2012 Agent-based modelling of the el farol bar problem *Simulation in Computational Finance and Economics: Tools and Emerging Applications*, ed B Alexandrova-Kabadjova *et al* (PA: IGI Global) pp 359-77.
[6] Schelling T C 1971 Dynamic models of segregation *J. Math. Sociol.* **1** 143-86
[7] Schelling T C 1978 *Micromotives and Macrobehavior* (NY: W. W. Norton & Company)
[8] Palmer R G, Arthur W B, Holland J H, LeBaron B and Tayler P 1994 Artificial economic life: A simple model of a stock market *Physica* D **75** 264-74
[9] LeBaron B 2002 Short-memory traders and their impact on group learning in financial markets *Proc. of the National Academy of Sciences* **99** 7201-06
[10] Pastore S, Ponta L and Cincotti S 2010 Heterogeneous information-based artificial stock market *New. J. Phys* **12** 053035
[11] Ponta L, Raberto M and Cincotti S 2011 A multi-assets artificial stock market with zero-intelligence traders *Europhys. Lett.* **93** 28002
[12] Gode D K and Sunder S 1993 Allocative efficiency of markets with zero-intelligence traders: market as a partial substitute for individual rationality *J. Polit. Sci.* **101** 119-37
[13] Weidlich W and Haag G 1983 *Concepts and Methods of a Quantitative Sociology* (Berlin: Springer)
[14] Weidlich W 2000 Sociodynamics: A systematic approach to mathematical modelling in the social sciences *Chaos. Soliton. Fract.* **18** 431-7
[15] Lux T 2012 Estimation of an agent-based model of investor sentiment formation in financial markets *J. Econ. Dyn. Control* **36** 1284-302
[16] Lux T 2009 Rational forecasts or social opinion dynamics? Identification of interaction effects in a business climate survey *J. Econ. Behav. Organ.* **72** 638-655
[17] Kirman A 1993 Ants, rationality, and recruitment *Q. J. Econ.* **108** 137-56
[18] Scharfstein D S and Stein J C 1990 Herd behavior and investment *Am. Econ. Rev.* **80** 465-79
[19] Day R H and Huang W 1990 Bulls, bears and market sheep *J. Econ. Behav. Organ.* **14** 299-329
[20] Shiller R J and Pound J 1989 Survey evidence on diffusion of interest among institutional investors *J. Econ. Behav. Organ.* **12** 47-66
[21] Kirman A and Teyssiere G 2002 Microeconomic models for long memory in the volatility of financial time series *Stud. Nonlinear Dyn.* E **5** 1-23
[22] Alfarano S, Lux T and Wagner F 2005 Estimation of agent-based models: The case of an asymmetric herding model *Comput. Econ.* **26** 19-49
[23] LeBaron B 2006 Agent-based computational finance *Handbook of computational economics* **2** 1187-233
[24] LeBaron B, Arthur W and Palmer R 1999 Time series properties of an artificial stock market *J. Econ. Dyn. Control* **23** 1487-516
[25] LeBaron B 2001 Evolution and time horizons in an agent-based stock market *Macroecon. Dyn.* **5** 225-54
[26] Anufriev M and Dindo P 2010 Wealth-driven selection in a financial market with heterogeneous agents *J. Econ. Behav. Organ.* **73** 327-58
[27] Chiarella C and He X 2002 *Handbook of Information Technology in Finance* ed D Seese *et al*
[28] LeBaron B 2011 Active and passive learning in agent-based financial markets *East. Econ. J.* **37** 35-43
[29] LeBaron B 2012 Heterogeneous gain learning and the dynamics of asset prices *J. Econ. Behav. Organ.* **83** 424-45
[30] Lux T and Marchesi M 1999 Scaling and criticality in a stochastic multi-agent model of a financial market *Nature* **397** 498-500
[31] Barabasi A L and Stanley H E 1996 *Fractal Concepts in Surface Growth* (MA: Cambridge University Press)
[32] Feng L, Li B, Podobnik B, Preis T and Stanley H E 2012 Linking agent-based models and stochastic models of financial markets *Proc. of the National Academy of Sciences* **109** 8388-93
[33] Lye R, Tan J P L and Cheong S A 2012 Understanding agent-based models of financial markets:

Agent-based models in financial market studies

A bottom-up approach based on order parameters and phase diagrams *Physica* A **391** 5521-31

[34] Gontis V and Kononovicius A 2014 Consentaneous agent-based and stochastic model of the financial markets *PLoS ONE* **9** e102201

[35] Ogilvy K W and Gray M A 1927 A contribution to the mathematical theory of epidemics *Proc. of the Royal Society* A **115** 700-21

[36] Lu Y, Wang J and Niu H 2015 Agent-based financial dynamics model from stochastic interacting epidemic system and complexity analysis *Phys. Lett.* A **14-5** 1023-31

[37] Gallegati M, Giulioni G and Kichiji N 2003 Complex dynamics and financial fragility in an agent based model *Adv. Complex Syst.* **6** 267-82

[38] Gatti D D, Guilmi C D, Gallegati M and Giulioni G 2007 Financial fragility, industrial dynamics, and business fluctuations in an agent-based model *Macroecon. Dyn.* **11** 62-79

[39] Gatti D D, Gallegati M and Russo A 2004 Technological innovation, financial fragility and complex dynamics *Tech. Rep.*

[40] Gatti D D, Guilmi C D, Gaffeo E, Giulioni G, Gallegati M and Palestrini A 2005 A new approach to business fluctuations: heterogeneous interacting agents, scaling laws and financial fragility *J. Econ. Behav. Organ.* **56** 489-512

IOP Conf. Series: Journal of Physics: Conf. Series **1039** (2018) 012023 doi:10.1088/1742-6596/1039/1/012023

ARIMA model to forecast international tourist visit in Bumthang, Bhutan

Choden and **Suntaree Unhapipat**

Department of Mathematics, Faculty of Science, Mahidol University, Bangkok

Email: chodenwangpo333@gmail.com

Abstract: The number of visitors to Bhutan increases each year making tourism industry the third largest revenue contributor to the nation. The country being the beneficiary of tourism it is very important to forecast tourist visit in the coming future. The monthly number of international visitors in Bumthang, Bhutan from January 2012 to December 2016 is modelled by using Box-Jenkins seasonal Autoregressive Integrated Moving Average (ARIMA) model. The best fitting ARIMA model is constructed based on Bayesian Information Criterion. The fitted seasonal ARIMA $(0, 0, 0) \times (1, 1, 0)_{12}$ is able to forecast international tourist visit of period Jan 2017-Jun 2017 with 91% accuracy. The forecasted result can be used as a tool to handle future challenges and to bring further development in tourism industry.

1. Introduction

Today tourism is one of the world's largest and fastest growing service industry. It accounted for 10 percent of global Gross Domestic Product (GDP), 7 percent of total exports and one in 11 jobs across the world in 2015. The international tourist arrival reached nearly 1.2 billion during the same year. Asia and Pacific region was reported as one of the fastest growing tourism region in the world with 5 percent growth rate, above 4.3 percent global average [1].

Like other nations Bhutan is also beneficiary of tourism industry, it begun in 1974 under the dynamic leadership of fourth king Jigme Singye Wangchuck with just 287 tourist visit [2]. It was introduced with primary objective of generating revenue especially foreign exchange, publishing country's unique culture and traditions to outside world and to contribute to country's socio-economic development.

Bhutan being very farsighted has been aware that an unrestricted flow of tourists can have negative impact on its pristine environment and rich and unique culture therefore adopted a policy of "High value-Low impact" tourism thus controlling the type and quantity of tourism right from the start. The philosophy of Gross National Happiness (GNH), adventure pursuit, pristine nature and spiritual values are also the reasons for people to visit the country [3].

Tourism sector in Bhutan is regarded as one of the most exclusive travel destination based on GNH values in the world, this is because of clearly established tourism policies, excellent tourism resources, developing and expanding private sectors, qualified and experienced personnel and established marketing channels and contacts. Today it is a vibrant business with high potential for growth and further development [4]. For this reason, we focus on modelling and forecasting the number of tourists visiting Bhutan.

1.1 Literature review
In 2001, Lim and McAleer [5] used Box-Jenkins method to forecast international travel demand for Australia by employing Autoregressive Integrated Moving Average (ARIMA) model. They created various alternative models from the dataset of period 1975-1989 based on three major tourist markets namely Malaysia, Singapore and Hong Kong. The study shows how best fitted model is selected and its application in the field of forecast. The authors also considered unit root test and seasonal unit root test before creating the relevant model which is very rare in many studies done before. This test will integrate seasonal data and converts non-stationary data to stationary as most tourism data are seasonal in nature. Over this entire test will improve validity of forecast made by the fitted models and makes judgment at each step of modelling process. The recommend fitted models were ARIMA (3, 1, 1) for Hong Kong, ARIMA (3, 1, 2) for Malaysia and ARIMA (4, 1, 0) for Singapore respectively. The forecast outcome of last two models was not as accurate as the one first one.

In 2008, Lim *et al* [6] forecasted number of nights spent by guest in New Zealand based on dataset from 1997 to 2007 using Holt-Winter triple exponential smoothing (model A) and Box Jenkins Autoregressive (AR), Moving average (MA) and Seasonal Autoregressive Moving Average (SARMA) models (Model B). The authors later compared these two models. After doing modeling procedures using Box- Jenkins method various Autoregressive Moving Average (ARMA) and SARMA models were created. The selection of model was done based on Akaike Information Criterion (AIC) and Schwarz Bayesian Criterion (SBC) and it was found that ARMA (2,1) and SARMA (2, 0, 1) × (0, 1, 1) are the best fitted models. The study reveals that model A shows the downfall of guest nights whereas forecast of guest nights shown by model B was quite positive. The study also tells that there are lot of researches were done in forecasting international tourist arrivals using both time series and regression method but in this particular study they said regression models cannot be used as guest nights can be both international and domestic. It is not a suitable method due to given complexities of different demand characteristics.

China being one of the top source tourism market for Australia; it has great impact on country's economy. Australia tries to maintain and promote Chinese tourist visit to their country and there are many studies done regarding to this context, among many in 2015, Ma *et al* [7] proposed a study to anticipate Chinese tourist visit to Australia by employing SARIMA model. They used monthly data set obtained from Australia Bureau Statistics and computations were done in STATA 13. ARIMA $(0, 0, 1) \times (0, 1, 1)_{12}$ was considered as best fitted model can forecast Chinese tourist visit three-period-ahead. The result confirms that February would be peak month every year for Chinese arrival to Australia and their visit will be seasonal for next two years. Based on the existing data, we are going to use the time series analysis.

2. Methodology

2.1 Data collection
For this study secondary data was used which was maintained by Tourism Council of Bhutan. The data consists of monthly international tourist visit in Bumthang from January 2012 to December 2016, "training dataset". Although tourist visit to almost all twenty provinces we focus on Bumthang province because of accessibility and availability of data and its importance to tourism demand in the country. Figure 1 shows Bhutan map shaded at Bumthang.

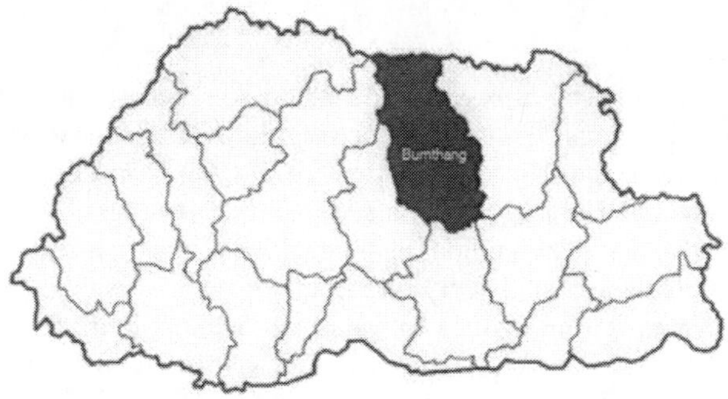

Figure 1. Bhutan map shaded at Bumthang.

2.2 Statistical analysis
In this study, the Box-Jenkins time series model is used to analyze international tourist visit to Bumthang. To apply Box-Jenkins model, we follow the following steps.

The first step is to *identify* all the tentative models. Identification consists of specifying appropriate AR, MA or ARMA and other order of the model. The identification is done by looking at Autocorrelation function (ACF) and Partial autocorrelation function (PACF) of the interested series.

The second step is to *estimate* the coefficients of the model. Coefficient of AR model can be estimated by least-square regression. The estimation of parameter usually requires more complicated iteration procedures but the computer programming automatically generate it.

Then the third step is to *check* the model. This step is also called diagnostic checking or verification. Two important elements in checking is to ensure that the residuals are white noise and estimated parameters are strictly significant.

Finally we *select* the best model among several competitive ARIMA models which might be suitable for the series thus we use Bayesian Information Criterion (BIC) for the model selection. The selected model is used for forecasting.

2.2.1 ARIMA model
If the process is not stationary, we have to take differencing term $(1-B)^d$ in the process. When the model ARMA (*p, q*) model on a time series which has been differenced *d* times we call this an ARIMA (*p, d, q*) model [8]. Thus the general ARIMA (*p, d, q*) written using backshift operator as:

$$\phi_p(B)(1-B)^d = \theta_0 + \theta_q(B)a_t$$

where $\phi_p(B) = 1-\phi_1 B-\phi_2 B^2 -\cdots-\phi_p B^p$ is the stationary AR operator,

$\theta_q(B) = 1-\theta_1 B-\theta_2 B^2 -\cdots-\theta_q B^q$ is the invertible MA operator and $\theta_0 = \mu(1-\phi_1-\phi_2-\cdots-\phi_p)$. is deterministic trend term.

2.2.2 Seasonal ARIMA
The seasonal ARIMA model includes both seasonal and non-seasonal factors in multiplicative model called ARIMA $(p,d,q)\times(P,Q,D)_s$ where *p* is non-seasonal AR order, *d* is non-seasonal differencing, *q* is non-seasonal MA order, *P* is seasonal AR order, *D* is seasonal differencing, *Q* is seasonal MA order and *s* is a seasonal period [9]. The general equation of the model is

$$\Phi_P(B^s)\phi_p(B)(1-B)^d(1-B^s)^D \dot{X}_t = \theta_q(B)\Theta_Q(B)^s a_t$$

where $\phi_p(B)$ is the non-seasonal autoregression component of order *p*, $\Phi_p(B^s)$ is the seasonal autoregression of order *P*, X_t is the current value of the time series examined, B is the backward shift operator $X_t(B^i) = X_{t-i}$, $(1-B)^d$ is the non-seasonal difference term, $(1-B^s)^D$ is seasonal difference term, $\theta_q(B)$ is the non-seasonal moving average of order q and $\Theta_Q(B)^s$ is the seasonal moving average of order Q [9].

3. Result analysis

3.1 Model identification

To use Box-Jenkins methodology, the series should be stationary. In this study, the graphical methods and normality test based on Kolmogorov and Shapiro test was conducted to check the stationary of the series in SPSS version 18. The test tests the null hypothesis that the series is normally distributed. At significant level α=0.05, we reject null hypothesis if P-value < α=0.05 or accept otherwise. From Table 1 we reject null hypothesis since p-value is less than α=0.05 i.e., series is not normally distributed. The graph of the series also shows strong dependence of variability on the level of series in which the data should be transformed to eliminate this dependence. To stabilize variance, we used log transformation. Figure 2 shows the plot of monthly international tourist visit in Bumthang. The monthly differenced logarithm transformed international tourist visit against time is plotted in Figure 3. The Table 2 shows the value of normality test of the logarithm transformed series where null hypothesis is accepted since P-value is greater than α=0.05 thus series are normally distributed.

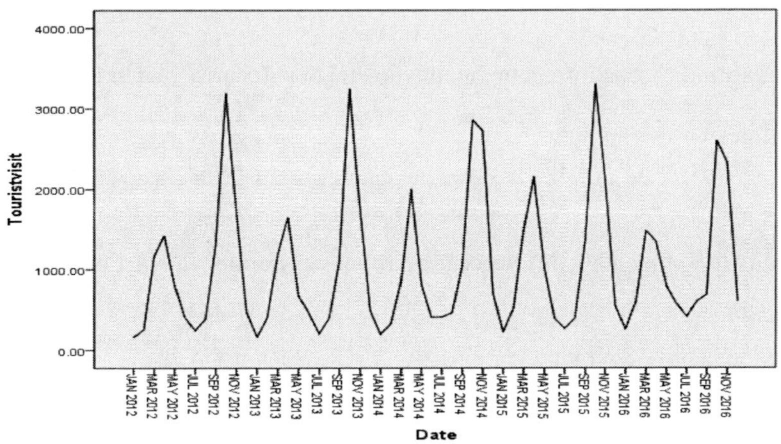

Figure 2. Monthly international tourist visit in Bumthang from Jan 2012- Dec 2016.

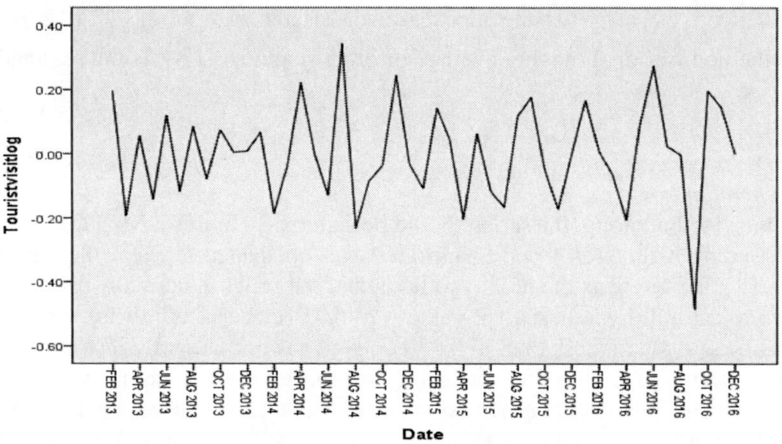

Figure 3. Monthly differenced logarithm transformed international tourist visit in Bumthang from Jan 2012- Dec 2016.

Table 1. Normality test of monthly international tourist visit to Bumthang.

	Kolmogorov-Smirnov			Shapiro-Wilk		
	Statistic	df	Sig.	statistic	df	Sig.
Tourist visit	0.183	60	0.000	0.840	60	0.000

Table 2. Normality test of monthly differenced logarithm transformed international tourist visit to Bumthang.

	Kolmogorov-Smirnov			Shapiro-Wilk		
	Statistic	df	Sig.	statistic	df	Sig.
Tourist visit	0.084	60	0.200	0.967	60	0.104

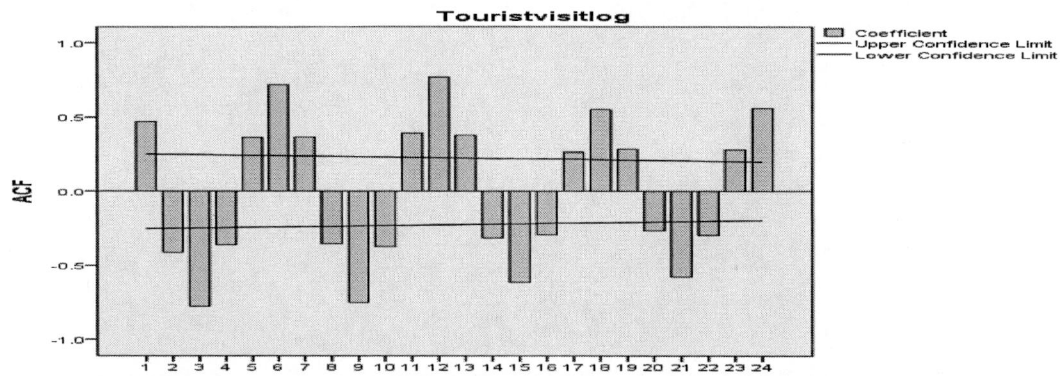

Figure 4. Sample ACF of monthly logarithm transformed international tourist visit in Bumthang.

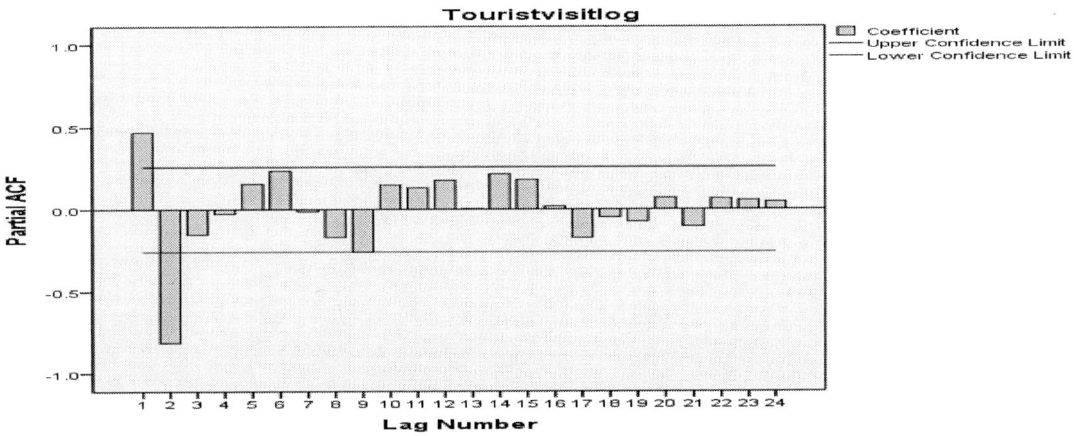

Figure 5. Sample PACF of monthly logarithm transformed international tourist visit in Bumthang.

Figure 4 and 5 shows sample ACF and PACF with 95% confidence limits. The ACF shows damp sine cosine wave and slow decay of the spikes indicating cyclic or seasonal movement of the correlation therefore a first seasonal differencing was done. Figure 6 and 7 shows sample ACF and PACF of monthly logarithm transformation of international tourist visit in Bumthang for first seasonal differencing at period 12 months. The sample ACF shows spike at lag 12 and PACF cuts off after lag 1. Therefore, the tentative models are ARIMA (0, 0, 0) × $(0, 1, 1)_{12}$, ARIMA (0, 0, 0) × $(1, 1, 0)_{12}$, ARIMA (0, 0, 0) × $(1, 1, 1)_{12}$ and ARIMA (0, 0, 0) × $(0, 1, 0)_{12}$.

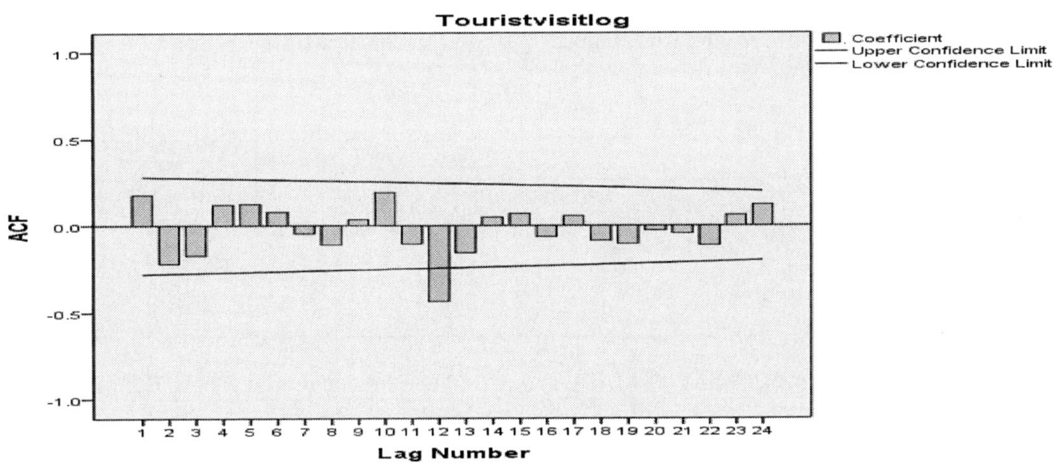

Figure 6. Sample ACF for the first seasonal differencing at period 12 months of logarithm transformed international tourist visit to Bumthang.

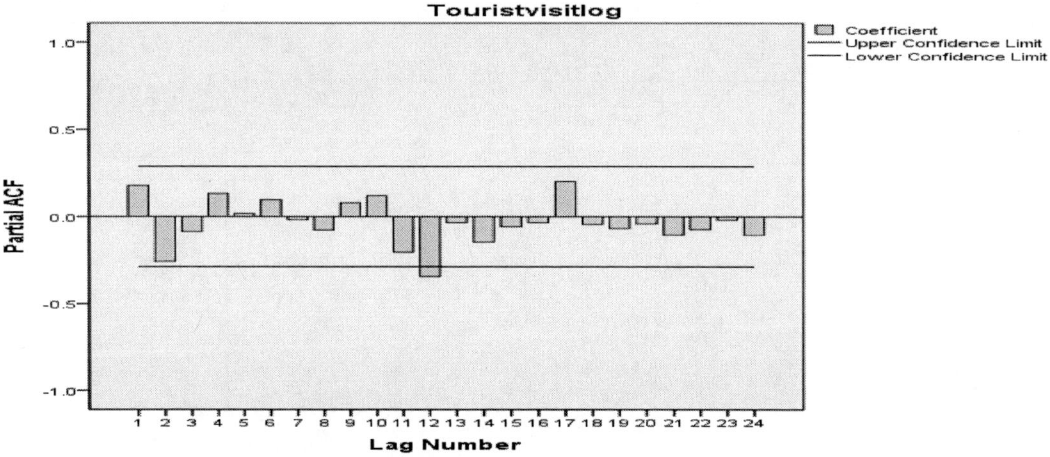

Figure 7. Sample PACF for the first seasonal differencing at period 12 months of logarithm transformed international tourist visit to Bumthang.

3.2 Model estimation and selection

To check model adequacy, we perform Q statistic test where test tests the null hypothesis that residuals from ARIMA model have no autocorrelation. At significant level α=0.05, we reject null hypothesis if P-value < α=0.05 or accept otherwise. The values of the Q statistic and P-values are given in Table 3 where we accept null hypothesis since P-value of tentative models is greater than α=0.05 i.e., residuals of ARIMA models have no autocorrelation thus the models are considered to be fitted models. To choose the best fitted model among the fitted models we consider the value of BIC, Root Mean Square Error (RMSE), Mean Absolute Percentage Error (MAPE), Mean Average Error (MAE) and R^2. The models are presented in Table 4. From the Table 4 the least value of BIC, RMSE, MAE, MAPE and highest R^2 is ARIMA (0, 0, 0) × $(1, 1, 0)_{12}$ indicating ARIMA (0, 0, 0) × $(1, 1, 0)_{12}$ as best model to forecast the monthly international tourist visit in Bumthang from January 2012 to December 2016.

Table 3. The Q statistic test for k = 24 of the tentative models for monthly international tourist visit to Bumthang.

Model	Q statistic	P-value
ARIMA (0, 0, 0) × $(0, 1, 1)_{12}$	14.02	0.67
ARIMA (0, 0, 0) × $(1, 1, 0)_{12}$	12.95	0.74
ARIMA (0, 0, 0) × $(1, 1, 1)_{12}$	11.01	0.81
ARIMA (0, 0, 0) × $(0, 1, 0)_{12}$	27.86	0.07

Table 4 Summary of BIC, RMSE, MAPE and R^2 of international tourist visit to Bumthang.

Model	R^2	RMSE	MAPE	MAE	BIC
ARIMA (0, 0, 0) × $(0, 1, 1)_{12}$	0.91	0.11	2.86	0.08	-4.35
ARIMA (0, 0, 0) × $(1, 1, 0)_{12}$	**0.92**	**0.10**	**2.85**	**0.08**	**-4.38**
ARIMA (0, 0, 0) × $(1, 1, 1)_{12}$	0.92	0.10	2.86	0.08	-4.36
ARIMA (0, 0, 0) × $(0, 1, 0)_{12}$	0.88	0.12	3.08	0.09	-4.13

3.3 Goodness of fit/ diagnostic checking

In time series modelling the selection of best fit model is directly related to how well the residuals analysis is performed. One of the assumptions of ARIMA model is that for a good model the residual should be white noise.

From Figure 8 the sample ACF and PACF of the model shows that the autocorrelation of the residual are all close to zero which mean they are uncorrelated and reveal no apparent patterns, hence the residual assume mean of zero and constant variance. Finally the P-value (0.74) for the Ljung-Box statistic clearly exceeds 5% for all lag orders. Thus the selected model ARIMA (0, 0, 0) × $(1, 1, 0)_{12}$ satisfies all the model assumptions and its parameter estimation is given in Table 5.

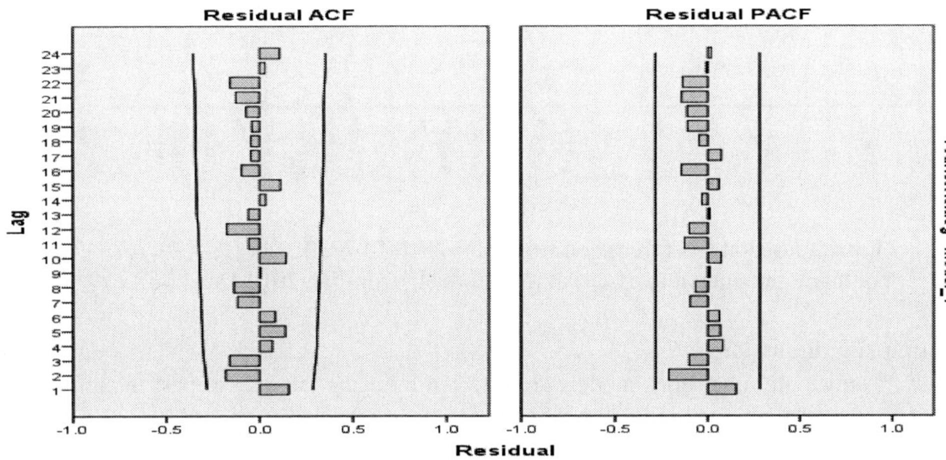

Figure 8. Sample ACF and PACF of the residuals of ARIMA (0, 0, 0) × $(1, 1, 0)_{12}$.

Table 5. Parameter estimation of appropriate ARIMA (0, 0, 0) × $(1, 1, 0)_{12}$.

Model	Estimate	Standard Error (SE)
Constant	0.03	0.01
Seasonal differencing	1	
MA seasonal lag 1	0.67	0.36

3.4 Forecasting using ARIMA (0, 0, 0) × $(1, 1, 0)_{12}$

The forecasted values with 95% forecast limit of the ARIMA (0, 0, 0) × $(1, 1, 0)_{12}$ of model for monthly international tourist visit in Bumthang are shown in Table 6 with standard error, lower and upper limit while actual forecast values are shown in Figure 9.

Table 6. Forecasted value for 6 months monthly international tourist visited for 2017.

Date	Tourist visited	Forecasted value	95% confidence level		Error
			Lower	Upper	
Jan-2017	201	251	159	398	-50
Feb-2017	369	550	347	871	-181
Mar-2017	779	1514	956	2399	-735
Apr-2017	1796	1950	1202	3090	-154
May-2017	915	1000	631	1585	-85
Jun-2017	610	501	316	794	109

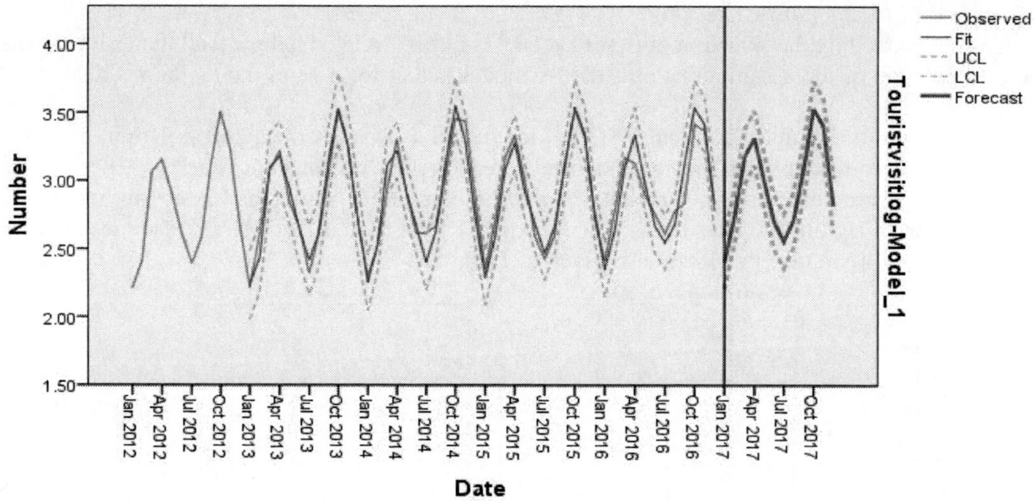

Figure 9. Actual and forecasted values for ARIMA (0, 0, 0) × $(1, 1, 0)_{12}$ of monthly international tourist visit to Bumthang from Jan 2012-Dec 2017.

4. Conclusion and discussion

In this study, a univariate time series models are selected by using the data of past monthly international tourist visit in Bumthang. The ACF and PACF of first seasonal differenced transformed data show seasonality at the period 12. Table 6 shows prominent forecasted values in March, April and May (spring in Bhutan) thus revealing seasonal movement of tourist in Bumthang. This is because the data collected was based on leisure tourist; it doesn't include official and personal tourists. The tourist visits Bumthang to experience cultural activities, visit spiritual sites, to enjoy treks on high mountains etc. These activities happens at fixed scheduled date every year usually in spring and autumn where weather is basically warm and less rainfall which makes it favorable for tourist to visit the place.

References

[1] Executive summary: Bhutan tourism monitor; annual report 2016. Available from www.tourism.gov.bt/about-tcb/about-tcb-2 [Assessed on 27th May, 2017]

[2] Tourism in Bhutan: Wikipedia. Available from htttps://en.wikipedia.org/wiki/Tourism-in-Bhutan. [Assessed on 25th May, 2017]

[3] Tourism Council of Bhutan. Available from www.tourism.gov.bt/about-tcb/about-tcb-2 [Assessed on 28th May, 2017]

[4] Tourism Council of Bhutan. Available from www.tourism.gov.bt/about-tcb/about-tcb-2 [Assessed on 28th May, 2017]

[5] Lim C, McAleer M. *Time series forecasts of international travel demand for Australia.* Tourism Management. 2002:23:389-396

[6] Lim C, Chang C, McAleer M. *Forecasting h(m)otel guest nights in New Zealand.* International Journal of Hospitality Management. 2009:28:228-235

[7] Ma E, Liu Y, Li J, Chen S. *Anticipating Chinese tourists' arrivals in Australia: A time series analysis.* Tourism Management Perspectives. 2016:17:50-58

IOP Conf. Series: Journal of Physics: Conf. Series **1039** (2018) 012023 doi:10.1088/1742-6596/1039/1/012023

[8] Nanthakumar L, Ibrahim Y. *Forecasting international tourism demands in Malaysia using Box – Jenkins's Sarima application.* University Malaysia Terengganus, Malaysia Terengganus, Kuala Terengganus. 2010; 3(2): 50 - 60
[9] Wei WWS. *Time series analysis*, Addison-Wesley, USA; (1990), p106

Acknowledgement

We would like to thank Thai International Cooperation Agency (TICA) for providing the financial support. We would also like to thank Tourism Council of Bhutan (TCB) for providing us latest available data.

Sensitivity of WRF microphysics schemes: case study of simulating a severe rainfall over Egypt

Muhammed Eltahan and Mohammed Magooda

Cairo University, Aerospace Engineering, Egypt

Abstract. In this work, sensitivity of ten microphysics schemes of Weather Research and Forecasting model were compared to determine which is more accurate. A case study of heavy rains over the Red Sea coast of Egypt on 26th October 2016 that caused damaging floods is the subject of this study. Three coastal areas were chosen, the city of Ras Ghareb, Wadi Elgemal National park, and Elba National park. The output precipitation from the model was compared to the satellite data obtained from Tropical Rainfall Measuring Mission (TRMM). Our analysis showed that the model has the capability to simulate the rainy events spatially and produce the cloud pattern. The results showed also that the model generally tends to overestimate the precipitation for all the ten schemes compared to TRMM data.

Keywords. Precipitation, WRF, Microphysics, TRMM

1. Introduction

Extreme weather events have negative impacts on communications and transportations, consequently resulting in catastrophic effects on distinct aspects of people's lives and economy, Figure 1.
Egypt, despite being mostly desert land, is susceptible to multiple severe rainfalls over its coastlines every year with effects extending sometimes over the entire country. Which urged for the need for prediction system to help authorities take preventive measures. Different numerical weather models alongside observations from both ground stations and satellite data are the base of such systems. Weather Research Forecast (WRF) model shows the capability to simulate such events and can be considered a corner stone of a prediction system [1]. WRF was also used before to simulate similar rainy event over Sinai Peninsula, Egypt on 18th January 2010 [2].
Microphysics schemes in numerical models play key role in simulating formation of cloud droplets, precipitation and land surface temperature (LST). It's also considered a key parameter in many hydrological, meteorological and environmental studies as it merges the interactions between the surface - atmosphere and energy fluxes between the atmosphere and the surface [3]. In this paper the effect of alternating microphysics schemes on simulating the rainfall that extended for two days was investigated and validated against satellite observations obtained from Tropical Rainfall Measuring Mission (TRMM) over three coastal regions.

IOP Conf. Series: Journal of Physics: Conf. Series **1039** (2018) 012024 doi:10.1088/1742-6596/1039/1/012024

Figure 1. Damages in Ras Ghareb city in 26th October 2016

2. Data preparation

Weather Research Forecasting (WRF) framework is an efficient framework for atmospheric simulation systems on different scales and one of the most used tools in both research and operational applications. The WRF Software Framework (WSF) works as an incubator for the dynamics solvers, physics packages that deals directly with the solvers, programs for initialization, chemistry and WRF Data Assimilation module (WRFDA). WSF contains two dynamics solvers (Both are Eulerian mass dynamical cores with terrain-following vertical coordinates), The first is the Advanced Research WRF (ARW) solver, and the second is the NMM (Non-hydrostatic Mesoscale Model), a solver developed at NCEP/ EMC. Those two dynamic solvers are the main Components of WRF used in this study.

2.1. Model Setup

WRF version 3.8 is used in this study to simulate meteorology over model domain with terrain shown in Figure 2. The model domain is defined as Lambert Projection extends from 24 E to 36 E (120 grid points) in the east-west direction and from about 21.5 N to 32.5 N (120 grid points) in the north-south direction at horizontal grid 10×10 Km^2. The vertical grid consists of 41 level from surface up to 10 hpa. The static geographical fields, such as terrain height, soil properties, vegetation fraction, land use and albedo, etc., are interpolated and prepared to the model domain by using geogrid program in the WRF preprocessing system (WPS).

The initial and lateral boundary conditions for the meteorological fields are obtained from the National Center for Environmental Predictions (NCEP). Final Analysis (FNL) fields available every 6 hours at a spatial resolution of 1°×1°. They were interpolated to the model domain by using ungrib and metgrid programs in the WRF preprocessing system (WPS).

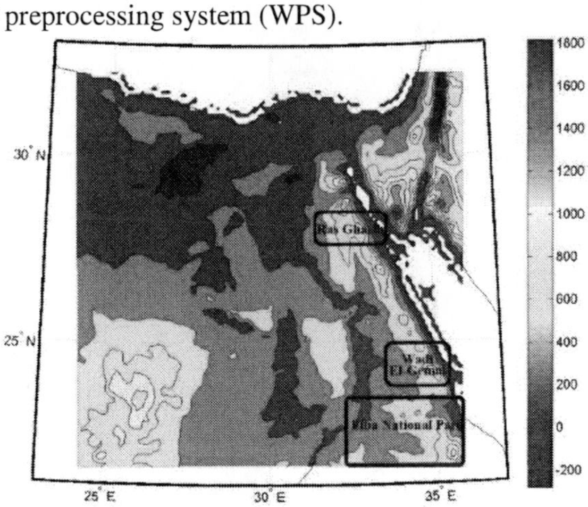

Figure 2. WRF Terrain map of Egypt showing the locations of the three selected domains for validation

2.2. Physical Parameterizations

The physical schemes used to configure the model are summarized in Table 1.

Table 1. Summary of model configuration

Scheme	Option number	Model
Microphysics	[1,2, 3, 4, 5, 6, 7, 8, 9,10]	Kessler [4]
		Lin (Purdue) [5]
		WRF single moment 3 (WSM3) [6]
		WRF single moment 5 (WSM5)
		Eta (Ferrier)
		WRF single Scheme 6b (WSM6) [7]
		Goddard [8]
		Thompson [9]
		Milbrandt-Yau Double Moment [10]
		Morrison Double Moment [11]
L Long-wave radiation	4	RRTMG [12]
S short wave radiation	4	RRTMG
Land surface	2	Unified Noah Land surface [13]
PBL model	1	Yonsei University
Surface Layer	1	MM5 similarity [14]
Cumulus	5	Grell 3D Ensemble Scheme [15]

3. Comparison and observations

3.1. Using maps

The spatial distribution for precipitation from ten different microphysics and TRMM data are shown in Figure 3. Generally, all the schemes successfully simulated the event with spatial distribution varying in the intensity of precipitation from one scheme to the other. TRMM always underestimated the precipitation with maximum reading of 35 mm over the Red sea, while the maximum precipitation from all the schemes reached up to 85 mm over different spatial domains depending on which scheme was used.

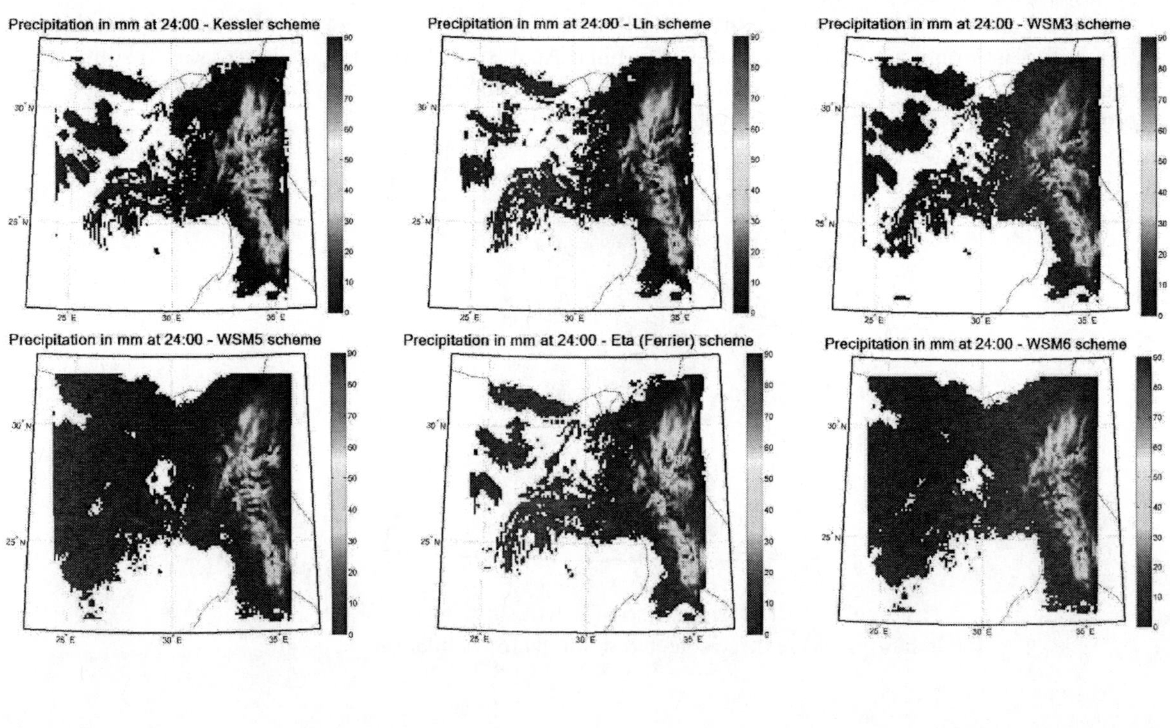

IOP Conf. Series: Journal of Physics: Conf. Series **1039** (2018) 012024 doi:10.1088/1742-6596/1039/1/012024

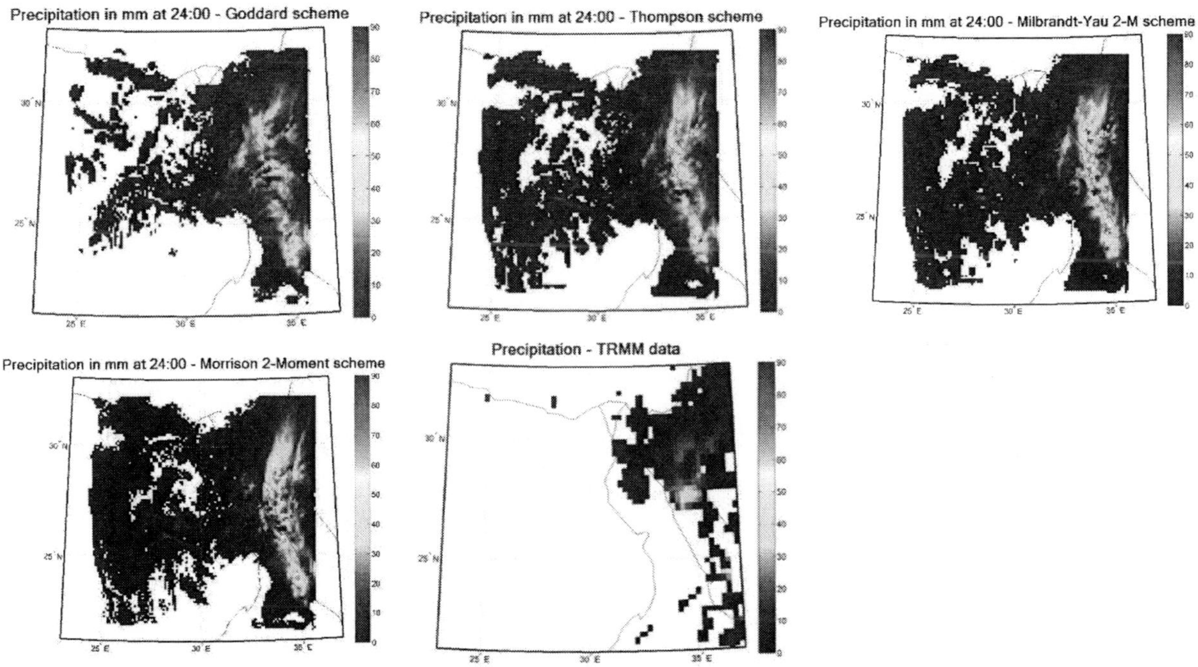

Figure 3. Spatial distribution for precipitation from TRMM satellite data and the ten microphysics over Egypt at midnight of the 26th October 2016

3.2. Averaging over three domains

Three domains over Ras Ghareb, Wadi Elgemal National park and Elba National park were selected to evaluate model performance with boundaries shown in Table 2.

Table 2. Selected domains with boundaries

	Upper boundary	Lower boundary	Left boundary	Right boundary
Ras Ghareb	29	27.5	32	34
Wadi Elgemal National park	24.83	24.1	33.9	36.46
Elba National park	23.9	21.97	33.9	37.6

The resulting average precipitation over the three domains using the same ten microphysics schemes were plotted against the TRMM average data for the exact same domains at the same time figure 4. All the schemes resulted in a high precipitation over both Elba National park and Wadi Elgemal National park while TRMM data underestimated the reading over both areas. Average Precipitation over Wadi Elgemal from TRMM was 12.716 mm. The Lin scheme was the closest to TRMM over wadi Elgemal by overestimating the precipitation by 14.2 mm while the farthest scheme was Eta scheme by 24.1 mm. The precipitation over Elba National park was 4.65. The closest scheme was Thompson which estimate the precipitation by 12.9. ETA scheme was the farthest one which evaluates it by 15.39 mm. While the average precipitation over Ras Ghareb from TRMM was 7.19 mm. The closest scheme to TRMM was Morrison double moment, overestimating precipitation by 8.5 mm while the farthest scheme was Kessler by almost 14.6 mm.

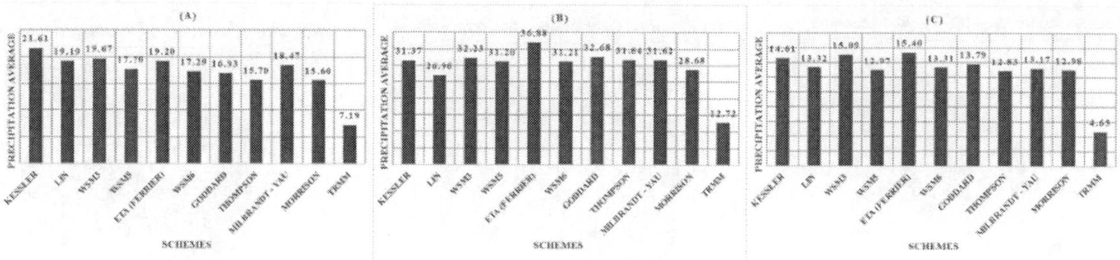

Figure 4. Average precipitation over three selected domains (A) Ras Ghareb (B) Wadi Elgemal (C) Elba National park from TRMM and the ten Microphysics schemes

3.3. Cloud pattern simulation

The model also successfully simulated the cloud pattern as shown in Figure 5 using Thompson scheme. This pattern is a function of the model height layer which is divided into three domains. Low cloud starts from the surface up to 0.66 eta level, mid cloud from 0.66 to 0.33 eta level and high cloud from 0.33 to 0 eta level. Clouds exist at all the height layers over Sinai, while there are generally clouds at low level over the coast of the red sea and on various levels at different spatial areas.

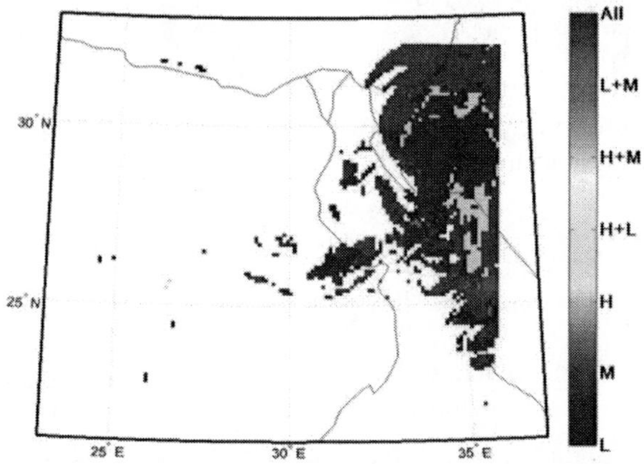

Figure 5. Cloud pattern over Egypt using Thompson scheme

4. Conclusions and future work

In this work the impact of several microphysics schemes on precipitation simulation was investigated. Performance of the WRF model was compared to TRMM satellite data in terms of precipitation. Three areas within the domains were selected to validate the model. WRF successfully simulated the spatial pattern of the precipitation, when compared to data obtained from TRMM, it was found that the model always overestimated the precipitation values.

Alternating between different microphysics schemes has direct impact on the spatial pattern of precipitation and on the distribution of the precipitation's density. The difference in the output among the ten microphysics schemes was very small, hence no specific scheme was alone very close to TRMM data over the three domains. Future works should include testing sensitivity of the model to changes in the preferences of the cloud scheme, also the performance of the model using data assimilation techniques should be investigated.

References

[1] Moustafa S. El-Sammany , 2010 "Forecasting of Flash Floods over Wadi Watier – Sinai Peninsula Using the Weather Research and Forecasting (WRF) Model ".

[2] Gamal El Afandi, Mostafa Morsy, and Fathy El Hussieny , 2013 "Heavy Rainfall Simulation over Sinai Peninsula Using the Weather Research and Forecasting Model" International

Journal of Atmospheric Sciences,Volume 2013 (2013), Article ID 241050, http://dx.doi.org/10.1155/2013/241050

[3] MC Anderson, JM Norman, GR Diak, WP Kustas, JR Mecikalski ,2011. "A two-source time-integrated model for estimating surface fluxes using thermal infrared remote sensing" Remote sensing of environment 60 (2), 195-216

[4] E. Kessler,1995,"On the continuity and distribution of water substance in atmospheric circulations " Atmospheric Research 38 (1995) 109-145 William C. Skamarock et al,2008 " A Description of the Advanced Research WRF Version 3"

[5] Lin, Yuh–Lang, Richard D. Farley, and Harold D. Orville, 1983: Bulk Parameterization of the Snow Field in a Cloud Model. J.Climate Appl. Met., 22, 1065–1092.

[6] Hong, Song–You, Jimy Dudhia, and Shu–Hua Chen, 2004: A revised approach to ice microphysical processes for the bulk parameterization of clouds and precipitation. Mon. Wea. Rev., 132, 103–120.

[7] Hong, S.–Y., and J.–O. J. Lim, 2006: The WRF single–moment 6–class microphysics scheme (WSM6). J. Korean Meteor. Soc., 42, 129–151.

[8] Tao, Wei–Kuo, Joanne Simpson, Michael McCumber, 1989: An Ice–Water Saturation Adjustment. Mon. Wea. Rev., 117, 231–235.

[9] Thompson, Gregory, Paul R. Field, Roy M. Rasmussen, William D. Hall, 2008: Explicit Forecasts of Winter Precipitation Using an Improved Bulk Microphysics Scheme. Part II: Implementation of a New Snow Parameterization. Mon. Wea. Rev., 136, 5095–5115.

[10] Milbrandt, J. A., and M. K. Yau, 2005: A multimoment bulk microphysics parameterization. Part I: Analysis of the role of the spectral shape parameter. J. Atmos. Sci., 62, 3051–3064.

[11] Morrison, H., G. Thompson, V. Tatarskii, 2009: Impact of Cloud Microphysics on the Development of Trailing Stratiform Precipitation in a Simulated Squall Line: Comparison of One– and Two–Moment Schemes. Mon. Wea. Rev., 137, 991–1007.

[12] Mlawer, Eli. J., Steven. J. Taubman, Patrick. D. Brown, M. J. Iacono, and S. A. Clough (1997), Radiative transfer for inhomogeneous atmospheres: RRTM, a validated correlated–k model for the longwave. J. Geophys. Res., 102, 16663–16682.

[13] Tewari, M., F. Chen, W. Wang, J. Dudhia, M. A. LeMone, K. Mitchell, M. Ek, G. Gayno, J. Wegiel, and R. H. Cuenca, 2004: Implementation and verification of the unified NOAH land surface model in the WRF model. 20th conference on weather analysis and forecasting/16th conference on numerical weather prediction, pp. 11–15.

[14] Paulson, C. A., 1970: The mathematical representation of wind speed and temperature profiles in the unstable atmospheric surface layer. J. Appl. Meteor., 9, 857–861.

[15] Grell, G. A, D. Devenyi, 2002: A generalized approach to parameterizing convection combining ensemble and data assimilation techniques. Geophys. Res. Lett., 29, 1693

IOP Conf. Series: Journal of Physics: Conf. Series **1039** (2018) 012025 doi:10.1088/1742-6596/1039/1/012025

Econometric Analysis of Influence of Monetary Policy on Macroeconomic Aggregates in Indian Economy

I V Tregub

Department of Data Analysis, Decision Making and Financial Technology, Faculty of Applied Mathematics and Information Technology, Financial University under the Government of the Russian Federation, Leningradsky Prospekt, 49, Moscow, Russia, 125993

Email: itregub@fa.ru

Abstract. This study attempts to examine the relationship between macroeconomic indicators of India and its monetary policy over the period 1980 to 2017. The paper studies the correlation between GPD, Balance of Payments, Inflation, Balance budget, Public and such independent variables as international reserves, money and quasi money, interest rate and reserve ratio with the help the linear regression using OLS and ANOVA test in Excel. From the analysis, it is clear that GDP, Public Debt and Inflation have the strongest dependence and significance in relation to monetary indicators.

1. Introduction

Nowadays India is a dynamically developing country, with strikingly high industry, labor resources and diverse mineral raw material resources, hugely progressing scientific and ethnical potential, rapidly growing service sector, which includes finances, banking and IT. India is among the leaders of Top-10 countries, with best indicators of industrial and agricultural development. The economy is versatile and composed of agriculture, crafts, manufacturing industry and textile.

Despite the fact, that two thirds of labor force is still employed directly or indirectly in agricultural business, the service sector expands at a fast pace and its role in Indian economy grows every year. India is also the main exporter of highly qualified personnel in IT, financial and programming spheres. Other branches, including heavy industry, pharmaceuticals, bio- and nanotechnology, telecommunications, construction, aviation industry and tourism also demonstrate high potential and rapid growth rates.

GDP growth of India has increased in 2014 to 7.3% from previous 6.9% and 5.1% the year before that. However, in 2016, GDP growth slowed and amounted to 7.1% (in compare with data for 2015, growth was 8%) [1]. As many developing countries, India aims high to increase its economic growth, keep inflation low, minimal public debt and other indicators of economic prosperity [2]. Therefore, the aim of the paper to study the influence of monetary policy on the growth of the Indian economy is important.

The paper is organized as follows. In Section 2, the methodology of creating of model will be given. The variables of the econometric model will be provided in Section 3.Then the data are modelled and analyzed in Section 4. In Section 5, we will draw a conclusion based on the analysis result.

2. Methodology of the model specification

In our work, we will use statistical methods of econometric modeling, which are based on four principles [3]. Using the first principle of the specification, we will include only linear equations in the model. We will estimate the coefficients of the model by the method of least squares [4], which is based on minimizing the sum of the squares of the residues. Next, we will check the premises of the Gauss-Markov theorem and the adequacy of the estimated model.

We will suppose that the monetary policy has a direct influence on GDP and the equation is of linear type. Following the 1st principle, we consider Keynesian monetary theory [5], which implies correlation between Reserve requirement, Money supply, Interest rate, Investment and GDP (X_{it}, i=1,…,4) and we will add the statement of the Lucas's theory [6]. We will be interested in the influence of the same set of monetary policy factors on various indicators of economic growth (Y_{jt}), so in general the model of econometric equations can be represented in the following general form:

$$\begin{cases} Y_{jt} = +a_1 \cdot X_{1t} + a_2 \cdot X_{2t} + a_3 \cdot X_{3t} + a_4 \cdot X_{4t} + \varepsilon_t; \\ E(\varepsilon_t) = 0;\ \sigma^2(\varepsilon_t) = const \end{cases} \quad (1)$$

Were ε_t is disturbance term normally distributed with a zero mean and a constant variance; Y_{jt} is a dependent variable; X_{it} – independent variable; a_{0i} – model coefficient; i = 1, …,4 – number of variable or coefficient; t – moment of time. There are no lag variables in our model.

3. Variables and Data

According to the approaches set forth in the Keynesian monetary theory, taking into account the statements of the Lucas's theory and the criticisms outlined in the scientific works [7] and [8], we have chosen as the instruments of monetary policy the following *independent variables*:

International reserves – are any kind of reserve funds that can be passed between the central banks of different countries. International reserves are an acceptable form of payment between these banks. The reserves themselves can either be gold or else a specific currency, such as the dollar or euro.

Money and quasi-money – comprise the sum of currency outside banks, demand deposits other than those of the central government, and the time, savings, and foreign currency deposits of resident sectors other than the central government.

Interest rate – the proportion of a loan that is charged as interest to the borrower, typically expressed as an annual percentage of the loan outstanding.

Requirment reserves ratio – a central bank regulation employed by most, but not all, of the world's central banks, that sets the minimum amount of reserves that must be held by a commercial bank.

The *dependent variables* are described as follows:

GDP – real growth rate (%) – the real economic growth rate measures economic growth, in relation to gross domestic product (GDP), from one period to another, adjusted for inflation - in other words, expressed in real as opposed to nominal terms.

Public debt (percentage of GDP) – Debt-to-GDP ratio is the ratio between a country's government debt (a cumulative amount) and its gross domestic product (GDP) (measured in years). A low debt-to-GDP ratio indicates an economy that produces and sells goods and services sufficient to pay back debts without incurring further debt.

Inflation rate (Consumer Price Index, %) – measures changes in the price level of a market basket of consumer goods and services purchased by households. he annual percentage change in a CPI is used as a measure of inflation. A CPI can be used to index (i.e., adjust for the effect of inflation) the real value of wages, salaries, pensions, for regulating prices and for deflating monetary magnitudes to show changes in real values.

Current account – the record of all economic transactions between the residents of the country and the rest of the world in a particular period (over a quarter of a year or more commonly over a year). Individuals, firms and government bodies make these transactions. Thus, the balance of payments includes all external visible and non-visible transactions of a country.

Budget balance (percentage of GDP) – the difference between national government revenues and expenditures, expressed as a percent of GDP. A positive (+) number indicates that revenues exceeded expenditures (a budget surplus), while a negative (-) number indicates the reverse (a budget deficit).

Data, related to macroeconomic indicators changes, is extracted from World Bank and covers 1995-2017, with 2017 statistics not included in calculations and used for comparison with further obtained results.

IOP Conf. Series: Journal of Physics: Conf. Series **1039** (2018) 012025 doi:10.1088/1742-6596/1039/1/012025

4. Evaluating and Results

For evaluating of the model coefficients with the aim of establishing the level of dependency of *GDP*, *Exchange rate* and *Growth rate*, *Public debt*, *Inflation rate, BoP* on international reserves, such as *money* and *quasi-money*, *interest rate* and *reserve ratio* and for making appropriate tests, we have used linear regression analysis and additional functions in Excel [9].

After the indicators are calculated we need to obtain F_{crit}, which represents critical value is a point on the test distribution that is compared to the test statistic to determine whether to reject the null hypothesis. If the absolute value of your test statistic is greater than the critical value, you can declare statistical significance and reject the null hypothesis. To find critical value we apply function 'F.INV.RT' and indicate probability (5%) and degrees of freedom, which in our case correspond to residual and regression values of ANOVA test.

The value of the R^2 is used to determine how strong or weak the regression equation is. If R^2 value lies between 0.8 and 1, then we can conclude that the regression equation is strong. However, if the value of the R^2 lies between 0 and 0.5, then we can say that the regression equation is a weak one.

The value of the F statistic is used to ascertain the overall significance of the GDP growth rate equation. We compare the value of the F statistic with the value of the critical value of F at a given significance level 5%. If the value of the F statistic is greater than the value of the F critical, then the overall equation is statistically significant.

The statistical significance of the parameters will be established. In testing for the statistical significance of the parameters, we use the t-test. It concludes that for a parameter to be statistically significant, the absolute value of the t-statistic should be greater than or equal to two. Using the t-test, the t-critical value is compared with the t-statistic at a given significance level (1%). If the t-statistic is greater than the t-critical value, then, the parameter in question is statistically significant. If otherwise, then, the parameter is not statistically significant.

The coefficients and the intercepts of independent variables are be obtained by using the ordinary least squares estimation technique. As a result of the evaluation we received the following models.

4.1. GDP - real growth rate (%) model

$$\begin{cases} \hat{Y}_{1t} = 6{,}10 + 0{,}018 \cdot \mathrm{ir} - 0{,}09 \cdot \mathrm{M2} - 0{,}31 \cdot \mathrm{i} + 0{,}48 \cdot \mathrm{rr} \\ R^2 = 0{,}42; F = 2{,}6; F_{crit} = 3{,}1 \end{cases} \tag{2}$$

Where

$\hat{Y}_1$ – estimated value of the real growth rate of Gross Domestic Product, (% to the previous period);
ir – international reserves (bln. USD);
$M2$ – money and quasi money (bln. USD);
i – interest rate (%);
rr – reserve ratio (%).

From the regression results one may conclude, that if the value of all the variables *ir, i, rr, M2* will be zero, then the GDP growth rate is 6.10, meaning that other variables also influence on the GDP growth rate. This is in complete agreement with the actually observed values of the growth rate of GDP. Actual GDP growth rate for 2016 was 7.1%.

Coefficients, attributed to International Reserves (*ir*) and Reserves Ratio (*rr*) are positive. This means that increasing the International Reserves on one bln. USD will cause GDP growth rate to increase only by 0,018% and 1% increase in Reserves Ratio will cause GDP growth rate to increase by 0,48%.

Coefficients, attributed to Money Aggregate and Inflation are negative, which means the otherwise relation with increase of them values and corresponding effect on GDP.

The value of the constant term and the slope coefficient under International Reserves and Money Aggregate are statistically significant. However, from the regression results of t-test the constants under *i* and *rr* are statistically insignificant (Tab.1.). Furthermore, according to the R^2 test obtained from the regression analysis, we may conclude, that the value $0,42$ means that about 42% of the total variation in the GDP growth rate is attributed to independent variables of the equation. It is statistically insignificant since $F < F_{crit}$. This means that there are other factors that can better explain the GDP growth rate of India.

4.2. Public debt (% of GDP) model

$$\begin{cases} \widehat{\text{Debt}} = 57{,}06 + 0{,}10 \cdot \text{ir} - 0{,}53 \cdot \text{M2} + 1{,}75 \cdot \text{i} + 0{,}71 \cdot \text{rr} \\ R^2 = 0{,}67; F = 7{,}15; F_{\text{crit}} = 3{,}1 \end{cases} \quad (6)$$

Where

$\widehat{\text{Debt}}$ – estimated value of Public debt, (percentage of GDP)

We may see that about 67 % of the total variation in Public debt of India is attributed to independent variables of the equation. It is statistically significant. Dependency of Public Debt is relatively high on independent variables.

The slope coefficients *ir,i,rr* are positive. Important dependence: a 1% increase in interest rate leads to 1,75% increase in Public Debt.

Coefficient of M2 is negative, which shows a logical reduction in debt when the money supply increases. Factors M2, *i, ir* are significant and *rr* is insignificant.

If all *ir, i, rr*, M2 are zero, then Public Debt (% of GDP) is 57,06% . The value of the constant term is statistically significant since its t-statistic is greater than 1.7.

4.3. Inflation rate (consumer prices) (%) model

$$\begin{cases} \widehat{\text{Infl}} = 1{,}61 + 0{,}00093 \cdot \text{ir} + 0{,}06 \cdot \text{M2} + 0{,}038 \cdot \text{i} + 0{,}36 \cdot \text{rr} \\ R^2 = 0{,}64; F = 6{,}39; F_{\text{crit}} = 3{,}1 \end{cases} \quad (7)$$

Where

$\widehat{\text{Infl}}$ – estimated value of Consumer Price Index, (%)

We may see that 64% of the total variation in Inflation rate of India is attributed to independent variables of the equation. It is statistically significant since $F > F_{crit}$. Linear equation is relatively strong. However, all the Coefficients are insignificant. This means that for the study of inflation within the framework of this model it is necessary to select other factors

4.4. BoP current account (% of GDP) model

$$\begin{cases} \widehat{\text{BoP}} = -3{,}44 - 0{,}013 \cdot \text{ir} + 0{,}021 \cdot \text{M2} + 0{,}21 \cdot \text{i} + 0{,}35 \cdot \text{rr} \\ R^2 = 0{,}59; F = 5{,}07; F_{\text{crit}} = 3{,}1 \end{cases}; \quad (8)$$

Where

$\widehat{\text{BoP}}$ – estimated value of Current account.

59 % of the total variation in Balance of Payments of India is attributed to independent variables of the equation. It is statistically significant since $F > F_{crit}$. Equation's relationship is medium. Coefficients M2, *i, rr* are positive. In the equation there is only one significant factor, this is International Reserves (*ir*). The negative coefficient under *ir* shows that increasing it by one bln. USD leads to a decrease BoP by 0.013 mln. USD, which is in complete agreement with the real data.

5. T-test

The critical value of Student's statistics for significance level 0.1 and df_1=20 is 1.72 (t_{crit}=1.72). As we can see from the table 1, several of intercept and the slope coefficients are less than t_{crit}. This means that the corresponding variables are insignificant and do not influence on macroeconomic indicators of Indian economy is weak.

Table 1. Results of t-test.

	Y_1	*Debt*	*Infl*	*BoP*	*Budg*
Intercept	1,70	3,70	0,39	-1,32	-3,45
ir	1,74	2,22	0,07	-1,78	-0,01
M2	-2,41	-3,30	1,50	0,78	0,02
i	-1,38	1,76	0,14	1,30	0,22
rr	0,81	0,28	0,53	0,82	0,36

In general, all the constricted models is completely relevant in economic terms and is important in economic investigations.

6. Conclusion

The research presented in this article made it possible to identify the impact of monetary policy factors on the macroeconomic indicators of the Indian economy. From our analysis, we strongly conclude that there is a positive strong dependence between GDP and Money and Quasi-money. At the same time, there is no great dependence between Public debt and interest rate. Inflation rate is just indifferent to international reserves. Current account does not depend on monetary policy nearly at all. What is more, strong significance between GDP and money supply has been proved, which indicates high sensibility of living costs and monetary indicators, in particular M2 and International reserves volume. On the other hand, GDP growth rate seems to be completely unrelated, as it under conditions of non-existence of monetary indicators, is just a 1% difference from actual growth rate of 2017. Hence, in terms of desirable increase in growth of GDP in India, policies should be concentrated on money supply and cash equivalents as well banking sphere as they strongly affect PPP. Furthermore, focus on retaining interest rate should also be considered as it greatly affects Public Debt. These conclusions might only be applicable in the short-term as more data and research are need to prove its sustainability in the long-run, which might imply monetary policy effectiveness only temporarily and a need to implement fiscal policy further.

References

[1] Paul Cashin *Report India 02'2017* – (International Monetary Fund) 2017
[2] Arestis P and Sawyer M 2005 *New Consensus Monetary Policy: an appraisal in The New Monetary Policy*, Cheltenham (UK: Edward Elgar)
[3] Turkina S N, Tregub I V 2014 *Menges model applied to the economy of the USA* Marketing i menedzhment innovacij **4** p 128-35Dougherty C *Introduction to Econometrics* – (Oxford University Press) 2004
[4] Dougherty C *Introduction to Econometrics* – (Oxford University Press) 2004
[5] Keynes J M 1936 *The General Theory of Employment, Interest and Money* (London: Macmillan)
[6] Lucas Robert E Jr 1984 *Money in a Theory of Finance* Carnegie Rochester Series on Public Policy 27 p 946
[7] Rogers C 2006 *Exogenous interest rates and modern monetary theory and policy* Essays in Honour of Basil J Moore (Cheltenham: Edward Elgar)
[8] Fisher I 1907 *The Rate of Interest; Its nature, determination and relation to economic phenomena*, (New York: Macmillan)
[9] Tregub I V On the applicability of the random walk model with stable steps for forecasting the dynamics of prices of financial tools in the Russian market 2016 *Journal of Mathematical Sciences* **5** p 716-21

IOP Conf. Series: Journal of Physics: Conf. Series **1039** (2018) 012026 doi:10.1088/1742-6596/1039/1/012026

Galois ring codes and their images under various bases

Virgilio Sison and Gio Carlo Delos Santos

Institute of Mathematical Sciences and Physics, University of the Philippines Los Baños, College, Laguna 4031, Philippines

E-mail: vpsison@up.edu.ph, gcdelossantos29@gmail.com

Abstract. In this paper we consider linear block codes B of length n over the Galois ring $GR(p^r, m)$ and obtain their images with respect to various bases of $GR(p^r, m)$ seen as a free module of rank m over the residue class ring $\mathbb{Z}_{p^r}$. Interesting new examples of dual, normal and self-dual bases of $GR(p^r, m)$ and their relationships are given. The image of B is a linear block code over $\mathbb{Z}_{p^r}$ of length mn and its generator matrix is formed row-wise by the images of $\beta_i G$, where $\{\beta_i\}_{i=1}^{m}$ is a chosen basis of $GR(p^r, m)$ and G is a generator matrix of B. Certain conditions in which the p^r-ary image is distance-invariant after a change in basis are investigated. Consequently a new quaternary code endowed with a homogeneous metric that is optimal with respect to certain known bounds is constructed.

1. Introduction

The motivation for this paper came from finding the p^r-ary image of a linear block code B of length n over the Galois ring $\mathcal{R} = GR(p^r, m)$ with respect to any m-basis of $\mathcal{R}$, and placing a bound on its minimum homogeneous distance in terms of the Hamming distance and other code parameters of B. The p^r-ary image is a linear block code of length mn over the residue class ring $\mathbb{Z}_{p^r}$, and its distance is with respect to the homogeneous metric defined on $GR(p^r, m)$ seen as a finite Frobenius chain ring. The main reference for this result is a previous paper [1] which considers for the most part the so-called polynomial basis of $GR(p^r, m)$. It is worthy to investigate the dual and normal bases to see how a change in basis affects the parameters of the p^r-ary image. For a thorough discussion of the bases of $GR(p^r, m)$ over $\mathbb{Z}_{p^r}$ the reader is referred to [2]. This paper constructed the dual basis using matrix algebra involving the generalized Frobenius automorphism. It showed that a basis of $GR(p^r, m)$ is self-dual if and only its automorphism matrix is orthogonal, and is normal if and only its automorphism matrix is symmetric.

The material is organized as follows: Section 2 gives the preliminaries and basic definitions while Section 3 gives the main results. Several new illustrative examples are also provided.

2. Preliminaries and definitions

A quick run-through of the theoretical requisites for this paper is presented in this section. This includes the structure of Galois rings as finite chain rings and Frobenius rings and the homogeneous weight that can be defined on them, the construction of Galois ring codes and their images with respect to a modular basis, and the notions of duality and normality of bases of Galois rings which are used in the derivation of optimal image codes. For a more thorough discussion of these topics the reader is referred to [1], [2], [3], [4], and [5].

2.1. Galois rings and homogeneous weight

A *Galois ring* is, in the most general sense, a finite commutative local ring with identity 1 such that its zero divisors together with the zero element form the unique maximal principal ideal $(p1)$ for some prime p. Consider the ring $\mathbb{Z}_{p^r}[x]$ of polynomials over the residue class ring $\mathbb{Z}_{p^r}$ of integers modulo p^r where r is a positive integer. Let $h(x)$ be a monic polynomial of degree m in $\mathbb{Z}_{p^r}[x]$ such that its image under the mod-p reduction map, extended coefficient-wise, is irreducible in $\mathbb{F}_p[x]$. Then the quotient ring $\mathbb{Z}_{p^r}[x]/(h(x))$ is a Galois ring with characteristic p^r and cardinality p^{rm}. Now let ω be a root of $h(x)$. It follows that $\mathbb{Z}_{p^r}[x]/(h(x))$ is ring isomorphic to the Galois extension $\mathbb{Z}_{p^r}[\omega]$ of $\mathbb{Z}_{p^r}$ in which the elements take the unique additive representation

$$z = a_0 + a_1\omega + \ldots + a_{m-1}\omega^{m-1} \tag{1}$$

where $a_i \in \mathbb{Z}_{p^r}$. Any Galois ring with characteristic p^r and cardinality p^{rm} is denoted by $GR(p^r, m)$. This ring is a finite chain ring of length r whose $r+1$ ideals $(p^i), i = 0, 1, \ldots, r$, with $p^{(r-i)m}$ elements, are linearly ordered by inclusion,

$$\{0\} = (p^r) \subset (p^{r-1}) \subset \ldots \subset (p) \subset \mathcal{R} \tag{2}$$

Observe that $GR(p, m)$ is the Galois field $\mathbb{F}_{p^m}$ and $GR(p^r, 1) = \mathbb{Z}_{p^r}$. Furthermore, $GR(p^r, m)$ can be seen as a free module of rank m over $\mathbb{Z}_{p^r}$ with the set

$$\mathcal{P}_m(\omega) = \{1, \omega, \omega^2, \ldots, \omega^{m-1}\} \tag{3}$$

as a free basis taken from the additive representation of $GR(p^r, m)$ in (1). The set $\mathcal{P}_m(\omega)$ is called the *standard* or *polynomial basis* of $GR(p^r, m)$. It is known that any other basis of $GR(p^r, m)$ has the same cardinality m.

For $GR(p^r, m)$ we apply the following homogeneous weight as given in [6] for finite chain rings.

$$w_{\mathtt{hom}}(x) = \begin{cases} 0 & \text{if } x = 0 \\ p^{m(r-1)} & \text{if } x \in (p^{r-1}) \setminus \{0\} \\ (p^m - 1)p^{m(r-2)} & \text{otherwise} \end{cases} \tag{4}$$

where (p^{r-1}) is the minimal ideal. The ring $GR(p^r, m)$ is Frobenius since it admits a generating character given by $\chi(z) = \xi^{b_{m-1}}$, where $\xi = \exp(2\pi i/p^r)$ for $z = \sum_{i=0}^{m-1} b_i\omega^i$. It is thus easy to compute from [3] and [7] that the average value of $w_{\mathtt{hom}}$ is equal to

$$\Gamma = (p^m - 1)p^{m(r-2)} \tag{5}$$

which is its minimum non-zero value.

2.2. Codes over $GR(p^r, m)$ and their p^r-ary images

A rate-k/n linear block code B of length n over $\mathcal{R} = GR(p^r, m)$ is an $\mathcal{R}$-submodule of $\mathcal{R}^n$ generated by a $k \times n$ matrix G over $\mathcal{R}$ whose rows span B such that no proper subset of the rows of G generates B. The rows of G need not be linearly independent. We say that B is free if the rows of G form a basis.

Let the set $\mathcal{B}_m = \{\beta_0, \beta_1, \ldots, \beta_{m-1}\}$ be a basis of $\mathcal{R}$ over $\mathbb{Z}_{p^r}$. We use the bijective map $\tau : \mathcal{R} \longrightarrow \mathbb{Z}_{p^r}^m$ given in [1] as follows

$$\tau(z) = (a_0, a_1, \ldots, a_{m-1}) \tag{6}$$

for $z = a_0\beta_0 + a_1\beta_1 + \ldots + a_{m-1}\beta_{m-1} \in \mathcal{R}$, $a_i \in \mathbb{Z}_{p^r}$. This map is extended coordinate-wise to $\mathcal{R}^n$. Thus, if $c \in B$ and $c = (c_0, c_1, \ldots, c_{n-1})$, $c_i = \sum_{j=0}^{m-1} a_{ij}\beta_j$, $a_{ij} \in \mathbb{Z}_{p^r}$, then

$$\tau(c) = (a_{00}, \ldots, a_{0,m-1}, \ldots, a_{n-1,0}, \ldots a_{n-1,m-1}) \tag{7}$$

in $\mathbb{Z}_{p^r}^{mn}$. The image $\tau(B)$ of B under τ with respect to $\mathcal{B}_m$ is called the p^r*-ary image* of B, and is obtained by simply substituting each element of $\mathcal{R}$ by the m-tuple of its coordinates over B. It is easy to prove that $\tau(B)$ is a linear block code of length mn over $\mathbb{Z}_{p^r}$. Since τ is bijective, $|B| = |\tau(B)|$. For the degenerate case $m = 1$, the block code B is a code over $\mathbb{Z}_{p^r}$ and the map τ is the identity map on B. We equip $\tau(B)$ with a homogeneous distance metric with respect to the weight $w_{\texttt{hom}}$ given in (4).

The following proposition in [1] shows how the minimum Hamming weight of B provides a bound for the minimum homogeneous distance of $\tau(B)$.

Proposition 2.1. *Let B be a linear block code of length n over $\mathcal{R} = GR(p^r, m)$ with minimum Hamming distance d, and $\tau(B)$ be the p^r-ary image of B with respect to a basis of $\mathcal{R}$ over $\mathbb{Z}_{p^r}$ with minimum homogeneous distance δ. Then*

$$\Gamma d \le \delta \le p^{r-1} md\,. \tag{8}$$

The image code $\tau(B)$ is *Type* α if $\delta = p^{r-1} d_{\tau(B)}$, where $d_\tau(B)$ is the Hamming distance of $\tau(B)$. The upper bound in (8) is sharpened further in the theorem below.

Theorem 2.2. *Let B be a linear block code of length n over $\mathcal{R} = GR(p^r, m)$ with minimum Hamming distance d, B_x the subcode of B generated by a codeword x with $w_H(x) = d$, and δ the minimum homogeneous distance of the p^r-ary image of B with respect to a basis of $\mathcal{R}$ over $\mathbb{Z}_{p^r}$. If B_x is free, then*

$$\delta \le \left\lfloor \frac{(p-1)p^{rm+r-2} md}{p^{rm} - 1} \right\rfloor . \tag{9}$$

The bound of Rabizzoni in [8, Theorem 1] becomes a specific case for Galois fields of Theorem 2.2. It was determined in [1] that the quaternary image of the 3-quasi cyclic rate-2/6 linear block code over $GR(4, 2)$ with generator matrix

$$\begin{pmatrix} 3+2\omega & 2+3\omega & \omega & \omega & 3+2\omega & 2+3\omega \\ \omega & 3+2\omega & 2+3\omega & 3+2\omega & 2+3\omega & \omega \end{pmatrix} \tag{10}$$

is a Type α code. Furthermore, a minimum-weight codeword $x = (1, 1 + 2\omega, 0, 3 + 2\omega, 3, 0)$ which generates a free code B_x with 16 elements makes the quaternary image to also reach the generalized Rabizzoni upper bound in (9). It should be emphasized that the basis used to get the quaternary image of this code via τ is the polynomial basis of $GR(4, 2)$ given by $\{1, \omega\}$, where $1 + \omega + \omega^2 = 0$.

2.3. Dual and normal bases of $GR(p^r, m)$

Referring to [2], we now consider bases of $\mathcal{R} = GR(p^r, m)$ other than the polynomial basis. Two bases $\{\alpha_1, \alpha_2, \ldots, \alpha_m\}$ and $\{\beta_1, \beta_2, \ldots, \beta_m\}$ are said to be dual if $T(\beta_i\alpha_j) = \delta_{ij}$, where δ_{ij} is the Kronecker delta and T is the generalized trace map from $\mathcal{R}$ down to $\mathbb{Z}_{p^r}$. Specifically, the basis $\{\beta_1, \beta_2, \ldots, \beta_m\}$ is self-dual if $T(\beta_i\beta_j) = \delta_{ij}$. It was proved that every basis has a unique dual

IOP Conf. Series: Journal of Physics: Conf. Series **1039** (2018) 012026 doi:10.1088/1742-6596/1039/1/012026

basis using the automorphism matrix $\Omega \in GL(m,\mathcal{R})$ relative to a basis $\{\beta_j\} = \{\beta_1, \beta_2, \ldots, \beta_m\}$ in terms of the generalized automorphism f on $\mathcal{R}$.

$$\Omega = \begin{pmatrix} \beta_1 & \beta_1^f & \beta_1^{f^2} & \ldots & \beta_1^{f^{m-1}} \\ \beta_2 & \beta_2^f & \beta_2^{f^2} & \ldots & \beta_2^{f^{m-1}} \\ \vdots & & & & \vdots \\ \beta_m & \beta_m^f & \beta_m^{f^2} & \ldots & \beta_m^{f^{m-1}} \end{pmatrix} \tag{11}$$

A normal basis of $GR(p^r, m)$ is a basis of the form $\{\alpha, \alpha^f, \alpha^{f^2}, \ldots, \alpha^{f^{m-1}}\}$ where $\alpha \in GR(p^r, m)$. The paper asserted that a basis $\{\beta_j\}$ with automorphism matrix Ω is self-dual if and only if Ω is an orthogonal matrix, and that $\{\beta_j\}$ is normal if and only if Ω is a symmetric matrix, or equivalently, the Frobenius automorphism f is the m-cycle $(\beta_1\, \beta_2 \ldots \beta_m)$.

3. Results and Discussion

We begin by giving new interesting examples of normal and dual bases of certain Galois rings. MAGMA routines are written to generate the bases of $GR(p^r, m)$, to derive the automorphism matrix and to test for symmetry and orthogonality. Several cases are considered including specifically the Galois fields $GR(p, m)$. We disregard the Galois ring $GR(p^r, 1)$ because it has a trivial basis.

Example 3.1. *We extend Example 3.8 and Example 3.15 of [2] and verify that none of the four distinct normal bases of $GR(4,2)$ is self-dual as Table 1 shows. Note that the dual of a normal basis is also normal.*

Table 1. Normal bases of $GR(4,2)$ and their duals

Normal Basis	Dual
$\{\omega, 3+\omega\}$	$\{2+\omega, 1+3\omega\}$
$\{1+\omega, 3\omega\}$	$\{3+\omega, 2+3\omega\}$
$\{2+\omega, 1+3\omega\}$	$\{\omega, 3+3\omega\}$
$\{3+\omega, 2+3\omega\}$	$\{1+\omega, 3\omega\}$

Example 3.2. *We extend Example 3.9 and Example 3.16 of [2] and find that there are eight distinct normal bases of $GR(4,3)$. Table 2 shows these normal bases and their duals. Observe that the first four normal bases are self-dual. For instance, consider the normal basis $\{3+\omega, 3+\omega^2, 1+3\omega+3\omega^2\}$. The automorphism matrix relative to this basis is the circulant matrix given by*

$$\Omega = \begin{pmatrix} 3+\omega & 3+\omega^2 & 1+3\omega+3\omega^2 \\ 3+\omega^2 & 1+3\omega+3\omega^2 & 3+\omega \\ 1+3\omega+3\omega^2 & 3+\omega & 3+\omega^2 \end{pmatrix} \tag{12}$$

The Frobenius automorphism on $GR(4,3)$ is a 3-cycle and Ω is symmetric. Further, Ω is orthogonal and the hence the basis is self-dual as well.

Table 2. Normal bases of $GR(4,3)$ and their duals

Normal Basis	Dual
$\{1+\omega, 1+\omega^2, 3+3\omega+3\omega^2\}$	$\{1+\omega, 1+\omega^2, 3+3\omega+3\omega^2\}$
$\{3+\omega, 3+\omega^2, 1+3\omega+3\omega^2\}$	$\{3+\omega, 3+\omega^2, 1+3\omega+3\omega^2\}$
$\{1+3\omega, 1+3\omega^2, 3+\omega+\omega^2\}$	$\{1+3\omega, 1+3\omega^2, 3+\omega+\omega^2\}$
$\{3+3\omega, 3+3\omega^2, 1+\omega+\omega^2\}$	$\{3+3\omega, 3+3\omega^2, 1+\omega+\omega^2\}$
$\{1+2\omega+\omega^2, 3+3\omega+\omega^2, 1+3\omega+2\omega^2\}$	$\{1+2\omega+3\omega^2, 3+\omega+3\omega^2, 1+\omega+2\omega^2\}$
$\{3+2\omega+\omega^2, 1+3\omega+\omega^2, 3+3\omega+2\omega^2\}$	$\{3+2\omega+3\omega^2, 1+\omega+3\omega^2, 3+\omega+2\omega^2\}$
$\{1+\omega+2\omega^2, 1+2\omega+3\omega^2, 3+\omega+3\omega^2\}$	$\{1+3\omega+2\omega^2, 1+2\omega+\omega^2, 3+3\omega+\omega^2\}$
$\{3+\omega+2\omega^2, 3+2\omega+3\omega^2, 1+\omega+3\omega^2\}$	$\{3+3\omega+2\omega^2, 3+2\omega+\omega^2, 1+3\omega+\omega^2\}$

Example 3.3. *Consider the Galois ring* $GR(8,2)=\mathbb{Z}_8[\omega]$*, where* ω *is the root of the monic basic irreducible polynonial* $x^2+x+1\in\mathbb{Z}_8[x]$*. The automorphism matrix with respect to the polynomial basis* $\mathcal{P}_2(\omega)=\{1,\omega\}$ *of* $GR(8,2)$ *is given by*

$$\Omega_P=\begin{pmatrix} 1 & 1 \\ \omega & 7+7\omega \end{pmatrix} \tag{13}$$

The inverse of Ω *is given by*

$$\Omega_P^{-1}=\begin{pmatrix} 3+5\omega & 5+2\omega \\ 6+3\omega & 3+6\omega \end{pmatrix} \tag{14}$$

so that $\{3+5\omega, 5+2\omega\}$ *is the dual of* $\mathcal{P}_2(\omega)$*. Table 3 gives the normal bases of* $GR(8,2)$ *and their duals. Note that none of the* 16 *distinct normal bases of* $GR(8,2)$ *is self-dual. But their duals are normal bases also. This can be seen through a simple determination of the automorphism matrix. For instance, the set* $\{\omega, 7+7\omega\}$ *is normal and its automorphism matrix is given by the circulant matrix*

$$\Omega_N=\begin{pmatrix} \omega & 7+7\omega \\ 7+7\omega & \omega \end{pmatrix} \tag{15}$$

which is symmetric but not orthogonal. In fact, its dual is $\{2+5\omega, 5+3\omega\}$.

We now specifically consider a free linear block code B over $GR(4,2)$ and $\tau(B)$ its quaternary image with respect to a basis $\{\beta_1,\beta_2\}$ over $\mathbb{Z}_4$. If B has rate-k/n, then $\tau(B)$ has rate-$2k/2n$. Consequently, if G is a generator matrix of B, then $\tau(B)$ is generated by the matrix

$$G[\tau(B)]=\begin{pmatrix} \tau(\beta_1 G) \\ \tau(\beta_2 G) \end{pmatrix} \tag{16}$$

This specific result is generalized in the following theorem.

Theorem 3.4. *Let* B *be a free rate-*k/n *linear block code of length* n *over* $GR(p^r,m)$ *with generator matrix* G *and* $\tau(B)$ *be the* p^r*-ary image of* B *with respect to the basis* $\{\beta_1,\beta_2,...,\beta_m\}$*. Then a generator of* $\tau(B)$ *is given by the* $mk\times mn$ *matrix*

$$\begin{pmatrix} \tau(\beta_1 G) \\ \tau(\beta_2 G) \\ \vdots \\ \tau(\beta_m G) \end{pmatrix}$$

Table 3. Normal bases of $GR(8,2)$ and their duals

Normal Basis	Dual
$\{\omega, 7+7\omega\}$	$\{2+5\omega, 5+3\omega\}$
$\{1+\omega, 7\omega\}$	$\{3+5\omega, 6+3\omega\}$
$\{2+\omega, 1+7\omega\}$	$\{5\omega, 3+3\omega\}$
$\{3+\omega, 2+7\omega\}$	$\{1+5\omega, 4+3\omega\}$
$\{4+\omega, 3+7\omega\}$	$\{6+5\omega, 1+3\omega\}$
$\{5+\omega, 4+7\omega\}$	$\{7+5\omega, 2+3\omega\}$
$\{6+\omega, 5+7\omega\}$	$\{4+5\omega, 7+3\omega\}$
$\{7+\omega, 6+7\omega\}$	$\{5+5\omega, \omega\}$
$\{3\omega, 5+5\omega\}$	$\{6+7\omega, 7+\omega\}$
$\{1+3\omega, 6+5\omega\}$	$\{3+7\omega, 4+\omega\}$
$\{2+3\omega, 7+5\omega\}$	$\{4+7\omega, 5+\omega\}$
$\{3+3\omega, 5\omega\}$	$\{1+7\omega, 2+\omega\}$
$\{4+3\omega, 1+5\omega\}$	$\{2+7\omega, 3+\omega\}$
$\{5+3\omega, 2+5\omega\}$	$\{7+7\omega, \omega\}$
$\{6+3\omega, 3+5\omega\}$	$\{7\omega, 1+\omega\}$
$\{7+3\omega, 4+5\omega\}$	$\{5+7\omega, 6+\omega\}$

Proof. Suppose the rows of G are $r_1, r_2, \ldots, r_k$. These rows form a basis for B. It can be shown by straightforward approach that the set

$$\{\tau(\beta_1 r_1), \ldots, \tau(\beta_1 r_k), \tau(\beta_2 r_1), \ldots, \tau(\beta_2 r_k), \ldots, \tau(\beta_m r_1), \ldots, \tau(\beta_m r_k)\}$$

is a basis for $\tau(B)$. Thus a generator matrix for $\tau(B)$ can have the elements of this basis as rows as shown below.

$$G[\tau(B)] = \begin{pmatrix} \tau(\beta_1 r_1) \\ \vdots \\ \tau(\beta_1 r_k) \\ \tau(\beta_2 r_1) \\ \vdots \\ \tau(\beta_2 r_k) \\ \vdots \\ \tau(\beta_m r_1) \\ \vdots \\ \tau(\beta_m r_k) \end{pmatrix} = \begin{pmatrix} \tau(\beta_1 G) \\ \tau(\beta_2 G) \\ \vdots \\ \tau(\beta_m G) \end{pmatrix} \tag{17}$$

This theorem shows that the generator matrix of the p^r-ary image is completely determined by the generator matrix G of the free linear block code B over $GR(p^r, m)$ and the chosen basis for $GR(p^r, m)$.

Referring back to the 3-quasi cyclic rate-2/6 linear block code B over $GR(4,2)$ with generator matrix given in (10) with Hamming distance $d = 4$, its quaternary image $\tau_{\mathcal{P}}(B)$ with respect to the polynomial basis $\mathcal{P}_2(\omega) = \{1, \omega\}$ is a rate-4/12 linear block code over $\mathbb{Z}_4$ with 256 codewords.

The generator matrix of the quaternary image $\tau_P(B)$ is given by

$$G[\tau_P(B)] = \begin{pmatrix} 1 & 0 & 0 & 0 & 3 & 1 & 0 & 2 & 3 & 2 & 1 & 1 \\ 0 & 1 & 0 & 0 & 3 & 2 & 2 & 2 & 2 & 1 & 3 & 0 \\ 0 & 0 & 1 & 0 & 3 & 3 & 3 & 2 & 0 & 2 & 1 & 3 \\ 0 & 0 & 0 & 1 & 1 & 0 & 2 & 1 & 2 & 2 & 1 & 2 \end{pmatrix} \tag{18}$$

Its Lee distance is $d_L = 8$ and Hamming distance $d_{\tau_P(B)} = 4$, making it a Type α code. With a suitable choice of minimum-weight word, the image code is also optimal with respect to the generalized Rabizzoni bound (9).

Example 3.5. *The dual basis of $\mathcal{P}_2(\omega)$ is the set $D = \{3+\omega, 1+2\omega\}$. Given the same $GR(4,2)$ code B above, the quaternary image $\tau_D(B)$ with respect to D is generated by the matrix*

$$G[\tau_D(B)] = \begin{pmatrix} 1 & 0 & 0 & 0 & 3 & 3 & 0 & 2 & 3 & 2 & 1 & 3 \\ 0 & 1 & 0 & 0 & 1 & 2 & 2 & 2 & 2 & 1 & 1 & 0 \\ 0 & 0 & 1 & 0 & 3 & 1 & 3 & 2 & 0 & 2 & 1 & 1 \\ 0 & 0 & 0 & 1 & 3 & 0 & 2 & 1 & 2 & 2 & 3 & 2 \end{pmatrix} \tag{19}$$

Although $\tau_D(B)$ is derived from the same code B via a change in basis, the quaternary image $\tau_D(B)$ is not equal to $\tau_P(B)$ and is an entirely new rate-4/12 linear block code over $\mathbb{Z}_4$ with 256 codewords. However, the distances are preserved. Its Lee distance is $d_L = 8$ and Hamming distance is $d_{\tau_D(B)} = 4$, which makes it a Type α code and a Rabizzoni-optimal code as well.

The code in Example 3.5 is the second known example of a $\mathbb{Z}_4$-code with such optimality. It should be observed further that the quaternary image of B with respect to the normal basis $N = \{\omega, 3+3\omega\}$ has exactly the same generator matrix as $\tau_P(B)$ and hence $\tau_P(B) = \tau_N(B)$. This example shows that it is possible to construct new good codes over $\mathbb{Z}_{p^r}$ from a given linear block code over $GR(p^r, m)$ with just a suitabe change in basis of $GR(p^r, m)$.

It may happen that the p^r-ary images of a linear block code B over $GR(p^r, m)$ are all the same under any basis of $GR(p^r, m)$ and thus all distances are retained, such as the case of the non-free rate-3/4 $GR(4,2)$-code generated by

$$G[B] = \begin{pmatrix} 1 & 1 & 1 & 3+2\omega \\ 0 & 2 & 0 & 2 \\ 0 & 0 & 2 & 2 \end{pmatrix} \tag{20}$$

However, there is a case where the image codes have varying homogeneous distances, such as the case of the non-free rate-2/4 $GR(4,3)$-code generated by

$$G[B] = \begin{pmatrix} 1 & 1 & 0 & \omega^2 \\ 0 & 2 & 2\omega^2 & 2+2\omega \end{pmatrix} \tag{21}$$

Example 3.6. *The polynomial basis of $\mathbb{F}_4 = \mathbb{F}_2[\omega]$ where $\omega^2+\omega+1 = 0$ is $\mathcal{P}_2(\omega) = \{1, \omega\}$ whose dual is $D = \{1+\omega, 1\}$, while a normal basis is $N = \{\omega, 1+\omega\}$. The following identities can be easily verified for any $v \in \mathbb{F}_4$.*

$$\tau_P(v) = \tau_D((1+\omega)v) = \tau_N(\omega v)$$

$$\tau_P(\omega v) = \tau_D(v) = \tau_N(1+\omega)v$$

Because of this result and by extension, the binary images of a linear block code over $\mathbb{F}_4$ with respect to $\mathcal{P}_2(\omega)$, D and N will be the same, so no change in basis is necessary. As a specific case, consider the rate-3/5 $\mathbb{F}_4$-linear block code over $\mathbb{F}_4$ generated by

$$G[B] = \begin{pmatrix} 1 & 0 & 0 & 1+\omega & \omega \\ 0 & 1 & 0 & \omega & 1+\omega \\ 0 & 0 & 1 & 1 & 1 \end{pmatrix} \tag{22}$$

This quaternary code is a $(5,3,3)$ maximum distance separable (MDS) code with 64 codewords. The image codes with respect to $\mathcal{P}_2(\omega)$, D and N are all $(10,6,3)$ binary codes with the same generator matrix given by

$$G[\tau(B)] = \begin{pmatrix} 1 & 0 & 0 & 0 & 0 & 0 & 1 & 1 & 0 & 1 \\ 0 & 1 & 0 & 0 & 0 & 0 & 1 & 0 & 1 & 1 \\ 0 & 0 & 1 & 0 & 0 & 0 & 0 & 1 & 1 & 1 \\ 0 & 0 & 0 & 1 & 0 & 0 & 1 & 1 & 1 & 0 \\ 0 & 0 & 0 & 0 & 1 & 0 & 1 & 0 & 1 & 0 \\ 0 & 0 & 0 & 0 & 0 & 1 & 0 & 1 & 0 & 1 \end{pmatrix} \tag{23}$$

This extends Example 1 in Section 3 of [9].

Acknowledgments
The first author gratefully acknowledges financial support from the UP Research Dissemination Grant and the UPLB Academic Development Fund.

References
[1] Solé P and Sison V 2007 Bounds on the minimum homogeneous distance of the p^r-ary image of linear block codes over the Galois ring $GR(p^r, m)$ *IEEE Trans. Inform. Theory* **53** 6 2270–73
[2] Sison V 2016 Bases of the Galois ring $GR(p^r, m)$ over the integer ring $\mathbb{Z}_{p^r}$ *Preprint* gr-qc/0401010
[3] Honold T 2001 A characterization of finite Frobenius rings *Arch. Math. (Basel)* **76** 406–15
[4] MacDonald B 1974 *Finite Rings with Identity* (Marcel Dekker)
[5] Wan Z-X 2003 *Lectures on Finite Fields and Galois Rings* (World Scientific)
[6] Greferath M and Schmidt S E 2001 Gray isometries for finite chain rings and a nonlinear ternary $(36, 3^{12}, 15)$ code *IEEE Trans. Inform. Theory* **45** 7 2522–24
[7] Greferath M and Schmidt S E 2000 Finite-ring combinatorics and MacWilliams Equivalence Theorem *J. Combin Theory Ser. A* **92** 17–28
[8] Rabizzoni P 1989 Relation between the minimum weight of a linear code over $GF(q^m)$ and its q-ary image over $GF(q)$ *Lecture Notes in Comput. Sci.* **388** 209–12
[9] Cerezo B R, Pasion A and Sison V 2006 Binary convolutional codes and linear block codes over $\mathbb{F}_4$ *Matimyas Matematika* **29** 1-2 9–17

IOP Conf. Series: Journal of Physics: Conf. Series **1039** (2018) 012027 doi:10.1088/1742-6596/1039/1/012027

The Frontier of Estimator Comparison between MLE and MEboot Estimation: Application for Optimization Management of Macroeconomics

S Wannapan[1] and C Chaiboonsri[2]
[1] Puay Ungphakorn Centre of Excellence in Econometrics, Faculty of Economics, Chiang Mai University, Thailand. Email: lionz1988@gmail.com.
[2] Puay Ungphakorn Centre of Excellence in Econometrics, Faculty of Economics, Chiang Mai University, Thailand.

Email: chukiat1973@gmail.com.

Abstract. One of the most difficult problems that many quantitative researchers have been trying to computationally solve is the parametric prediction. Interestingly, is the maximum likelihood estimator really the best estimator for data predictive estimating in the recent moment? This question leads the authors to conduct the mathematically experimental study by using data generating processes (DGP), entropy calculating, and cross-entropy analyses for seeking the best estimator between the maximum likelihood method (MLE) and maximum entropy bootstrapping approach (MEboot). Furthermore, the experimental solution would be employed to the real application in a macro econometrical research. Consequently, the empirical results in this paper can be the sensible tool for mathematicians or even economists to improve the data prediction in time-series analyses.

1.Introduction

Historically, it is like saying that to let data stories presenting moments in the future is not proper anymore. When we are focusing on behavioural information in the future, data collections should be observed from data in the future. However, the problem is that it is extremely difficult to find the future observations which contain the conditions of robustness and consistency. The former condition regards to provide an alternative and potentially powerful explanation of the generalization ability of data distributions. The latter condition is about to re-prove the statistical consistency of the linear classifier [1] such as the autoregressive regression (AR model) or even the generalized autoregressive conditional heteroskedasticity (GARCH process). Consequently, mathematical simulated experiments such as Data Generating Processes (DGP) relied on [2] and [3], Entropy analyses [4], even Cross-Entropy analyses (CE) [5] are employed in this paper to seek the best experimental solution for this issue and apply it to an empirical application.

Statistically, multivariate linear models are estimated by mostly employing the maximum likelihood estimator (MLE) to seek a parametric solution for several decades. For example, [6], [7], [8], and [9]. However, is MLE the most precise estimator? Interestingly, the maximum entropy bootstrapping estimator (MEboot) introduced by [10] can be the potential tool that evaluates linear models more precisely, but it is rare that this estimator is empirically employed to both experimental analyses and real applications. For

IOP Conf. Series: Journal of Physics: Conf. Series **1039** (2018) 012027 doi:10.1088/1742-6596/1039/1/012027

example, [11] and [12]. Accordingly, the comparison of these two estimator in mathematical experiments conducted in this paper can indicate the sensible way to improve a forecasting research, especially in macroeconomics.

2. The Objective and Scope of Research

The scope of the research is divided into two sections. First, it is the experiment of mathematical calculations using two simulated observation sets (Data Generating Process: DGP), including the sets of 500 and 1,000 random samples. Afterward, the comparison between the maximum likelihood estimator (MLE) and maximum entropy bootstrapping estimator (MEboot) is originated by applying the entropy and cross-entropy methods. Second, the section of the empirical application for macroeconomic time-series variables is conducted to confirm that the results from the experimental comparison can improve the parametric estimation of the empirical research based on the Markovitz optimization management by [13]. Additionally, Bayesian statistics relied on [14], [15], and [16] was employed to estimate the econometric models. For instance, ADF unit-root testing [17], Markov switching modelling [18]. The yearly observations employed in this section are Thailand macroeconomic factors such as gross domestic products (GDP), inflation rates, interest rates, exchange rates, and money supplies. These are collected during 1965 to 2015.

3. The Overall of Methodologies

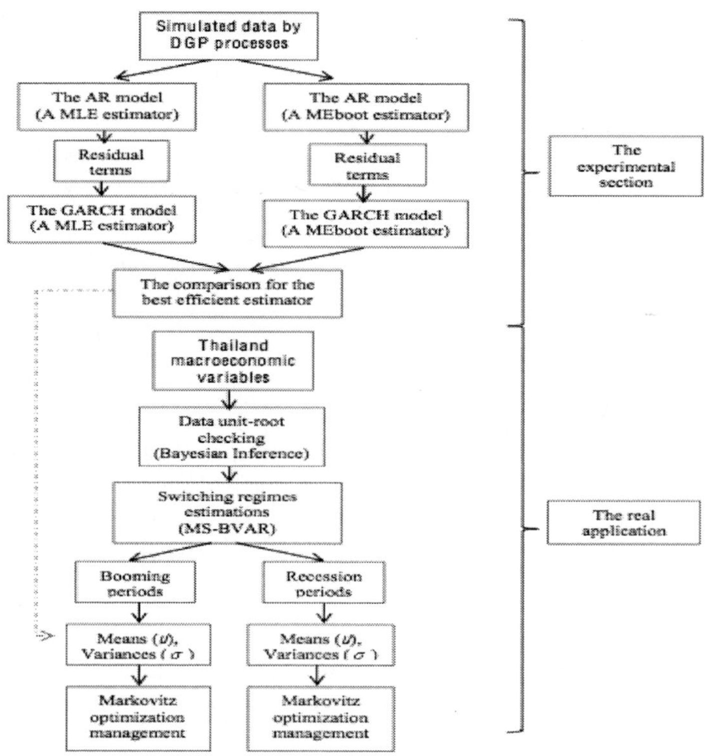

Figure 1. The conceptual framework of the research

3.1 *The results of the mathematically experimental analysis.*
The section of the computationally experimental analysis in this paper is originally proposed to investigate the best parametric estimator between the MLE and MEboot estimations. Simulated data sets were estimated by using the maximum likelihood estimator (MLE) and maximum entropy bootstrapping approach (MEboot). From the AR-GARCH model, residual terms were observed and transformed into the cumulative distribution function (CDF), which is a real-valued random variable. The CDF terms were then employed to calculate on the entropy analysis, including the general entropy and cross-entropy method. Empirically, the results of the cross-entropy comparison were shown in Table 1. In the case of 500 random observations, there is obvious that the MEboot estimation more provides the precise estimator rather than using the MLE approach. In other words, the CE results stated the value of cross-entropy calculating on the residual terms observed from the MEboot estimation (82.0112) is the nearest entropy value to the overall entropy (65.9109). Similarly, in the case of 1,000 random samples, the result still confirms that the MEboot estimator can give the precise parameter rather than the counterpart. Obviously, the value of cross-entropy calculating on the residual terms observed from the MEboot estimation (161.1296) is the shortest distance value from the overall value (65.4534). Accordingly, it is sensible to conclude that the MEboot estimator will be chosen to empirically apply to the realistic time-series indexes.

Table 1. The Data Generating Processes (DGP) in the AR-GARCH model based on the simulated random sampling and the Entropy analyses.

Details	AR-GARCH (n = 500)			AR-GARCH (n = 1,000)		
	Overall Entropy	MLE (CE analysis)	MEboot (CE analysis)	Overall Entropy	MLE (CE analysis)	MEboot (CE analysis)
Estimated entropy values (calculating from residual terms generated by AR-GARCH)	65.9109	83.2470	82.0112[a]	65.4534	161.2539	161.1296[a]

Noted: [a] the nearest CE value to the overall entropy value

3.2 The Empirical Application in the Real Economic Research
The ADF test based on Bayesian inference analyses the null hypothesis that a time-series data is non-stationary I(1) against the alternative I(0). The result was represented in Table 3. The results of the selecting prior for the switching estimation based on Bayesian inference and all of divided regimes, defining as boom and recession periods, were shown in Table 2. Empirically, the optimization approach for portfolio managements is employed to clarify the proportion of Thailand economic factors. Descriptively, Table 3 shows average values, standard errors and variances of five variables, which are successfully estimated from the AR-GARCH model using the MEboot estimator. As Markowitz's mean-variance optimization by minimizing standard errors, the empirical result shows that the optimal approach can efficiently maximize the expected growth of the whole economic system and minimize the risky value of the whole economic expansion simultaneously. The details in Table 4 are also consonant. In booming periods, the sharp ratio is a double increment from 0.090 to 0.169. Similarly, in recession periods, the ratio is a little bit improvement from 0.333 to 0.410. As a result, this can be implied the overall economic system in both booming and recessing periods is improved when we employ the optimal management approach.

Table 2. The MSBVAR estimation for Thailand macroeconomic factors during 1976 to 2015.

Variables	Stage	Duration (Forecasting by MSBVAR)
Thailand economic system (including GDP, inflation, interest rate, exchange rate, money supply)	Economic Boom periods	17 years
	Economic Recession periods	22 years

Table 3. Descriptive information estimated from the AR-GARCH model with the MEboot estimator.

Stage	Details	GDP I(0) (%)	Inflation I(0) (%)	Interest rate I(0) (%)	Exchange rate Bath per US dollar I(0) (%)	Money supply I(0) (%)
Boom	Mean from the AR model (u)	0.485	0.322	0.246	-0.185	0.504
	Variances from the GARCH model (σ)	0.432	0.387	0.298	0.223	0.449
Recession	Mean from the AR model (u)	0.413	0.134	0.568	0.339	0.449
	Variances from the GARCH model (σ)	0.409	0.161	0.508	0.018	0.518

Noted: I(0) represents the trends of collected variables are stationary

Table 4. the mean-variance portfolio optimization based on minimizing standard errors

Stage	Details	Neutral Economic Cycling movement	Optimal policy management for Economic Cycling movement
Boom periods	The expected value of whole system growth, (u): mean	0.275	0.504
	The risky value of whole system growth (σ): variance	3.041	2.991
	The sharp ratio of whole system (u/σ)	**0.090**	**0.169**
Recession periods	The expected value of whole system growth, (u): mean	0.381	0.568
	The risky value of whole system growth (σ): variance	1.142	1.385
	The sharp ratio of whole system (u/σ)	**0.333**	**0.410**

For Table 5, in the booming periods, money supplies especially are the factor needed to be the intensively controllable indicator. The latent value, which equals 0.46, also implied that the economic system is not efficient since there is some information which is elusive. In the case of recession periods, on the other hand, the factor of interest rates is the predominant. The latent value, which is equivalent to 0.43, stated that the economic system is not going to be efficient in the downsizing situations.

Table 5. The proportion of systematically economic optimization.

Stage	Details	Macroeconomic variables				
		GDP	Inflation	Interest rate	Exchange rate	Money supply
Boom periods	Optimal portions of growth (forecasting interval value)	0.36 (8.00- 10.90)	0.31 (4.94-6.46)	0.28 (6.37-8.17)	0.14 (2.96-3.37)	0.37 (17.73-24.27)
	Summary			**1.46**		
	Latent value			**0.46**		
Recession periods	Optimal portions of growth (forecasting interval value)	0.29 (3.61-4.67)	0.23 (3.65-4.50)	0.33 (6.13-8.15)	0.28 (3.72-4.75)	0.30 (10.03-13.05)
	Summary			**1.43**		
	Latent value			**0.43**		

4. Conclusion

The succession of mathematical modelling comparisons to investigate the suitable estimator for data predicting is already clarified in this paper. Experimentally, the DGP approach in the AR-GARCH model indicates the empirical result that the parametric estimator calculated by the MEboot method can efficiently provide a precise parameter rather than using the MLE estimation. This is strongly confirmed by the outcomes of cross-entropy (CE) calculations. Applying in the Markovitz optimization management, this application shows the estimated results empirically represented that the major macroeconomic indexes should be emphasized in booming periods are money supplies, GDP, and inflation rates, respectively. On the other hand, the predominant factors should be intensive concerned are interest rates, money supplies, and GDP, respectively.

References

[1] Xu H, Caramanis C, and Mannor S 2009 Robustness and regularization of support vector machines *Journal of Machine Learning Research* Vol 10 pp 1485-1510.
[2] Nguyen H T 2014 *Probability for statistics in econometrics* Center of Excellence in Econometrics, Faculty of Economics, Chiang Mai University, Thailand.
[3] Sirca S J, and Omladic M 2017 The JLS model with ARMA/GARCH errors *ARS MATHEMATICA CONTEMPORANEA* Vol 13 pp 63-79.
[4] Vinod H D 2013 Maximum entropy bootstrap simulations for variance estimation Available at SSRN: https://ssrn.com/abstract=2295723.
[5] Vázquez E F 2011 Updating weighting matrices by cross-entropy *Investigaciones Regionales* Vol 21 pp 53-69.
[6] Abidin S, Reddy K, and Zhang C 2014 Intensity of price and volatility spillover effects in Asia-Pacific basin equity markets *Australasian Accounting, Business and Finance Journal* Vol 8 pp 3-18.
[7] Ferenstein E and Gasowski M 2004 Modelling stock returns with AR-GARCH processes *SORT* Vol 28 pp 55-68.
[8] Lama A, Jha G K, Paul R K, and Gurung B 2015 Modelling and forecasting of price volatility: an application of GARCH and EGARCH models *Agricultural Economics Research Review* Vol 28 pp 73-82.
[9] Mousazadeh S and Cohen I 2011 AR-GARCH in presence of noise: parameter estimation and its application to voice activity detection *IEEE TRANSACTIONS ON AUDIO, SPEECH, AND LANGUAGE PROCESSING* Vol 9 pp 916-926.
[10] Vinod H and Lopez-de-Lacalle J 2009. Maximum entropy bootstrap for time series: the meboot R package *Journal of Statistical Software* Vol 29.
[11] Chaitip P and Chaiboonsri C 2013 AEC's demand for ICT: maximum entropy bootstrap approach in panel data *Procedia Economics and Finance* Vol 5 pp 125-132.
[12] Chaiboonsri C and Chaitip P 2013 A boundary analysis of ICT firms on Thailand stock market: a maximum entropy bootstrap approach and Highest Density Regions (HDR) approach *Int. J. Computational Economics and Econometrics* Vol 3 pp 14-26.
[13] Cornuejols G and Tütüncü R 2006 *Optimization methods in finance*. Carnegie Mellon University Press USA.
[14] Chen C W S, Chen S Y, and Lee S 2013 Bayesian unit root test in double threshold heteroskedastic models *Comput Econ* Vol 42 pp 471-490.
[15] Jeffreys H 1961 Theory of probability 3ed New York: Oxford University Press.
[16] Kass R E and Raftery A E 1995 Bayes factors *Journal of the American Statistical Association* Vol 90 pp 773-795.
[17] Said S E, Dickey D 1984 Testing for unit roots in autoregressive moving-average models with unknown order *Biometrika* Vol 71 pp 599–607.
[18] Hamilton J 1989 A new approach to the economic analysis of nonstationary time series and business cycle *Econometrica* Vol 57 pp 357-384.

Designing Efficient Configurable QCA Nano Circuit for Morphological Operations in Image Processing

Shiv Bhusan Tripathi, Aron Narzary, Rahul Toppo, Mrinal Goswami and Bibhash Sen

National Institute of Technology, Durgapur, India

E-mail: (jadnaps95, aron2766narzary, rsitoppo, mrinal.goswami21)@gmail.com, bibhash.sen@cse.nitdgp.ac.in

Abstract. Morphological operations like erosion and dilation are important operations in various video and image processing applications. This research presents a configurable Quantum-dot Cellular Automata (QCA) nano circuit for morphological applications in image processing. QCA is a field-coupled emerging nano-technology which can be applied in various image processing operations. The implemented QCA design utilizes two cascaded 5-input majority voters where dilation and erosion can be performed with the help of a control line. The proposed design is compared with the previous results which demonstrate substantial improvement in terms of area, delay and cell count.

1. Introduction

The size of transistor is scaling down towards nanometre scale to improve the performance. But, due to physical limitations of CMOS technology, further scale down is facing several problems such as power consumption, electro-migration failures, interconnectivity, lithography problems etc. [1]. Therefore, immense research is going on to find a replacement of current Complementary Metal Oxide Semiconductor (CMOS) technology. Several emerging technologies such as Carbon Nanotubes, Silicon Nanowires, Quantum-dot Cellular Automata, Single-Electron Transistors are emerging as a possible replacement of current CMOS. The Quantum-dot Cellular Automata (QCA) promises higher circuit densities, faster speed, low power consumption over the traditional CMOS technology [2]. The integration densities of QCA technology can touch 1012 cells/cm^2 and the circuit switching frequency can be close to terahertz [3]. QCA is a field-coupled nano technology where information travels with the help of local interaction between the electrons. Thus, QCA is considered as one of the most propitious nanoelectronic technologies [2, 4].

Now-a-days, the most important requirement of any computational system is processing speed and low power consumption. Image processing is a very challenging domain in very-large-scale integration (VLSI) technology due to its huge requirement in computational power. Till date, CMOS based VLSI technology is extensively used to improve the quality of image processing systems. In this regard, several implementations are reported such as nonlinear filters, mathematical morphological operations, multi-functional applications etc.

The complete paper is ordered in the following way: an ephemeral discussion of morphological operations in image processing is presented in section 2. Section 3 will instigate some of the

IOP Conf. Series: Journal of Physics: Conf. Series **1039** (2018) 012028 doi:10.1088/1742-6596/1039/1/012028

basics of QCA. The proposed designs are discussed in section 5 followed by conclusion in section 6.

2. Basics of Morphological Operation in Image Processing

Morphological operations act as a tool to extract image components that are useful for further representation and description of the image region. Here, non-linear operations are performed related to the shape or morphology of the image. It does not concern about exact pixel values, but concern about the relative ordering of pixel values. Some of these operations are - Erosion, Dilation, Opening, Closing etc. Erosion and Dilation are the basic operations in morphological image processing. The basic morphological operations are well suited for binary images with only white and black pixels. The inputs of these morphological operations are - the binary image and the binary structuring element. The background and foreground regions are assumed as white and black pixels respectively. The basic morphological operations (erosion and dilation) can be performed on the binary image with the help of a binary structuring element. The structuring element is superimposed over the input image to coincide the origin of the structuring element with the input image.

2.1. Erosion on binary image

In erosion, there are two inputs, the first input is the image that is going to be eroded and another one is the structuring element of known origin, using which erosion is applied to the input image and an eroded image is produced as output. The origin of structuring element visit every pixel of the input image. If the structuring element is totally contained in input image then that location is marked in the processed image. It can be mathematically defined as

$$I \ominus S = \{(i,j) : S(i,j) \subset I\}$$

Here S is the 3×3 structuring element, I is the input image and $\ominus$ is the erosion operator. The above equation suggests that S is translated by (i, j) and whether that translated S is contained in I (Figure 1).

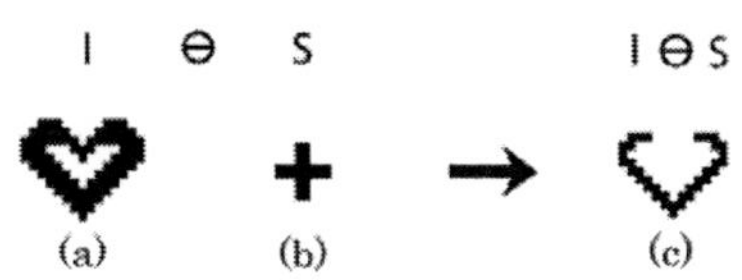

Figure 1. (a) Input Image (I) (b) Structuring Element (S) (c) Eroded image

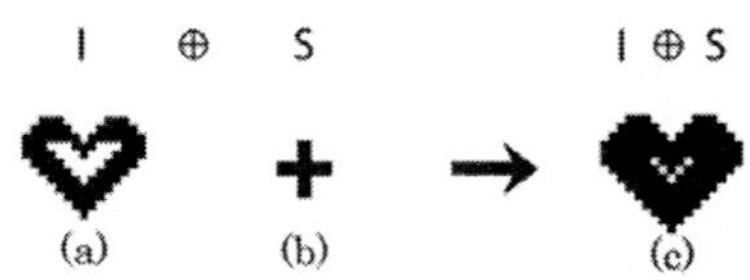

Figure 2. (a) Input Image (I) (b) Structuring Element (S) (c) Dilated image

2.2. Dilation on Binary image

In dilation, there are also two inputs, first one is the image that is going to be dilated and another one is the reflected structuring element about its origin. In dilation, as in erosion, the origin of reflected structuring element visit every pixel of the input image and if at that position reflected structuring element is partially contained in input image then that location is marked in the processed image. The mathematical representation of dialation is:

$$I \oplus S = \{(i,j) : (\hat{S})_(i,j) \cap I \neq \phi\}$$

Here S is 3×3 the structuring element, I is the input image, $\oplus$ is the dilation operator and $\hat{S}$ is the reflection of S about its origin (Figure 2).

3. Basics of Quantum-dot Cellular Automata

A QCA cell has four quantum dots with two free electrons. Based on their position, binary logic 0 and logic 1 can be performed. A 3-input majority voter is used as logic gates which can be represented as M(A,B,C) = AB + BC + CA (Figure 3). By applying fixed polarization at one of the input cells, it can operate as an AND or OR gate. NOT gate can be implemented using two different ways which is shown in Figure 3. Moreover, to increase the programmability feature of the voter, 5-input majority voter is extensively utilized now-a-days [5]. The logic function of 5-input majority voter is: Maj(A, B, C, D, E) = ABC + ABD + ABE + ACD + ACE + ADE + BCD + BCE + BDE + CDE. In this work, we have used 5-input majority voter which is proposed in [5], to implement our designs. QCA clocking has four periodic phases with a $\Pi/2$ phase difference between contiguous phases (Figure 4). The four different phases are: switch, hold, release, relax.

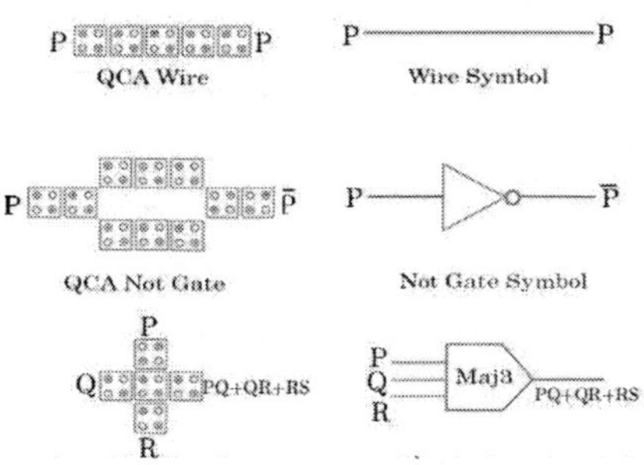

Figure 3. Fundamental Gates

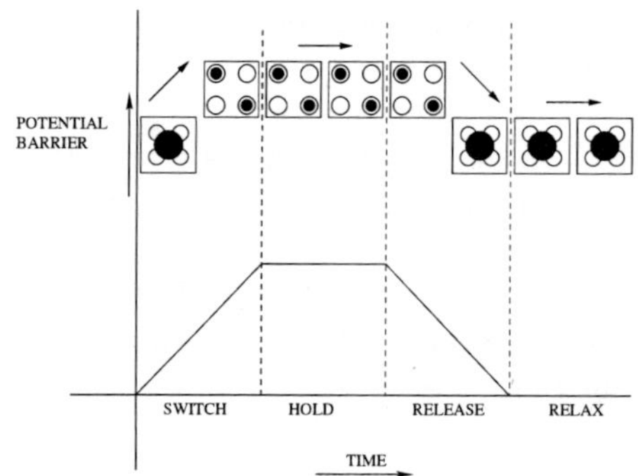

Figure 4. QCA Clocking

4. Related Work

The design architecture and fault detection techniques are extensively investigated in QCA [6–8]. However, QCA based designs for image processing applications are not studied in great detail. Few interesting QCA based image processing attempts are discussed in [3, 9, 10]. Image thresholding using QCA is attempted in [9]. In [10], two separate designs for morphological erosion and dilation are proposed. A configurable design for morphological erosion and dilation operation is first attempted in [3] but the design covers huge area, consume more cells and also clock with respect to the design proposed in this work.

5. Proposed QCA Designs

5.1. Proposed Non-Configurable Design

A 3×3 structuring element is considered to perform the proposed erosion and dilation operation. First, the structuring element is placed on the input image. For erosion, structuring element should be contained in the input image and for dilation, it should be partially contained. Then, all five-pixel value of the binary image that falls under the marked pixels of 3×3 structuring element is extracted. If all the black pixels of the structuring element coincide with black pixels on the input image, record black the pixel of the output image that corresponds to the

IOP Conf. Series: Journal of Physics: Conf. Series **1039** (2018) 012028 doi:10.1088/1742-6596/1039/1/012028

structuring elemnt. If we take AND operation of these pixels then the resultant image will be the eroded image of the input image. Similarly, if we take OR operation of these pixels then the resultant image will be the dilated image of the input image. These two operations can be implemented in hardware as shown in Figure 5. The proposed design consists of two cascaded 5-input majority gates where controlling l_1, l_2, l_3 and l_4 erosion and dilation can be performed. If $l_1 = l_2 = l_3 = l_4 = -1$ then it perform erosion and if $l_1 = l_2 = l_3 = l_4 = 1$ then it will perform dilation. A comparison between the previous attempt and the proposed work is presented in Table 1 which signifies the superiority of the proposed design.

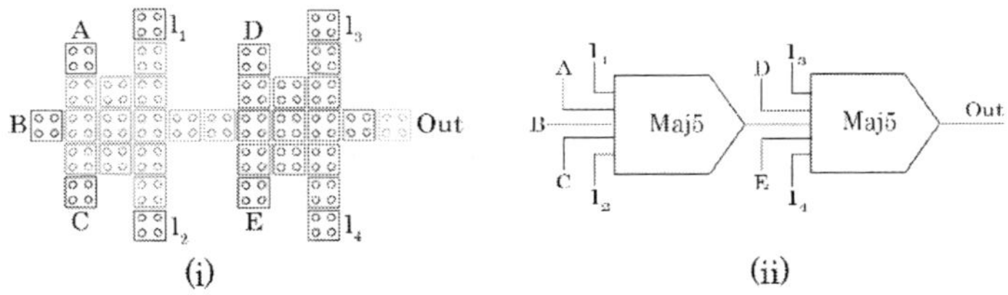

Figure 5. QCA layout and logical diagram of proposed non-configurable design

Table 1. Comparison of proposed non-configure design with Mardiris non-configure design

Design	Cell Count	Clock Cycle	Area (μm^2)
Mardiris design [10]	35	3	0.043
Proposed design	35	2	0.030
Improvement %	-	33.33	30.23

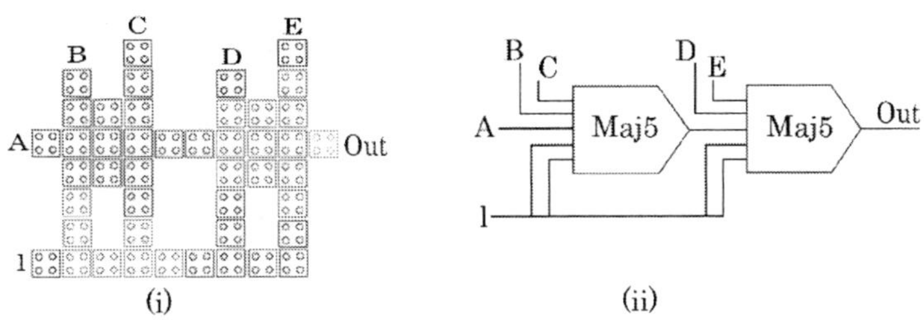

Figure 6. QCA layout and logical diagram of proposed configurable design

5.2. Proposed Configurable Design For Erosion and Dilation

As it is discussed in the previous section, erosion and dilation can be done using AND and OR gates, a new configurable design for both erosion and dilation is proposed as shown in Figure 6. The design consists of six inputs (A, B, C, D, E and l) and one output (Out). If l = 1 then the design performs dilation otherwise it performs erosion operation. The simulation outcomes are displayed in figure 7 and 8. In the Figure 7, the input l is fixed at polarization P= -1 where as in Figure 8, the input l is fixed at polarization P= 1. A detailed comparison between previous design and the proposed design is shown in Table 2. The proposed configurable design of erosion and dilation is more area efficient, cost-effective as well as faster than the previous design.

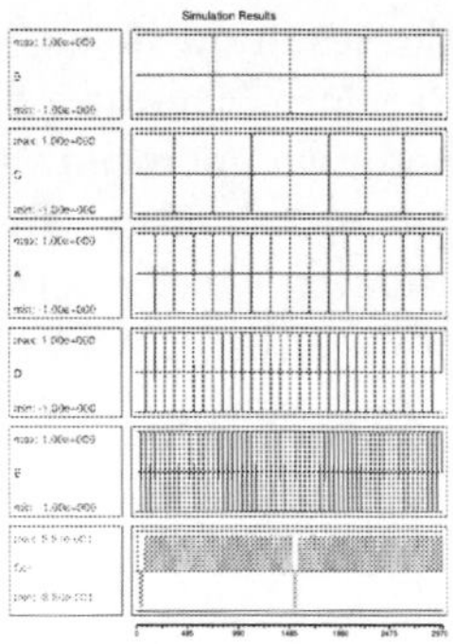

Figure 7. Simulation of QCA based Dilation

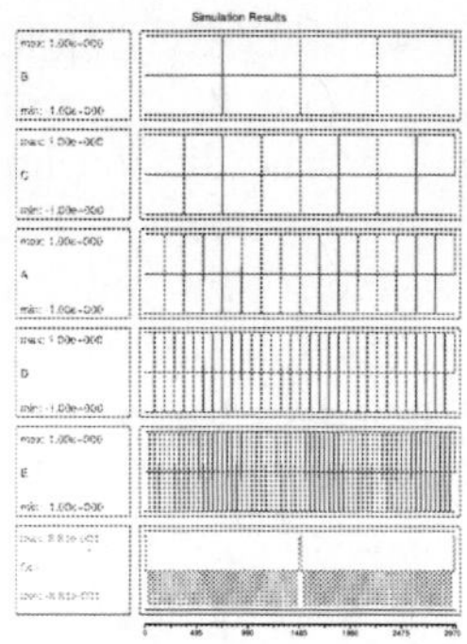

Figure 8. Simulation of QCA based Erosion

Table 2. Comparison of Proposed Configurable design with Mardiris Configurable design

Design	**Cell Count**	**Clock Cycle**	**Area** (μm^2)
Mardiris design [3]	51	4	0.05
Proposed design	45	3	0.03
Improvement %	11.76	25	40

5.3. Simulation Parameters
QCADesigner (Version 2.0.3) [11] coherence vector simulation engine is utilized to substantiate the proposed designs using all the default parameters.

6. Conclusion
In this work, a novel configurable QCA design is discussed for the implementation of erosion and dilation operations for binary images. The configurable QCA design utilized two cascaded 5-input majority voter. The simulation outcomes substantiate the correct operation of the proposed design. The proposed design is more area efficient (11.76%), cost-effective (25%) as well as faster (40%) than the previous design.

References
[1] International technology roadmap for semiconductors (itrs) URL `http://www.itrs2.net/`
[2] Lent C S, Tougaw P D, Porod W and Bernstein G H 1993 *Journal of Nanotechnology* **4** 49–57
[3] Mardiris V and Chatzis V 2016 *Journal of Engineering Science and Technology Review* **9** 25–30
[4] Goswami M, Sen B, Mukherjee R and Sikdar B K 2017 *Microelectronics Journal* **60** 1 – 12
[5] Akeela R and Wagh M 2011 *NSTI Nanotech* **2** 978 – 981
[6] Nath R K, Sen B and Sikdar B K 2017 *IET Circuits, Devices Systems* **11** 201–208
[7] Nath R K, Sen B, Daga R, Chakraborty N, Tibrewal H and Sikdar B K 2015 *2015 19th International Symposium on VLSI Design and Test* pp 1–5
[8] Chattopadhyay S, Tripathi S B, Goswami M and Sen B 2016 *2016 20th International Symposium on VLSI Design and Test (VDAT)* pp 1–2
[9] Sen B, Anand A S, Adak T and Sikdar B K 2011 *2011 International Conference on Innovations in Information Technology* pp 356–360
[10] Mardiris V and Chatzis V 2014 *Image Processing Algorithms Implementation Using Quantum Cellular Automata* (Cham: Springer International Publishing) pp 65–84
[11] Walus K, Dysart T J, Jullien G A and Budiman R A 2004 *IEEE Transactions on Nanotechnology* **3** 26–31

IOP Conf. Series: Journal of Physics: Conf. Series **1039** (2018) 012029 doi:10.1088/1742-6596/1039/1/012029

Density functional theory study of uni- and bi-metallic small Pt and Ni clusters for fuel conversion applications

Alexandra B. Santos-Putungan and Darwin Barayang Putungan*

Physics Division, Institute of Mathematical Sciences and Physics, University of the Philippines Los Baños, College, Los Baños, Laguna 4031 Philippines

Email: dbputungan@up.edu.ph

Abstract. In this work, we employed density functional theory calculations to probe the binding characteristics of C and OH on small uni-metallic (Pt_3 and Ni_3) and bi-metallic (Pt_2Re and Ni_2Fe) clusters for energy conversion applications. Both uni-metallic and bi-metallic clusters were found to possess relatively good HOMO-LUMO gaps, indicating that the clusters are electronically stable against excitations and high energy environment. It is found that the introduction of foreign atoms on uni-metallic clusters impacts C and OH binding characteristics, which could be useful in the design and fine-tuning of novel nanocatalysts for energy conversion applications such as CO/CO_2 methanation and hydrogenation reactions. It is further inferred that Pt_3 and Pt_2Re clusters bind C and OH moderately, and could be utilized in further studies concerning energy conversion.

1. Introduction

One of the main problems that the world would encounter in the coming years is energy production and usage. It is generally known that the world's fossil fuel reserves are finite, and thus it will be depleted in the next few decades. Although there is a strong drive in the proliferation of non-conventional and renewable energy sources such as from light and wind, still, the current technology needed for it to fully replace fossil fuel is not matured enough and needed additional tweaking and improvements before it can be realized as a major energy source. It is therefore of importance to find alternative ways of producing conventional fuels that satisfy the following requirements: (1) sustainable and renewable, in a sense that fuel can be produced in a close cycle and that the fuel usage by-products can be converted back, (2) production should be efficient, and can meet current demands, and (3) should be safe and practical for mass production and distribution [1–3]. For such requirements to be met, the reactions accompanying fuel conversion should be at its optimum, which can be attained through proper choice of catalysts. One of the widely used method in producing the so-called "artificial" fossil fuel is steam reforming, in which raw carbon-rich materials such as coke is made to react with steam at high temperature and pressure to produce methane, propane, and other higher-order hydrocarbons [4–9]. Its by-product, called syngas that is composed of H_2 and CO gases, can be recycled to produce hydrocarbon fuels. Syngas can also be produced from biomass materials, making syngas conversion one of the most studied reaction in the field of catalysis [10–12].

In this regard, this work focuses on the theoretical study of transition metal atomic clusters as potential catalysts for syngas conversion. Specifically, this work tries to assess the effect of alloying these clusters to create bi-metallic atomic clusters on important parameters such as species binding on certain sites. This is important as the binding of intermediate species on the catalysts could dictate the efficiency of the reaction, and thus proper tuning of this could prove useful in the design of catalysts.

From [13], it was illustrated in the volcano plot that certain transition metal surfaces were found to have good reactivity when it comes to syngas conversion. Syngas conversion proceeds by breaking CO into C and O, and then hydrogenation of C into CH_4, O to OH and subsequent conversion to H_2O. Thus, C and OH binding on the catalyst surface would determine the efficiency of the reaction, which is the subject of this study.

2. Methodology

The calculations were carried out in the paradigm of density functional theory (DFT) via the Quantum Espresso package suite [14]. Three-atom clusters were considered in this work: the uni-metallic Pt_3, Ni_3, and the bi-metallic Pt_2Re, and Ni_2Fe. These clusters were chosen to analyze the effect of alloying those transition metals that were found to either bind strongly or weakly to C and OH, as was presented in [4]. As for calculations details, Perdew-Burke-Ernzerhof (PBE) exchange-correlation functionals within the generalized gradient approximation (GGA) were utilized, together with norm-conserving pseudopotentials. Kinetic energy cut-off of 60 Ry is used, and the sampling of the Brilliouin zone is done via a 1 x 1 x 1 k-point mesh representing the real-space unit cell with more than 10 Å vacuum in all spatial directions. Electronic convergence is set to 10^{-5} Ry and atoms were allowed to relax until the forces on each is below 10^{-3} Ry/a.u.

Optimized geometries, as well as charge distributions, HOMO-LUMO gap and OH and C binding energies are presented and analyzed. The binding energy of OH and C on the clusters in defined as

$$E_{binding} = E_{system} - E_{cluster} - E_{OH/C} \quad (1)$$

3. Results and Discussions

Figure 1 shows the optimized geometries of the metal clusters, together with the corresponding bond lengths. All structures possess the isosceles triangle geometry, except for Ni_3 which possess an equilateral triangle geometry. Using Bader charge analysis, it was found that for uni-metallic clusters such as Pt_3 and Ni_3, the electron distribution among the atoms is relatively uniform. On the other hand, for the bi-metallic ones which include Pt_2Re and Ni_2Fe, the foreign atoms Re and Fe act as electron donor. The Re atom in Pt_2Re donates around 0.56 e to the Pt atoms, while Fe in Ni_2Fe donates around 0.41 e to Ni atoms. To analyse the electronic structure of the metal clusters, their HOMO-LUMO gaps were calculated, as presented in Table 1. The HOMO-LUMO gap can be used to draw conclusions on the stability and reactivity of the clusters [15,16]. As can be observed from the given values, all metallic clusters have relatively fine HOMO-LUMO gaps, with Ni_3 having the largest gap among the clusters. Clusters with large HOMO-LUMO gaps are said to be stable under excitations or high energy environment. In catalysis perspective, this is quite a good indication that the catalyst will still function close to ideal operations, as the chance of it being used up in the reaction, and exhibit the so-called "site-poisoning" is lower.

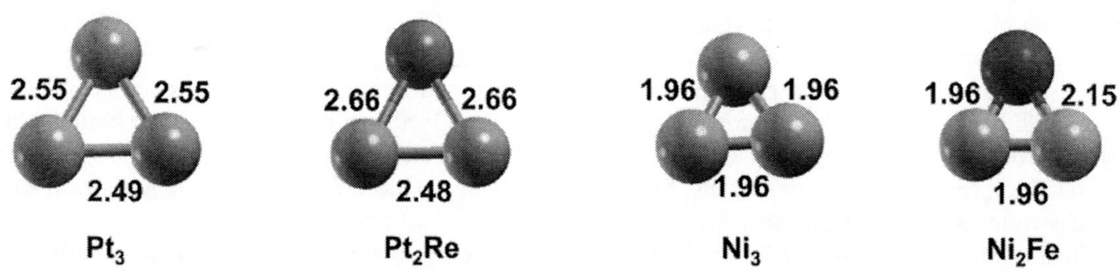

Figure 1. Optimized geometries of the three-atom uni-metallic and bi-metallic clusters. Bond lengths are reported in units of Å.

For the calculation of the binding energies of OH and C, the optimized geometries and bond lengths for C and OH adsorption are shown in Figures 2 and 3 respectively. It can be inferred that both C and

OH binds with shorter bond lengths to Ni_3 and Ni_3Fe. To analyse charge distribution, charge difference isosurfaces were plotted for C and OH adsorption (Figure 4). Here, red regions denote electron rich regions, while blue ones denote electron deficient regions. The isosurfaces indicate that the binding between the clusters and C and OH has an ionic nature. To substantiate on these, we calculated the charge transfer to C and OH to the clusters via Bader charge analysis (Table 1). It is found that C and OH gains substantial amounts of electrons from the clusters (except for the case of Ni_3, in which C loses a minimal 0.15*e* of charge to Ni). These charge transfers become more pronounced for bimetallic clusters. The calculated binding energies is also listed in Table 1, and a plot of the C binding energy (E^C) against that of OH (E^{OH}), and the plot of C binding energy (E^C) against that of OH (E^{OH}) is shown in Figure 5.

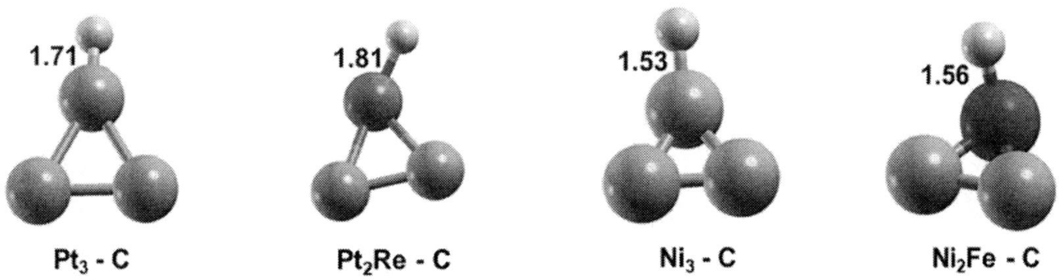

Figure 2. Optimized geometries of adsorbed C on uni-metallic and bi-metallic clusters. Bond lengths are reported in units of Å.

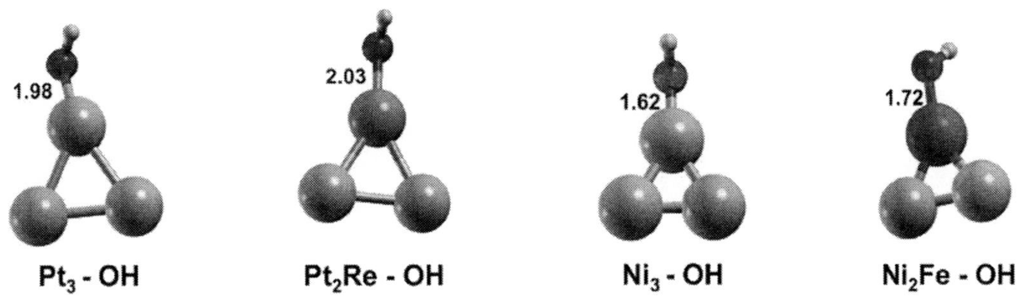

Figure 3. Optimized geometries of adsorbed OH on uni-metallic and bi-metallic clusters. Bond lengths are reported in units of Å.

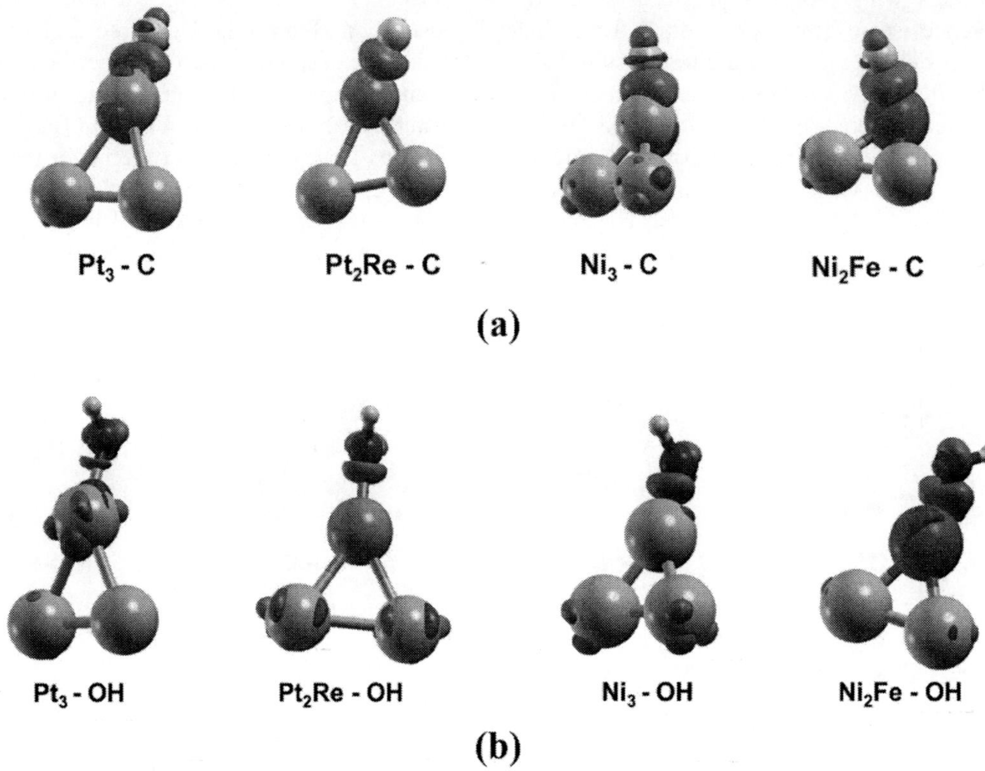

Figure 4. Charge difference isosurfaces for (a) C adsorption and (b) OH adsorption on uni-metallic and bi-metallic clusters. The isosurface level is set to 0.02*e* $Å^{-3}$.

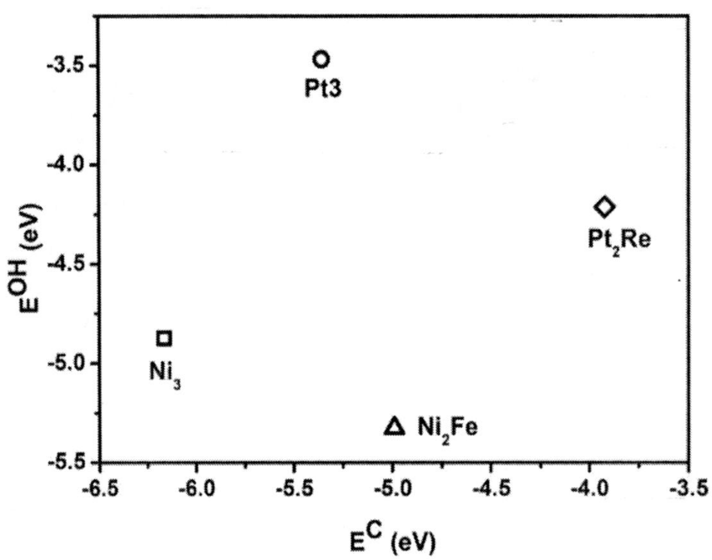

Figure 5. Plot of the C binding energy against that of OH for each small metallic cluster.

Table 1. HOMO-LUMO gap, C and OH charge transfer and binding energies

	HOMO-LUMO gap (eV)	Charge transfer to C (*e*)	Charge transfer to OH (*e*)	C binding energy (eV)	OH binding energy (eV)
Pt_3	0.51	0.176	1.42	-5.36	-3.47
Pt_2Re	0.31	0.685	1.60	-3.92	-4.21
Ni_3	0.87	-0.149	1.37	-6.17	-4.88
Ni_2Fe	0.22	0.162	1.48	-4.99	-5.33

For proper reactions to proceed, for instance in the case of methanation or CO/CO_2 hydrogenation reactions, the binding of intermediate species on the catalyst must be at the optimum level [13,17,18]. Analyzing each C and OH adsorption pair, it can be observed that Ni_3 binds C and OH quite too strong, and thus it can be inferred that the cluster would not be a good catalyst for syngas conversion. For the bimetallic case, Ni_2Fe, it appears that C adsorption energy is reduced, although OH binding is further enhanced. On the other hand, Pt_3 and Pt_2Re clusters seem to have good binding characteristics with respect to C and OH. In the uni-metallic case, there is moderate OH binding, although C binding is still quite high. Replacing one Pt with Re significantly lowered C binding energy, with a modest increase in OH binding. In general, Ni_3 and Ni_2Fe clusters adsorb C and OH quite strongly as compared to that of Pt_3 and Pt_2Re. This work has demonstrated that small bi-metallic clusters could be employed to fine tune C and OH binding energies, which could be utilized in the design of useful catalysts for energy conversion applications.

4. Summary and Conclusion

The work carried out theoretical calculations in the paradigm of DFT to assess the catalytic potential of select small metal clusters: uni-metallic Pt_3 and Ni_3 clusters, and the bi-metallic clusters Pt_2Re and Ni_2Fe clusters. Results showed that the clusters are structurally and electronically stable. C and OH binding were found to be too strong on Ni_3 and Ni_2Fe, while these are quite reduced in the case of Pt_3 and Pt_2Re especially in the case of C adsorption. Due to the moderate adsorption characteristics of C and OH on Pt_3 and Pt_2Re clusters, it is inferred that these could be employed in the fine tuning of reaction intermediates' binding properties for syngas conversion and energy conversion applications as a whole. Moreover, this work has demonstrated the effectivity of introducing foreign atoms to form bimetallic clusters in improving binding characteristics of C and OH.

References

[1] Dunn S 2002 *Int. J. Hydrogen Energy* **27** 235–64
[2] Maeda K and Domen K 2010 *J. Phys. Chem. Lett.* **1** 2655–61
[3] Ponrouch A., Monti D, Boschin A., Steen B, Johansson P and Palacín M R 2015 *J. Mater. Chem. A* **3** 22–42
[4] Dong C, An L, Yang Y and Zhang J 2010 *Asia-Pacific Power Energy Eng. Conf. APPEEC* 0–3
[5] Van Herwijnen T, Van Doesburg H and De Jong W A 1973 *J. Catal.* **28** 391–402
[6] Hu D, Gao J, Ping Y, Jia L, Gunawan P, Zhong Z, Xu G, Gu F and Su F 2012 *Ind. Eng. Chem. Res.* **51** 4875–86
[7] Kambolis A, Schildhauer T J and Kröcher O 2015 *Chim. Int. J. Chem.* **69** 608–13
[8] Dry M E 2004 *Appl. Catal. A Gen.* **276** 1–3
[9] Hammer B and Nørskov J K 2000 *Adv. Catal.* **45** 71–129
[10] Remediakis I N, Abild-Pedersen F and Nørskov J K 2004 *J. Phys. Chem. B* **108** 14535–40
[11] Lo J M H and Ziegler T 2007 *J. Phys. Chem. C* **111** 11012–25
[12] Domazetis G, Raoarun M and James B D 2007 *Energy and Fuels* **21** 2531–42
[13] Nørskov J K, Abild-Pedersen F, Studt F and Bligaard T 2011 *Proc. Natl. Acad. Sci. U. S. A.* **108** 937–43
[14] Giannozzi P, Baroni S, Bonini N, Calandra M, Car R, Cavazzoni C, Ceresoli D, Chiarotti G L, Cococcioni M, Dabo I, Dal Corso A, De Gironcoli S, Fabris S, Fratesi G, Gebauer R,

IOP Conf. Series: Journal of Physics: Conf. Series **1039** (2018) 012029 doi:10.1088/1742-6596/1039/1/012029

Gerstmann U, Gougoussis C, Kokalj A, Lazzeri M, Martin-Samos L, Marzari N, Mauri F, Mazzarello R, Paolini S, Pasquarello A, Paulatto L, Sbraccia C, Scandolo S, Sclauzero G, Seitsonen A P, Smogunov A, Umari P and Wentzcovitch R M 2009 *J. Phys. Condens. Matter* **21**

[15] Gupta U, Ulises Reveles J, Melko J J, Khanna S N and Castleman A. W 2009 *Chem. Phys. Lett.* **467** 223–9

[16] Joswig J O, Roy S, Sarkar P and Springborg M 2002 *Chem. Phys. Lett.* **365** 75–81

[17] Andersson M P, Abild-Pedersen F, Remediakis I N, Bligaard T, Jones G, Engbæk J, Lytken O, Horch S, Nielsen J H, Sehested J, Rostrup-Nielsen J R, Nørskov J K and Chorkendorff I 2008 *J. Catal.* **255** 6–19

[18] Nørskov J K, Bligaard T, Rossmeisl J and Christensen C H 2009 *Nat. Chem.* **1** 37–46

IOP Conf. Series: Journal of Physics: Conf. Series **1039** (2018) 012030 doi:10.1088/1742-6596/1039/1/012030

Application of log-linear models to pneumonia patients: a case study of Jigme Dorji Wangchuck National Referral Hospital (JDWNRH) in Bhutan

Sonam Zam and Montip Tiensuwan
Department of Mathematics, Faculty of Science, Mahidol University, Bangkok 10400, Thailand

Email: moon7su@gmail.com

Abstract: Pneumonia is a lung infection or swelling (inflammation) of the tissue in one or both lungs caused by bacteria *Streptococcus pneumonia.* Pneumonia claims the lives of the world's most vulnerable children. The aim of this study is to find the association between pneumonia variables of pneumonia patients. The recorded data of 4190 JDWNRH pneumonia patients from 2005 to 2016 were used. A test of independence was conducted for all pneumonia patients and individual gender to identify the association between any two variables using chi-square test and Cramer's V statistic. The variables with maximum Cramer's V were formulated using two-dimensional log-linear models to obtain estimate parameters, expected frequencies and standardized residuals. Three-dimensional log-linear models were applied to test the association among any three variables. The analysis of deviance is used to select the best models. The result showed that most paired variables were significantly related at p-value less than 0.01. For all data, ward has a strong association with pneumonia types, age, status of last contact and region. In addition, length of stay has association with status of last contact.

1. Introduction

Google defines 'pneumonia' as a lung infection or swelling (inflammation) of the tissue in one or both lungs [1]. The bacteria *Streptococcus pneumonia* is one of the most common causes of pneumonia throughout the world. This bacterium was discovered by a French microbiologist Louis Pasteur and an American microbiologist George Sternberg in 1881 [2]. The second major cause of pneumonia is the bacteria known as *Haemophilus influenza*, discovered by Pfeiffer in 1892 [3].The symptoms of pneumonia are cough, fever, sweating, shivering, loss of appetite, headache, nausea, joint and muscle pain and a hard time breathing. For most people, pneumonia can be treated at home. It often disappears in 2 to 3 weeks yet older people, babies, and people with other diseases can become very ill. They may need to be in the hospital. Some of the effective preventive measures are exclusive breastfeeding for the first 6 months of life, an effective and inexpensive way to treat childhood pneumonia is through the use of antibiotics and vaccines, and using clean cooking stove.

Therefore, the purpose of this study is to identify the associated factors among personal and clinical variables of pneumonia patients using log-linear models.

2. Methodology

2.1Data collection

In this study, recorded data of 4190 (2323 males and 1867 females) of pneumonia patients from 2005 to 2016 were collected from Jigme Dorji Wangchuck National Referral Hospital (JDWNRH), Bhutan. JDWNRH is renowned across Bhutan for giving the most essential health management and assessment services. Not only the hospital looks into the referred cases by 20 provinces' hospital, but also keenly serve non-referred cases from the neighboring cities and states with the best of its efforts. The data consisted of personal variables and clinical variables which are shown in Table 1. The personal data includes three variables: age, region and gender. The clinical data includes five variables: pneumonia type, ward, number of visits, length of stay and status of last contact.

Table 1: Categories and levels of personal and clinical variables included in the models

Category	Divisions
Personal	
Age (years)	≤10, 11-20, >20
Gender	Male, Female
Region	1.Southern, Eastern, Western, Central, Others 2.Southern, Eastern, Western, Others (only for three-dimensional log-linear models of male and female)
Clinical	
Pneumonia type	Pneumonia-unspecified organism, Others
Ward	1.Pediatric, Medical, Others 2.Pediatric, Others (only for three-dimensional log-linear models of male and female)
Hospital visits (days)	≤ 2, > 2
Length of stay (days)	≤ 15, 16-30, >30
Status of last contact	1.Recovered, Death, Others 2.Recovered, Others (only for three-dimensional log-linear models of male and female)

Some of the categories have more than one divisions, it is to reduce the value being blank or zero.

From Table 2, the personal data show that more than 50 percent of the patients were male and about 54 percent of the pneumonia patients were children aged 10 and below.

From clinical data, most of the patients were diagnosed with pneumonia-unspecified organism. Among wards, most of the patients were admitted in pediatric ward and least patients were admitted in eye and physical ward. About 98.7 percent of the patients visited hospital less than or equals to two days with majority of the patients being admitted in the hospital for less than or equals to 15 days. Status of last contact tells us that with less number of deaths, around 96.1 percent of the patients recovered from pneumonia.

2.2 Statistical Analysis

In this study, we carried out statistical analysis in three stages. A test of independence was conducted to learn about the association between personal and clinical variables, and among personal and clinical variables incorporated with Cramer's V statistic. Cramer's V measures the strengths of association between two categorical variables after chi-square has determined significance. The value ranges from 0

IOP Conf. Series: Journal of Physics: Conf. Series **1039** (2018) 012030 doi:10.1088/1742-6596/1039/1/012030

to 1. A value close to 0 indicates a weak association between variables and close to +1 shows a strong association. The Cramer's V can be obtained by

$$V = \sqrt{\frac{\chi^2}{n\,(\mathrm{v}-1)}}, \tag{1}$$

where χ^2 is derived from chi-squared test, v is the smaller of the number of rows and columns, and n is a sample size. This statistic, Cramer's V can attain the maximum of 1 for tables of any dimension.

Table 2: Descriptive data for personal data and clinical data classified by gender

Variables	Male		Female		Total	
	Number	%	Number	%	Number	%
Personal data						
Age (years)						
≤ 10	1274	54.8	972	52.1	2246	53.6
11-20	576	24.8	511	27.4	1087	25.9
>20	473	20.4	384	20.6	857	20.5
Region						
Southern	891	38.4	709	38.0	1600	38.2
Eastern	451	19.4	338	18.1	789	18.8
Western	486	20.9	464	24.9	950	22.7
Central	456	19.6	344	18.4	800	19.1
Others*	39	1.7	12	0.6	51	1.2
Clinical data						
Pneumonia type						
Pneumonia, Unspecified organism	2143	92.3	1716	91.9	3859	92.1
Others**	180	7.7	151	8.1	331	7.9
Ward						
Pediatric	1796	77.3	1408	75.4	3204	76.5
Medical	187	8.0	153	8.2	340	8.1
Others***	340	14.6	306	16.4	646	15.4
Hospital visits						
≤ 2	2293	98.7	1842	98.7	4135	98.7
> 2	30	1.3	25	1.3	55	1.3
Length of stay (days)						
≤ 15	2220	95.6	1787	95.7	4007	95.6
16-30	82	3.5	68	3.6	150	3.6
> 30	21	0.9	12	0.6	33	0.8
Status of last contact						
Recovered	2232	96.1	1797	96.3	4029	96.2
Death	86	3.7	66	3.5	152	3.6
Others****	5	0.2	4	0.2	9	0.2

Note: *such as India, China, Australia & Bangladesh. **such as pneumonia due to influenza virus, Streptococcus pneumonia, etc.***such as neonatal, eye, etc.****such as absconded & referred out.

2.2.1 Two-dimensional log-linear models

The model given is applied to study the association between any two variables. The saturated or maximal model is written as

$$LogE\left(Y_{jk}\right)=\mu+\alpha_j+\beta_k+\left(\alpha\beta\right)_{jk} \quad , \tag{2}$$

where μ is the grand average of the logarithms of the expected frequencies, α_j, β_k are the main effects of the factors, and $(\alpha\beta)_{jk}$ is the interaction effect of two factors with degrees of freedom $1+(J-1)+(K-1)+(J-1)(K-1)=JK$.

2.2.2 Three-dimensional log-linear models
Like Two-dimensional models, three-dimensional models are applied to study the association between any three variables. The full model is given by

$$LogE\left(Y_{jkl}\right)=\mu+\alpha_j+\beta_k+\gamma_l+\left(\alpha\beta\right)_{jk}+\left(\alpha\gamma\right)_{jl}+\left(\beta\gamma\right)_{kl}+\left(\alpha\beta\gamma\right)_{jkl} \tag{3}$$

where μ is the grand average or overall mean, $\alpha_j, \beta_k, \gamma_l$ are the main effects of the factors, $(\alpha\beta)_{jk}, (\alpha\gamma)_{jl}, (\beta\gamma)_{kl}$ are the interaction effects of two factors, and $(\alpha\beta\gamma)_{jkl}$ is the interaction effects of three factors, with $1+(J-1)+(K-1)+(L-1)+(J-1)(K-1)+(J-1)(L-1)+(K-1)(L-1)+(J-1)(K-1)(L-1)=JKL$ degrees of freedom.

2.2.3 Goodness of fit
Goodness of fit test is used to determine whether a log-linear model used matches the real sceneries.

H_0: The smaller model fits the data well.

H_1: The larger model fits the data well.

Chi-squared statistic and log-likelihood ratio statistic are used to measure goodness of fit for contingency tables as followings:

1. Chi-square statistic is

$$\chi^2=\sum_{j=1}^{J}\sum_{k=1}^{K}\frac{(O_{jk}-E_{jk})^2}{E_{jk}} \tag{4}$$

The test is to reject H_0 if $\chi^2>\chi^2_{\alpha,(J-1)(K-1)}$ and accept H_0 otherwise, with an approximated significance level α, and degrees of freedom $(J-1)(K-1)$.

2. The log-likelihood ratio statistic or deviance is

$$G^2=2\sum_j\sum_k O_{jk}\log\frac{O_{jk}}{E_{jk}} \tag{5}$$

with $(J-1)(K-1)$ degrees of freedom. Thus, the test is to reject H_0 if $D>\chi^2_{\alpha,(J-1)(K-1)}$ and accept H_0 otherwise.

3. Results and discussion
In this section, we present the results from testing associations between two distinct pneumonia variables for male and female patients using chi-square statistic and Cramer's V. Table 3 illustrates the summary of the relationship between the personal and clinical variables which gave maximum Cramer's V for male and female pneumonia patients at p-value < 0.01.
Table 3 shows that ward and status of last contact, length of stay and status of last contact, age and ward, and region and ward have strong association for both the genders. However, pneumonia type and ward had association only for male patients. It also indicates that ward is having a strong relationship with personal variables as well as clinical variables.

Table 3: Summary of the association between variables in personal data and clinical data which produced maximum Cramer's V value

Variable	Chi-square	d.f.	p-value	Cramer's V
Male				
Pneumonia type and ward	14.920**	2	0.001	0.080^
Ward and status of last contact	142.520**	4	<0.001	0.175^
Length of stay and status of last contact	19.586**	4	0.001	0.065^
Age and ward	955.964**	4	<0.001	0.454^
Region and ward	266.466**	8	<0.001	0.239^
Female				0.080^
Ward and status of last contact	60.995**	4	<0.001	0.128^
Length of stay and status of last contact	34.662**	4	<0.001	0.096^
Age and ward	729.264**	4	<0.001	0.442^
Region and ward	184.321**	8	<0.001	0.222^

Note: ** significant at p-value <0.01, ^maximum Cramer's V value for each variable.

Two-dimensional log-linear models

From Table 3, we choose the two variables which had an association and gave the maximum Cramer's V-value to formulate the two-dimensional log-linear models, and acquired 9 models: 5 for male and 4 for female, in the form $Log[E(Y_{jk})] = \mu + \alpha_j + \beta_k + (\alpha\beta)_{jk}$. We present only one model corresponding to that of age and ward for male patients in Table 4.

Table 4: The frequencies (percent) of age and ward for male patients

Ward	Age			Total
	≤10	11-20	>20	
Pediatric	1161(49.98)	494(21.27)	141(6.07)	1796(77.31)
Medical	0(0.00)	9(0.39)	178(7.66)	187(8.05)
Others*	113(4.86)	73(3.14)	154(6.63)	340(14.64)
Total	1274(54.84)	576(24.80)	473(20.36)	2323(100.00)

Note: * such as neonatal, eye, etc.

In addition, we have also fitted the model under H_0 and the maximal model under H_1 and obtained χ^2=955.964, the log-likelihood ratio statistic (D) = 847.053 and degrees of freedom (d.f.) = 4, hence at a significance level 0.05 we reject H_0 and select the maximal or saturated model. The estimates of the main effects of interaction terms of the saturated model are given in Table 5.

Table 5: Estimates of parameters of saturated model for age and ward for male patients

Parameter	Model under H_0	Saturated model	Parameter	Model under H_0	Saturated model
μ	7.057	6.893	α_1	0.000	0.000
α_2	-33.004	-2.262	α_3	-2.330	-1.664
β_1	0.000	0.000	β_2	-0.855	-0.794

The other models were obtained in the same way. The log-likelihood ratio statistic, chi-square value and degrees of freedom for the associated variables for individual gender are represented in Table 6.

Table 6: The log-likelihood ratio statistic, chi-square value and degrees of freedom for associated variables on male and female patients

Associated variables	D	χ^2	d.f.	*p*-value
Male				
Ward and pneumonia type	12.260	14.920	2	0.001
Ward and status of last contact	112.415	142.520	4	<0.001
Ward and age	847.053	955.964	4	<0.001
Region and ward	254.654	266.466	8	<0.001
Female				
Ward and status of last contact	50.316	60.995	4	<0.001
Length of stay and status of last contact	15.478	34.662	4	<0.001
Ward and age	656.510	729.264	4	<0.001
Region and ward	185.508	184.321	8	<0.001

Three-dimensional log-linear models

In this section, we determine whether or not any association is the same for both the gender for all patients and age for individual gender. Farr *et al.* [4], found that crude death rates from pneumonia for male is more than for female, that is why gender has been selected as one in each set of three variables for all patients. We selected age as one in each set of three variables for individual gender because pneumonia is the infectious cause of death in children under the age of 5 [5]. The other two variables for each set are age and region, pneumonia type and ward, ward and status of last contact, length of stay and status of last contact, age and ward, and region and ward for all patients and pneumonia type and ward, ward and status of last contact, length of stay and status of last contact, age and ward, region and ward for male patients, and ward and status of last contact, length of stay and status of last contact, age and ward, and region and ward for female patients. From these, three-dimensional log-linear models were formulated using the values of deviance difference and deviance difference degrees of freedom to obtain the best models. The results are shown in Table 7, however we present only one model in detail that is for the ward, status of last contact and age for female patients but other models are also obtained in the same manner.

Table 7: Analysis of deviance for the ward, status of last contact and age for female patients

Model	Parameter	Deviance	d.f.	Deviance difference	d.f. difference
(1)	Full model: $\mu, \alpha_j, \beta_k, \gamma_l, (\alpha\beta)_{jk}, (\alpha\gamma)_{jl}, (\beta\gamma)_{kl}, (\alpha\beta\gamma)_{jkl}$	0	0	-	-
(2)	Delete $(\alpha\beta\gamma)_{jkl}$ from (1)	16.649	2	16.649*	2
(3)	Delete $(\alpha\beta)_{jk}$ from (2)	28.507	3	11.858*	1
(4)	Delete $(\alpha\gamma)_{jl}$ from (2)	544.848	5	528.199*	3
(5)	Delete $(\beta\gamma)_{kl}$ from (2)	618.532	7	601.883*	5

Note: * significant at *p*-value < 0.01

According to the chi-square, the best model is (2), that is $Log[E(Y_{jkl})] = \mu + \alpha_j + \beta_k + \gamma_l + (\alpha\beta_{jk}) + (\alpha\gamma)_{jl} + (\beta\gamma)_{kl}$, $j = 1,2;\ k = 1,2;\ l = 1,2$, which shows that, for female patients, age is related with ward and status of last contact, while ward is related to status of last contact. The observed frequency, expected frequency and standardized residual of ward, status of last contact and age are shown in Table 8.

Table 8: Observed, expected frequency and standardized residual of ward, status of last contact and age for female patientS

Cell *j*	*k*	*l*	Observed (Expected under H_1)	Expected under H_0	Standardized Residual
1	1	1	853	854.5144	-0.0518
1	2	1	25	23.4856	0.3125
2	1	1	93	91.4856	0.1583
2	2	1	1	2.5144	-0.9550
1	1	2	418	417.1820	0.0400
1	2	2	0	0.8180	-0.9044
2	1	2	92	92.8180	-0.0849
2	2	2	1	0.1820	1.9175
1	1	3	111	99.4583	1.1573
1	2	3	1	12.5417	112.0000
2	1	3	230	241.5417	-0.7426
2	2	3	42	30.4583	2.0913

The best model of three-dimensional log-linear models for each set of three variables is summarized in Table 9. According to the three-dimensional log-linear models, the best models for each set of three variables are not the same. The relationship among variables in each set contributed for choosing the best model.

From Table 8, there is an interaction of three effects in models (1), (6), (7), (8), (10), (11), (13). Model (1) shows that the association between age and region is not the same for both the gender. Model (2) illustrates that the distribution of gender is same for the ward and pneumonia type. According to the best model, model (3), the association between ward and status of last contact is the same for both genders. As per model (4), the association between length of stay and status of last contact is the same for both genders. Likewise, model (5) depicts association of gender being depended on ward and age. Model (6) shows that the association between ward and region is not the same for both genders. As per model (7), the dispersal of age is not the same for pneumonia type and ward. Model (8) and model (11) demonstrates that the association of ward and status of last contact is not equivalent for all ages. According to the best model, model (9) and model (12), length of stay and status of last contact depended on age. Similarly, models (10) and (13), the association between age and ward is not the same for region.

4. Conclusion

In our study, we have shown only the relative variables which give the maximum Cramer's V value in each table of association. There are numerous studies in the world that have investigated on pneumonia. A special feature of this study is being the first research to be carried out in studying pneumonia patients using log-linear model in Bhutan and this approach led to a valuable result which can be used by experts in Bhutan to battle against pneumonia.

This study showed that most paired variables were significantly related at p-value less than 0.01. For all patients and individual gender, ward has a strong association with pneumonia types, age, status of last contact and region. In addition, length of stay has association with status of last contact. From the numerical results, we can see that most of the pneumonia patients are residing in southern region of Bhutan while there were less patients from other countries like Australia, Bangladesh, India and China receiving treatment in Bhutan.

Table 9: Summary of the best model of three-dimensional log-linear models

Variables	The best model
(1) Age (α_j), region (β_k), and gender (γ_l)	$Log[E(Y_{jkl})] = \mu + \alpha_j + \beta_k + \gamma_l + (\alpha\beta)_{jk} + (\alpha\gamma)_{jl} + (\beta\gamma)_{kl} + (\alpha\beta\gamma)_{jkl}$, $j = 1, 2, 3; k = 1, 2, 3, 4, 5; l = 1, 2$
(2) Ward (α_j), pneumonia type (β_k), and gender (γ_l)	$Log[E(Y_{jkl})] = \mu + \alpha_j + \beta_k + \gamma_l + (\alpha\beta)_{jk} + (\alpha\gamma)_{jl} + (\beta\gamma)_{kl}$, $j = 1, 2, 3; k = 1, 2; l = 1, 2$
(3) Ward (α_j), status of last contact (β_k), and gender (γ_l)	$Log[E(Y_{jkl})] = \mu + \alpha_j + \beta_k + \gamma_l + (\alpha\beta)_{jk} + (\alpha\gamma)_{jl} + (\beta\gamma)_{kl}$ $j = 1, 2, 3; k = 1, 2, 3; l = 1, 2$
(4) Length of stay (α_j), status of last contact (β_k) and gender (γ_l)	$Log[E(Y_{jkl})] = \mu + \alpha_j + \beta_k + \gamma_l + (\alpha\beta)_{jk} + (\alpha\gamma)_{jl} + (\beta\gamma)_{kl}$ $j = 1, 2, 3; k = 1, 2, 3; l = 1, 2$
(5) Ward (α_j), age (β_k), and gender (γ_l)	$Log[E(Y_{jkl})] = \mu + \alpha_j + \beta_k + \gamma_l + (\alpha\beta)_{jk} + (\alpha\gamma)_{jl} + (\beta\gamma)_{kl}$ $j = 1, 2, 3; k = 1, 2, 3; l = 1, 2$
(6) Ward (α_j), region (β_k), and gender (γ_l)	$Log[E(Y_{jkl})] = \mu + \alpha_j + \beta_k + \gamma_l + (\alpha\beta)_{jk} + (\alpha\gamma)_{jl} + (\beta\gamma)_{kl} + (\alpha\beta\gamma)_{jkl}$, $j = 1, 2, 3; k = 1, 2, 3, 4, 5; l = 1, 2$
(7) Pneumonia type (α_j), ward (β_k), and age (γ_l)	$Log[E(Y_{jkl})] = \mu + \alpha_j + \beta_k + \gamma_l + (\alpha\beta)_{jk} + (\alpha\gamma)_{jl} + (\beta\gamma)_{kl} + (\alpha\beta\gamma)_{jkl}$, $j = 1, 2; k = 1, 2; l = 1, 2, 3$
(8) Ward (α_j), status of last contact (β_k), and age (γ_l)	$Log[E(Y_{jkl})] = \mu + \alpha_j + \beta_k + \gamma_l + (\alpha\beta)_{jk} + (\alpha\gamma)_{jl} + (\beta\gamma)_{kl} + (\alpha\beta\gamma)_{jkl}$, $j = 1, 2; k = 1, 2; l = 1, 2, 3$
(9) Length of stay (α_j), status of last contact (β_k) and age (γ_l)	$Log[E(Y_{jkl})] = \mu + \alpha_j + \beta_k + \gamma_l + (\alpha\beta)_{jk} + (\alpha\gamma)_{jl} + (\beta\gamma)_{kl}$ $j = 1, 2; k = 1, 2; l = 1, 2, 3$
(10) Ward (α_j), region (β_k) ,and age (γ_l)	$Log[E(Y_{jkl})] = \mu + \alpha_j + \beta_k + \gamma_l + (\alpha\beta)_{jk} + (\alpha\gamma)_{jl} + (\beta\gamma)_{kl} + (\alpha\beta\gamma)_{jkl}$, $j = 1, 2; k = 1, 2, 3, 4; l = 1, 2, 3$
(11) Ward (α_j), status of last contact (β_k), and age (γ_l)	$Log[E(Y_{jkl})] = \mu + \alpha_j + \beta_k + \gamma_l + (\alpha\beta)_{jk} + (\alpha\gamma)_{jl} + (\beta\gamma)_{kl} + (\alpha\beta\gamma)_{jkl}$, $j = 1, 2; k = 1, 2; l = 1, 2, 3$
(12) Length of stay (α_j), status of last contact (β_k) and age (γ_l)	$Log[E(Y_{jkl})] = \mu + \alpha_j + \beta_k + \gamma_l + (\alpha\beta)_{jk} + (\alpha\gamma)_{jl} + (\beta\gamma)_{kl}$ $j = 1, 2; k = 1, 2; l = 1, 2, 3$
(13) Ward (α_j), region (β_k), and age (γ_l)	$Log[E(Y_{jkl})] = \mu + \alpha_j + \beta_k + \gamma_l + (\alpha\beta)_{jk} + (\alpha\gamma)_{jl} + (\beta\gamma)_{kl} + (\alpha\beta\gamma)_{jkl}$, $j = 1, 2; k = 1, 2, 3, 4; l = 1, 2, 3$

The prevalence of pneumonia was found to be highest among children and male patients. Since pneumonia is the number one infectious diseases killer of children, pediatric ward had the highest number of patients being admitted with more being male. About 98.7 percent of the patients visited hospital less than or equals to two days with majority of the patients being admitted in the hospital for less than or equal to 15 days. Status of last contact tells us that with less number of deaths, around 96.1 percent of the patients recovered from pneumonia.

Bhutan Multiple Indicator Survey (2010) [6], reveals that about 7% of the children are reported to have had the symptoms of Pneumonia and 74% of children are said to have received treatment at the health care unit. Rest 26% have not received any treatment, though 26% seems small, the battle with pneumonia has not ended. Nevertheless, with awareness campaign on importance of breast feeding, living in clean surroundings, using antibiotics and vaccines, there are reduction in pneumonia patients with each year.

Though the frequency of hospital visits influences the severity of pneumonia, it could not find an association with any of the personal and clinical variables using log-linear models. The log-linear model only determines the association between the variables [7]. Therefore, another interesting study would be on pneumonia variables using logistic regression models.

References

[1] Definition of pneumonia. Available from *http://www.webmd.com/lung/tc/pneumonia-topic-overview#1* [Assessed 1st June, 2017].
[2] Causes of pneumonia. Available from *http://www.emedicinehealth.com/bacterial_pneumonia/page2_em.htm* . [Assessed 1st June, 2017].
[3] Causes of pneumonia. Available from *http://www.lung.org/lung-health-and-diseases/lung-disease-lookup/pneumonia/what-causes-pneumonia.html?referrer=https://www.google.co.th/* . [Assessed 1st June, 2017].
[4] Farr MB, Bartlett RLC, Wadsworth J, Miller LD. Risk factors for community acquired pneumonia diagnosed upon hospital admission. Respiratory medicine (2000) 94, 954-963.
[5] Pneumonia, the cause of death in children *http://www.who.int/mediacentre/factsheets/fs331/en/* . [Assessed 4th June, 2017]
[6] Bhutan multiple indicator survey. Available from *http://www.nsb.gov.bt/publication/files/pub1he155wy.pdf*. [Assessed 8th June, 2017].
[7] Tiensuwan M, Yimprayoon P, Lenbury Y. Application of log-linear models to cancer patients: A case study of data from the national cancer institute, Thailand. Mahidol University. 2005: 1283-1291.

Acknowledgements
The authors would like to thank Thailand International Cooperation Agency (TICA) for the scholarship and record section of Jigme Dorji Wangchuck National Referral Hospital (JDWNRH), Ministry of Health in Bhutan for supporting us with the pneumonia data.

IOP Conf. Series: Journal of Physics: Conf. Series **1039** (2018) 012031 doi:10.1088/1742-6596/1039/1/012031

Application of logistic regression models to cancer patients: a case study of data from Jigme Dorji Wangchuck National Referral Hospital (JDWNRH) in Bhutan

Choney Zangmo and **Montip Tiensuwan**
Department of mathematics, Faculty of Science, Mahidol University, Bangkok 10400, Thailand

Email: zangmo_c@yahoo.com

Abstract: Cancer is an uncontrolled disease caused by a damage to the cell's DNA. Cancer rates are increasing every year and is the biggest concern in many countries including Bhutan. Bhutan being a very small country loses around 100 to 200 patients every year due to cancer. This study involved 3013 cancer patients who are treated at Jigme Dorji Wangchuck National Referral Hospital (JDWNRH) from 2010 to 2016. Those patients who cannot be treated in the country are referred out to other countries for treatment. This study is to identify the factors that affect the survival of all cancer patients as well as male and female patients separately. This has been done using logistic regression model where the dependent variable is the patient's last status. The best fitted model was obtained from the analysis of deviance. The test showed that the last status of the patient and the variables of personal and clinical data are mostly significant at p-value less than 0.05. The factors affecting the last status of cancer are age, length of stay and the cancer site. Cancer site is one of the main factors that affect the survival of both male and female patients.

1. Introduction

Cancer is a disease caused by an uncontrolled division of abnormal cells in the parts of the body. Cancer has become more than just a global health problem and is also a cause of deep suffering to the individual patients, their families and to the community at large. Cancer affects everyone which represents a tremendous burden to the family and country. According to world Health Organization, cancer has found to be the second leading cause of death throughout the globe and is accounted for death of 8.8 million people in 2015 worldwide [1].

In Taiwan, pancreatic cancer is the eighth leading cause of death. Approximately 800 people die of pancreatic cancer yearly [2] and prostate cancer is the second leading cause of death in the world [3]. The exact cause of cancer is unknown but there are some preventable risk factors which can be avoided such as using tobacco, alcohol, unhealthy foods and unhealthy lifestyles [4]. Cancer of the oral cavity, pharynx, larynx, and esophagus in Italy showed smokers have higher risk to cancer. However, for cancer of esophagus, alcohol consumption showed higher risk to cancer than smoking [5].

Cancer can be very expensive, devastating, worrisome and could possibly lead to death. However, early detection increases the survival figures [6-7]. It has become a major concern around the world. In Bhutan, the number of cancer cases has been increasing over the years and there were 473 cases reported in 2016. Female patients are mostly affected by cancer than male patients. To create awareness of such disease, Bhutan has come up with programs such as health campaign, banning of tobacco, pap-smear screenings among women and vaccination to minimize cancer risk [8].

Therefore, the purpose of this study is to apply logistic regression models to identify the cancer variables associated with patient's last status for all cancer patients and for individual gender.

2. Methodology

2.1 Data collection

The subjects on this study were cancer patients treated at Jigme Dorji Wangchuck National Referral Hospital (JDWNRH) in Thimphu, Bhutan from 2010 to 2016. The hospital is located in the capital of the country. A total of 3013 cancer patients with 1657 females and 1356 males were collected from the record section of JDWNRH through the application form. The cancer data was classified into two categories as follows:

Part I Personal data: This part consists of only two variables i.e. the age and gender

Part II Cancer / clinical data: This part consists of four variables including visits, length of stay, cancer site, and patient's last status.

Table 1 provides the categories and levels of personal and clinical variables in the models.

Table 1. Categories and levels of personal and clinical variables included in the models

Category	Level
Personal	
Gender	male, female
Age (years)	0-25, 26-50, 51- 75, >75
Clinical	
Length of stay (days)	≤ 5, >5
Visits (times)	≤ 5, >5
Cancer site	lip, oral cavity and pharynx, digestive organs, respiratory organs, female genital organs, male genital organs, breast, bone and skin, others (brain, throat, neuroendocrine tumors and urinary tract)
Patients last status	alive, death (binary logistic regression) alive, death, uncertain (multinomial logistic regression)

For the binary logistic regression, we use the dependent variable as death and alive. For multinomial logistic regression, we used the dependent variables as alive, death or uncertain.

Data were entered into excel and has been analyzed using IBM SPSS 18 (SPSS Ins).

2.2 Statistical Analysis

In this study, the statistical analyses are given as follows.

2.2.1 Probability distribution

In logistic regression models we have to construct design variables. The partial method involves setting all of the design variables equal to zero for the reference group, and then setting a simple design variable equal to one for each of the other groups. If the independent variable is declared to be in k categories, then the number of the design variables is equal to k-1. This is illustrated in Table 2.

Next, We define $X = 1$ if the outcome is a success and $X = 0$ if the outcome is a failure, then the probability of X is given as $\Pr(X = 1) = \pi$ and $\Pr(X = 0) = 1 - \pi$ where $0 \leq \pi \leq 1$.

IOP Conf. Series: Journal of Physics: Conf. Series **1039** (2018) 012031 doi:10.1088/1742-6596/1039/1/012031

Table 2 Illustration of the coding of the design variables using the partial method

Level	Design variables			
	X_1	X_2	...	X_{k-1}
1	0	0	...	0
2	1	0	...	0
3	0	1	...	0
⋮	⋮	⋮	⋱	⋮
k	0	0	...	1

The logit is defined as the logarithm of the odds ratio $\left(\frac{\pi_i}{1-\pi_i}\right)$ which is given by an equation

$$\text{Logit}\,(\pi_i) = \log\left(\frac{\pi_i}{1-\pi_i}\right) = \beta_0 + \beta_1 x_1 + \ldots + \beta_s x_{is}\,.$$

The fitted value is $\pi_i = \dfrac{e^{\log it(\pi_i)}}{1+e^{\log it(\pi_i)}}$.

2.2.2 Goodness of fit

To measure the goodness of fit model, we use log likelihood ratio statistic (deviance).

$$D = 2\sum o \log\frac{o}{e}$$

where o denotes the observed frequencies and e denotes the expected frequencies.
From the value of D, we can test the hypothesis of the model which are

H_0: The modified model fits the data well.

H_1: The modified model does not fit the data well.

At significant level α = 0.05, we reject the H_0 if $D > \chi^2_{\alpha,\ (\text{degrees of freedom})}$ or accept H_0 otherwise.
The standard procedure of choosing the significant value is $\alpha = 0.05$.

3. Results and Discussion

3.1 Binary logistic regression

Binary logistic regression is used when the dependent variable is dichotomous. The number and percentage of clinical and personal data by gender are shown in Table 3. From Table 3, according to the age groups, most of the male and female patients affected by cancer ranges from 51-75 years of age. Most of the hospital visits were less than or equal to five times which is about 47.2% for male and 52.8% for female. Majority of the patients were admitted in the hospital for more than five days. Regarding the patients last status, more patients recovered from cancer. Death occurred more to male cancer patients (56%) than female patients (44%). In contrast majority of the female patients are diagnosed with cancer. The death percentage of the male patients was higher than the female patients although females were the ones who were mostly diagnosed with cancers. For the cancer site, most of the male patients suffered from digestive organ cancer. For female, the genital organs cancer had the highest number of occurrences. There are few numbers of male patients who were diagnosed with genital organs cancer.

Table 3 The number and percentage of personal and clinical data by gender

Variables	Male		Female	
	Number	Percent	Number	Percent
Age Groups (Years)				
0-25	55	38.2	89	61.8
26-50	165	26.8	450	73.2
51-75	715	46.6	819	53.4
>75	283	60.9	182	39.1
Visits				
≤ 5	1186	44.1	1505	55.9
> 5	32	47.8	35	52.2
Length of stay (Days)				
≤ 5	490	40.3	726	59.7
> 5	728	47.2	814	52.8
Patient status				
Alive	881	40.9	1275	59.1
Death	337	56	265	44
Cancer Site				
Lip, oral cavity and pharynx	79	62.7	47	37.3
Digestive organs	707	58.9	493	41.1
Respiratory organs	134	59.8	90	40.2
Female genital organs	0	0	511	100
Male genital organs	53	100	0	0
Breast	0	0	109	100
Bones and skin	54	54.5	45	45.5
Others*	191	43.8	245	56.2
Total	1218	44.2	1540	55.8

Note: *such as brain, throat, neuroendocrine tumor and urinary tract.

First we describe the relationship between the dependent and the independent variables. We find the significant values and the values for each variable is recorded. Backward elimination is used to delete the variables that are not significant and only keep the variables that are significant. The *p*-values for gender (X_1), age (X_2), length of stay (X_3), and cancer site (X_5-X_{11}) are 0.046, 0.007, 0.000 and 0.000, respectively and they are all significant. For the variable visits (X_4), the value is not significant with *p*-value 0.563. Through the backward elimination method, the variable visit is removed. The result is shown in Table 4. It shows the significant values for all cancer variables. For cancer site, some categories are not significant. We cannot delete the categories which are not significant because the cancer site is a design variable.

Using the variables that are significant, we conducted an interaction test. All the interaction conducted was found to be significant and the result is shown in Table 5. Thus, these variables were considered in the logistic regression analysis.

Table 4 Significant values of all cancer patients

Variables		*p*-value
Gender	Male	
	Female	0.046
Age		0.007
Length of Stay	≤ 5	-
	> 5	0.000
Cancer site	Lip, oral cavity and pharynx	0.000
	Digestive organs	0.059
	Respiratory organs	0.083
	Female genital organs	0.007
	Male genital organs	0.113
	Breast	0.026
	Bones and skin	0.227
	Others*	0.506
Constant		0.000

Note: *such as brain, throat, neuroendocrine tumors and urinary tract.

Table 6 Significant values of cancer data for the individual gender

Variables		*p*-value	
		Male	Female
Age		0.541	0.001
Length of Stay	≤ 5	-	-
	> 5	0.000	0.040
Cancer site	Lip, oral cavity and pharynx	0.002	0.000
	Digestive organs	0.065	0.494
	Respiratory organs	0.177	0.281
	Female genital organs	-	0.037
	Male genital organs	0.210	-
	Breast	-	0.041
	Bones and skin cancer	0.336	0.338
	Others*	0.845	0.643
Constant		0.000	0.000

Note: *such as brain, throat, neuroendocrine tumors and urinary tract

Table 5 Analysis of deviance for the determined variables and p-value for possible interaction of interest to be added to the main effects only model for all cancer patients

Interaction	Deviance	Deviance difference	d.f	d.f. difference	*p*-value
Main effects only	**1507.006**	-	**1470**	-	-
Gender × Age	1504.758	2.248**	1469	1	0.000
Gender × Length of stay	1505.378	1.628**	1469	1	0.000
Gender × Cancer site	1505.805	1.201**	1466	4	0.000
Age × length of stay	1505.243	1.763**	1469	1	0.000
Age × Cancer site	1501.948	5.058**	1463	7	0.000
Length of stay × Cancer site	1490.064	16.942**	1463	7	0.000

Note: ** Significant at p-value < .01
Main effects only: Age, Gender, Length of stay and Cancer site

Since the variable gender is significant, we consider the model for individual gender. For the male cancer patients, length of stay (X_2) and cancer site (X_4-X_8) are significant with p-value < 0.01. The variable age (X_1) and visits (X_3) has p-value 0.541 and 0.730 and they are not significant but age being one of the most biologically important variables and interacts with other variable, we have to make a decision concerning variable age in the model. Thus, we have included the variable age in the model.

For the female cancer patients, the results of fitting the model of female cancer patients with the analysis of deviance for the patient's last status contains the age (X_1), length of stay (X_2) and the cancer site (X_4 -X_9) which has the p-value 0.001, 0.040 and 0.000, respectively. Again the variable visits(X_3) is not significant with p-value 0.746. This variable has been removed using backward elimination which is shown in Table 6.

IOP Conf. Series: Journal of Physics: Conf. Series **1039** (2018) 012031 doi:10.1088/1742-6596/1039/1/012031

Table 7 shows the deviance and *p*-value for possible interaction added in the main effects only model for male cancer patients.

Table 7 Analysis of deviance for the determined variables and *p*-value for possible interaction of interest to be added to the main effects only model for male cancer patients

Interaction	Deviance	Deviance difference	d.f	d.f. difference	*p*-value
Main effects only	**698.103**	-	**622**	-	-
(Age × Length of stay) + Age	697.290	0.813**	620	2	0.000
(Age × Cancer site) +Age	693.806	4.297**	616	6	0.000
Length of stay × Cancer site	692.270	5.833**	616	6	0.000

Note: ** Significant at *p*-value < .01
Main effects only: Age, Length of stay and Cancer site

Table 8 shows the deviance and *p*-value for possible interaction added in the main effects only model.

Table 8 Analysis of deviance for the determined variables and *p*-value for possible interaction of interest to be added to the main effects only model for female cancer patients

Interaction	Deviance	Deviance difference	d.f	d.f. difference	*p*-value
Main effects only	**803.047**	-	**843**	-	-
Age × length of stay	795.860	7.187**	842	1	0.000
Age × Cancer site	797.003	6.044**	837	6	0.000
length of stay × Cancer site	786.509	16.538**	837	6	0.000

Note: ** Significant at p-value < .01
Main effects only: Age, length of stay and Cancer site

The interaction terms included into this model shows the possibility to describe the effects of the associated variables to the patient's last status. Table 8 shows the individual interaction added to the main effects only model. It shows all the interaction tests to be significant. When all the main effects only model and interaction terms are added together, we get the result which is shown in Table 9. We see that the variables age, length of stay, cancer site and interaction terms between cancer site and length of stay are significant at *p*-value < 0.05. Since the interaction term between cancer site and length of stay is significant, the variable cancer site is included in the model.

The odd of age covariate is 1.064. This shows that with every increase in age, the odds of death increases by 1.064 times. The female patients admitted in hospital for more than 5 days have about 29.067 times higher odds of death than those patients who are not admitted and the odds ratio is relatively very high. The group of female patients who has cancer located in the digestive organs, respiratory organs, female genital organs, breast, bone and skin and some other sites have higher odds to death than the female patients who are affected by lip, oral cavity and pharynx cancer (reference group). That is because the odds of death of these groups are greater than the groups of patients who have cancer located in lip, oral cavity and pharynx. The group of female patients who are admitted in hospital for more than 5 days and has digestive organs cancer has 0.081 times less odds to death than the patients with the cancer located in lip, oral cavity and pharynx. A female patient with cancer located in female reproductive organs who is admitted in the hospital for more than 5 days has about 0.142 times less odds to death.

Table 9 Estimated coefficients, standard errors, Wald's test and odds ratio with 95% C.I. of the best model for female cancer patients

Variable	Estimated coefficients	Estimated Std. Error	Wald's test	*p*-value	Exp(B)	95% C.I.for EXP(B)	
						Lower	Upper
X_1	0.062	0.031	3.945	0.047	1.064	1.001	1.13
X_2	3.377	1.341	6.341	0.012	29.278	2.114	405.515
X_4	4.643	2.431	3.648	0.056	103.865	0.886	12178.027
X_5	2.313	2.716	0.725	0.394	10.106	0.049	2073.456
X_6	2.791	2.461	1.286	0.257	16.292	0.131	2025.326
X_7	3.540	2.877	1.514	0.219	34.453	0.123	9676.525
X_8	4.198	2.768	2.301	0.129	66.558	0.293	15101.951
X_9	3.881	2.426	2.559	0.11	48.454	0.417	5625.991
$X_1 \times X_2$	-0.017	0.009	3.404	0.065	0.983	0.965	1.001
$X_4 \times X_1$	-0.042	0.031	1.783	0.182	0.959	0.903	1.02
$X_5 \times X_1$	-0.009	0.035	0.066	0.797	0.991	0.925	1.062
$X_6 \times X_1$	-0.035	0.032	1.185	0.276	0.966	0.907	1.028
$X_7 \times X_1$	-0.065	0.041	2.484	0.115	0.937	0.864	1.016
$X_8 \times X_1$	-0.060	0.037	2.633	0.105	0.942	0.876	1.013
$X_9 \times X_1$	-0.039	0.031	1.575	0.210	0.962	0.905	1.022
$X_4 \times X_2$	-2.513	1.199	4.392	0.036	0.081	0.008	0.85
$X_5 \times X_2$	-1.860	1.295	2.062	0.151	0.156	0.012	1.971
$X_6 \times X_2$	-1.955	1.228	2.536	0.111	0.142	0.013	1.57
$X_7 \times X_2$	-0.931	1.476	0.398	0.528	0.394	0.022	7.116
$X_8 \times X_2$	-1.319	1.542	0.732	0.392	0.267	0.013	5.488
$X_9 \times X_2$	-1.685	1.244	1.835	0.176	0.185	0.016	2.123
Constant	-6.879	2.426	8.042	0.005	0.001		

Note: X_1: Age, X_2: Length of stay, X_4-X_9: Cancer site

The estimated coefficients, standard errors, Wald's test and odds ratio (95% confidence interval) of the best model for all cancer patients and male cancer patients were obtained in the similar manner as obtained for female cancer patients. The patient's visits to hospital has found to be not significant for all cancer patients and for the individual gender. The factors affecting the patient's last status of cancer are age, length of stay and the cancer site. Cancer site and length of stay are the main factors that affect the survival of all cancer patients as well as the individual gender.

3.2 Multinomial logistic regression

Multinomial logistic regression is used when the dependent or the outcome variable has more than two categories.

According to Table 10, more than 70% of the patients recovered from cancer. Death occurred more to male cancer patients than female patients but majority of the females are diagnosed with cancer. Very few patients have the status uncertain and we do not know whether the patients are dead or alive. The patients are either referred out or not recovered in the uncertain category. The patients with digestive organs cancer has the highest number of death and about 17.5% of the patients with bone and skin cancer is uncertain to death.

Table 10 The number and percentage of personal and clinical data by patient's last status

Variables		Patients last status			Total
		Alive	Death	Uncertain	
Gender	Male	881(65)	337 (24.9)	138 (10.2)	1356
	Female	1275(76.9)	265 (16)	117 (7.1)	1657
Age group	0-25	123(66.5)	21 (11.4)	41 (22.2)	185
	26-50	518(78.2)	97(14.7)	47 (7.1)	662
	51-75	1183 (71.6)	351 (21.2)	119 (7.2)	1653
	>75	332 (64.7)	133 (25.9)	48 (9.4)	513
Length of stay	≤ 5	1015 (79.5)	201 (15.8)	60 (4.7)	1276
	> 5	1141 (65.7)	401(23.1)	195 (11.2)	1737
Visits	≤ 5	2099 (71.3)	592 (20.1)	254 (8.6)	2945
	> 5	57 (83.8)	10 (14.7)	1 (1.5)	68
Cancer site	Lip, oral cavity and pharynx	101 (67.3)	25 (16.7)	24 (16)	150
	Digestive organs	851 (65.5)	349 (26.8)	100 (7.7)	1300
	Respiratory organs	156 (62.2)	68 (27.1)	27 (10.8)	251
	Female genital organs	467 (88.3)	44 (8.3)	18 (3.4)	529
	Male genital organs	47 (85.5)	6 (10.9)	2 (3.6)	55
	Breast	102 (87.9)	7 (6)	7 (6)	116
	Bones and skin	86 (71.7)	13 (10.8)	21 (17.5)	120
	Others*	346 (71.6)	90 (20)	56 (8.5)	492

Note: *such as brain, throat, neuroendocrine tumor and urinary tract. The number in parenthesis are in percentage.

Table 11 represents the parameter estimates for the patient's last status and cancer data of all cancer patients. For this analysis, the reference group is death of patients due to cancer. Alive of patient's status relative to death of patient's status and uncertain of patient's status relative to death of patient's status are used to estimate the parameters of the model. For the patients last status alive, the variables age, gender, length of stay and cancer site are significant with p-values 0.010, 0.040, 0.000 and (0.490, 0.060, 0.140, 0.000, 0.040, 0.000, 0.050), respectively. Similarly, for the patients last status uncertain, the variables age, visits, length of stay and cancer site are significant with p-values 0.000, 0.000, 0.010 and (0.020, 0.130, 0.790, 0.790, 0.950, 0.220, 0.010), respectively. Gender and visits are not significant, so we do not consider the model for individual gender. Thus, the model considers the factors age, length of stay and cancer site as the best fit model. Moreover, cancer mostly affected the elder patients.

The first half of the table describes the factors associated with the alive of patients. When looking at the age, the odds ratio of age is 0.992. For the patients admitted in the hospital with less than or equal to 5 days, the odds for the alive of patients relative to death of patients is 1.535. For the cancer site, the patients diagnosed with lip, oral cavity and pharynx cancer has 1.196 times chances to live. Patients with digestive organs cancer has the log odd value decreased by 0.27 when comparing the alive of patients to the death of patients and the odds ratio is 0.764. For patients with female genital organs cancer, male genital organs cancer, breast cancer and bone and skin cancer, the odds of being alive is 2.540, 2.614, 3.328 and 1.905, respectively when compared to the death of patients.

IOP Conf. Series: Journal of Physics: Conf. Series **1039** (2018) 012031 doi:10.1088/1742-6596/1039/1/012031

Table 11 Parameter Estimates for patients last status and cancer data of all cancer patients

last status	Variables		B	Std. Error	Wald	*p*-value	Exp(B)	95% C I for Exp(B) Lower Bound	95% C I for Exp(B) Upper Bound
Alive	Intercept		1.603	0.208	59.55	0.000			
	Age		-0.01	0.003	6.624	0.010	0.992	0.986	0.998
	Visits		0.024	0.043	0.301	0.580	1.024	0.941	1.114
	Gender	Male	-0.22	0.104	4.329	0.040	0.806	0.658	0.988
		Female							
	length of stay	≤ 5	0.428	0.102	17.73	0.000	1.535	1.257	1.873
		> 5							
	Cancer Site	Lip, oral cavity and pharynx	0.179	0.257	0.484	0.490	1.196	0.722	1.98
		Digestive organs	-0.27	0.145	3.459	0.060	0.764	0.575	1.015
		Respiratory organs	-0.29	0.196	2.213	0.140	0.747	0.509	1.097
		Female genital organs	0.932	0.206	20.57	0.000	2.54	1.698	3.8
		Male genital organs	0.961	0.459	4.39	0.040	2.614	1.064	6.423
		Breast	1.202	0.413	8.487	0.000	3.328	1.482	7.471
		Bones and skin	0.645	0.323	3.979	0.050	1.905	1.011	3.589
		Others*							
Uncertain	Intercept		1.218	0.315	14.9	0.000			
	Age		-0.02	0.004	23.7	0.000	0.979	0.97	0.987
	Visits		-0.51	0.135	14.09	0.000	0.603	0.463	0.785
	Gender	Male	0.068	0.166	0.166	0.680	1.07	0.773	1.481
		Female							
	length of stay	≤ 5	-0.46	0.177	6.638	0.010	0.633	0.448	0.897
		> 5							
	Cancer site	Lip, oral cavity and pharynx	0.803	0.345	5.405	0.020	2.232	1.134	4.392
		Digestive organs	-0.35	0.229	2.331	0.130	0.705	0.451	1.104
		Respiratory organs	-0.08	0.303	0.072	0.790	0.922	0.509	1.67
		Female genital organs	-0.09	0.343	0.068	0.790	0.914	0.467	1.791
		Male genital organs	-0.05	0.848	0.004	0.950	0.948	0.18	4.998
		Breast	0.705	0.571	1.523	0.220	2.023	0.661	6.194
		Bones and skin	1.098	0.400	7.537	0.010	2.997	1.369	6.561
		Others*							

The reference category is: Death.
Note: *such as brain, throat, neuroendocrine tumors and urinary tract.

The second half of the table describes the factors associated with the uncertain category of patient's last status. For the factor age, the log odds value for the uncertain of patients compared to the death of patients is decreased by 0.02. The odds ratio is 0.979. The log odds value for the patients visiting the hospital is decreased by 0.51. So the odds for uncertain of patients compared to death of patients is 0.603. For the patients admitted in the hospital with less than or equal to 5 days, the odds for the uncertain of patients is 0.633 times less than the death of patients. For the patients diagnosed with lip, oral cavity and pharynx cancer, breast cancer and bone and skin cancer, the odds of uncertain relatively to death is 2.232, 2.023 and 2.997, respectively.

4. Conclusion

According to the results, we obtained the best model for all cancer, male and female cancer patients with the dependent variable as the patient's last status. If we randomly sample one patient and get information for the factors in the model, we can predict the status of last contact of the patient. From each model we obtain the estimated logit to get the fitted value.

In the binary logistic regression, for all cancer patients, the factors which effect the patient's last status are age, length of stay and cancer site. We perform in the same manner to the female and male cancer patients. Age being one of the most important factors, we considered the variable age in the model. Interaction terms were included in the model using the analysis of deviance to assess their significance. For multinomial logistic regression also, the factors age, length of stay and cancer site were highly related to the patient's last status. Therefore, these factors were considered into the model.

Finally, we conclude that, the logistic regression models that can identify the factors which affect the status of last contact of the cancer patients are also used to predict the status of cancer patients. Thus, further interesting studies could be applying this method to be used with other diseases such as diabetes, cardiovascular disease, HIV, etc. and also applying other statistical method to analyze the data and compare the results.

References

[1] World Health Organization. Available from http://www.who.int/mediacentre/factsheetVfs297/en/. [Assessed 5th, October, 2017].
[2] Chang Chun-Lang, Hsu Ming-Yuan. The study that applies artificial intelligence and logistic regression for assistance in differential diagnostic of pancreatic cancer. *Expert Systems with Applications 36* 2009; 10663-10672
[3] Yuri Prahara, Rochadi Sungsang, Danarto Raden. A Device for Predicting Prostate Cancer Risk: A Logistic Regression. *Journal of Prostate Cancer* 2016; 1-5
[4] Cancer facts in Bhutan-Bhutan cancer society. Available from http://www.bhutancancer.org/cancer-facts-in-bhutan/. [Assessed 11th, October, 2017]
[5] Franceschi Silvia, Talamini Renato, Barra Salvatore, E.Baron Anna, *et al.* Smoking and Drinking in Relation to Cancer of the Oral Cavity, Pharynx, Larynx, and Esophagus in Northern Italy. *Cancer Research* 1990; 6502-6507
[6] Yusuff H, Mohamad N, Ngah KU, Yahaya SA. Breast cancer analysis using logistic Regression. 2012; 14-22
[7] Madhu B *et al.* A Multinomial Logistic Regression Analysis to Study The Influence Of Residence And Socio-Economic Status On Breast Cancer Incidence In Southern Karnataka. *International Journal of Mathematics and Statistics Invention (IJMSI).E-ISSN: 2321 – 4767 P-ISSN: 2321 – 4759. Volume 2 Issue 5 2014*
[8] Stomach cancer in Bhutan- world life expectancy. Available from http://www.worldlifeexpectancy.com/bhutan-stomach -cancer [Assessed 7th, October, 2017]

Acknowledgements

We would like to thank Thailand International Cooperation Agency (TICA) for the scholarship and the record section of Jigme Dorji Wangchuck National Referral Hospital (JDWNRH) in Bhutan for providing and supporting us with cancer data.

IOP Conf. Series: Journal of Physics: Conf. Series **1039** (2018) 012032 doi:10.1088/1742-6596/1039/1/012032

Risk factors for cardiovascular diseases (CVDs) patients in Bhutan

Yeshi Dorji and Montip Tiensuwan

Department of Mathematics, Faculty of Science, Mahidol University, Bangkok, Thailand

Email:yeepee89@gmail.com

Abstract. This paper aims to identify a survival curve involving risk factors for cardiovascular diseases patients in Bhutan. There are 5,206 cardiovascular diseases patients recorded in Jigme Dorji Wangchuck National Referral Hospital (JDWNRH) medical record department as of January 2006 to December 2016. In addition, it examines a relationship between survival time and some risk factors by using univariate and multivariate Cox's proportional hazard models. Application of Log-rank test and Cox's proportional hazard model compares the survival curves and find the covariates that affects the survival time. The results showed that, during the follow up a total of 844 (16.2%) CVD patients died which 216 (13.3%) was young-aged, 201 (13.7%) a middle-aged and 427 (20.2%) an old-aged. The survivorship curves are plotted using Kaplan Meier method and identified some risk factors using univariate and multivariate Cox's model. The risk factors were identified as, namely, age, sex, frequency of hospital visits, CVD subtype, number of diagnosis, ward, hypertension, diabetes, myocardial infarction, and cerebral infarction. Further, risk factors for each age group are illustrated at the significant p-value <0.05.

1. Introduction

Cardiovascular Diseases (CVDs) are the number one cause of death globally and to the less economically developed country like Bhutan. Every year, more number of people die of cardiovascular disease than from any other cause. Cardiovascular Disease (CVD) is the condition that affects the functioning of heart organ and flow of blood in the circulatory system [1]. The competitive risk factors like diabetes, obesity, and the metabolic syndrome contributes to the ascending trend of CVD patients and its mortality. Thus, in the face of public health and individual's economy, it is a huge burden requiring a renewed approach towards prevention of risk factors [2-4].

Survival models in finding risk factors for cardiovascular patients would be effective in lengthening mortality, treatment and lifestyle modification for CVD patients [5-7]. Various studies conducted to frame a risk factors model for short and long terms risk [8]. The association between CVD sub-types and age groups are significant [9-11]. Kaplan Meier and Cox's models have widely used to study the relation between CVD's patients and associated risk factors [12-14].

The aim of this study is to compare the survival function for each age group within covariates and examine the relationship between survival time and risk factors by Cox's proportional hazard models.

2. Methods

2.1. *Study Participants*

The JDWNRH is the medical center of Bhutan keeping the update of the statistical data available regarding patients suffering from any disease. Country is yet to cope with technology and trend regarding the system of automatic (computerization) like other developed nations. It's low financial and economic background keeps Bhutan in peace with neither hunger of it nor craving under the philosophy of GNH (Gross National Happiness). Thus, country's data is statistically in-qualitative as most of the variables/factors are lacking in the package. However, in this registry, 1,618 were registered as young-aged (<45 years) patients, 1,471 as middle-aged (45-65 years) and 2,117 as an old-aged (>65 years) from January 2006 to December 2016. None of the patients were excluded from the study.

2.2. *Risk factors*

To identify the risk factors after onset of CVD which further affects the survival time, we considered some conventional variables available in the data, namely, age, sex, frequency of hospital visits, CVD subtypes, number of diagnosis, ward, hypertension, diabetes mellitus, myocardial infarction and cerebral infarction. Some other risk factors like, HDL and LDL cholesterol, BMI, current smoking, and physical activity were absent in the data. Its absence has weighted the data low. However, available variables were used to find its significance to the survival time/time to event. Age of patients were categorized into <45 years as young aged/less risk patient, 45-65 years as middle aged/moderate risk patients, and >65 years as an old aged/high risk patient. The cardiovascular disease is generally considered as an old age disease. The CVD subtype, hypertension, diabetes mellitus, myocardial infarction and cerebral infarction were extracted from the diagnosis 1, 2, 3, and 4 of the data based on ICD-10 coding for CVD.

2.3 *Statistical Analysis*

All the participants (n = 5,206) were used throughout the data analysis and interpretation. Kaplan-Meier (KM) method was used to estimate the cumulative distribution function of each risk factor, and also compared and plotted the survivorship within each level of all risk factors. The KM method remained helpful in finding risk factors which affect the survival time of CVD patients.

The KM estimator of the survivorship function at time t is obtained by the equation below

$$\hat{s}(t_j) = \frac{number\ of\ surviving\ past\ t_j}{number\ of\ subject\ \text{in}\ the\ study},$$

where t_j are ordered failure times for j=0, 1, 2,…,k; where k denotes the number of distinct times at which some subject failed.

A non-parametric test (Log-rank test) is used for comparing the survival time between different groups of variables and risk factors. It is very much powerful for non-overlapping survival curves. When survival curves are compared, we use the following testing hypothesis:

H_0 : No difference between survival curves.

H_1 : Difference between survival curves.

The most popular testing method is called the Log-rank test. The Log-rank statistic for two groups are given by

$$\frac{(O_i - E_i)^2}{Var(O_i - E_i)}; i = 1, 2, \ldots, k,$$

where

$$(O_i - E_i) = \sum_j (m_{ij} - e_{ij}),$$

and m_{ij} is the observed counts for the ith group at the jth ordered failure time, e_{ij} is the expected counts for the ith group at the jth ordered failure time.

The univariate and multivariate Cox regression allowed us to analyse the effects on survival of several prognostic variables in a survival model. It assumed the effects of different prognostic variables on survival as constant over time. The Cox's PH model is a special type of regression model, which can be used to analyse data that contains censored observations. The hazard function formula for the Cox's PH model is written as

$$h(t,\mathbf{X}) = h_o(t)\exp(\sum_{i=1}^{p}\beta_i x_i),$$

where **X** is a set of covariates $(x_1, x_2, \ldots, x_p)$,

$h_0(t)$ is the baseline hazard function that is a function of *t*, but does not involve the x's

β_i is the coefficient of x_i.

All statistical analysis was performed with SPSS Version 18.

3. Results

The cumulative survival functions of all CVD patients are considered and is shown in Figure 1. The cumulative survival of all CVD patients were more than 80%. It indicates that patients are going to survive some more years if the death is not caused due to other factors. The CVD patient's data were then used to plot cumulative survival functions and compare survival functions for each age group covariates. Referring to Figure 2, it shows that patients less than or equal to 45 years have better survival prognosis than other two middle and an old aged groups. Though middle aged groups survival function is comparatively less significant with young aged group, the curves do not overlap. The log-rank statistics and corresponding p-value on Table 1, indicate that the three categories of patients differ to each other in survival function.

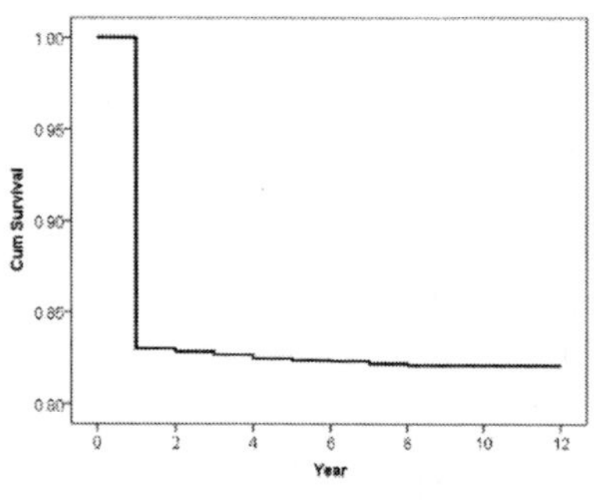

Figure 1 Cumulative survival functions for CVD patients.

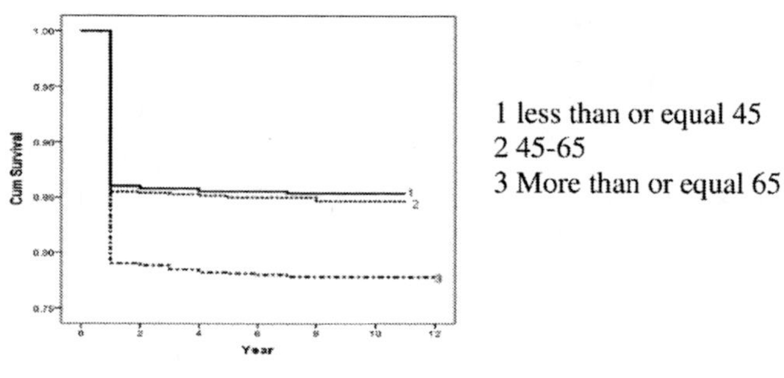

Figure 2 Cumulative survival functions for age group covariates of CVD patients.

Table 1 Test statistic for equality of survival distribution.

Covariate	Log-rank statistic	d.f.	p-value
Age group for CVD patients	40.003**	2	0.000

**significant at p-value <0.01

Table 2 Baseline characteristics of participants grouped in three age categories: number (%)

	Less than 45 yrs (<45)	45-65 yrs	More than 65 yrs (>65)
Gender			
Female	797 (49.3)	625 (42.5)	958 (45.3)
Frequency Visit			
1 Time	1429 (88.3)	1339 (91.0)	1907 (90.1)
2 Times	153 (9.5)	104 (7.1)	168 (7.9)
3 Times	36 (2.2)	28 (1.9)	42 (1.9)
CVD Subtype			
ARF	73 (4.5)	21 (1.4)	8 (0.4)
CRHD	188 (11.6)	69 (4.7)	31 (1.5)
HpD	177 (10.9)	258 (17.5)	447 (21.1)
IHD	63 (3.9)	85 (5.8)	173 (8.2)
PHD & DPC	73 (4.5)	36 (2.4)	55 (2.6)
Other HD	241 (14.9)	229 (15.6)	616 (29.1)
CrD	208 (12.8)	337 (22.9)	579 (27.4)
DAAC	22 (1.4)	16 (1.1)	22 (1.0)
DVLL	548 (33.9)	415 (28.2)	172 (8.1)
Others	25 (1.5)	5 (0.3)	14 (0.7)
No. of Diagnostics			
1 Disease	1128 (69.7)	849 (57.7)	994 (46.9)
2 Diseases	382 (23.6)	451 (30.7)	721 (34.1)
3 Disease	90 (5.6)	134 (9.1)	295 (13.9)
4 Disease	18 (1.1)	37 (2.5)	107 (5.1)
Ward			
Medical	494 (30.5)	640 (43.5)	1141 (53.9)
Surgical	478 (29.5)	411 (27.9)	257 (12.1)
Pvt Cabin	79 (4.9)	147 (9.9)	355 (16.8)
Others	567 (35.0)	273 (18.6)	364 (17.2)
Hypertension	44 (2.7)	120 (8.2)	233 (11.0)
Diabetes	18 (1.1)	47 (3.2)	92 (4.3)
Myocardial Infarction	26 (1.6)	47 (3.2)	77 (3.6)
Cerebral Infarction	54 (3.3)	97 (6.6)	210 (9.9)

Note: ARF- Acute rheumatic fever, CRHD-chronic rheumatic heart disease, HpD-hypertensive disease, IHD-Ischemic heart disease, PHD & DPC-pulmonary heart disease and disease of pulmonary circulation, Other HD- other forms of heart disease, CrD-cerebrovascular disease, DAAC- disease of arteries, arterioles and capillaries, DVLL-disease of veins, lymphatic vessels and lymph nodes, others- other unspecified disorders of circulatory system.

Therefore, we divided the data into three groups as follows: young aged CVD patients, middle aged CVD patients and old aged CVD patients. In addition, we used type and risk factor in clinical data, and all covariates in personal data to plot and compare survival functions for each group. Each covariates association and comparison of Log-rank test with its p-value are given in Tables 3, 4 and 5.

Table 2 shows the baseline characteristics of participants grouped into three age categories. 49.3%, 42.5% and 45.3% were female patients of young aged, middle aged and old aged group respectively. Patients visiting hospital ranged to the maximum of nine times and for the convenience, it was grouped into one time, two times and more than three times. Some patients experienced the event during nine times of visit, some during one time only and some still survived even after nine times visit. The visit frequency to the hospital for young patients found to be 88.3%, 9.5% and 2.2%, respectively. For middle aged patients, 91%, 7.1% and 1.9%, respectively. An old-aged patient's visit was 90.1%, 7.9% and 1.9%, respectively. Above figures depicts that the frequency of CVD patients visiting hospital one time is more and concludes that patients have less complications. Thus, more the frequency of visit, more the complications of the disease.

CVD is not just one disease; it is anything related to the defect of circulatory system. Therefore, the disease is grouped into 10 according to the ICD-10-CM diagnosis codes. After onset of disease, some patients carry multiple diagnosis, up to four diagnosis. Some patients had diabetes mellitus and hypertension, where we considered hypertension and diabetes mellitus as another separate covariate. Their associated diagnostics was supposed to affect the mortality rate of patients and evaluated in Tables 3, 4 and 5 for all covariates within age groups.

Mean age of total participants was 57 years and 41% were an old aged. Some prognostic variables were associated to the survival time of all three categories of age. However, gender remained statistically insignificant for all groups of age. The follow up duration of the study was 11 years from January 2006 to December 2016. During that period, 844 events occurred, of which 216 were young-aged, 201 middle aged and 427 an old-aged participant. Tables 3, 4 and 5 shows the log-rank test of three groups of participants respectively.

Table 3 Test statistics for equality of survival distributions for young-aged CVD patients.

Covariate	Log-rank statistics	d.f.	p-value
Gender	0.002	1	0.963
Frequency of visit	4.013	2	0.134
CVD sub-types	284.927**	9	0.000
No. of diagnostics	118.162**	3	0.000
Ward	192.510**	3	0.000
Hypertension	2.281	1	0.131
Diabetes	0.075	1	0.784
Myocardial infarction	12.726**	1	0.000
Cerebral infarction	4.442*	1	0.035

*significant at p-value <0.05, **significant at p-value <0.01.

Young aged CVD patients of Bhutan have a strong association with CVD sub-types, number of diagnostics, ward, myocardial infarction and cerebral infarction covariates (p-value <0.05). The gender is not significant with p-value 0.963, which does not have association at all. It means, it has a probability error of 96.3%, where male and female does not have different survival time. In addition, frequency of visit, hypertension and diabetes does not have association with young-aged CVD patients as shown in Table 3 as their p-value is greater than 0.05.

Table 4 Test statistics for equality of survival distributions for middle-aged CVD patients.

Covariate	Log-rank statistics	d.f.	p-value
Gender	1.724	1	0.189
Frequency of visit	10.244**	2	0.006
CVD sub-types	161.223**	9	0.000
No. of diagnostics	41.879**	3	0.000
Ward	152.686**	3	0.000
Hypertension	9.805**	1	0.002
Diabetes	0.450	1	0.502
Myocardial infarction	0.558	1	0.455
Cerebral infarction	7.375**	1	0.007

**significant at p-value <0.01.

Regarding the middle-aged patients (45-65 years), covariates like frequency of visit, CVD sub-types, number of diagnostics, ward, hypertension and cerebral infarction have p-value less than 0.05. It indicates that the middle-aged patient's survival time will depend on those covariates. Some variables like, gender, diabetes and myocardial infarction does not make a difference in survival time as their significant level is more than 0.05 as shown in Table 4.

IOP Conf. Series: Journal of Physics: Conf. Series **1039** (2018) 012032 doi:10.1088/1742-6596/1039/1/012032

Table 5 Test statistics for equality of survival distributions for old-aged CVD patients.

Covariate	Log-rank statistics	d.f.	p-value
Gender	2.280	1	0.131
Frequency of visit	0.002	2	0.999
CVD sub-types	126.554**	9	0.000
No. of diagnostics	18.964**	3	0.000
Ward	60.921**	3	0.000
Hypertension	0.779	1	0.377
Diabetes	0.538	1	0.463
Myocardial infarction	4.522*	1	0.033
Cerebral infarction	3.473	1	0.062

*significant at p-value <0.05, **significant at p-value <0.01.

Table 5 shows the test statistics for equality for survival distributions for old-aged CVD patients. Where CVD sub-types, number of diagnostics, ward and myocardial infarction associates and depicts that, an old-age CVD patient's survival time would depend on those factors. Some covariates like gender, frequency of visit, hypertension, diabetes, and cerebral infarction does not have relation with the old-age CVD patients. It's survival time wont depend on those variables.

Thus, some of the considered risk factors have shown an association with test statistics at significance level <0.05. The risk of mortality increases significantly with age. CVD sub-types, number of diagnostics and ward were statistically significant for all groups of age and they are identified to be the significant risk factors for CVD patients. When survivor functions are plotted using KM method, every plot showed no overlapping and each groups are different in all categorical variables. Moreover, the overall hazard ratio of the all-categorical variables grouped into three age groups are summarized in Table 6.

All the covariates were entered into the Cox's proportional hazard model. The table shows the general hazard ratio of the all covariates in three groups. The covariates which are not related to survival time at significance level 0.05 are gender, number of diagnostics and cerebral infarction. Hazard ratio explains the comparison of effect of each covariate on survival time of patients.

The subjects who have visited 2 times and more than 3 times have hazard ratio 1.02 and 1.04 for young aged patients, 0.49 and 0.47 for middle aged patients with 1.03 and 1.09 for old aged patients respectively. Hazard ratio for young aged and old aged patients is greater than subjects who visited hospital one time. So chances of death for 2 times visit and 3 times visit for young aged patients increases at a rate of 1.02 and 1.04 times respectively and old aged patients follows the same. In the similar manner, some CVD sub types, number of diagnostics, ward, hypertension, diabetes, myocardial infarction and cerebral infarction had hazard ratio greater than reference (value 1). A young aged patient who has diabetes are associated with increased death rate of 2.28.

4. Discussion

We use Kaplan-Meier model to plot the cumulative survival curves of each covariate and test the difference of cumulative survival curves within each covariate by using log-rank statistics. We found that younger CVD patients have better survival prognosis than other two middle and old aged patients. For each group of patients, the gender covariates do not have difference in survival time.

For univariate Cox's proportional hazard model, we use one covariate for one model to find the association. Next, we use all covariates which we get from the univariate Cox's proportional hazard model in multivariate Cox's proportional hazard model.

Table 6 Multivariable-adjusted hazard ratios for all three groups of age: number (%)

	Less than 45 yrs (<45)	45-65 yrs	More than 65 yrs (>65)
	HR (95% CI)	HR (95% CI)	HR (95% CI)
Frequency Visit			
1 Time	1.00 (reference)	1.00 (reference)	1.00 (reference)
2 Times	1.02 (0.51-2.01)	0.49 (0.26-0.96)*	1.03 (0.53-2.0)
3 Times	1.04 (0.47-2.29)	0.47 (0.20-1.08)	1.09 (0.52-2.27)
CVD Subtype			
ARF	1.00 (reference)	1.00 (reference)	1.00 (reference)
CRHD	0.82 (0.17-4.09)	0.48 (0.04-5.34)	1.23 (0.00-1.9E29)
HpD	0.96 (0.22-4.18)	0.39 (0.05-3.28)	551.5 (0.00-1.9E20)
IHD	0.42 (0.09-1.99)	0.41 (0.05-3.06)	446.5 (0.00-1.5E20)
PHD & DPC	1.57 (0.31-7.87)	0.42 (0.04-4.05)	935.8 (0.00-3.1E20)
Other HD	2.52 (0.58-10.92)	1.25 (0.15-10.06)	707.7 (0.00-2.4E20)
CrD	4.21 (1.03-17.23)*	1.86 (0.26-13.56)	2050.7 (0.00-6.8E20)
DAAC	3.46 (0.83-14.46)	1.38 (0.19-10.14)	2395.7 (0.00-7.9E20)
DVLL	2.20 (0.39-12.17)	0.00 (0.00-6.3E101)	2528.3 (0.00-8.5E20)
Others	0.04 (0.00-0.49)*	0.04 (0.00-0.46)**	268.0 (0.00-9.0E19)
No. of Diagnostics			
1 Disease	1.00 (reference)	1.00 (reference)	1.00 (reference)
2 Diseases	0.76 (0.28-2.09)	1.32 (0.53-3.31)	0.59 (0.38-0.95)
3 Disease	1.66 (0.60-4.56)	1.96 (0.79-4.88)	1.09 (0.69-1.71)
4 Disease	2.33 (0.82-6.65)	1.97 (0.75-5.14)	0.93 (0.57-1.52)
Ward			
Medical	1.00 (reference)	1.00 (reference)	1.00 (reference)
Surgical	0.28 (0.20-0.39)**	0.28 (0.21-0.38)**	0.54 (0.43-0.68)**
Pvt Cabin	0.22 (0.09-0.48)**	0.22 (0.11-0.43)**	0.42 (0.27-0.65)**
Others	0.06 (0.01-0.42)*	0.25 (0.13-0.47)**	0.44 (0.32-0.61)**
Hypertension	1.16 (0.57-2.35)	1.27 (0.81-2.01)	1.81 (1.31-2.49)**
Diabetes	2.28 (0.56-9.34)	1.18 (0.47-2.97)	1.14 (0.68-1.91)**
Myocardial Infarction	0.28 (0.09-0.79)*	0.87 (0.19-3.92)	0.42 (0.21-0.83)*
Cerebral Infarction	1.16 (0.61-2.22)	1.24 (0.75-2.05)	1.29 (0.92-1.82)

*significant at p-value <0.05, **significant at p-value <0.01.

The risk factors that affect the survival time are the frequency of hospital visits, CVD subtypes, no. of diagnosis, ward, hypertension, myocardial infarction and cerebral infarction. A participant who visited hospital not so frequently has better survival prognosis than those who visited more than one time. It depicts the severity of the disease or need of better treatment. Other unidentified cardiovascular disease has less survival time due to inaccuracy or delay in detection of the disease. Thus, further delaying treatment.

A patient with more number of disease has poor prognosis. Some participants who were treated in the surgical ward showed a better prognosis in survival as immediate surgery saves lives. Participants treated in other wards like ears and nose (ENT) ward, psychiatric ward, orthopaedic ward etc., showed poor survival time. Hypertension affected the survival time of only middle-aged patients. Myocardial infarction and cerebral infarction showed difference in survival time commonly to young patients.

Further, study could be done in more exclusively with more detailed variables like cholesterol, marital status, BMI, current smoking status.

References

[1] Steinbaum RS, (2017, July 25). *Heart Disease Home*. Retrieved from Heart and Cardiovascular

Diseases : http://www.webmd.com/heart-disease/guide/diseases-cardiovascular#1

[2] Hata J, et al. (2016). Risk prediction models for mortality in patients with cardiovascular disease: The biobank japan project. *Journal of epidemiology,* s71-s76.

[3] Pearson A T, et al. (2002). Consensus Panel Guide to Comprehensive Risk Reduction for Adult Patients Without Coronary or Other Astherosclerotic Vascular Diseases. *AHA Scientific Statement,* 388-399.

[4] Yusuf S, et al. (2001). Part II: Variations in Cardiovascular Disease by Specific Ethnic Groups and Geographic Regions and Prevention Strategies. *Clinical Cardiology: New Frontiers.*2855-2864.

[5] Freedman S D, et al. (1998). The Relation of Overweight to Cardiovascular Risk Factors Among Children and Adolescents. *The Bogalusa Heart Study.* 1175-1182.

[6] Berenson S G, et al. (1998). Association between multiple cardiovascular risk factors and athersclerosis in children and young adults. *The New England Journal of Medicine.* 1650-1656.

[7] Kannel B W, et al. (1979). Diabetes and Glucose Tolerance as Risk Factors for Cardiovascular Disease. *The Framingham Study.* 120-126.

[8] Turner CR et al. (1998). Risk factors for coronary artery disease in non-insulin dependent diabetes mellitus: United Kingdom Prospective diabetes study (UKPDS:23). *British Medical Journal (BMJ),* 823-828.

[9] Conroy MR et al. (2003). Estimation of ten-year risk of fatal cardiovascular disease in Europe:the SCORE project. *European Heart Journal,* 987-1003.

[10] Lloyd-Jones MD, et al. (2006). Prediction of Lifetime Risk for Cardiovascular Disease by Risk factor Burden at 50 Years of Age. *Epidemology*, 791-798.

[11] Soares PG, et al. (2016). Progression of Mortality due to Diseases of the Circulatory System and Human Development Index in Rio de Janeiro Municipalities. *Original Article, Epidemology* , 314-322.

[12] Wilson WFP et al. (1998). Prediction of coronary heart disease using risk factor categories. *Circulation: Special report,* 1837-1847.

[13] Blacher J et al. . (2001). Artrial Calcifications, Arterial Stiffness, and Cardiovascular Risk in End-Stage Renal Disease. *Hypertension*, 938-942.

[14] Assmann G et al. (2002). Simple Scoring Scheme for Calculating the Risk of Acute Coronary Events Based on the 10_year Follow-Up of the Prospective Cardiovaascular Munster (PROCAM) Study. *Circulation*, 310-315.

Acknowledgements

We would like to thank Thailand International Cooperation Agency (TICA) for the scholarship and Record Section of Jigme Dorji Wangchuck National Referral Hospital (JDWNRH) in Bhutan for providing and supporting with possible required CVDs patient's data.

Time series analysis of diabetes patients: A case study of Jigme Dorji Wangchuk National Referral Hospital in Bhutan

Thukten Singye and Suntaree Unhapipat

Department of Mathematics, Faculty of Science, Mahidol University, Bangkok 10400, Thailand.

Email: thuktensingye@gmail.com

Abstract: Diabetes has become a concern in Bhutan with its growing number of patients reported to the hospital. Increasing number of patient is likely to result in rising demand for the medical emergencies. Due to having only three referral hospitals in Bhutan,it is important to forecast the future incidences and prepare with proper resource planning. The monthly number of Diabetes patients obtained from Jigme Dorji Wangchuk National Referral Hospital (JDWNRH) is fitted by autoregressive integrated moving average (ARIMA) model. Such dataset starting from January 2006 to December 2016 is divided into two sub datasets; training dataset (January 2006 – December 2015) and validation dataset (January 2016 – December 2016). Using ARIMA, several models were evaluated based on the Bayesian Information Criterion (BIC) and Ljung-Box Q statistics. ARIMA(0, 1, 1) is the best model to describe and predict the future trends of Diabetes incidences on considering the simplest parsimonious lowest order model. Therefore, the proposed model will help to plan appropriately and allocate resources for emergencies.

1. Introduction

Diabetes is a serious life-long disease in which levels of glucose in the blood and urine become too high because the body's ability to produce or respond to the hormone insulin released by the pancreas gets impaired and cannot move the glucose into the cells. There are two main types of diabetes which are Type 1 and Type 2. These two different types of diabetes have different effects and can be caused by different factors. However, they both can cause serious health complications and need to be treated and managed properly.

Type 1 diabetes is a condition when the pancreas fails to produces insulins because the insulins producing cells are damaged or destroyed, so this causes the elevation of glucose level in the blood. Type 2 diabetes is a condition when the body does not produce enough insulin or there is extremely less insulin to move the glucose to the cells moreover it is also a case when the insulin fails to work properly. Therefore, the sugar level in the blood is risen.

Bhutan is a land lock country situated in the heart of the great Himalaya where its maximum proportion of population depends on the agriculture. However, shift onto the sedimentary lifestyle give raise to the diseases like diabetes, heart disease, obesity and some other in the recent years. It is a great concern for country like Bhutan whose population is just about 808,000 as per the *United Nations, Department of*

Economic and Social Affairs, Population Division (2017)[1]. Out of which 47 percent of population falls under the age group of 25 to 59, so it is in fact alarming to know that this working group of people are the victim of these diseases.

Diabetes Mellitus (DM) is currently a major problem prevailing in the country, which results in the lifelong complication and heavy health expenditure. As per WHO International Diabetes Federation, they estimated 422 million adults were living with diabetes in 2014. Therefore, it showed an increase in associated risk factors such as being overweight or obese.

Unfortunately, the country has only three-referral hospital, few district hospitals and BHU (Basic Health Units). However, they are not that equipped with the infrastructures, equipment and with skilled labor. So ultimately, pressure will come to the referral hospitals because sophisticated laboratory tests are essential to differentiate between type 1 diabetes and type 2 diabetes. To understand whether it requires insulin injections for survival or is it where the body cannot properly use the insulin it produces. Type 2 diabetes are most common of people with diabetes and it had occurred maximum with the adults and few children in our dataset. Therefore, in this paper we will find the best-fit models to forecast the future trends for diabetes patients as a whole and type 2 diabetes patient differently and forecast them subsequently.

Recently, there are a number of literatures using time series to model a clinical data. The followings are some time series application for forecasting number of patients in many diseases. However, very few papers are available for diabetes incidence researches using the time series model. In 2005, Earnest *et al.* under BioMed Central (BMC) health services research uses autoregressive integrated moving average (ARIMA) models to predict and monitor the number of beds occupied during a severe acute respiratory syndrome (SARS) outbreak in tertiary hospital Singapore. They found that the ARIMA (1, 0, 3) model was able to describe and predict the number of beds occupied during the SARS outbreak well. The mean absolute percentage error (MAPE) for the training set and validation set were 5.7% and 8.6% respectively, which they found was reasonable for use in the hospital setting. Furthermore, the model also provided three-day forecasts of the number of beds required. Total number of admissions and probable cases admitted on the previous day were also found to be independent prognostic factors of bed occupancy [2]. In 2011, Liu, *et al.* studied the forecasting incidence of hemorrhagic fever with renal syndrome in China. By using the ARIMA model, they worked on Chinese hemorrhagic fever with renal syndrome (HFRS) data during 1978 to 2008. The best model is selected based on minimum value of Akaike Information criterion (AIC) and Ljung-Box test were used. They assess their validity of their proposed model using the MAPE between the observed and fitted HFRS incidence. Finally, they used the fitted ARIMA model to forecast the incidence of the years 2009 to 2011 [3]. In 2014, Dan *et al.* used SARIMA model to forecast malaria mortality rate. In this study, they used Box-Jenkins methodology to build ARIMA for malaria mortality rate for the period 1996 to 2013 with a total of 216 data point. They did the one-year ahead forecast. The best model is selected based on minimum values of AIC and Schwarz Bayesian Information Criterion (SBC). The model $\text{ARIMA}(1,1,1)\times(0,0,1)_{12}$ was recommended to be the best model for forecasting the malaria mortality rate for the coming period of January 2014 to December 2014 [4]. In 2015, Appiah *et al.* has also done time series analysis of Malaria cases in Ejisu Juaben Municipality and found ARIMA (2, 1, 1), autoregressive process of order 2, differencing of order 1 and moving average of order 1 best fit for the secondary data. Using the obtained model, they forecasted for the next two years from 2014 and 2016 [5]. In 2016, Pan *et al.* used the ARIMA model for forecasting the patient number of epidemic disease. They have used actual data of every day patient number of epidemic disease between January and August 2014, in total 223 days, which they obtained from real life CDC (center of disease control). They identify time series model of ARIMA (7, 1, 0) best fit with the forecast accuracy 92.1% [6]. In 2017, Villani *et al.* BMC Health Services Research forecasted the prehospital EMS demand for the emergencies using the time series modelling. Four-one thousand four hundred and fifty-four prehospital diabetic emergencies who attended within the period from January 2009 until December 2015. The SARIMA (0, 1, 0) x $(0, 1, 0)_{12}$ model provided the best fit, with a MAPE of 4.2% [7]. Therefore, time series model is an appropriate model dealing with the number of patients. We will use time series to model and predict the diabetes patient dataset obtained from JDWNRH of Bhutan by using Box-Jenkins approach.

2. Data collection and methodology

2.1 Data collection

We are using the monthly dataset of diabetes patient reported to the referral hospital from 2006 to 2016 for this study. The dataset is retrieved from Jigme Dorji Wangchuk National Referral Hospital (JDWNRH), Bhutan. Such data set is divided into two sets; training data (January 2006 – December 2015) and validation data (January 2016 – December 2016). Model generation is done based on the training data set.

2.2 Methodology

We used Box-Jenkins approach to forecast the number of diabetes patients of JDWNRH for next two years using the existing data from 2006 until 2016. The Box-Jenkins forecasting models are based on statistical concepts and principles as follows.

2.2.1 Identification

This is a stage of determining necessity of differencing, to produce stationarity and defining the order of seasonal and non-seasonal Autoregressive (AR) and Moving average (MA) operators for the series. Overall in this step appropriate structure Autoregressive Integrated Moving Average (ARIMA) and the order of the model is specified. Therefore, autocorrelation function (ACF) and partial autocorrelation function (PACF) plays a vital role in the identifying the AR and MA order.

2.2.2 Estimation

The second step is to estimate the coefficients of the model. Coefficient of AR models can be estimated by least-square regression. The parameter estimation usually requires more complicated iteration procedure but using the computer programing automatically generates it. In this study, we use the Statistical Package Social Sciences (SPSS 18) software to estimate the coefficient.

2.2.3 Diagnostic checking

In this step, we verify whether the estimates of the parameters are significant. Then we ensured the residual plots is a white noise. Once the model is fitted, we need to do diagnostic checking by studying the ACF of residual. If all the autocorrelation and partial correlation are small then the model is working fine but if some of the autocorrelations are large, the values of autoregressive order p or moving average order q are adjusted and the model is re-estimated. The checking of residuals and adjusting the values p and q continues until the resulting residuals contain no additional structure. Once appropriate model is selected, the program may be used to generate forecasts and associated probability limits.

2.2.4 Model selection

Finally, the best ARIMA model is selected by using the Bayesian Information Criterion (BIC) and the selected model is used for forecasting the future diabetes patients of the JDWNRH.

2.3 Box-Jenkins method

2.3.1 The Autoregressive AR (p) process

The AR model is a value of time series after regression on previous values from that same time series. It expresses time series as a linear function of past values. The AR model of order *p*, denoted by AR (*p*), is

$$y_t = \phi_0 + \sum_{i=1}^{p} \phi_i Y_{t-i} + \epsilon_t \quad (1)$$

where ϕ_0 is a constant, $\phi_1, \phi_2, \phi_3, \ldots, \phi_p$ are the model parameters and $\epsilon_t \sim N(0, \sigma^2)$ is a white noise error.

2.3.2 The Moving-Average MA(q) process

The MA model of order *q*, denoted by MA (*q*), is given as

$$y_t = \epsilon_t + \sum_{j=1}^{q} \theta_j \epsilon_{t-j} \quad (2)$$

where $\epsilon_t, \epsilon_{t-1}, \epsilon_{t-2}, \ldots, \epsilon_{t-q}$ are white noise error and $\theta_1, \theta_2, \theta_3, \ldots, \theta_q$ are model parameters

2.3.3 The Autoregressive Moving Average (ARMA) process

If the data perfectly stationary, we can simply model in the format in the model below. By using the linear combination of AR and MA models, given in the equation (1) and (2), the ARMA model of order (*p, q*) can be written as

$$y_t - \phi_1 y_{t-1} - \phi_2 y_{t-2} - \cdots - \phi_p y_{t-p} = \epsilon_t - \theta_1 \epsilon_{t-1} - \theta_2 \epsilon_{t-2} + \cdots - \theta_q \epsilon_{t-q} \quad (3)$$

where $\phi's$ (phis) are the autoregressive parameters to be estimated, the $\theta's$ (thetas) are the moving average parameters to be estimated, the *Y*'s are the original time series values and the $\epsilon's$ a residuals or errors which are assumed to follow the normal probability distribution.

In order to make the writing of model easier, Box-Jenkins use the backshift operator. The backshift operator *B* can change the time period *t* to time period *t* - 1. For example, $BY_t = Y_{t-1}$, $B^2Y_t = Y_{t-2}$ and so on. So the above model can be rewritten as:

$$\left(1 - \phi_1 B - \cdots - \phi_p B^p\right) Y_t = \left(1 - \theta_1 B - \cdots - \theta_q B^q\right) \epsilon_t \quad (4)$$

Further can be abbreviated as:

$$\phi_p(B) Y_t = \theta_q(B) \epsilon_t \quad (5)$$

where $\phi_p(B) = \left(1 - \phi_1 B - \cdots - \phi_p B^p\right)$ and $\theta_q(B) = \left(1 - \theta_1 B - \cdots - \theta_q B^q\right)$.

2.3.4 The ARIMA (p, d, q) forecasting equation

However, it is impossible to have the data always stationary. Therefore, to make the process we have to do differencing in non-stationary series so the differenced series $(1-B)^d Y_t$ must be add in the process ARMA (*p, q*).

Where the a_t is a sequence of identically distributed uncorrelated deviates, referred to as "white noise".

In many situations where differencing is employed, a non-zero constant term θ_0 will not be required. For brevity, the equation is generally written as:

$$\phi_p(B)(1-B)^d Y_t = \theta_0 + \theta_q(B)\epsilon_t, \qquad (6)$$

Where

$$y_t = a_t + \sum_{i=1}^{p} \phi_t Y_{t-1} + \epsilon_t + \sum_{j=1}^{q} \theta_t \epsilon_{t-1}$$

and $(1-B)^d Y_t$ is a differenced series and ϵ_t's is a sequence of identically distributed uncorrelated deviates, referred to as "white noise"[8].

3. Results and discussion

3.1 Identification

The monthly Diabetes patients reported to the hospital from 2006 to 2016 is plotted in the Figure 1. The data plot display trends, which indicates that the mean is not stationary. However, it is necessary to check variance stationarity first. To investigate the stationarity of variance, Box-Cox transformation is used. Figure 1 shows the time series plot of the monthly number of diabetes patients from JDWNRH, Bhutan.

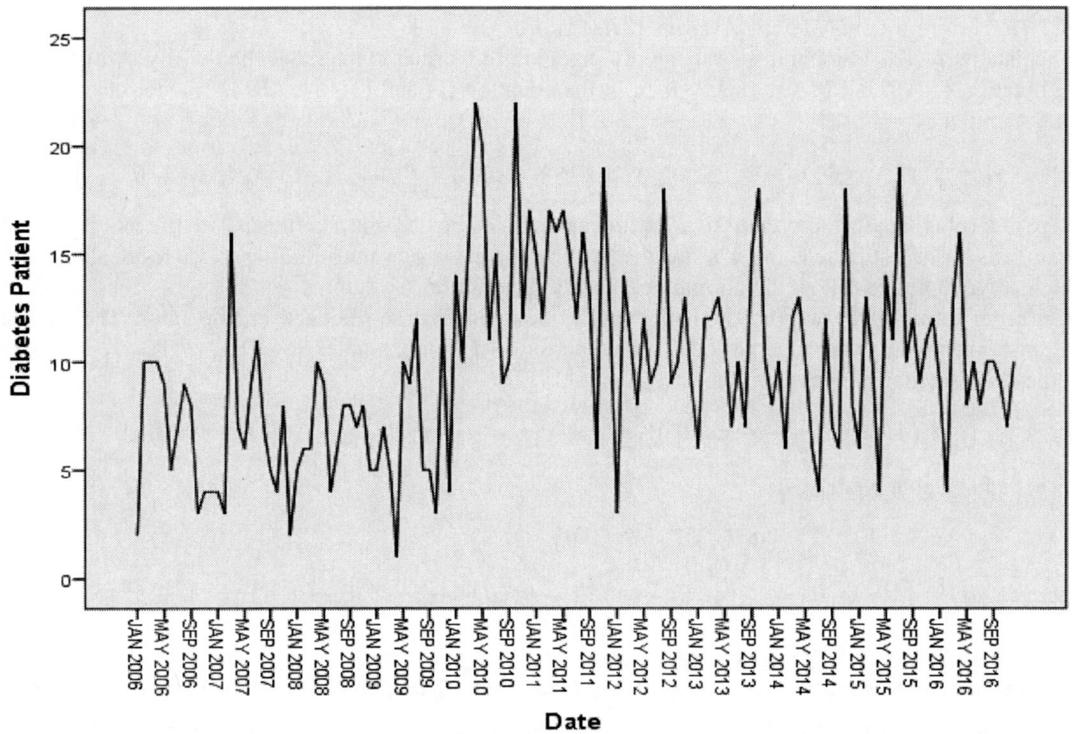

Figure 1. Monthly number of diabetes patients from JDWNRH (January 2006- December 2016)

3.1.1 Tests for stationarity

It is necessary to check the stationarity of the variance. We can use Shapiro-Wilk and Kolmogorov-Smirnov[a] for the normality test. According to the result providing in Table1, we can conclude that the variance was not constant (*p*-value < 0.05). Hence, root transformation which is Box-Cox power transformation is used to stabilize the variance as shown in the Table 2. The transformed series is plotted in Figure 2.

Table 1. Tests of Normalityof the empirical data.

	Kolmogorov-Smirnov[a]			Shapiro-Wilk		
Diabetes	Statistic	df	P-value	Statistic	df	P-value
Patients	.094	120	**.011**	.972	120	**.014**

a. Lilliefors Significance Correction

Table 2. Tests of Normality of the transformed data.

	Kolmogorov-Smirnov[a]			Shapiro-Wilk		
Root transformatio	Statistic	df	P-value	Statistic	df	P-value
n	.078	120	**.070**	.972	120	**.496**

a. Lilliefors Significance Correction

The power transformation analysis shows that the need of root transformation for the data. Further, to make the mean stationary, first order differencing is used. The monthly root transformed diabetes data with first order differencing is plotted in the Figure 2. Notice that the root transformed with differenced series is mean stationary and constant variance.

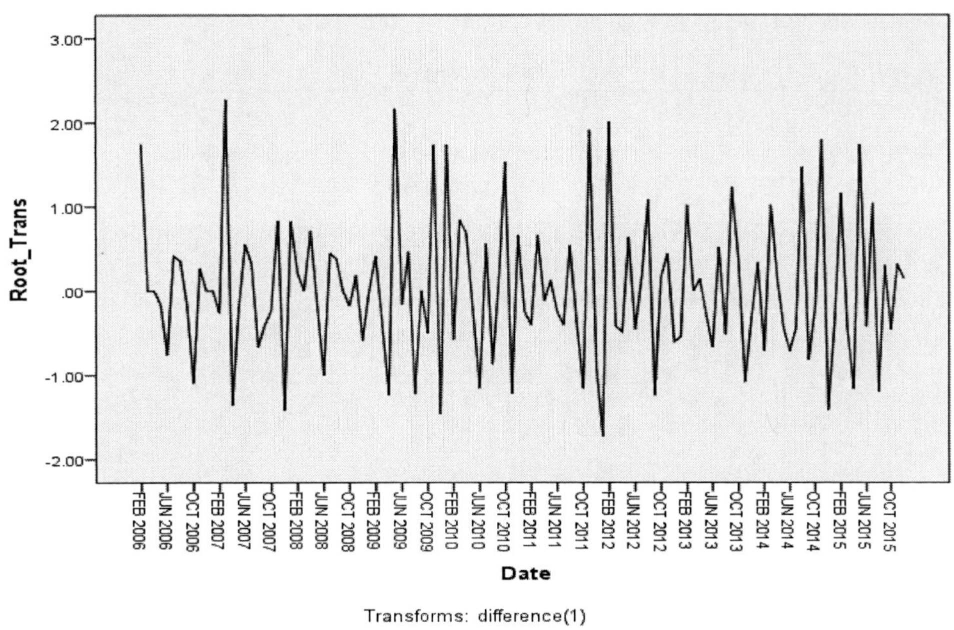

Figure 2. Time series plot of Diabetes patients after the first differencing.

IOP Conf. Series: Journal of Physics: Conf. Series **1039** (2018) 012033 doi:10.1088/1742-6596/1039/1/012033

3.1.2 Graphical representation of autocorrelations

From the ACF plots below (Figure 3), the value of its function is less than one. The parameter of any given function is less the one, it means that there is stationarity.

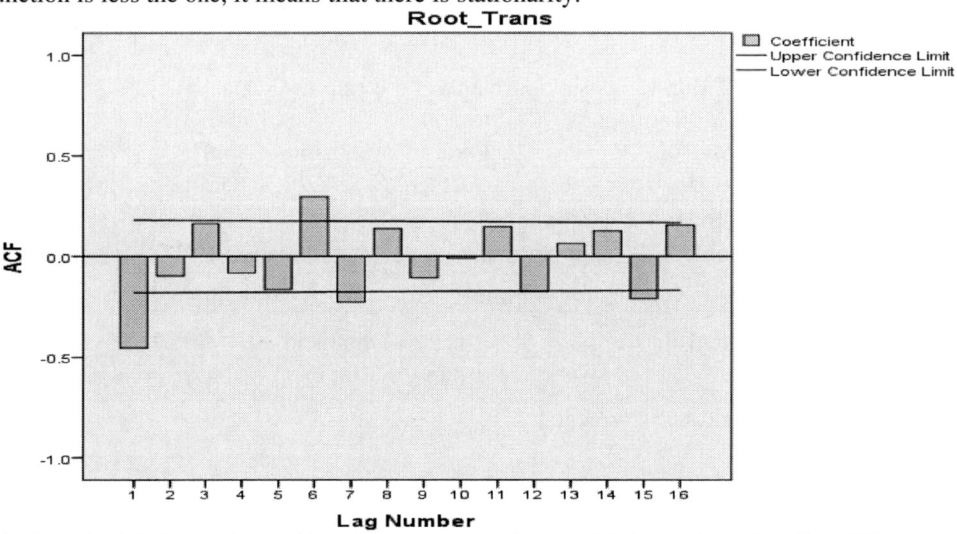

Figure 3. Sample ACF plot of monthly square root transformed Diabetes data after first differencing

The plot shows a negative spike at lag 1 and there is no other spikes showing that it is over differenced. Further the plot decays to zero on the both sides of the mean. From the Figure 3 it can be concluded that an appropriate model is moving average process of order 1, i.e., MA(1).

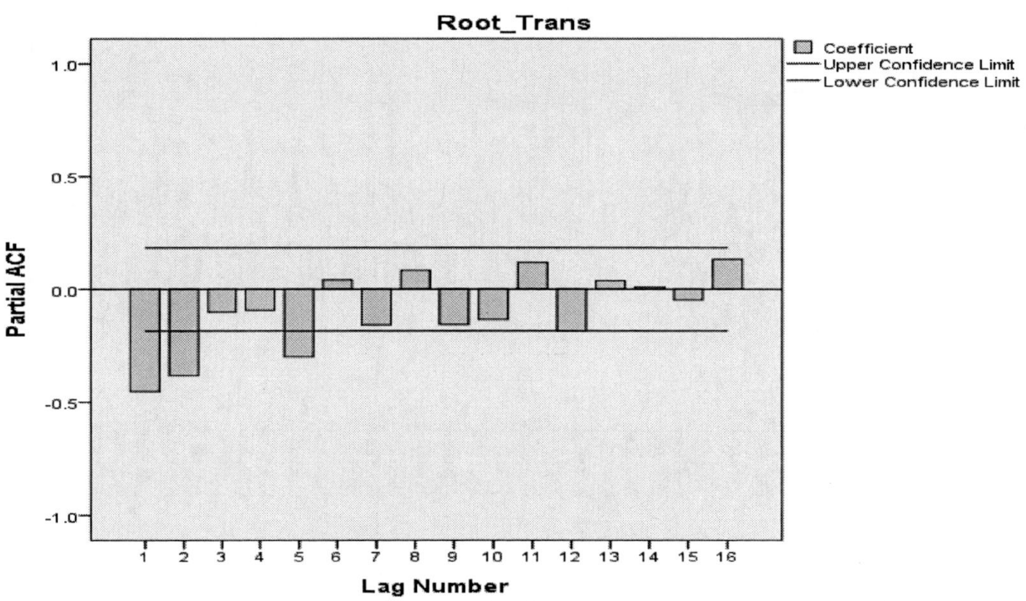

Figure 4. Sample PACF of monthly square root transformed Diabetes data after first differencing

Observing the plots of ACF and PACF plot above, it could be noticed that the ACF displays the sharper cuts off compared to the PACF. According to the PACF plot in Figure 4, we use the autoregressive process of order 2, i.e., AR(2).

Based on the Ljung-Box Q statistics, the models with significant p-values are selected and shown in Table 3 below. The resultant BIC providing in Table 4 shows that the most appropriate model to fit diabetes patients' data is ARIMA(0, 1, 1). We are going to estimate parameters of the model in the next subsection.

Table 3. Competitive models for the monthly diabetes patients based on the Ljung-Box Q statistics test

Model	Q statistic	P-value
1. ARIMA (0, 1, 0)	79.624	0.000
2. ARIMA (1, 1, 0)	50.858	0.000
3. ARIMA (0, 1, 1)	**21.669**	**0.198**
4. ARIMA (1, 1, 1)	**21.069**	**0.176**
5. ARIMA (2, 1, 0)	**23.806**	**0.094**
6. ARIMA (2, 1, 1)	**19.379**	**0.197**

Table 4 RMSE and BIC of all possible models resulting in Table 5.

Model	RMSE	BIC
ARIMA (0, 1, 1)	**0.678**	**-0.736**
ARIMA (1, 1, 1)	0.670	-0.639
ARIMA (2, 1, 0)	0.701	-0.551
ARIMA (2, 1, 1)	0.671	-0.598

3.2 Diagnostic checking and parameter estimation

Diagnostic checking is also known as the verification. It is to do with the testing of the goodness of fit test statistics of a model. We study the ACF and PACF of the residual plot to see it is white noise. If all the autocorrelation and partial correlation are small then the model is working fine but if some of the autocorrelations are large, the values of p or q are adjusted and the model is re-estimated. The checking of residuals and adjusting the values of p and q continues until the resulting residuals contain no additional structure.

Form Figure 5 the sample ACF and PACF of the model shows that the autocorrelations of the residual are all close to zero which mean they are uncorrelated, hence the residual assume mean of zero and constant variance. Finally, the p-value (0.198) for the Ljung-Box statistic clearly exceeds 5% for all lag orders. Thus, the selected model ARIMA (0, 1, 1) satisfies all the model assumptions.

Table 5 shows the estimated parameter of ARIMA(0, 1, 1). Therefore, the estimated model is $y_t = \epsilon_t - 0.825\epsilon_{t-1}$.

Table 5. Parameter estimates of model ARIMA(0, 1, 1)

				Estimate	SE	t	P-value
ARIMA(0, 1, 1)	Square Root Transformation	Difference		1			
		MA	Lag 1	.825	.054	15.297	.000

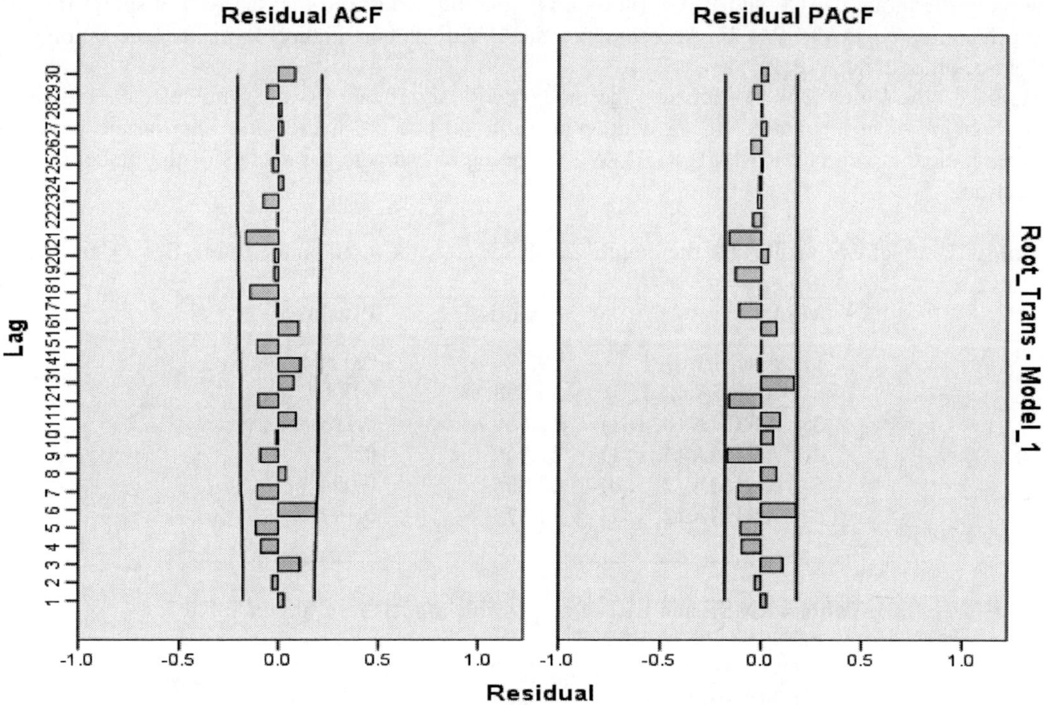

Figure 5. Sample ACF and PACF of the residual of ARIMA (0, 1, 1)

3.3 Forecasting using ARIMA (0, 1, 1)

The forecast values with 95 percent forecast limit of the ARIMA (0, 1, 1) of model for monthly diabetes patients are shown in Table 6 with standard error, lower and upper limit and its actual forecasted values.

Table 6. Forecast updates for 1 year- monthly Diabetes patient reported to the hospital

Date	Diabetes patient	Forecasted value	95% confidence limit		Error
			Lower	Upper	
Jan 2016	9	10.89	3.92	21.44	-1.89
Feb 2016	4	10.89	3.84	21.62	-6.89
Mar 2016	13	10.89	3.76	21.81	2.11
Apr 2016	16	10.89	3.69	22.00	5.11
May 2016	8	10.89	3.61	22.18	-2.89
Jun 2016	10	10.89	3.53	22.37	-0.89
Jul 2016	8	10.89	3.46	22.56	-2.89
Aug 2016	10	10.89	3.39	22.66	-0.89
Sep 2016	10	10.89	3.35	22.85	-0.89
Oct 2016	9	10.89	3.28	23.04	-1.89
Nov 2016	7	10.89	3.20	23.23	-3.89
Dec 2016	10	10.89	3.13	23.43	-0.89

4 Conclusion

In this paper, we studied the Diabetes Patients using the ARIMA modelling and forecasting method. Box-Jenkins was used to study the monthly Diabetes patients reported to the JDWNRH in Bhutan for the period of January 2006 to December 2016. Our main intension for this study is to forecast the monthly diabetes case for the coming period of January 2017 to December 2018. Table 7 below shows the forecast for next two years and Figure 6 is the scatter plot, it determine potential relationships among Actual and Forecast value. List of tentative ARIMA models are developed based on the Ljung-Box Q statistic test, BIC and after testing the significant of the estimated parameters, ARIMA(0, 1, 1) model was seen best fit for forecasting the Diabetes patient.

Table 7. Forecast for two years- monthly Diabetes patients of ARIMA(0, 1, 1) Model.

	2017			2018		
Months	Lower limit	forecast	Upper limit	Lower limit	forecast	Upper limit
January	3.06	10.89	23.52	2.43	10.89	25.50
February	3.03	10.89	23.72	2.37	10.89	25.70
March	2.96	10.89	23.91	2.34	10.89	25.81
April	2.89	10.89	24.11	2.28	10.89	26.01
May	2.86	10.89	24.21	2.34	10.89	26.11
June	2.79	10.89	24.40	2.19	10.89	26.32
July	2.722	10.89	24.60	2.16	10.89	26.42
August	2.69	10.89	24.70	2.10	10.89	26.63
September	2.62	10.89	24.90	2.07	10.89	26.73
October	2.56	10.89	25.00	2.02	10.89	26.83
November	2.53	10.89	25.20	1.99	10.89	27.04
December	2.46	10.89	25.40	1.96	10.89	27.14

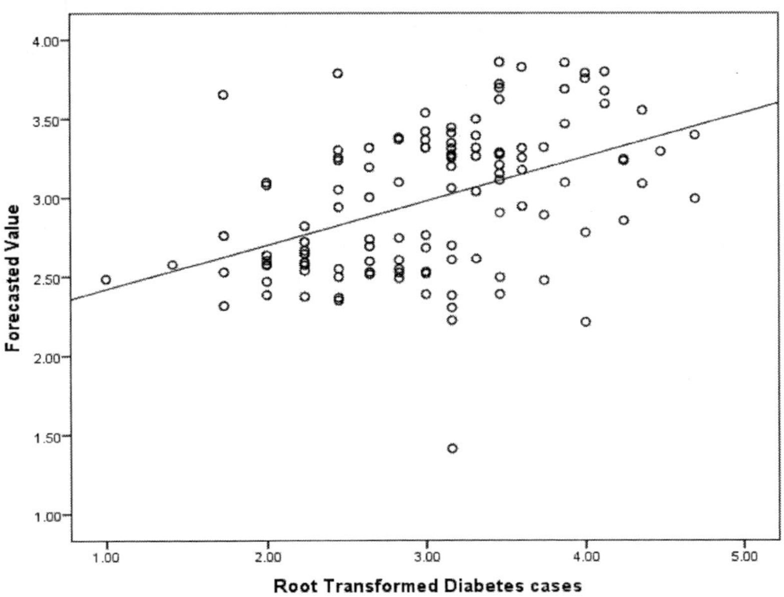

Figure 6. Scatter plot for Actual and Forecast value

References

[1] United Nation: New York. (2017). Department of Economic and Social Affairs. Retrieved from World Population Prospects: https://esa.un.org/unpd/wpp/Publications/Files/WPP2017_KeyFindings.pdf.
[2] Earnest, A., Chen, M.I., Ng, D., & Sin, L.Y. (2005, May 11). Using autoregressive integrated moving average (ARIMA) models. Retrieved from BMC Health Services Research: https://bmchealth servres.biomedcentral.com/track/pdf/10.1186/1472-6963-5-36?site=bmchealthservres.biomed central.com.
[3] Liu Q. Liu X., Jiang B. and Yang W. (2011). Forecasting incidence of hemorrhagic fever with renal syndrome in China using ARIMA model.
[4] Dan D. E., Jude O and Idochi O. (2014). Modelling and Forecasting Malaria Mortality Rate using SARIMA Models (A Case Study of Aboh Mbaise General Hospital, Imo State Nigeria). Science Publishing Group.
[5] Sampson, A. T., Henry, O., & Nabubie, I. B. (2015, June 06). Time Series Analysis of Malaria cases in Ejisu-Juaben Municipality. Retrieved from International journal of scintific and technology research volume 4. : http://www.ijstr.org/final-print/june2015/Times-Series-Analysis-Of-Malaria-Cases-In-Ejisu-juaben-Municipality.pdf
[6] Pan, Y., Zhang, M., Chen, Z., Zhou, M., & Zhang, Z.(2016). An ARIMA Based Model for Forecasting the Patient. Retrieved from Shenzhen University: http://ieeexplore.ieee.org/stamp/stamp.jsp?arnumber=7538560
[7] Villani, M., Earnest, A., Nanayakkara, N., Smith, K., Courten, B., & Zoungas, S. (2017). Time Series Modelling to forecast Prehospital EMS demand for diabetic emergencies. Retrieved from BMC Health Services Research: https://bmchealthservres.biomedcentral.com/track/pdf/10.1186/s12913-017-2280-6?site=bmchealthservres.biomedcentral.com.
[8] Wei W.S. (1990). Time series analysis, Addison-Wesley, USA.

Acknowledgements

We would like to thank Thai International Cooperation Agency (TICA) for providing the financial support. We also would like to thank Jigme Dorji Wangchuk National Referral Hospital (JDWNRH), Bhutan for providing and letting us use latest available data.

Stepwise Regression Modeling of the Retardance in Dextran-Citrate Coated Ferrofluids

J F Lin[1], J J Sheu[2] and Y Y Yang[1]

[1] Department of Industrial Design, Far East University, No. 49, Zhonghua Rd., Xinshi Dist., Tainan City 74448, Taiwan R.O.C.
[2] Department of Mechanical Engineering, Southern Taiwan University of Science and Technology, No. 1, Nantai St., Yungkang Dist., Tainan City 71005, Taiwan R.O.C.

E-mail: jacklin@mail.feu.edu.tw

Abstract. Double-layer coating citrate (citric acid, CA) and dextran on Fe_3O_4 ferrofluids (FFs) have been conducted for biomedical application such as hyperthermia. In a previous study, the magneto-optical retardance of dextran-citrate (DC) coated FFs was measured by a Stokes polarimeter. At this work, instead of using the multiple regression (MR) analysis, we adopted the stepwise regression to model the retardance of DC coated FFs in a systematic way. From the results of stepwise regression by MATLAB, the developed model had highly predictable ability owing to F of 667.365 and had a high correlation coefficient R of 0.99989. The average absolute error of predicted retardances to measured retardances was just 0.85%. Using the genetic algorithm (GA) in MATLAB, the optimized parametric combination was determined as [0.075, 75.856, 0.323, 0.65] corresponding to Fe_3O_4 concentration, coating temperature, CA mass, and dextran mass. The maximum retardance was found as 119.653°, close to those of [0.075, 75.794, 0.322, 0.65] and 119.658° obtained by Fmincon function and a relative error of -0.004%. Above all, the second order stepwise regression method was useful to model the retardance of DC coated FFs, and the maximum global retardance was determined by GA.

1. Introduction

Citrate (citric acid, CA) or dextran coated Fe_3O_4 ferrofluids (FFs) have been conducted for biomedical application such as hyperthermia [1, 2]. It has been known that the surface of the magnetic nanoparticles (MNPs) can be stabilized in an aqueous dispersion by the adsorption of CA or dextran. Effect of synthesis conditions on the properties of CA- and dextran-coated iron oxide nanoparticles was discussed [3-5]. The stabilized biocompatible CA coated FF was successfully produced with high retardance and magnetization in the previous study [4]. The retardance values (magneto-optical property as the phase retardation of linearly birefringent medium, such as quartz and certain liquid crystals) of FFs were measured by a developed Stokes polarimeter [4]. In the recent work, the dextran-citrate (DC) coated FFs have been developed, and we investigate not only magnetic retardance but also heating performance of the DC coated FFs for the hyperthermia application [2].

For the modeling and prediction of the retardance in FFs, multiple regression (MR) and optimization of retardance in CA coated FFs were executed by Taguchi method and Microsoft Excel, and the F value of regression model was large enough [6]. However, the modeling of retardance executed by Microsoft Excel was not in a systematic way. Further, the stepwise regression executed by MATLAB was successful for modeling the retardance of CA coated FFs [7]. The regression model had highly predictable ability and had a high correlation coefficient of one. Moreover, genetic algorithm (GA) solver (in MATLAB® Optimization Toolbox) was used to find the global maximum retardance value efficiently.

In this study, the second order stepwise regression executed by MATLAB was successfully used to model the retardance of DC coated FFs. The regression model had highly predictable ability and had a high correlation coefficient of 0.99989. Above all, two optimization techniques, including Fmincon function and GA solver (both in MATLAB® Optimization Toolbox), were used to find the local and global maximum retardance value, respectively. Then the DC coated FFs could be more useful in practical applications such as hyperthermia.

2. Method

In this study, the methods including uniform design method, stepwise regression method, Fmincon function and genetic algorithm were briefly introduced.

2.1. Uniform design method

Uniform design (UD) method is based on uniform distribution in number theory. It makes experiment points uniformly scattered in the range of experiment parameters for getting more information by less experiment [2]. The optimal synthesis condition of DC coated Fe_3O_4 FFs with high retardance was determined by the UD and the test results of retardance were shown as table 1 [2].

Table 1. Uniform design and test results of retardance.

	Level of parameter				
Test No.	Fe_3O_4 (g/ml)	Temp. (℃)	CA (g)	Dextran (g)	Retardance (deg)
1	1(6/80)	2(50)	4(0.30)	7(0.65)	27.8628
2	2(6/80)	4(60)	8(0.36)	5(0.60)	13.5872
3	3(6/90)	6(70)	3(0.24)	3(0.55)	12.2155
4	4(6/90)	8(80)	7(0.36)	1(0.55)	31.8867
5	5(6/100)	1(45)	2(0.24)	8(0.65)	7.1405
6	6(6/100)	3(55)	6(0.30)	6(0.60)	26.5201
7	7(6/110)	5(65)	1(0.24)	4(0.60)	30.3393
8	8(6/110)	7(75)	5(0.30)	2(0.55)	20.0765
9	9(6/120)	9(85)	9(0.36)	9(0.65)	56.7898

2.2. Stepwise regression

While dealing with large number of independent variables, it is of significant importance to determine the best combination of these variables to predict the dependent variable [8]. Stepwise regression method is a routine statistic technique used for variable selection and is a combination of forward and backward procedures. The method starts with an initial model which includes a subset of all the candidate variables. And then, a so-called stepping procedure which iteratively altering the model established at the previous step by adding or removing variables in accordance with the stepping criteria is used in order to add significant variables or remove insignificant variables.

2.3. Fmincon function and genetic algorithm

The Fmincon function (Find MINimum of CONstraint nonlinear multivariable function, a deterministic gradient-based method) was in-built in MATLAB and used for optimization [9]. It uses sequential quadratic programming (SQP) method for optimization. SQP methods represent the state of the art in nonlinear programming methods. Based on the works of Biggs, Han and Powell; an SQP method mimics Newton's method for constraint optimization. It is an iterative method of starting from some initial point and converging to a constrained local minimum [9]. It was noted that different algorithms could be chosen, including interior point, active set, and trust region reflective. And this technique has the disadvantage that the solution found could be a local minimum of the function depending on the initial estimates of the parameters. Moreover, the genetic algorithm (GA) technique is based on the natural process of evolution to solve optimization and search problems. There are three main operators in GA which are reproduction, crossover and mutation [10]. GAs were pioneered by Holland in the 1960s and 70s and have emerged as a highly effective technique for solving a wide variety of optimization problems [10]. Above all, the global maximum could be found by GA.

3. Results and discussions

3.1. Results of stepwise regression

Results of the second order stepwise regression by MATLAB are shown in figure 1, the model has highly predictable ability owing to F of 667.365 and has a high correlation coefficient R of 0.99989, which is suggested that this regression model is statistically significant and gives useful information about the response variable. The average absolute error of predicted retardances to measured retardances is less as 0.85%. The regression equation is given as:

$$Retardance=-415.933+1760.06X_3-0.176898X_2^2-8857.9X_3^2-300.709X_4^2+25.0825X_1X_2+52.1548X_2X_3+12.4852X_2X_4 \quad (1)$$

As shown in figure 1 and table 2, we can see the meaningfulness degrees of input variables. This degree of meaningfulness is determined via being value of p-val below 0.05. Thus meaningfulness ranking of input variables is determined as 7(X_3^2), 13(X_2X_4), 6(X_2^2), 12(X_2X_3), 9(X_1X_2), 3(X_3), and 8(X_4^2). As shown in figure 1, the RMSE (Root Mean Square Error) value is 0.609. Further, the R^2 value is determined as 0.999786 (close to 1). It is indicated that the ability of prediction in the second order stepwise regression is excellent.

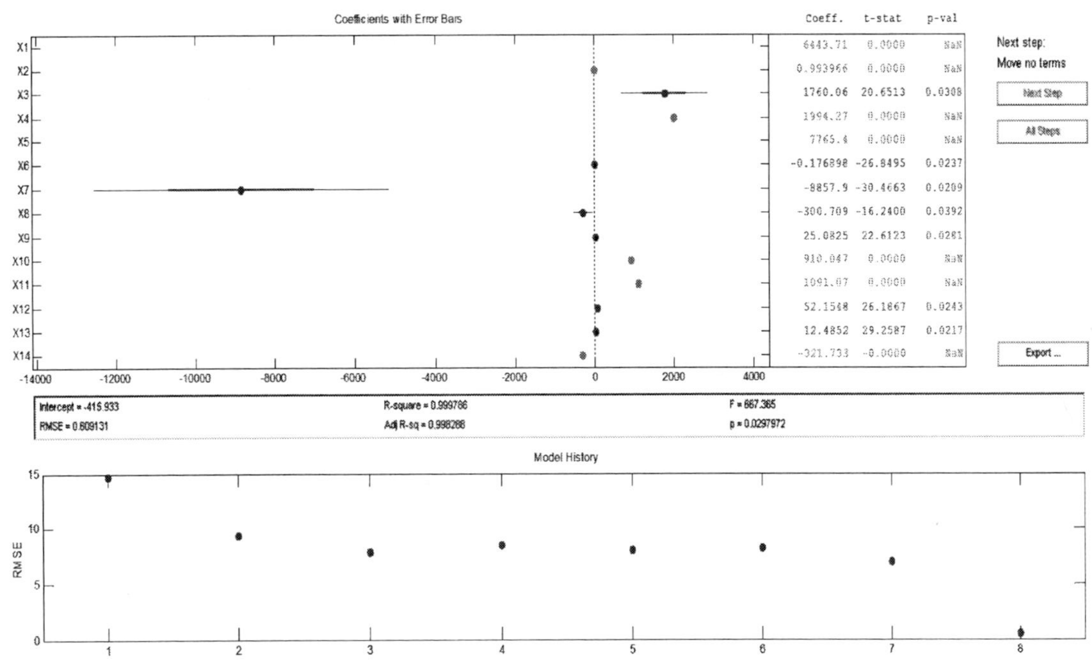

Figure 1. Results of the second order stepwise regression.

The significance t-test of each variable is listed in table 2. It is seen that the significant influence of each term on the retardance of DC coated FFs is given by: $X_3^2 > X_2X_4 > X_2^2 > X_2X_3 > X_1X_2 > X_3 > X_4^2$. This indicates that the effect of CA mass (X_3) on the retardance is of the most significance and followed by dextran mass (X_4), coating temperature (X_2), and Fe_3O_4 concentration (X_1). The relative influence of cross correlations on the retardance of DC coated FFs (from examining the magnitude of the *P* value or the significance level of the cross items in *t*-test) is given by: $X_2X_4 > X_2X_3 > X_1X_2$. The item of coating temperature (X_2) seems significant in the second order stepwise regression; the X_2 plays an important role in the influence of retardance judging from the X_2X_4, X_2^2, X_2X_3, and X_1X_2.

Table 2. Results of the second order stepwise regression.

	Coefficient	*t*-test	*P* value
X_3	1760.06	20.6513	0.0308
$X_2^2\,(X_6)$	-0.176898	-26.8495	0.0237
$X_3^2\,(X_7)$	-8857.9	-30.4663	0.0209
$X_4^2\,(X_8)$	-300.709	-16.2400	0.0392
$X_1X_2\,(X_9)$	25.0825	22.6123	0.0281
$X_2X_3\,(X_{12})$	52.1548	26.1867	0.0243
$X_2X_4\,(X_{13})$	12.4852	29.2587	0.0217

3.2. Results of optimization using Fmincon fuction and genetic algorithm

In this study, the start points are chosen as [0.05, 45, 0.24 0.55], [0.05625, 55, 0.27, 0.575], [0.0625, 65, 0.30, 0.60], [0.06875, 75, 0.33, 0.625], and [0.075, 85, 0.36, 0.65] are input into the program in MATLAB Optimization Toolbox subsequently. It is noted that @FitFun830 is obtained by changing the sign of equation (1) and is used to find the minimum value. Hence, the maximum retardance for the five different start points is obtained the same as 119.658° for [0.075, 75.794, 0.322, 0.65]. A graph output using Fmincon function is shown as figure 2 for start points are set as [0.0625, 65, 0.30, 0.60]. The Fmincon function method is efficient and solves the problem directly. It is known that the local maximum or minimum could be found by Fmincon function, and the global extreme could be further found by using GA technique or a hybrid of the GA and Fmincon optimization technique.

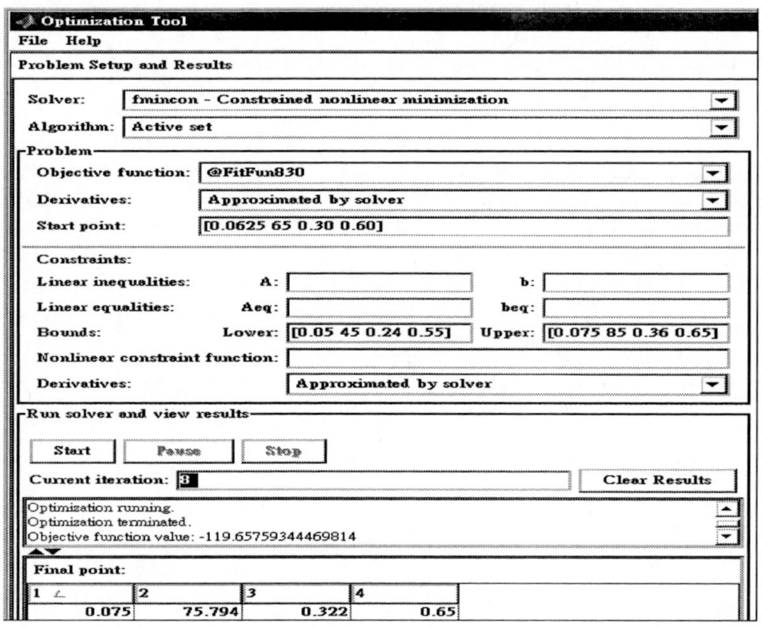

Figure 2. A graph output using the fmincon in MATLAB® Optimization Toolbox.

Further, the optimized parametric combination and maximum retardance of DC coated FFs are determined as [0.075, 75.856, 0.323, 0.65] and 119.653° (the highest retardance in 12 runs of GA and as shown in figure 3), close to that of [0.075, 75.794, 0.322, 0.65] and 119.658° obtained by Fmincon function and a relative error of -0.004% in retardance. The corresponding graph output using GA in

MATLAB® optimization is shown in figure 3. It is noted that the Fitness function as @FitFun830 is obtained by changing the sign of equation (1) and is used to find the minimum value.

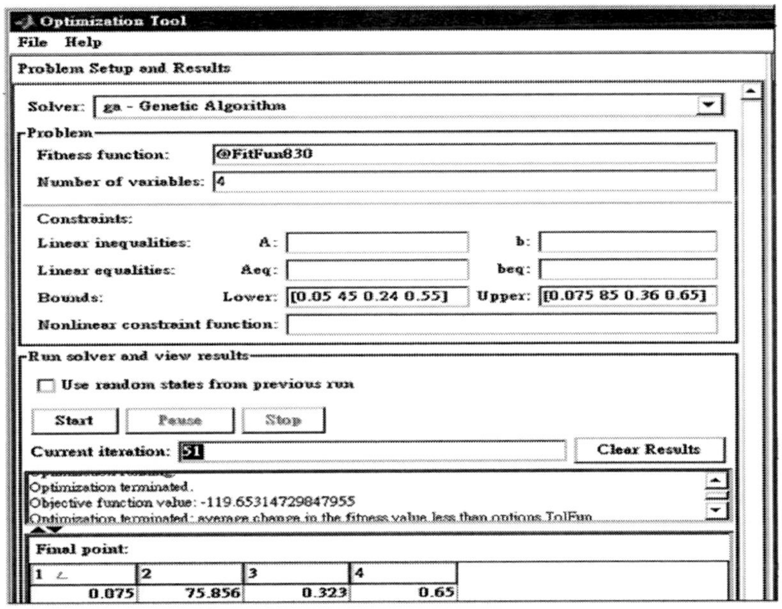

Figure 3. A graph output using the GA in MATLAB® optimization.

4. Conclusions

The second order stepwise regression method was successfully used to model the retardance of DC coated FFs. According to the results of the second order stepwise regression, the *F* value was obtained as high as 667.365 and had a correlation coefficient of 0.99989; the developed regression model had highly predictable ability. Using the genetic algorithm in MATLAB for 12 runs, the optimized parametric combination with the highest retardance value was determined as [0.075, 75.856, 0.323, 0.65]. The maximum retardance was found as 119.653°, close to those of [0.075, 75.794, 0.322, 0.65] and 119.658° obtained by Fmincon function in MATLAB and a relative error of -0.004%.

References

[1] Cheraghipour E et al 2012 *J. Biomed. Sci. Eng.* **5** 715
[2] Lin J F et al 2017 *AIMS Mater. Sci.* **4** 231
[3] Li L et al 2013 *Microel. Eng.* **110** 329
[4] Lin J F et al 2014 *J. Magn. Magn. Mater.* **372** 147
[5] Hong R Y et al 2009 *Chem. Eng. J.* **150** 572
[6] Lin J F et al 2015 *Proc. SPIE* **9302** 930229
[7] Lin J F and Qiu X R 2017 *IOP Conf. Materials Science and Engineering* **215** 012003.
[8] Cevik A et al 2010 *Adv. Eng. Softw.* **41** 611
[9] Sreeraj P et al 2013 *J. Mech. Eng. Research* **5** 154
[10] Lin W L et al 2009 *IEEE J. L. T.* **27** 4136

Acknowledgements

The financial support provided by the Ministry of Science and Technology in Taiwan under MOST 105-2221-E-269-001 and MOST 106-2221-E-269-001 is gratefully acknowledged.

Optimal control for the transmission dynamics of malaria disease model

Bundit Unyong
Department of Mathematics, Faculty of Science and Technology,
Phuket Rajabhat University, Phuket, Thailand

Email: bunditun@gmail.com

Abstract. In this work a modified SEIR-SVEVIV model that described the dynamics transmission of malaria disease was proposed and analyzed. The standard method is used to analyze the behaviors of the proposed model. The results shown that there were two equilibrium points; disease free and endemic equilibrium point. The qualitative results are depended on a basic reproductive number (R_0). We obtained the basic reproductive number by using the next generation method technique and finding the spectral radius. Routh-Hurwitz criteria is used for determining the stabilities of the model. If $R_0 < 1$, then the diseases free equilibrium point is local asymptotically stable: that is the disease will died out, but if $R_0 > 1$, then the endemic equilibrium is local asymptotically stable. After that the SEIR-SVEVIV model is modified from the first model by adding the optimal control functions that includes two times – dependent control functions with one minimizing the contract between the susceptible human and the infected vector and the other, minimizing the population of the infected human. The result from the numerical solutions of the models are shown and compared for supporting the analytic results.

1. Introduction

Malaria is caused by parasites that are transmitted to people through the bites of infected female mosquitoes. *P. falciparum* is the most deadly malaria parasite and the most prevalent in Africa, where malaria cases and deaths are heavily concentrated. The first symptoms of malaria – fever, headache, chills and vomiting – usually appear between 10 and 15 days after the mosquito bite. Without prompt treatment, *P. falciparum* malaria can progress to severe illness and death. WHO recommends a multi-pronged strategy to prevent, control and eliminate malaria. Key interventions include: the use of insecticide-treated mosquito nets and indoor residual spraying, diagnostic testing, and treatment of confirmed cases with effective anti-malarial medicines. In recent years, these measures have dramatically lowered the malaria burden in many settings. Malaria transmission continues in many countries around the world however, and causes hundreds of thousands of deaths each year (WHO) [1]. Malaria situation in Thailand week 1-43 on 1 Sep 2017, from1 Jan 2017 to 1 Sep 2017. No. of Thai malaria cases is about 7,902 and No. of foreigner malaria cases is 2,976.By total that decrease 35.31% from the year 2016 [2]. Mathematical models have become an important tool for understanding the spread and control of disease because of this disease is caused by virus, therefore no drug can cure this disease specificially [3-7]. This paper is organized respectively as follows. In section 2, we present an SEIR-SVEVIV model for malaria. The standard method is used to analyze the behaviors of the proposed model. The analysis of optimization problem is presented in section 3. In section4, we give a numerical appropriate method and the simulation corresponding results. Finally, the conclusions are summarized in section 5.

2. Material and method

2.1 Material

In this paper, we study the transmission of malaria diseases through mathematical modell. By using the standard method to analyze the behaviours of the model which was adopted from [8]. The population consist of two groups: human population N and population vector N_v. Human population be divided into four disease-state compartments: susceptible individuals (S), people who can catch the disease; exposed individuals (E), people whose body is a host for the infectious agent but are not yet able to transmit the disease; infectious individuals (I), people who have the disease and can transmit the disease; recovered individuals (R), people who have recovered from the disease. Population vector or mosquitoes N_v are divided into three groups of mosquitoes: the susceptible mosquitoes population S_v, the exposed mosquitoes population E_v and the infected mosquitoes population I_v. In this study, we assumed that there are numbers of people in the populations that have already infected by the virus while others have not. It is also assumed that the transmission of the virus continues in the population but number of mosquitoes as the vector is constant. People and mosquitoes are categorized in one group at a time. Then we obtained the transmission model as shown by a system of ordinary differential equations as follows.

Human Population;

$$\frac{dS}{dt} = A - \left(\frac{b\beta SI_{\upsilon}}{1+\upsilon I_v}\right) - \mu S + \varphi R \qquad (1)$$

$$\frac{dE}{dt} = \left(\frac{b\beta SI_{\upsilon}}{1+\upsilon I_v}\right) - (\mu+\alpha)E \qquad (2)$$

$$\frac{dI}{dt} = \alpha E - (\mu+\gamma+\delta)I \qquad (3)$$

$$\frac{dR}{dt} = \gamma v I - (\mu+\varphi)R \qquad (4)$$

Vector Population;

$$\frac{dS_v}{dt} = B - \left(\frac{b\beta_v S_v I}{1+\upsilon_v I}\right) + \mu_v S_v \qquad (5)$$

$$\frac{dE_v}{dt} = \left(\frac{b\beta_v S_v I}{1+\upsilon_v I}\right) - (\mu_v+\alpha_v)E_v \qquad (6)$$

$$\frac{dI_v}{dt} = \alpha_v E_v - (\mu_v+\delta_v)I_v \qquad (7)$$

Where; $S+E+I+R=N=\frac{A}{\mu}$ and $S_v+E_v+I_v=N_v=\frac{B}{\mu_v}$,

$S(t)$ is the susceptible human population at time t

$E(t)$ is the exposed human population at time t

$I(t)$ is the infected human population at time t

$R(t)$ is the recovered human population at time t

N is the total number of human population ,

$S_v(t)$ is the susceptible mosquitoes population at time t

$E_v(t)$ is the exposed mosquitoes population at time t

$I_v(t)$ is the infected mosquitoes population at time t
N_v is the total number of mosquitoes population ,
b is average bite of mosquitoes that potentially infected
A is the constant recruitment rate of the human,
B is the constant recruitment rate of the mosquito,
μ is the death rate of human population ,
μ_υ is the death rate of mosquitoes population ,
β is the probability that a bite by an infectious mosquito results in transmission of disease to human
β_v Probability that a bite results in transmission of parasite to a susceptible mosquito
α is the rate of progression of humans from the exposed state to the infectious state
α_v is the rate of progression of mosquitoes from the exposed state to the infectious state
γ is the recovery rate for humans from the infectious state to the recovered state
δ is the disease-induced death rate of human
δ_v is the disease-induced death rate of mosquito
φ is the rate of loss of immunity in humans
υ is an antibody produced by human in response to the incidence of infection caused by mosquito
υ_v is an antibody produced by mosquito in response to the incidence of infection caused by human
u is the contract rate between the infectious mosquito and susceptible human
v is the rate at which infectious human are treated at each time period

2.1.1 Basic properties of the model
Positivity and Boundedness of the Solution
Theorem 1. If the initial data $S \geq 0, E \geq 0, I \geq 0, R \geq 0$ and $S_v \geq 0, E_v \geq 0, I_v \geq 0$ then the solution $(S, E, I, R, S_v, E_v, I_v)$ of the malaria control model (1) are non-negative for all t > 0. Therefore, $\limsup_{t\to\inf} N(t) \leq \frac{A}{\mu}, \limsup_{t\to\infty} N_v(t) \leq \frac{B}{\mu_v}$, Where $S+E+I+R = N = \frac{A}{\mu}, S_v + E_v + I_v = N_v = \frac{B}{\mu_v}$

The Invariant Region
Theorem 2. The region $\Omega = \Omega_h \cup \Omega_v \subset R^4 \times R^3$ is positively invariant for the model with non-negative initial condition in R^7 ,where

$$\Omega_h = \{(S,E,I,R) \in R^4 \mid S+E+I+R \leq \frac{A}{\mu}\}, \Omega_v = \{(S_v,E_v,I_v) \in R^3 \mid S_v + E_v + I_v \leq \frac{B}{\mu_v}\}.$$

The equilibrium points for $(S^*, E^*, I^*, R^*, S_v^*, E_\upsilon^*, I_\upsilon^*)$ are found by setting the right hand side of each equations (1-7) equal to zero. We obtained two equilibrium points as follows;

$$S = \frac{A+\varphi R}{(\mu + \frac{b\beta I_v}{1+\nu I_v})}, E = \frac{(\frac{b\beta S I_v}{1+\nu I_v})}{(\mu+\alpha)}, I = \frac{\alpha E}{(\mu+\gamma+\delta)}, R = \frac{\gamma I}{\mu+\varphi},$$

$$S_v = \frac{B}{\frac{b\beta_v I}{1+\nu_v I} + \mu_v}, E_v = \frac{\frac{b\beta_v S_v I}{1+\nu_v I}}{\alpha_v + \mu_v}, I_v = \frac{(\alpha_v E_v)}{\mu_v + \delta_v}$$

2.1.1.1 Disease Free Equilibrium Point (E_0) :

IOP Conf. Series: Journal of Physics: Conf. Series **1039** (2018) 012035 doi:10.1088/1742-6596/1039/1/012035

In the absence of the disease in the community, there are $I = 0$ and $I_v = 0$, we obtained $E_0(S, E, I, R, S_v, E_v, I_v)$ where

$$S = \frac{A}{\mu}, E = 0, I = 0, R = 0, S_v = \frac{B}{\mu_v}, E_\upsilon = 0, I_\upsilon = 0$$

2.1.1.2 Endemic Equilibrium Point (E_1):

In case the disease is presented in the community, $I > 0$ *and* $I_\upsilon > 0$, we obtained, $E_1(S^*, E^*, I^*, R^*, S_v^*, E^*_\upsilon, I^*_\upsilon)$ where;

$$S^* = \frac{A + \varphi R^*}{(\mu + \frac{b\beta I_v^*}{1 + \nu I_v^*})}, E^* = \frac{(\frac{b\beta S^* I_v^*}{1 + \nu I_v^*})}{(\mu + \alpha)}, I^* = \frac{\alpha E^*}{(\mu + \gamma + \delta)}, R^* = \frac{\gamma I^*}{\mu + \varphi}, S_v^* = \frac{B}{\frac{b\beta_v I^*}{1 + \nu_v I^*} + \mu_v}, E_v^* = \frac{\frac{b\beta_v S_v^* I^*}{1 + \nu_v I^*}}{\alpha_v + \mu_v}, I_v^* = \frac{(\alpha_v E_v^*)}{\mu_v + \delta_v}$$

2.1.2 Basic Reproductive Number (R_0)

We obtained a basic reproductive number by using the next generation method [9]. By rewriting the equations (1) – (7) in matrix form;

$$\frac{dX}{dt} = F(X) - V(X) \tag{16}$$

Where $F(X)$ is the non-negative matrix of new infection terms and $V(X)$ is the non-singular matrix of remaining transfer terms.

And setting;

$$F = \left[\frac{\partial F_i(E_0)}{\partial X_i}\right], V = \left[\frac{\partial V_i(E_0)}{\partial X_i}\right]$$

(17)

for all $i, j = 1, 2, 3, 4, 5, 6, 7$ be the Jacobean matrix of $F(X)$ and $V(X)$ at E_0 .The basic reproductive number (R_0) is the number of secondary case generate by a primary infectious case (Van den Driessche and Watmough, 2002). It can be evaluated through the formula;

$$\rho\left(FV^{-1}\right) \tag{18}$$

Where FV^{-1} is called the next generation matrix and $\rho\left(FV^{-1}\right)$ is the spectral radius (largest eigenvalues) of FV^{-1}. Then we get the reproduction number R_0 where,

$$R_0 = \frac{(b^2 \beta \alpha \alpha_v AB)}{\mu(\mu + \alpha)(\gamma + \delta + \mu)(\delta_v + \mu_v)(\alpha_v + \mu_v)} \tag{19}$$

Finally, Routh-Hurwitz criteria is used for determining the stabilities of the model. If $R_0 < 1$, then the disease free equilibrium point is local asymptotically stable: that is the disease will died out, but if $R_0 > 1$, then the endemic equilibrium is local asymptotically stable. In this paper, we use optimal control this method as part of control measures for malaria disease. From the system of equations (1-

7), we include two controls *u* and *v* that represent, respectively, the effort rate that reduces the contract between the infectious vector and the susceptible individuals and the rate at with infectious human are treated at each time period. The mathematical system with controls is given by the nonlinear differential equations subject to non-negative initial conditions as the following;

$$\begin{aligned}
&\frac{dS}{dt} = A - \left(\frac{ub\beta SI_v}{1+\upsilon I_v}\right) - \mu S + \varphi R && (1)\\
&\frac{dE}{dt} = \left(\frac{ub\beta SI_v}{1+\upsilon I_v}\right) - (\mu+\alpha)E && (2)\\
&\frac{dI}{dt} = \alpha E - (\mu+\gamma+\delta+v)I && (3)\\
&\frac{dR}{dt} = \gamma v I - (\mu+\varphi)R && (4)\\
&\frac{dS_v}{dt} = B - \left(\frac{b\beta_v S_v I}{1+\upsilon_v I}\right) + \mu_v S_v && (5)\\
&\frac{dE_v}{dt} = \left(\frac{b\beta_v S_v I}{1+\upsilon_v I}\right) - (\mu_v+\alpha_v)E_v && (6)\\
&\frac{dI_v}{dt} = \alpha_v E_v - (\mu_v+\delta_v)I_v && (7)
\end{aligned} \qquad (20)$$

Where, $S(0) > 0, E(0) \geq 0), I(0) \geq 0, R(0) \geq , S_v(0) > 0, E_\upsilon(0) \geq 0$ and $I_\upsilon(0) \geq 0$

2.2 method

2.2.1 Optimal control for the transmission dynamics of malaria disease model

In this section we use the optimal control theory to analyze the behavior of the system of equations (20). The objective one is to minimize the contract between susceptible human and the infected vector and the other is to minimize the population of the infected human. Mathematically, for a fixed terminal time t_f , the problem is to minimize the objective functional;

$$J(u,v) = \int_0^{t_f} \left[S(t) + I(t) + \frac{B_1}{2}u^2(t) + \frac{B_2}{2}v^2(t) \right] dt \qquad (21)$$

The parameter $B_1 \geq 0$ and $B_2 \geq 0$ denote weights that balance the size of the terms for a fixed terminal time t_f. Hence we are interested in finding an optimal control pair u^* and v^*,such that:

$$J(u^*,v^*) = \min\{J(u,v) : (u,v) \varepsilon U\} \qquad (22)$$

Where, $U = \{(u,v) \mid 0 < u(t) < 1, 0 < v(t) < 1, t \in [0,t_f], \text{u and v are Lebesgue measurable}\}$

Next, applying the Pontryagin's Maximum Principle (Kirschner et al.,1997),we derive necessary conditions for our optimal control and corresponding state variables, including the two control functions. Therefore we have seven corresponding adjoint variables where λ_1 corresponds to S , λ_2 corresponds to E , λ_3 corresponds to I, λ_4 corresponds to R , λ_5 corresponds to S_v, λ_6 corresponds to E_v and λ_7 corresponds to I_υ.

The Hamiltonian equation is formed by allowing each of the adjoint variables correspond to each of the state variables accordingly and combining the result with the objective functional as below:

$$H = S(t) + I(t) + \frac{B_1}{2}u^2(t) + \frac{B_2}{2}v^2(t) + \sum_{i=1}^{7}\lambda_i f_i \tag{23}$$

Where f_i is the right hand side of the differential equation of the ith state variables.
The adjoint equations are formed by taking the derivative of the Hamiltonian with respect to each of the state variables as follow; By applying the Pontryagin's maximum principle [10] and the existence result of optimal control from [11] we obtain the following theorem:

Theorem 3 There exists an optimal control $(u^*, v^*) \in U$,and corresponding solution $S^*, E^*, I^*, R^*, S_v^*, E^*_\upsilon$ and I_υ^* that minimize $J(u,v)$ over U .And there exists adjoint functions $\lambda_1, \lambda_2, \lambda_3, \lambda_4, \lambda_5, \lambda_6$ and λ_7 verifying;

$$\lambda_1' = -1 + \lambda_1(\mu + \frac{ub\beta I_v}{1+vI_v}) - \lambda_2(\frac{ub\beta I_v}{1+vI_v}) \ , \lambda_2' = \lambda_2(\mu + \alpha) - \lambda_3\alpha \ ,$$

$$\lambda_3' = \lambda_3(\mu + \gamma + \delta + v) - \lambda_4\gamma v + \lambda_5(\frac{(1+v_vI)(b\beta_vS_v) + b\beta_vS_vIv_v}{(1+v_vI)^2})$$
$$-\lambda_6(\frac{(1+v_vI)(b\beta_vS_v) - b\beta_vS_vIv_v}{(1+v_vI)^2}) + \lambda_7(\mu_v + \delta_v) - 1$$

$$\lambda_4' = -\lambda_1\varphi + \lambda_4(\mu + \varphi) \ , \ \lambda_5' = \lambda_5(\frac{b\beta_vI}{1+v_vI} + \mu_\upsilon) - \lambda_6(\frac{b\beta_vI}{1+v_vI}),$$

$$\lambda_6' = \lambda_6(\alpha_v + \mu_v) - \lambda_7\alpha_\upsilon, \ \lambda_7' = (\lambda_1 - \lambda_2)(\frac{(1+vI_v)(ub\beta S) - (ub\beta SI_vv)}{(1+vI_v)^2}) + \lambda_7(\mu_v + \delta_v)$$

With the transversality conditions; $\lambda_1(t_f) = \lambda_2(t_f) = \lambda_3(t_f) = \lambda_4(t_f) = \lambda_5(t_f) = \lambda_6(t_f) = \lambda_7(t_f) = 0$, and the optimize control (u^*, v^*) is given by

$$u^* = \min\left\{1, \max\left\{0, \frac{((\lambda_1 - \lambda_2)(b\beta I_v)S^*}{B_1(1+vI_v)}\right\}\right\}$$
$$v^* = \min\left\{1, \max\left\{0, \frac{(\lambda_3 - \lambda_4)(I^* - \gamma)}{B_2}\right\}\right\}.$$

Proof.

The existence of optimal control can be proved by using the results from [3]. The adjoint equations and transversality conditions can be obtained by using Pontryagin's Maximum Principle such that,

$$\lambda_1' = -1 + \lambda_1(\mu + \frac{ub\beta I_v}{1+vI_v}) - \lambda_2(\frac{ub\beta I_v}{1+vI_v}) \ , \ \lambda_2' = -\frac{\partial H}{\partial E} = \lambda_2(\mu+\alpha) - \lambda_3\alpha \ ,$$

$$\lambda_3' = \lambda_3(\mu+\gamma+\delta+v) - \lambda_4\gamma v + \lambda_5(\frac{(1+v_vI)(b\beta_vS_v)+b\beta_vS_vIv_v}{(1+v_vI)^2}) - \lambda_6(\frac{(1+v_vI)(b\beta_vS_v)-b\beta_vS_vIv_v}{(1+v_vI)^2})$$
$$+\lambda_7(\mu_v+\delta_v)-1$$

$$\lambda_4' = -\frac{\partial H}{\partial R} - \lambda_1\varphi + \lambda_4(\mu+\varphi) \ , \ \lambda_5' = -\frac{\partial H}{\partial E_\upsilon} = \lambda_5(\frac{b\beta_vI}{1+v_vI} + \mu_\upsilon) - \lambda_6(\frac{b\beta_vI}{1+v_vI}),$$

$$\lambda_6' = -\frac{\partial H}{\partial I_\upsilon} = \lambda_6(\alpha_v + \mu_v) - \lambda_7\alpha_v, \ \lambda_7' = (\lambda_1 - \lambda_2)(\frac{(1+vI_v)(ub\beta S) - (ub\beta SI_v v)}{(1+vI_v)^2}) + \lambda_7(\mu_v+\delta_v),$$

The optimal control pair (u^*, v^*) are obtained by finding the derivative of the Hamiltonian equation with respect to the control variables, equating to zero, and solving equation. Then we get;

$$\frac{\partial H}{\partial u} = B_1u(t) - \frac{\lambda_1 b\beta SI_v}{1+vI_v} + \frac{\lambda_2 b\beta SI_v}{1+vI_v} = 0 .$$

Then the optimal value for u is; $u^* = \dfrac{(\lambda_1-\lambda_2)(b\beta I_v^* S^*)}{B_1(1+vI_v^*)}$. And $\dfrac{\partial H}{\partial v} = B_2 v(t) - \lambda_3 I + \lambda_4\gamma I = 0$,

Hence the optimal value for v is; $v^* = \dfrac{(\lambda_3-\lambda_4)(I^*)(1-\gamma)}{B_2}$

By the bounds in U of the control, the optimal control pair (u^*, v^*) is given by

$$u^* = \min\left\{1, \max\left\{0, \frac{(\lambda_1-\lambda_2)(b\beta I_v^* S^*)}{B_1(1+vI_v^*)}\right\}\right\}, v^* = \min\left\{1, \max\left\{0, \frac{(\lambda_3-\lambda_4)(I^*)(1-\gamma)}{B_2}\right\}\right\}. \quad (24)$$

For supporting analytic results we need to resolve the optimal control model numerically.

3. Results

In this section we present the numerical simulations obtained by solving numerically from the following optimality system;

$$\frac{dS}{dt} = A - (\frac{u^* b\beta SI_v}{1+\upsilon I_v}) - \mu S + \omega R \quad (1), \qquad \frac{dS_\upsilon}{dt} = B - (\frac{b\beta_v S_v I}{1+\upsilon_v I}) + \mu_v S_v \quad (5)$$

$$\frac{dE}{dt} = (\frac{u^* b\beta SI_v}{1+\upsilon I_v}) - (\mu+\alpha)E \quad (2), \qquad \frac{dE_\upsilon}{dt} = (\frac{b\beta_v S_v I}{1+\upsilon_v I}) - (\mu_v + \alpha_v)E_v \quad (6)$$

$$\frac{dI}{dt} = \alpha E - (\mu+\gamma+\delta+v^*)I \quad (3), \qquad \frac{dI_\upsilon}{dt} = \alpha_v E_v - (\mu_v+\delta_v)I_v \quad (7)$$

$$\frac{dR}{dt} = \gamma v^* I - (\mu+\varphi)R \quad (4),$$

$$\lambda_1' = -\frac{\partial H}{\partial S} - 1 + \lambda_1(\mu + \frac{ub\beta I_v}{1+\nu I_v}) - \lambda_2(\frac{ub\beta I_v}{1+\nu I_v}) \ , \ \lambda_2' = -\frac{\partial H}{\partial E} = \lambda_2(\mu+\mu) - \lambda_3\alpha \ ,$$

$$\lambda_3' = \lambda_3(\mu+\gamma+\delta+v) - \lambda_4\gamma v + \lambda_5(\frac{(1+v_v I)(b\beta_v S_v) + b\beta_v S_v I v_v}{(1+v_v I)^2})$$

$$-\lambda_6(\frac{(1+v_v I)(b\beta_v S_v) - b\beta_v S_v I v_v}{(1+v_v I)^2}) + \lambda_7(\mu_v + \delta_v) - 1,$$

$$\lambda_4' = -\frac{\partial H}{\partial R} - \lambda_1\varphi + \lambda_4(\mu+\varphi) \ , \lambda_5' = -\frac{\partial H}{\partial E_\upsilon} = \lambda_5(\frac{b\beta_\upsilon I}{1+v_v I} + \mu_\upsilon) - \lambda_6(\frac{b\beta_\upsilon I}{1+v_v I}),$$

$$\lambda_6' = -\frac{\partial H}{\partial I_\upsilon} = \lambda_6(\alpha_v + \mu_v) - \lambda_7\alpha_\upsilon \ , \lambda_7' = (\lambda_1 - \lambda_2)(\frac{(1+\nu I_v)(u^* b\beta S) - (u^* b\beta S I_v \nu)}{(1+\nu I_v)^2}) + \lambda_7(\mu_v + \delta_v),$$

With $S(0)=S_0, E(0)=E_0, I(0)=I_0, R(0)=R_0, S_v(0)=S_{v_0}, E_v(0)=E_{v_0} and I_v(0)=I_{v_0}$ and $\lambda_i(t_f)=0,(i=1,2,3,4,5,6,7)$

Since ,there were initial condition for the state variables and terminal conditions for the adjoint variabless and the optimality system is two-point boundary value problem, with separated boundary conditions at $t=0$ and t_f. Then we use the semi-implicit finite difference method to solve the optimality system (20). We partition the interval $[t_0, t_f]$ at the point $t_i = t_0 + ih(i=0,1,2,...,n)$, where h is the time step such that $t_n = t_f$. And we define the state and adjoint variable; $S(t), E(t), I(t), R(t), S_\upsilon(t), E_\upsilon(t), I_\upsilon(t), \lambda_1, \lambda_2, \lambda_3, \lambda_4, \lambda_5, \lambda_6, \lambda_7$ and the control u and v in terms of nodal points $S_i, E_i, I_i, R_i, S_{v_i}, E_{v_i}, I_{v_i}, \lambda_1^i, \lambda_2^i, \lambda_3^i, \lambda_4^i, \lambda_5^i, \lambda_6^i, \lambda_7^i, u^i$ and v^i. After that we use the combination of forward and backward difference approximation. The simulations at endemic state were carried out using the following values taken from table1, with initial condition; S(0)=200,E(0)=40,I(0)=20,R(0)=0,S_v(0)=2000,E_υ(0)=40 and I_υ(0)=60. and results show below

IOP Conf. Series: Journal of Physics: Conf. Series **1039** (2018) 012035 doi:10.1088/1742-6596/1039/1/012035

Table 1. Parameters values used in numerical simulation at endemic state.

Parameters	Description	Value
μ	Death rate of human populations	0.0000548 day^{-1}
A	Recruitment rate of susceptible human	0.000215
μ_v	Death rate of vector populations	0.0667 day^{-1}
B	Constant inflow of infective rate	0.07
b	The average bite of mosquitoes that potentially infected	0.02
φ	the rate of loss of immunity in humans	0.09
δ	The disease-induced death rate of human	0.001
δ_v	Proportional rates of mosquitoes exposed to the virus infection	0.01
γ	the duration of infection in the body	0.05
β	Probability that a bite by an infectious mosquito results in transmission of disease to human	0.1
β_v	Probability that a bite results in transmission of parasite to a susceptible mosquito	0.09
N	Number of human populations	10000
N_v	Number of mosquitoes populations	20000
α	The rate of progression of humans from the exposed state to the infectious state	0.0588
α_v	The rate of progression of mosquitoes from the exposed state to the infectious state	0.0556
u	The effort rate that reduces the contract between the infectious vector and the susceptible individuals	$0 < u < 1$
v	The rate at with infectious human are treated at each time period.	$0 < v < 1$

3.1 Numerical results for a modified SEIR-SVEVIV *model for each state variable with and without control* u and v.

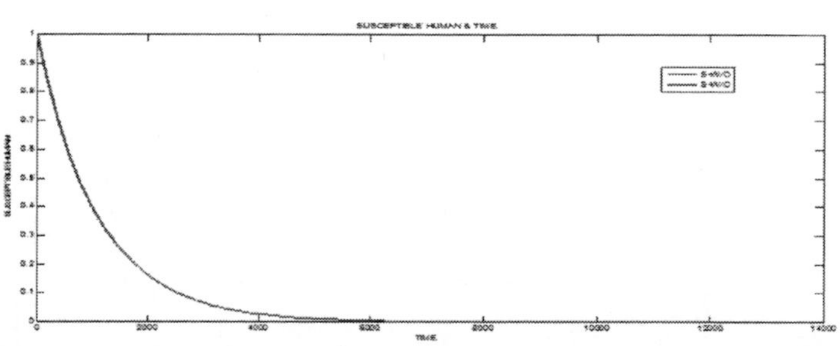

Figure 1 Represent time series of susceptible individuals (S) with and without controls. It's show the number of susceptible individuals (S) with controls decreased faster than the number of susceptible individuals (S) without controls.

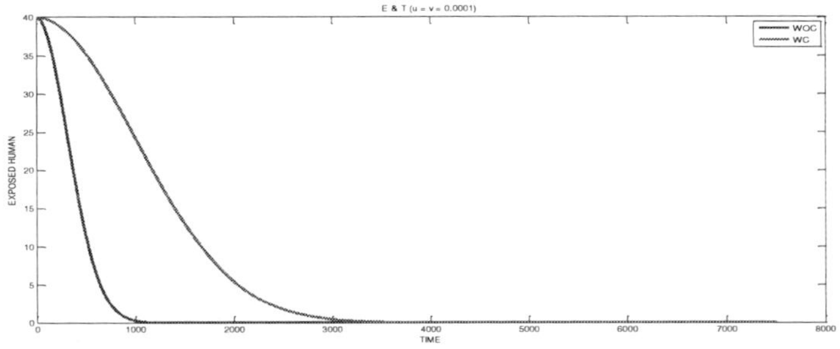

Figure 2 Represent time series of exposed individuals (E) with and without controls. It's show the number of exposed individuals (S) with controls are increased faster than the number of exposed individuals (S) without controls.

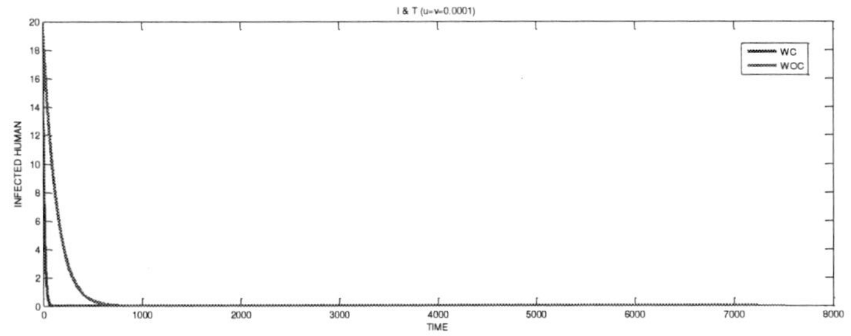

Fig. 3 Represent time series of infectious individuals (I) with and without controls. It's show that the number of infected individuals (I) with controls are decreased rapidly.

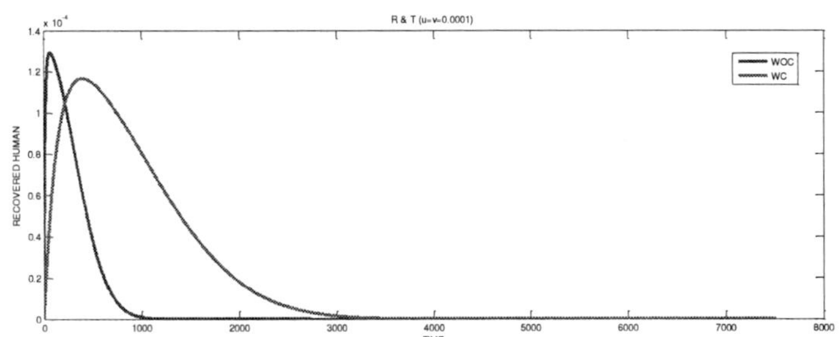

Fig. 4 Represent time series of recovered human (R) with and without controls. It's show the number of recovered human (R) with controls are increased more faster than the number of recovered human (R) without controls.

4. Conclusions

In this paper, a modified SEIR-SVEVIV model for transmission dynamics of malaria disease was proposed and analyzed in order to better understand the transmission and spread of the malaria disease and tried to find an effective strategy for its prevention. The qualitative results are depended on a basic reproductive number (R_0). We obtained the basic reproductive number by using the next generation method technique and finding the spectral radius. Routh-Hurwitz criteria is used for determining the stabilities of the model. If $R_0 < 1$, then the diseases free equilibrium point is local asymptotically stable:

that is the disease will died out, but if $R_0 > 1$, then the endemic equilibrium is local asymptotically stable. To reduce the contract between the susceptible human and the infected vector and the other, minimizing the population of the infected human. The optimal control theory has been applied. By using the Pontryagin's maximum principle, the explicit expression of the optimal controls was obtained. Simulation results indicate that the numbers of infectious individuals (I) are decreased after control, but the number of recovered human (E) increased after control.

References

[1] WHO. *Fact sheets: Dengue and dengue haemorrhagicfever.*http://www.who.int/mediacentre/factsheets/fs117/en/ (2014).
[2] http://www.thaivbd.org/n/histories/view/2662
[3] Derouich, M. and Boutayeb, A., Dengue Fever: Mathematical modelling and computer simulation. *Applied Mathematics and Computation.* **177**(2), 528–544 (2006).
[4] Derouich, M. A., Boutayeb and Twizell, E. H., A Model of Dengue Fever. *Brunel Universty,England.* p. 1–10 (2003).
[5] Pongsumpun, P., Transmission model for dengue disease with and without the effect of extrinsic incubation period. *KMITL Sci. Tech.* J. Thailand **6**, 74–82 (2006).
[6] Pongsumpun, P. and K. Patanarapelert., Infection risk to travelers going to dengue fever endemic regions. *Mahidol University*, Thailand **35**, 155–159 (2004).
[7] Pongsumpun, P. and Tang,I., Transmission of dengue hemorrhagic fever in age structured population. *Mathematical and Computer Modelling* **37**, 949–961 (2003).
[8] Ang, K. C. and Li, Z., Modeling the spread of dengue in Singapore. In Conference Proceedings for The International Congress on Modeling and Simulation 1999, Hamilton, New Zealand, December 1999, Vol. 2, p. 555–560 (2002).
[9] Van den Driessche, P and Watmough,J (2002). Reproductive numbers and sub-threshold endemic equilibria for compartment models of disease transmission.Math. Biosci.;180;29-48
[10] L. S. Pontryagin, V. G. Boltyanskii, R. V.Gamkrelidze, E. F. Mishchenko, The Mathematical Theory of Optimal Processes, Wiley, New York, 1962.
[11] H.Raj joshi, S. Lenhart, M Y Li, L. Wang, Optimal control methods applied to disease models-Contemporary Mathematics, Volume 410, 2006

Acknowlegments

The authors are grateful to the Department of Mathematics, Faculty of Science and Technology and Phuket Rajabhat University, Thailand for providing the facilities to carry out the research.

IOP Conf. Series: Journal of Physics: Conf. Series **1039** (2018) 012036 doi:10.1088/1742-6596/1039/1/012036

The Role of Family on the Transmission Model of Methamphetamine

Surapol Naowarat[1] and Nuengruedee Kumat[2]

[1] Department of Mathematics, Faculty of Science and Technology, Suratthani Rajabhat University, Surat Thani ,84100, Thailand.

[2]Department of Mathematics, Faculty of Science and Technology,Suratthani Rajabhat University, Surat Thani ,84100, Thailand.

Email: su1963n@gmail.com

Abstract: In this paper, we proposed and analyzed the mathematical model of methamphetamine epidemics with the role of family was taken into account. The objective of this research was to investigate the role of family on the transmission model of methamphetamine epidemics. The standard method of differential equations was used to analyze the proposed model, to find the drug free equilibrium points and drug present equilibrium point of the model and to determine the stability of the proposed model for both equilibrium points. The drug reproductive number (R_0) was found by applying the next generation matrix. The analytic solution and numerical solution were carried out. The results shown the proposed model which represented by five differential equations, consist of five subgroups that were the susceptible , light methamphetamine users, hard methamphetamine users, clients of health services in treatment and recovered individuals. We found that there were two equilibrium points, drug free equilibrium point and drug present equilibrium point. From the numerical simulation, when the effective of the role of family $k = 0.9$, the drug reproductive number $R_0 = 0.4928 < 1$, which mean that the methamphetamine epidemic was not occurred. Whenever the effective of the role of family $k = 0.1$, the drug reproductive number $R_0 = 3.7296 > 1$, which mean that the methamphetamine epidemic was occurred. Both drug free equilibrium points and drug present equilibrium points were local asymptotically stable. We concluded that if we increase the effective of the role of family (k) in the human population, then the methamphetamine users will decrease.

1. Introduction

In recent year, Western Cape Province of South Africa, there has been a dramatic increase in treatment demand for drugs such as dagga, mandrax, cocaine, heroin and methamphetamine (MA). MA patients defined heroin as drug users who used MA as a primary or secondary drug of abuse, increase from 121 to 376 patients from second half of 2003 to the first half of 2004. More than 50% of the patients were under the age of 20 [1]. MA is a highly addictive stimulant drug. The common effects of intoxication are increased energy and self-confidence, euphoria, heightened libido and appetite suppression. Prolonged use is usually characterized by severe weight loss, mood swing, violent behavior and body organ disorders [2]. Kalula and Nyabadza[3] extended the model by White and Comiskey [4] to model substance abuse in South Africa in order to qualitatively investigate the dynamics of substance abuse and predict drug abuse trends. Morris and Parry [5] reported that the high rates of increase of HIV infections in communities with high MA use. MA use has been linked to risky sexual behavior and sexually transmitted infections. Families and friends play an incredibly critical role in motivation individuals with drug problems to enter treatment [6]. Rittisak and Naowarat [7] proposed and analyzed a mathematical model to study the dynamics of giving up smoking with effect of education campaign. It found that with an increase in the education campaign, the number of everyday smokers will decrease. Nyabadza, Njagarah and Smith [8] studied the modeling the dynamic of crystal meth abuse in the presence of drug-supply chains in South Africa by considering a model for MA use that tracks drug-supply chains, and accounts for rehabilitation and amelioration for the addicted. There was a unique drug free equilibrium. The sensitivity analysis is applied for this model. It found that the parameters with the most control over the epidemic are the quitting rate of light-drug users and the person-person contact rate between susceptible individuals and MA uses. This study is motivated by the recent work of [1] , by adding the parameter the role of family of [6,9] for those study. So, the objective for this study is to propose and analyze the MA epidemics with the role of family.

This paper is organized as follows. In section 2, the deterministic model is presented to describe the role of family. In section 3, the model analysis is carried out. In section 4, numerical results are illustrated our analytic results. The conclusion is presented in section 5.

2. Material and method

2.1 Material

To formulate the model , we consider the total human population at time t, denoted by N , is divided into five disjoint classes namely susceptible individuals at risk of using MA class S , light MA users class I_1, hard MA users class I_2 , clients of health care services in treatment class T , and recovered individuals class, R . Hence, $S+I_1+I_2+T+R=N$. The possible changes in the life of a MA user can be tracked by the schematic representation in Fig. 1

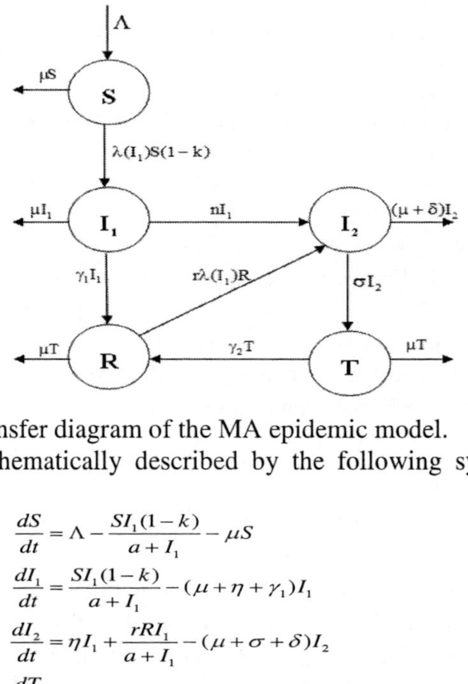

Fig. 1. Transfer diagram of the MA epidemic model.

Based on Fig. 1, the model is mathematically described by the following system of differential equations:

$$\begin{aligned}
\frac{dS}{dt} &= \Lambda - \frac{SI_1(1-k)}{a+I_1} - \mu S \\
\frac{dI_1}{dt} &= \frac{SI_1(1-k)}{a+I_1} - (\mu+\eta+\gamma_1)I_1 \\
\frac{dI_2}{dt} &= \eta I_1 + \frac{rRI_1}{a+I_1} - (\mu+\sigma+\delta)I_2 \\
\frac{dT}{dt} &= \sigma I_2 - (\mu+\gamma_2)T \\
\frac{dR}{dt} &= \gamma_1 I_1 + \gamma_2 T - r\frac{RI_1}{a+I_1} - \mu R
\end{aligned} \qquad (1)$$

Where Λ is the recruitment rate of susceptible individuals, μ is the natural dead rate, η is the relative infectivity of I_2 when compare to I_1, r is the level of relapse to being a hard MA user, σ is the uptake rate into treatment programs, δ is the remove from the hard MA users' class that include drug related,

γ_1 is the recovery rate for light MA users, γ_2 is the recovery rate for MA users under treatment, a is the density of human population and k is the effective of the role of family.

2.2 *Methods*

From the proposed model, by setting the right-hand sides of the equations (1) to zero. We obtained two equilibrium points,

2.2.1 Drug Free Equilibrium(DFE) denoted by $E_0(S,I_1,I_2,T,R)$ **:** In this case when $I_1=0$, we obtained

$$S=\frac{\Lambda}{\mu},\ I_2=0, T=0, R=0, \text{ that is } E_0(\frac{\Lambda}{\mu},0,0,0,0).$$

2.2.2 Drug Present Equilibrium (DPE) denoted by $E_1(S^*,I_1^*,I_2^*,T^*,R^*)$ **:** In the case when $I_1^*>0$, we obtained

$$S^*=\frac{\Lambda(a+I_1^*)}{A_1I_1^*+a\mu},\ I_2^*=\frac{B_1I_1^{3*}+B_2I_1^{2*}+naA_5I_1^*}{B_3I_1^{2*}+B_4I_1^*+B_5},\ T^*=\frac{\sigma I_1^{2*}}{\mu+\gamma_2},\ R^*=\frac{A_2I_1^{2*}+A_3I_1^*+(\gamma_2a\sigma+\gamma_2\sigma I_1^*)I_2^*}{A_4I_1^*+A_5} \text{ and } I_1^*=\frac{-C_2\pm\sqrt{C_2^2-4C_1C_3}}{2C_1},$$

Where $A_1=(1-k)+\mu, A_2=\gamma_1(\mu+\gamma_2),\ A_3=a\gamma_1(\mu+\gamma_2),\ A_4=(\mu+r)(\mu+\gamma_2),\ A_5=a\mu(\mu+\gamma_2),\ B_1=nA_4+rA_2,\ B_2=nA_5+rA_3+anA_4,$ $B_3=(\mu+\sigma+\delta)A_4-\gamma_2\sigma r,\ B_4=(\mu a+\sigma a+\delta a)A_4+(\mu+\sigma+\delta)A_5-\gamma_2a\sigma r,\ B_5=(\mu+\sigma+\delta)aA_5, C_1=(\mu+\eta+\gamma_1)A_1,$ $C_1=(\mu+\eta+\gamma_1)A_1,\ C_2=a(\mu+\eta+\gamma_1)A_1+a\mu(\mu+\eta+\gamma_1)-\Lambda(1-k),\ C_3=\mu a^2(\mu+\eta+\gamma_1)-\Lambda a(1-k).$

2.2.3 Drug Reproductive Number (R_0)

The drug reproductive number is defined as the number of secondary infections generated by a typical infected in an otherwise disease free population in his/her whole infectious period. The drug reproductive Number (R_0) for our model is computed using the method described in [10] and changing the notation as in [10] the matrices F and V are given by

$$F=\begin{bmatrix}0\\ \frac{I_1S(1-k)}{a+I_1}\\ 0\\ 0\\ 0\end{bmatrix}, V=\begin{bmatrix}-\Lambda+\mu S\\ (\mu+\eta+\gamma)I_1\\ (\mu+\sigma+\delta)I_2-\eta I_1-r\lambda(I_1)R\\ -\sigma I_2+(\mu+\gamma_2)T\\ -\gamma_1I_1-\gamma_2T+r\lambda(I_1)R+\mu R\end{bmatrix}.$$

Now, the matrix F and V evaluated at disease free equilibrium point are given by

$$F=\begin{bmatrix}0&0&0&0&0\\ 0&\frac{\Lambda(1-k)}{\mu}&0&0&0\\ 0&0&0&0&0\\ 0&0&0&0&0\\ 0&0&0&0&0\end{bmatrix}, V=\begin{bmatrix}\mu&\frac{\Lambda(1-k)}{a\mu}&0&0&0\\ 0&\mu+\eta+\gamma&0&0&0\\ 0&-\eta&\mu+\sigma+\delta&0&0\\ 0&0&-\sigma&\mu+\gamma_2&0\\ 0&-\gamma_1&0&-\gamma_2&\mu\end{bmatrix}.$$

And the matrix FV^{-1} is given by

$$FV^{-1}=\begin{bmatrix}0&0&0&0&0\\ 0&\frac{\Lambda(1-k)}{a\mu(\mu+\eta+\gamma)}&0&0&0\\ 0&0&0&0&0\\ 0&0&0&0&0\\ 0&0&0&0&0\end{bmatrix}.$$

So, the drug reproduction number (R_0) which is the spectral radius of the matrix FV^{-1} is given by

$$R_0=\frac{\Lambda(1-k)}{a\mu(\mu+\eta+\gamma_1)}.$$

2.2.4 The Stability of the Model

At the DFE, if all the eigenvalues of the Jacobean matrix of the system have negative real parts, then the DFE is locally asymptotically stable.

Theorem 1 The DFE for the system (1) is locally asymptotically if $R_0<1$.

Proof. The Jacobian matrix (J_0) of the model (1) at the DFE is given by

$$J_0 = \begin{bmatrix} -\mu & \frac{\Lambda(1-k)}{a\mu} & 0 & 0 & 0 \\ 0 & \frac{\Lambda(1-k)}{a\mu} - (\mu+\eta+\gamma_1) & 0 & 0 & 0 \\ 0 & \eta & -(\mu+\sigma+\delta) & 0 & 0 \\ 0 & 0 & \sigma & -(\mu+\gamma_2) & 0 \\ 0 & \gamma_1 & 0 & \gamma_2 & -\mu \end{bmatrix}.$$

The eigenvalues of the Jacobian matrix are the solutions of the characteristic equation $|J_0 - \lambda I| = 0$,

$$\begin{vmatrix} -\mu-\lambda & \frac{\Lambda(1-k)}{a\mu} & 0 & 0 & 0 \\ 0 & \frac{\Lambda(1-k)}{a\mu} - (\mu+\eta+\gamma_1)-\lambda & 0 & 0 & 0 \\ 0 & \eta & -(\mu+\sigma+\delta)-\lambda & 0 & 0 \\ 0 & 0 & \sigma & -(\mu+\gamma_2)-\lambda & 0 \\ 0 & \gamma_1 & 0 & \gamma_2 & -\mu-\lambda \end{vmatrix} = 0 .$$

Expanding the determinant into a characteristic equation, we have

$$(\lambda+\mu)(\lambda+\mu)(\lambda+\mu+\gamma_2)(\lambda+\mu+\sigma+\delta)(\lambda+\frac{\Lambda(1-k)}{a\mu}-(\mu+\eta+\gamma_1))=0 .$$

Thus, all the eigenvalues of the Jacobian matrix have negative real parts, implying that the DFE point is locally asymptotically stable $(R_0 < 1)$.

Theorem 2 The DPE for the system (1) is locally asymptotically if $R_0 > 1$

Proof. The Jacobian matrix (J_1) of the model (1) at the DPE is given by

$$J_1 = \begin{bmatrix} -D_1 & D_3 & 0 & 0 & 0 \\ D_2 & -D_4 & 0 & 0 & 0 \\ 0 & D_5 & -D_7 & 0 & D_9 \\ 0 & 0 & \sigma & -D_8 & 0 \\ 0 & D_6 & 0 & \gamma_2 & -D_{10} \end{bmatrix}.$$

Where $D_1 = \frac{I_1^*(1-k)}{a+I_1^*}+\mu,\ D_2 = \frac{I_1^*(1-k)}{a+I_1^*},\ D_3 = \frac{aS^*(1-k)}{(a+I_1^*)^2},\ D_4 = \frac{aS^*(1-k)}{(a+I_1^*)^2}+(\mu+\eta+\gamma_1),\ D_5 = \eta+\frac{arR^*}{(a+I_1^*)^2},$

$D_6 = \gamma_1+\frac{arR^*}{(a+I_1^*)^2},\ D_7 = \mu+\sigma+\delta,\ D_8 = \mu+\gamma_2,\ D_9 = \frac{rI_1^*}{a+I_1^*},\ D_{10} = \frac{rI_1^*}{a+I_1^*}+\mu$.

The eigenvalues of the Jacobian matrix are the solutions of the characteristic equation $|J_1 - \lambda I| = 0$,

$$\begin{vmatrix} -D_1-\lambda & D_3 & 0 & 0 & 0 \\ D_2 & -D_4-\lambda & 0 & 0 & 0 \\ 0 & D_5 & -D_7-\lambda & 0 & D_9 \\ 0 & 0 & \sigma & -D_8-\lambda & 0 \\ 0 & D_6 & 0 & \gamma_2 & -D_{10}-\lambda \end{vmatrix} = 0.$$

Expanding the determinant into a characteristic equation, we have

$$\lambda^5 + M_1\lambda^4 + M_2\lambda^3 + M_3\lambda^4 + M_4\lambda + M_5 = 0 .$$

Where $M_1 = D_1+D_4+D_7+D_8+D_{10}$, $M_2 = D_{10}(D_7+D_8)+(D_1+D_4)(D_7+D_8+D_{10})$, $M_5 = \sigma\gamma_2 D_9(D_1D_4+D_2D_3)$,

$M_3 = D_7D_8D_{10}+D_7D_8+(D_1+D_4)(D_7D_8+D_7D_{10}+D_8D_{10})+\sigma\gamma_2 D_9$, $M_4 = (D_1+D_4)(D_7D_8D_{10}+\sigma\gamma_2 D_9)$.

By applying Routh Hurwitz Criteria. All the eigenvalues of the Jacobian matrix have negative real parts,
if the coefficient of characteristic equation is satisfied this conditions:

1. $M_1 > 0, M_2 > 0, M_3 > 0, M_4 > 0, M_5 > 0.$
2. $M_1M_2M_3 > M_3^2 + M_1^2M_4$,
3. $(M_1M_4 - M_5)(M_1M_2M_3 - M_3^2 - M_1^2M_4) > M_5(M_1M_2 - M_3)^2 + M_1M_2^5$,

So, it implied that the DPE is locally asymptotically stable $(R_0 > 1)$.

3. Results

From the numerical simulation, we presented the numerical results as shown in Fig. 2 and Fig. 3

3.1 Stability of drug free state: Using the values of parameters for simulation as followed, $\Lambda = 0.028$, $\mu = 0.025$, $\eta = 0.008$, $r = 3.000$, $\sigma = 0.09$, $\delta = 0.02$, $\gamma_1 = 0.01$, $\gamma_2 = 0.02$, $a = 0.01$, $k = 0.90$.We obtained the values of eigenvalues and the drug reproductive number as follows

$$\lambda_1 = \lambda_2 = -0.025, \lambda_3 = -0.055, \lambda_4 = -0.04288, R_0 = 0.3696 < 1, E_0 = (1.12, 0, 0, 0, 0)\ .$$

Since all eigenvalues are to be negative and the drug reproductive number is less than one, the DFE, E_0 will be locally asymptotically stable, as shown in Fig. 2

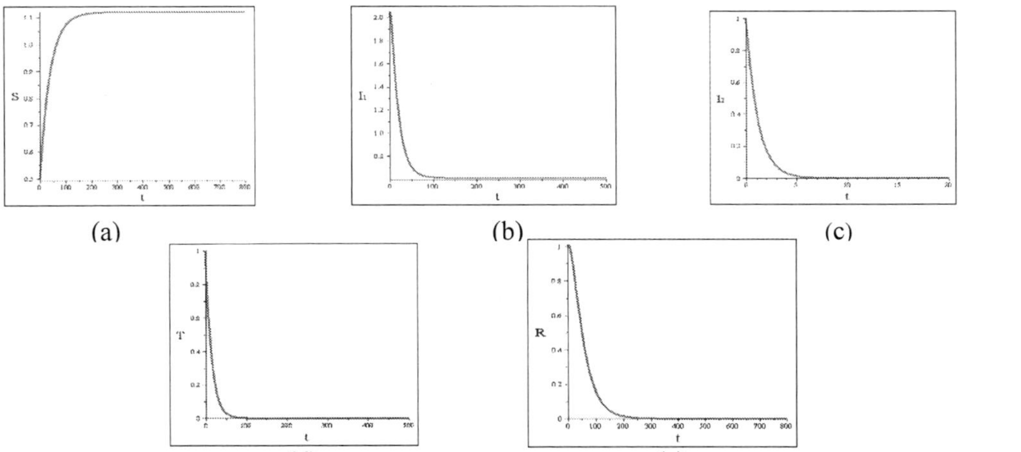

(a) (b) (c) (d) (e)

Fig. 2. The time series of (a) susceptible individuals at risk of using MA S, (b) light MA users I_1, (c) hard MA users I_2 ,(d) clients of health care services in treatment T, and recovered individuals, R with the values of parameters used from the text. We can see that the solutions converge to the DFE $(1.12, 0, 0, 0, 0)$

3.2 Stability of drug present state:

We changed the values of the effective of role family $k = 0.01$ and kept the other values of parameters to be those given in text. We obtained the values of eigenvalues and the drug reproductive number as follows

$$\lambda_1 = -0.0446964, \lambda_2 = -0.910689, \lambda_3 = -2.97826, \lambda_4 = -0.045205 - 0.03414i, \lambda_4 = -0.045205 + 0.03414i, R_0 = 3.7296 > 1,$$
$$E_1 = (0.0306070, 0.736034, 0.102036, 0.166967, 0.0026646).$$

Since all eigenvalues are to be negative and the drug reproductive number is greater than one, the DPE , E_1 will be locally asymptotically stable, as shown in Fig. 3

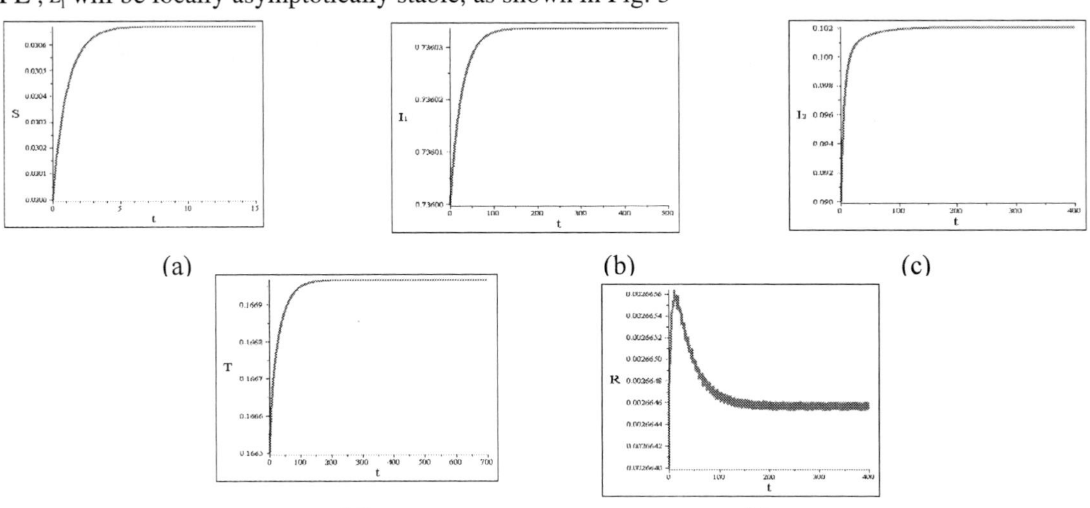

(a) (b) (c) (d) (e)

Fig. 3.The time series of (a) susceptible individuals at risk of using MA class S ,(b) light MA users I_1 ,(c) hard MA users I_2 ,(d) clients of health care services in treatment T , and recovered individuals , R with the values of parameters used from the text. We can see that the solutions converge to the DPE.

4. Conclusions

In this paper, we developed and analyzed a mathematical model for the MA epidemics. The role of family was taken into account. We analyzed the model by using the standard method. The drug reproductive is obtained through the use of next generation method and calculated the spectral radius of next generation matrix to obtain the drug reproductive number, $R_0 = \frac{\Lambda(1-k)}{a\mu(\mu+\eta+\gamma_1)}$. The drug reproductive number is the threshold condition for determining the stability of the model which are shown in Fig. 2and Fig. 3. Our simulation results shown that R_0 will increase when the effective of role of family decrease. We found the values of $R_0 = 0.3696,\ 3.7296$ when $k = 0.90,\ k = 0.01$, respectively. It seems that the light MA users and hard MA users will decrease when the effective of role of family is increased.

References

[1] Parry, C. 2005. *Substance abuse trends in the Western Cape: Summary Alcohol and Drug Abuse.* Medical Research Council, pp 195-213.

[2] Kapp, C. 2008. *Crysmeth boom adds to South Africa's health challenges* pp 193.

[3] Kalula, A. S. and Nyabadza, F. 2012. *A theoretical model for substance abuse in the presence of Treatment, S. Afr. J. Sci.* **108**(3-4), pp 1-12.

[4] White E and Comiskey C (2007)*Heroin epidemics treatment and ODE modelling.* Math. Biosci. **2007**(208), pp 312-324.

[5] Morris K and Parry C(2006) *South African methamphetamine boom could fuel further HIV* Lancet. **2006**(6) pp 471.

[6] Giord, S. 2017. *Family involvement is important in substance abuse treatment,* Psych Central. Jul 2017.

[7] Rittisak, T. and Naowarat, S. 2015. *Stability analysis and qualitative behavior of giving up smoking model with education campaign,* AJBAS, **2015**(9) pp 46-53.

[8] Nyabadza,F.Njagarah,J.B.H.and Smith,R.J. 2012. *Modelling the dynamic of crystal meth('Tik') abuse in the presence of drug-supply chains in South Africa*. Bull. Math. **2012**(1) pp 24-48.

[9] Mushanyuzy, J. Nyabadza, F. Muchatibayaz, G. and Stewartz, A.G.R. 2016. *The role of family in initiating methamphetamine abuse treatment: insights through a mathematical model.* J. Appl Computat Math **2016**(5), pp 1-10.

[10] Driessche P V, and J Watmough J 2002 *Reproduction numbers and sub-threshold endemic equilibria for compartmental models of disease transmission*. Math Biosci, **2002**(180), pp 29-48.

Acknowledgment

The authors would like to express their sincere thanks to the Department of Mathematics, Faculty of Science and Technology, Suratthani Rajabhat University, Thailand.

Asymptotic Solutions of the Planar Squeeze Flow of a Herschel-Bulkley Fluid

Pavan Kumar Singeetham and Vishwanath Kadaba Puttanna
National Institute of Technology Karnataka, Mangalore, India

Email: shastryvishwanath@gmail.com

Abstract. In this study, we present the analysis of the squeeze flow of a Herschel-Bulkley fluid between parallel plates that are approaching each other with a constant squeeze motion. The classical lubrication analysis predicts the existence of a central unyielded zone bracketed between near-wall regions. This leads to the well-known squeeze flow paradox for viscoplastic fluids. Since the kinematic arguments show that there must be a finite velocity gradient even in the unyielded zone, thereby precluding the existence of such regions. This paradox may, however, be resolved within the framework of a matched asymptotic expansions approach where one postulates separate expansions within the yielded and apparently unyielded (plastic) zones. Based on the above technique, we circumvent the paradox, and develop complete asymptotic solutions for the squeeze flow of a Herschel-Bulkley fluid. We derive expressions for the velocity, pressure and squeeze force. The effects of the yield threshold on the pseudo-yield surface that separates the sheared and plastic zones, and squeeze force for different values of non-dimensional yield stress have been investigated.

1. Introduction

The squeeze film phenomenon occurs for the close approach of a pair of surfaces, and conforms to the classical lubrication paradigm. This approach leads to a sharp growth in the pressure within the narrow gap (between the surfaces), this growth being proportional to the fluid viscosity. While, squeeze flow problems have been analysed extensively for Newtonian fluids, we here consider the same for viscoplastic fluid between planar geometry. Bingham, Casson and Herschel-Bulkley are some of the most commonly used models to represent viscoplastic fluids. More details on these models can be found in Bird et.al [1] and Barnes [2] review articles. Here in this study, we use the Herschel-Bulkley model to represent the viscoplastic fluid, which is a generalized model of non-Newtonian fluids. For such fluids, flow occurs when the stress in the gap exceeds critical yield stress.

Practically, this type of flow behaviour occurs in many areas, such as slurries and suspensions, muds, clays, certain polymer solutions, lavas, etc. Therefore, such materials have applications in different fields, ranging from the oil, gas industries to the concrete used for construction. The non-Newtonian behaviour of fresh concrete, fly ash, mining slurries and cement flows has been verified by conducting field and laboratory experiments [3-5]. However, the Herschel-Bulkley model is more suitable to represent the true rheological behaviour of viscoplastic fluids over a sufficiently wide range of shear rates. The flow behaviour of viscoplastic materials is important to analyse, due to the existence of yield surface which separates yielded and unyielded regions.

The geometry with small aspect ratio, having viscoplastic material as a fluid medium have a long history. Some of the researchers have analysed the squeeze flow of viscoplastic fluid and the presence of yield surface. But most of the earlier works [6-9] debate on the existence of true unyielded plug regions in the problem of squeeze flow of viscoplastic fluids. The classical lubrication theory leads to the well-known "squeeze flow paradox", since the velocity component of the unyielded plug region varies in the flow direction. This implies that true unyielded plug regions cannot exist. However the paradox can be resolved by using numerical simulations or approximate solutions.

Some of the investigators have used discretized approximations such as finite difference or finite element methods to analyse the squeeze flow of viscoplastic materials and verified with the experimental results [10-15]. Several researchers have used the technique of asymptotic expansions to analyse the viscoplastic squeeze flow problems. Initially Walton and Bittleston [16] have used this technique to study the Bingham plastics in a narrow eccentric annulus. Balmforth and Craster [17], later Frigaard and Ryan [18] and Putz et al. [19] have used asymptotic expansions to resolve the squeeze flow paradox and developed the consistent solution for thin-layer problems. Recently, Muravleva [20] has analysed the squeeze flow of a Bingham fluid in a planar geometry using numerical simulations and asymptotic expansions [17].

In the present work, we study the squeeze flow problem of a viscoplastic fluid in a planar geometry using the Herschel-Bulkley model by applying the technique of asymptotic expansions and also suggest corrections to resolve the squeeze flow paradox.

2. Mathematical formulation

The objective is to develop a consistent solution for the squeeze flow of an incompressible viscoplastic fluid for a planar geometry using Herschel-Bulkley fluid model, which resolves the squeeze flow paradox. Figure 1 shows the schematic of the squeeze flow problem. The gap width $2H^*$ between the plates of length $2L^*$ is filled with a viscoplastic material, while the plates approach each other with a constant velocity v_s^*.

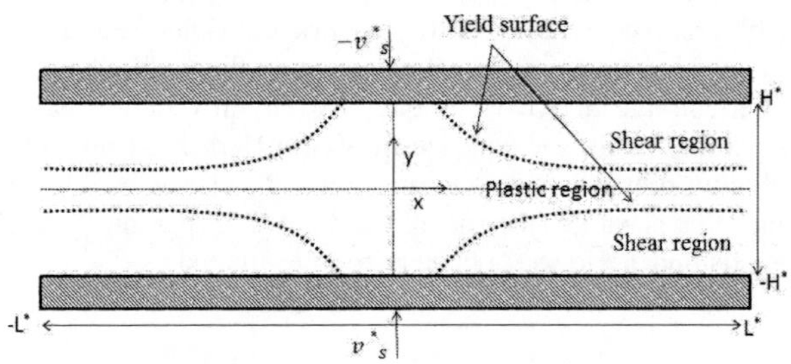

Figure 1. Schematics of the Core formation.

The governing equations of the problem in two dimensional form is as follows:

$$\frac{\partial u^*}{\partial t^*}+u^*\frac{\partial u^*}{\partial x^*}+v^*\frac{\partial u^*}{\partial y^*}=-\frac{\partial p^*}{\partial x}+\frac{\partial \tau^*_{xx}}{\partial x^*}+\frac{\partial \tau^*_{xy}}{\partial y}, \tag{1}$$

$$\frac{\partial v^*}{\partial t^*}+u^*\frac{\partial v^*}{\partial x^*}+v^*\frac{\partial v^*}{\partial y^*}=-\frac{\partial p^*}{\partial y}+\frac{\partial \tau^*_{xy}}{\partial x^*}+\frac{\partial \tau^*_{yy}}{\partial y}, \tag{2}$$

$$0=\frac{\partial u^*}{\partial x^*}+\frac{\partial v^*}{\partial y^*}. \tag{3}$$

The constitutive equation for Herschel-Bulkley fluid model in three dimensional form is given by

$$\tau^*_{ij}=\left[K\left|\dot{\gamma}^*\right|^{n-1}+\frac{\tau_0}{\left|\dot{\gamma}^*\right|}\right]\dot{\gamma}^*_{ij} \quad \text{for } \left|\tau^*\right|>\tau_0,$$

$$\dot{\gamma}^*_{ij}=0 \quad \text{for } \left|\tau^*\right|\le\tau_0.$$

In the present work, (u^*, v^*) is the velocity of the fluid, p^* is the pressure, τ^*_{xx}, τ^*_{xy}, τ^*_{yx} and τ^*_{yy} are the components of deviatoric stress tensor. To non-dimesionalize the governing equations, we have used different scales in both x and y directions, with plate half-length L^* as the horizontal length scale and half thickness H^* as the vertical length scale respectively. v^*_s is taken as characteristic velocity in the transverse direction, the velocity in x- direction is scaled with u^*_c and time with L^*/u^*_c. Here pressure is scaled with $\frac{K^*L^*}{H^*}\left(\frac{u^*_c}{H^*}\right)^n$. The both shear and extensional stress components are scaled with $K^*\left(\frac{u^*_c}{H^*}\right)^n$ and $K^*\left(\frac{u^*_c}{H^*}\right)^{n-1}\frac{u^*_c}{L^*}$ respectively. Here $\mathrm{Re} = \frac{\rho^*(u^*_c)^2}{K^*}\left(\frac{H^*}{u^*_c}\right)^n$ is the Reynolds number, where ρ^* and K^* are the density and consistency coefficient of the viscoplastic material.

Assuming the effect of fluid inertia is negligible, equations (1)-(3) can be written in dimensionless form as:

$$0 = -\frac{\partial p}{\partial x} + \varepsilon^2 \frac{\partial \tau_{xx}}{\partial x} + \frac{\partial \tau_{xy}}{\partial y}, \tag{4}$$

$$0 = -\frac{\partial p}{\partial y} + \varepsilon^2\left(\frac{\partial \tau_{xy}}{\partial x} + \frac{\partial \tau_{yy}}{\partial y}\right), \tag{5}$$

$$0 = \frac{\partial u}{\partial x} + \frac{\partial v}{\partial y}. \tag{6}$$

Here $\varepsilon = \frac{H^*}{L^*}$ is called the aspect ratio. The dimensionless form of constitutive equations are given by

$$\tau_{ij} = \left[|\dot{\gamma}|^{n-1} + \frac{N}{|\dot{\gamma}|}\right]\dot{\gamma}_{ij} \text{ for } |\tau| > N,$$

$$\dot{\gamma}_{ij} = 0 \text{ for } |\tau| \le N.$$

And the strain rate tensor $\dot{\gamma}_{ij}$ is given by

$$\dot{\gamma}_{xx} = 2\frac{\partial u}{\partial x},\ \dot{\gamma}_{xy} = \dot{\gamma}_{yx} = \frac{\partial u}{\partial y} + \varepsilon^2\frac{\partial v}{\partial x},\ \dot{\gamma}_{yy} = 2\frac{\partial v}{\partial y}.$$

The second invariants $\bar{\bar{\dot{\tau}}}$ and $\bar{\bar{\dot{\gamma}}}$ are denoted by τ and γ respectively, i.e. ,

$$\tau = \sqrt{\tau^2_{xy} + \varepsilon^2\tau^2_{xx}},\ \dot{\gamma} = \sqrt{\gamma^2_{xy} + \varepsilon^2\gamma^2_{xx}}$$

In the above equations dimensionless Herschel-Bulkley number N is defined by $N = \frac{\tau_0}{K^*}\left(\frac{H^*}{u^*_c}\right)^n$, where τ_0 is the yield stress.

The above equations (4)-(6) are solved by applying appropriate boundary conditions given below.

At $y=1 \Rightarrow u=0, v=-1$, at $y=-1 \Rightarrow u=0, v=1$, at $x=1 \Rightarrow \sigma_{xx}=-p+\varepsilon^2\tau_{xx}=0, \tau_{xy}=0$, in the plane of symmetry $y=0 \Rightarrow v=0, \tau_{xy}=0$, and on the plane symmetry $x=0 \Rightarrow u=0, \tau_{xy}=0$.

2.1. *Asymptotic expansions*

The equations (4)-(6) along with the boundary conditions are solved by introducing asymptotic expansions.

$$u(x,y)=u^0(x,y)+\varepsilon u^1(x,y)+\varepsilon^2 u^2(x,y)+\cdots, \qquad v(x,y)=v^0(x,y)+\varepsilon v^1(x,y)+\varepsilon^2 v^2(x,y)+\cdots,$$

$$p(x,y)=p^0(x,y)+\varepsilon p^1(x,y)+\varepsilon^2 p^2(x,y)+\cdots, \qquad \tau(x,y)=\tau^0(x,y)+\varepsilon\tau^1(x,y)+\varepsilon^2\tau^2(x,y)+\cdots.$$

Substituting these expansions in equations (4)-(6) and comparing the leading order terms (i.e. ε^0 terms), we get

$$0=-\frac{\partial p^0}{\partial x}+\frac{\partial \tau_{xy}^0}{\partial y}, \tag{7}$$

$$0=-\frac{\partial p^0}{\partial y}, \tag{8}$$

$$0=\frac{\partial u^0}{\partial x}+\frac{\partial v^0}{\partial y}. \tag{9}$$

2.1.1. *Shear region*

Solving equation (8), we have $p^0=p_0(x)$ and from equation (7) along with the boundary condition, we have

$$\tau_{xy}^0=\frac{\partial p^0}{\partial x}y. \tag{10}$$

And we have $\tau^0=\left|\tau^0\right|$ and $\dot{\gamma}^0=\left|\frac{\partial u^0}{\partial y}\right|$, one can write leading order stress tensor components as

$$\tau_{xy}^0=\left(\left|\frac{\partial u^0}{\partial y}\right|^n+N\right)\operatorname{sgn}\left(\frac{\partial u^0}{\partial y}\right), \tag{11}$$

$$\tau_{xx}^0=2\left(\left|\frac{\partial u^0}{\partial y}\right|^{n-1}\left|\frac{\partial u^0}{\partial x}\right|+N\frac{\frac{\partial u^0}{\partial x}}{\left|\frac{\partial u^0}{\partial y}\right|}\right). \tag{12}$$

Here velocity vanishes on the surface of the plate and maximum at the center in the domain x>0 and y>0. When the material is squeezed out, velocity becomes positive (u>0) implies $\frac{\partial u^0}{\partial y}<0$.

Solving equation (11) with appropriate boundary condition, we have

$$u^0 = \frac{n}{n+1}\left(\frac{N}{y_0}\right)^{1/n}\left[(1-y_0)^{1+1/n} - (y-y_0)^{1+1/n}\right]. \tag{13}$$

Now consider the first order approximations (i.e. ε^1 terms), we get

$$0 = -\frac{\partial p^1}{\partial x} + \frac{\partial \tau^1_{xy}}{\partial y}, \tag{14}$$

$$0 = -\frac{\partial p^1}{\partial y}, \tag{15}$$

$$0 = \frac{\partial u^1}{\partial x} + \frac{\partial v^1}{\partial y} \tag{16}$$

and $\tau^1_{xy} = n\left|\frac{\partial u^0}{\partial y}\right|^{n-1}\frac{\partial u^1}{\partial y}$, $u^1_{y=1} = 0$. After integrating equations (14) and (15), we obtain

$$p^1 = p_1(x),\ \tau^1_{xy} = y p_1^{'}(x) + g(x). \tag{17}$$

Solving equation (17) with the boundary condition, we get

$$u^1(x,y) = \frac{1}{n+1}\left(\frac{N}{y_0}\right)^{1/n-1}\left[\begin{array}{l} g(x)(n+1)\left\{(y-y_0)^{1/n} - (1-y_0)^{1/n}\right\} \\ + p_1^{'}(x)\left\{(ny_0+y)(y-y_0)^{1/n} - (1+ny_0)(1-y_0)^{1/n}\right\}\end{array}\right], \tag{18}$$

here g(x) is the unknown function of integration.

2.1.2. *Plastic region*

At leading order, for each x in [0, 1], we have $\tau^0_{xy} = y p_0^{'}(x)$. The value of τ^0_{xy} attains a maximum at the plane y=1 and vanishes at the plane y=0. Hence, there exists a plane $y = y_0$ at which $\left|\tau^0_{xy}\right| = N$, where $\partial u^0 / \partial y = 0$. Hence, at leading order $y_0 = -N / p_0^{'}(x)$ which is the position of the pseudo-yield surface. For $y \in [0, y_0]$, we have $\tau_0 < N$ and $\dot{\gamma}^0 = 0$. Therefore, one can write the plug velocity from equation (13), i.e.,

$$u^0 = \frac{n}{n+1}\left(\frac{N}{y_0}\right)^{1/n}(1-y_0)^{1+1/n}. \tag{19}$$

To determine the pseudo-yield surface $y = y_0(x)$, one can use integral form of equation (9), i.e.,

$$\int_0^1 u^0(x,y)dy = x. \tag{20}$$

Now substituting equations (13) and (19) into (20) leads an algebraic equation for the yield surface, $y_0(x)$,

$$\frac{(1-y_0)^{2+1/n}}{2+1/n} + \frac{n+1}{n}\left(\frac{y_0}{N}\right)^{1/n} x - (1-y_0)^{1+1/n} = 0. \tag{21}$$

The algebraic equation (21) can be solved by using any of the numerical method to obtain the yield surface, $y_0(x)$. From the equation (19), it can be observed that $u^0(x) = u_0(y_0(x))$ is purely a function of x such that, $\frac{\partial u_0}{\partial x} \neq 0$ where as $\frac{\partial u_0}{\partial y} = 0$.

From above relations, it is obvious that the leading order velocity $u_0(x)$ changes in x direction which implies that the plug region is not a true plug region. This is the basis of the lubrication paradox for yield stress fluids. This problem arises due to the absence of diagonal components of the stress.

This paradox may, however, be resolved within the framework of a matched asymptotic expansions approach where one postulates separate expansions within the yielded and apparently unyielded (plastic) zones. The yielded zones conform to the lubrication paradigm with the shear stress being much greater than all other stress components. On the other hand, the shear and extensional stresses are comparable in the 'plastic region', with the overall stress magnitude being asymptotically close to but just above the yield threshold.

These shear and plastic regions are separated by an interface represented by a smooth pseudo-yield surface $y = y_0(x)$. Let us consider the domain near the center plane of thickness $0 < y < y_0$. Below the fake yield surface, $y = y_0(x)$, the asymptotic expansion described above breaks down. To find the appropriate solution in this region, which incorporates changes in horizontal velocity component we modify u velocity as follows:

$$u(x,y) = u^0(x) + \varepsilon u^1(x,y) + \varepsilon^2 u^2(x,y) + \cdots$$

Using these expansions, we can find stress components

$$\tau_{xx}^{-1} = \frac{2N}{\dot{\gamma}^0}\frac{\partial u^0}{\partial x},\ \tau_{yy}^{-1} = \frac{2N}{\dot{\gamma}^0}\frac{\partial v^0}{\partial y},\ \tau_{xy}^{0} = \frac{N}{\dot{\gamma}^0}\frac{\partial u^1}{\partial y},\ \text{where } \dot{\gamma}^0 = \sqrt{\left(\frac{\partial u^1}{\partial y}\right)^2 + 4\left(\frac{\partial u^0}{\partial x}\right)^2}.$$

From these components, we obtain

$$\tau_{xx}^{-1} = N\sqrt{1 - \frac{y^2}{y_0^2}} \text{ and } \tau_{xy}^{0}\dot{\gamma}^0 = N\frac{\partial u^1}{\partial y}.$$

Solving for $\frac{\partial u^1}{\partial y}$, we get

$$u^1 = 2u_0'(x)\sqrt{y_0^2 - y^2} + u_1^*(x), \tag{22}$$

where $u_1^*(x)$ is an unknown function of integration.

For the first order approximation we have the following equations:

$$0 = -\frac{\partial p^1}{\partial x} + \frac{\partial \tau_{xy}^1}{\partial y} + \frac{\partial \tau_{xx}^{-1}}{\partial x}, \tag{23}$$

$$0 = -\frac{\partial p^1}{\partial y} + \frac{\partial \tau_{xx}^{-1}}{\partial y}, \tag{24}$$

$$0 = -\frac{\partial u^1}{\partial x} + \frac{\partial v^1}{\partial y}. \tag{25}$$

IOP Conf. Series: Journal of Physics: Conf. Series **1039** (2018) 012037 doi:10.1088/1742-6596/1039/1/012037

Solving above equations (23)-(24), we get

$$\tau_{xy}^{1}(x, y) = y\psi'(x) + p_0''(x)\left(y_0^2 \sin^{-1}(y/y_0) - y\sqrt{y^2 - y_0^2} \right), \tag{26}$$

here $\psi(x)$ is an unknown function of integration.

2.2. *Matching*

Using the technique of matching, one can calculate the unknown functions of integration. Since shear stress is continuous at $y = y_0(x)$, one can obtain

$$p_1(x) = \psi(x),\ g(x) = Ny_0'(x)\frac{\pi}{2}. \tag{27}$$

Similarly from continuity of velocity at $y = y_0(x)$, we get

$$u_1^*(x) = \frac{-1}{n+1}\left(\frac{N}{y_0}\right)^{1/n-1}(1-y_0)^{1/n}\left((n+1)g(x) + p_1'(x)(ny_0+1)\right). \tag{28}$$

Substituting above integral constants in equations (13), (18), (19) and (22), we get the velocity distribution in both shear an plastic regions. Further, using the equation of continuity equation (6), one can calculate the unknown value $p_1'(x)$:

$$p_1'(x) = \frac{\pi(2n+1)(1+3n+2n^2)(1+ny_0)y_0 N\left(\frac{N}{y_0}\right)^{1/n}}{(1-y_0)^{1/n}(2n^2y_0^2 + 2ny_0 + n + 1)^2}. \tag{29}$$

3. Pressure Distribution and Squeeze force

One of the important aspects of squeeze flow problem is squeeze force. To get this, we need to know the pressure distribution at the plate along the flow direction. The pressure gradient in shear region is given by $\frac{\partial p^s}{\partial x} = \frac{\partial p^0}{\partial x} + \varepsilon\frac{\partial p^1}{\partial x}$.

Substituting $p_0'(x)$ and $p_1'(x)$, integrating above expression, we get the pressure distribution in shear region as follows:

$$p^s(x) = C_1 + \frac{-nN(1+2ny_0)(1-y_0)^{1+1/n}\left(\frac{N}{y_0}\right)^{1/n}}{(1+n)(1+2n)y_0} - \varepsilon\left[(1+2n)\pi\left(\frac{1}{4n}\log\left(1+n+2ny_0+2n^2y_0^2\right) + \frac{\tan^{-1}\left(\frac{1+2ny_0}{\sqrt{1+2n}}\right)}{2n\sqrt{1+2n}}\right)\right]. \tag{30}$$

Similarly, one can obtain pressure distribution in plastic region as follows:

$$p^p(x) = C_1 + \frac{-nN(1+2ny_0)(1-y_0)^{1+1/n}\left(\frac{N}{y_0}\right)^{1/n}}{(1+n)(1+2n)y_0} - \varepsilon\left[(1+2n)\pi\left(\frac{1}{4n}\log\left(1+n+2ny_0+2n^2y_0^2\right) + \frac{\tan^{-1}\left(\frac{1+2ny_0}{\sqrt{1+2n}}\right)}{2n\sqrt{1+2n}}\right)\right] - \varepsilon N\sqrt{1-\frac{y^2}{y_0^2}}$$

(31)

The constant C_1 can be calculated using boundary condition $p=0$ at $x=1$.

IOP Conf. Series: Journal of Physics: Conf. Series **1039** (2018) 012037 doi:10.1088/1742-6596/1039/1/012037

Further the squeeze force can be calculated by using the relation:

$$F = 2\int_0^1 p(x)dx \tag{32}$$

where $p(x)$ can be substituted by using equations (30) and (31).

4. Results and Discussion

The core thickness along the principal flow direction for different values of Herschel-Bulkley number (N) and power law index (n) have been computed and the results are shown in figures 2 and 3. From these figures, we can observe that core thickness decreases along the horizontal direction from center plane ($x = 0$) to the edge of the plate. The rate of change of core thickness increases with increase in Herschel-Bulkley number. Hence the core thickness in general for fluids with higher Herschel-Bulkley numbers is more than the fluids with lower Herschel-Bulkley numbers. Further, as power law index increases core thickness decreases and also the rate of change of core thickness increases with increase in power-law index.

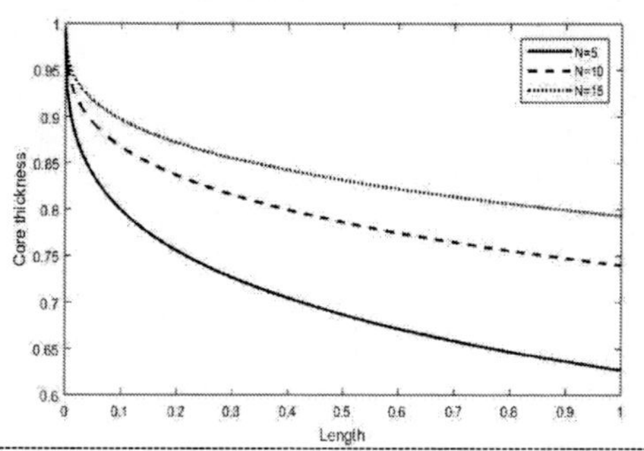

Figure 2. Core thickness formation at n=0.5.

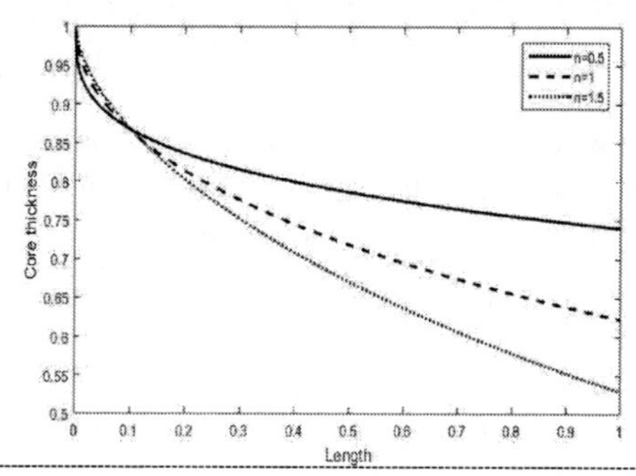

Figure 3. Core thickness formation at N=10.

Figures 4 and 5 show the results of the squeeze force for various values of aspect ratio (ε), power-law index (n) and Herschel-Bulkley number (N). From these figures, we can observe that, squeeze force increases considerably with the increase in Herschel-Bulkley number for a particular value of aspect ratio. Also, squeeze force increases with increasing power-law index. Further, as aspect ratio increases squeeze force decreases for large Herschel-Bulkley numbers and the rate of decrease in squeeze force is marginal with increase in aspect ratio.

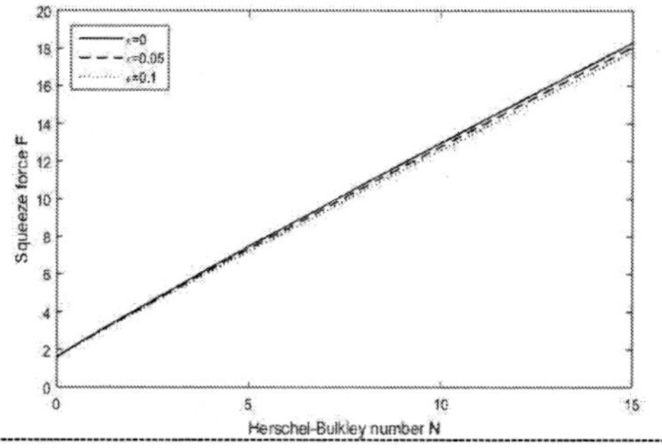

Figure 4. Squeeze force distribution at n=0.5.

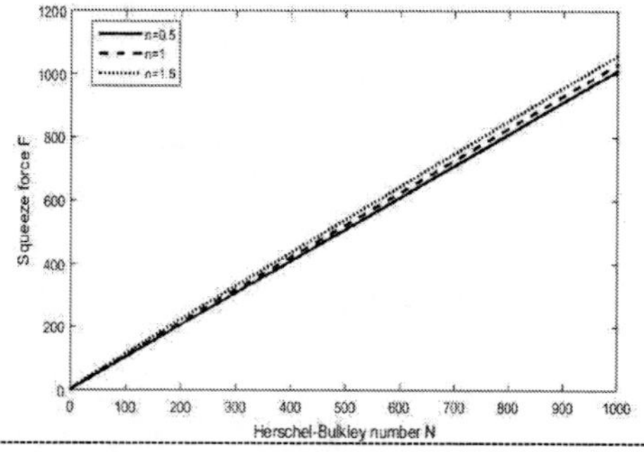

Figure 5. Squeeze force distribution at $\varepsilon = 0.05$.

5. Conclusions

The squeeze flow behaviour between two parallel plates lubricated by a Herschel-Bulkley fluid with constant squeeze motion is theoretically analysed using the technique of a matched asymptotic expansions. We obtain consistent asymptotic solutions which are free from "squeeze flow paradox".

- The shape of the core thickness along the horizontal direction is determined numerically for various values Herschel-Bulkley number and power-law index.
- It is found that core thickness increases with increase in Herschel-Bulkley number where as core thickness decreases with increase in power-law index.
- Expressions for velocity and pressure are calculated analytically for various values of Herschel-Bulkley number, power-law index and aspect ratio.
- Squeeze force results for various values of aspect ratio, Herschel-Bulkley number and power-law index have been calculated numerically.
- The rate change of squeeze force for different values of Herschel-Bulkley number increases with increase in power-law index. Further, squeeze force decreases with increase in aspect ratio.

References

[1] R B Bird, G C Dai and B J Yarusso 1983 The rheology and flow of viscoplastic material *Rev Chem Eng.* **1** 1-70
[2] H A Barnes 1999 The yield stress - a review *J Nonnewton Fluid Mech.* **81** 133-178
[3] H Alehossein 2009 Viscous, cohesive, non-Newtonian, depositing, radial slurry flow *Int J Miner Process.* **93** 11-19
[4] H Alehossein, B Shen, Z Qin and C Huddlestone-Holmes 2012 Flow analysis, transportation, deposition of frictional viscoplastic slurries and pastes in Civil and Mining Engineering *J Mater Civ Eng.* **24** 644-657
[5] H Alehossein and Z Qin 2013 Mechanics of granular-frictional-visco-plastic fluids in civil and mining engineering *AIP Conference Proceeding* **1558** 196-199
[6] G G Lipscomb and M M Denn 1984 Flow of Bingham fluids in complex geometries *J Nonnewton Fluid Mech.* **14** 337-346
[7] E J O'Donovan and R I Tanner 1984 Numerical study of the Bingham squeeze film problem *J Nonnewton Fluid Mech.* **15** 75-83.
[8] D K Gartling and N Phan-Thien 1984 A numerical simulation of a plastic fluid in parallel-plate plastometer *J Nonnewton Fluid Mech.* **14** 347-360
[9] S D R Wilson 1993 Squeezing flow of a Bingham material *J Nonnewton Fluid Mech.* **47** 211-219
[10] M J Adams, I Aydin, B J Briscoe and S K Sinha 1997 A finite element analysis of the squeeze flow of an elasto-viscoplastic paste material *J Nonnewton Fluid Mech.* **71** 41-57
[11] T W Chan and D G Baird 2002 An evaluation of a squeeze flow rheometer for the rheological characterization of a filled polymer with a yield stress *Rheol Acta.* **41** 245-256
[12] G H Meetan 2005 Flow of soft solids squeezed between planar and spherical surfaces *Rheol Acta* **44** 563-572
[13] J Zhang, K E Khayat and A P Noronha 2005 Three dimensional lubrication flow of a Herschel-Bulkley fluid *Int J Numer Methods Fluids* **50** 511-530
[14] J D Sherwood, D Durban 1998 Squeeze flow of a Herschel-Bulkley fluid *J Nonnewton Fluid Mech.* 77 115-121
[15] K P Vishwanath and A Kandasamy 2010 Inertia effects in circular squeeze film bearing using Herschel-Bulkley lubricants *Appl Math Model.* **34** 219-227
[16] I C Walton and S H Bittleston 1991 The axial flow of a Bingham plastic in a narrow eccentric annulus *J Fluid Mech.* **222** 39-60
[17] N Balmforth and R Craster 1999 A consistent thin layer theory for Bingham plastics *J Nonnewton Fluid Mech.* **84** 65-81
[18] I Frigaard and D Ryan 2004 Flow of a visco-plastic fluid in a channel of slowly varying width *J Nonnewton Fluid Mech.* **123** 67-83
[19] A Putz, I A Frigaard and D M Martinez 2009 On the lubrication paradox and the use of regularization methods for lubrication flows *J Nonnewton Fluid Mech.* **163** 62-77
[20] L Muravleva 2015 Squeeze plane flow of viscoplastic Bingham material *J Nonnewton Fluid Mech.* **220** 148-161

IOP Conf. Series: Journal of Physics: Conf. Series **1039** (2018) 012038 doi:10.1088/1742-6596/1039/1/012038

Structure-preserving algorithm for thermal conduction of local thermal nonequilibrium saturated porous medium

Xuemei Liu[1,2]

[1] School of Mechanics, Civil Engineering and Architecture, Northwestern Polytechnical University, Xi'an, Shaanxi, 710072, P.R. China
[2] Department of Engineering Mechanics, Chang'an University, Xi'an, Shaanxi, 710064, P.R. China

Email: liuxuemei@chd.edu.cn

Abstract. Based on the two-energy equation model and constitutive equations of local thermal nonequilibrium saturated porous media, one-dimensional thermal conduction equations of solid skeleton and pore fluid are established. Introducing orthogonal variables, a first order generalized multi-symplectic structure-preserving form of thermal conduction equations is derived as well as the errors of generalized multi-symplectic conservation law and local momentum. The temperature profiles of two phases are obtained and the effect of the heat exchange coefficient between two phases on the process from local thermal nonequilibrium to local thermal equilibrium is revealed numerically. According to the relative error between numerical solution and analytical solution derived from the separating variables method, it can be concluded that this structure-preserving scheme is a valid scheme with high accuracy. Numerical errors of generalized multi-symplectic conservation law and local momentum are presented for two typical types of the thermal conduction in one-dimensional saturated porous media. According to these numerical results, it can be concluded that this structure-preserving algorithm has long-time numerical stability and good conservation properties.

1. Introduction

In recent decades, the problems of heat transfer in porous media have been investigated widely in a variety of fields, such as composite material manufacturing processes[1,2], heat and moisture motion in porous media[3-5], heat and mass transfer process of biological tissues[6,7], and so on [8,9]. Simacek P and Advani S G[1,2] examined the temperature distribution during non-isothermal saturated flow through the porous media. Sajjadi M and Azaiez J[5] examined the effect the heat transfer coefficient on the melting process in the melting porous media. Zhu G M *et al* [7] pointed out that the model for an unsaturated porous medium can be used to describe the flow and heat transfer of sweat in the skin layer. In these fields, model of local thermal equilibrium is usually assumed in a representative element volume (REV). In order to reveal the heat exchange between two phases, the model of local thermal nonequilibrium in porous media has attracted researchers' more attentions[10-12]. Nield D A and Kuznetsov A V[10] investigated the effect of local thermal nonequilibrium among the particle, fluid, and solid-matrix phases of a porous media saturated by a nanofluid analytically. Dehghan M *et al* [11] analytically studied the local thermal non-equilibrium effect in the developed region of forced convection in the saturated porous media at several boundary conditions. However, there are few researches on preserving-structure property in the thermal conduction problem.

As we known, structure-preserving algorithm has applied widely and its high precision and excellent stability are outstanding extremely[13-18]. Aydın A and Karasözen B[13] considered the multi-symplectic six-point scheme for the integration of coupled nonlinear Schrödinger equations. Wang Y Sh *et al* [14] discussed the concept of local structure-preserving algorithms for partial differential equations. Zhang S Y and Deng Z Ch[15] demonstrated the structure-preserving algorithms can preserve canonical character of system real solution. Mcdonald F *et al* [17] investigated the property of preserving travelling wave solutions for multi-symplectic discretizations. In view of the dissipative system, Hu W P *et al* [19-21] investigated the generalized multi-symplectic structure-preserving algorithm and indicated excellent long-time numerical behavior and good conservation property of this method in several various fields.

Based on theory of porous media and two-energy equation model, thermal conduction equations of one-dimensional local thermal nonequilibrium saturated porous medium are established firstly in this paper. Secondly, introducing canonical variables, structure-preserving form for heat conduction equations are constructed. Then, a midpoint box discretization scheme is obtained. Finally, the thermal conduction process associated with heat exchange coefficient between two phases is simulated numerically. Meanwhile, the numerical errors of generalized multi-symplectic conservation law and local momentum are also investigated.

2. Thermal conduction model of local thermal nonequilibrium saturated porous medium

An incompressible elastic saturated porous medium ϕ^s is considered, which is completely filled with inviscid pore fluid ϕ^f and occupies the domain V in the initial state. Based on the theory of porous media, each constituent is regarded as a continuum by using $\mathbf{u}^i=\mathbf{u}^i(\mathbf{X},t)$, $(\mathbf{X},t)\in V\times[0,+\infty)$ to represent its own motion. Energy balance equation is given as follows [12],

$$\rho^i \dot{\varepsilon}_i^i - \boldsymbol{\sigma}^i \cdot \mathbf{D}^i - \rho^i r^i + \operatorname{div} \mathbf{q}^i = \hat{e}^i - \hat{\mathbf{p}}^i \cdot \dot{\mathbf{u}}^i, \quad i = s, f \tag{1}$$

where ρ^i, $\boldsymbol{\sigma}^i$, $\boldsymbol{q}^i$ and ε^i are macroscopic quality density, macroscopic stress tensor, heat flux vector and specific internal energy of the constituent ϕ^i, respectively. $\mathbf{D}^i$ is macroscopic strain rate tensor which satisfies $\mathbf{D}^i = (\operatorname{grad}\dot{\mathbf{u}}^i + \operatorname{grad}^{\mathrm{T}}\dot{\mathbf{u}}^i)/2$. $\hat{\mathbf{p}}^i$ and $\hat{e}^i$ are the interaction force and heat exchange between two phases , which satisfy $\hat{\mathbf{p}}^s + \hat{\mathbf{p}}^f = \mathbf{0}$ and $\hat{e}^s + \hat{e}^f = 0$, respectively. $\dot{(\cdots)}_i$ denotes the material time derivative with the motion of constituent ϕ^i. Assuming small deformation of solid phase and small velocity of pore fluid, neglecting thermal expansion of solid skeleton, additional thermal conduction and thermal convection caused by the relative motion between two constituents, the constitutive equations are expressed as follows (See reference12).

$$\begin{cases} \boldsymbol{\sigma}^s = -n^s p\,\mathbf{I} + \boldsymbol{\sigma}^{se}, \boldsymbol{\sigma}^f = -n^f p\,\mathbf{I} + \boldsymbol{\sigma}^{fe}, \boldsymbol{\sigma}^{se} = 2\mu^s\mathbf{E}^s + \lambda^s(\mathbf{E}^s\cdot\mathbf{I})\mathbf{I}, \quad \boldsymbol{\sigma}^{fe}=0, \\ \hat{\mathbf{p}}^f = p\operatorname{grad} n^f - S_v w^f, \hat{e}^f = \hat{\mathbf{p}}^f\cdot\dot{\mathbf{u}}^s - e_\Theta(\theta^f - \theta^s), \mathbf{q}^s = -k^{ss}\operatorname{grad}\theta^s, \mathbf{q}^f = -k^{ff}\operatorname{grad}\theta^f \end{cases} \tag{2}$$

where $\boldsymbol{\sigma}^{se}$ and $\boldsymbol{\sigma}^{fe}$ are the effective stresses tensors of the solid skeleton and the porous fluid, respectively. p is the effective pore pressure of the fluid and θ^i is the temperature of constituent ϕ^i. e_Θ is the heat exchange constant between two constituents and S_v is the viscous dissipation constant. w^f is the seepage velocity of the pore fluid. μ^s and λ^s are the macroscopic Lame constants of the solid skeleton. k^{ff} is the ordinary thermal conduction coefficient of the constituent ϕ^i.

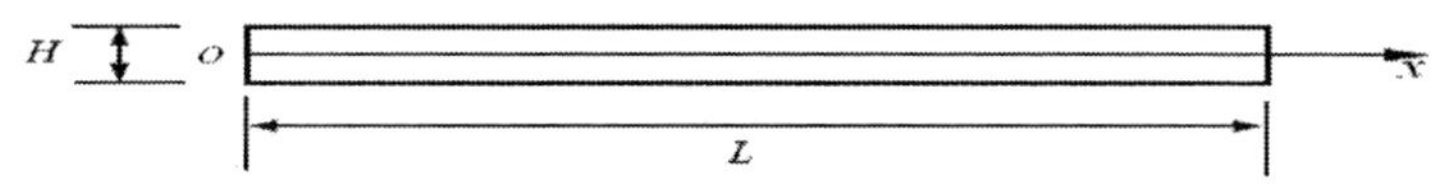

Figure 1. Thermal conduction of local thermal nonequilibrium saturated porous media.

In view of one-dimensional saturated porous media shown as figure 1, substituting the constitutive equations into equation (1), two-energy equations are simplified as follows,

$$\begin{cases} \rho^s c^s \partial_t \theta^s - k^{ss} \partial_{xx} \theta^s - e_\Theta(\theta^f - \theta^s) = 0 \\ \rho^f c^f \partial_t \theta^f - k^{ff} \partial_{xx} \theta^f + e_\Theta(\theta^f - \theta^s) = 0 \end{cases} \tag{3}$$

in which the partial heat sources of two constituents is neglected (see Reference 12). In equation (3), introducing the following dimensionless constants and variables:

$\bar{\rho}^s = \rho^s/(\rho^s+\rho^f), \bar{\rho}^f = \rho^f/(\rho^s+\rho^f), \bar{c}^s = c^s/(c^s+c^f), \bar{c}^f = c^f/(c^s+c^f), \bar{k}^{ss} = k^{ss}/(k^{ss}+k^{ff}), \bar{k}^{ff} = k^{ff}/(k^{ss}+k^{ff}),$

$\bar{e}_\Theta = H^2 e_\Theta/(k^{ss}+k^{ff}), \bar{\theta}^s = \theta^s/\Theta_0, \bar{\theta}^f = \theta^f/\Theta_0, \bar{x} = x/L, \bar{t} = t/T, T = H^2(\rho^s+\rho^f)(c^s+c^f)/(k^{ss}+k^{ff}), \lambda = H/L,$

dimensionless thermal conduction equations can be obtained below,

$$\begin{cases} \rho^s c^s \partial_t \theta^s - k^{ss} \lambda^2 \partial_{xx} \theta^s - e_\Theta(\theta^f - \theta^s) = 0 \\ \rho^f c^f \partial_t \theta^f - k^{ff} \lambda^2 \partial_{xx} \theta^f + e_\Theta(\theta^f - \theta^s) = 0 \end{cases} \tag{4}$$

where a dimensional mark is still used to express a dimensionless quantity.

3. Local conservation laws of the local thermal nonequilibrium saturated porous medium

Introducing canonical momenta: $\partial_x\theta^s=\varphi$, $\partial_x\theta^f=\phi$ and defining $\mathbf{z}=[\theta^s, \theta^f, \varphi, \phi]^T$ as the state variable , equation (4) can be rewritten as the coupled first-order Hamilton partial differential equations,

$$\mathbf{M}\partial_t z + \mathbf{K}\partial_x z = \nabla_z S(z) \tag{5}$$

where $S(z) = \lambda^2(k^{ss}\varphi^2 + k^{ff}\phi^2)/2 + e_\Theta\{\theta^f\theta^s - [(\theta^s)^2 + (\theta^f)^2]/2\}$ is the Hamiltonian function.

$$\mathbf{M} = \begin{pmatrix} \rho^s c^s & 0 & 0 & 0 \\ 0 & \rho^f c^f & 0 & 0 \\ 0 & 0 & 0 & 0 \\ 0 & 0 & 0 & 0 \end{pmatrix}, \mathbf{K} = \begin{pmatrix} 0 & 0 & -k^{ss}\lambda^2 & 0 \\ 0 & 0 & 0 & -k^{ff}\lambda^2 \\ k^{ss}\lambda^2 & 0 & 0 & 0 \\ 0 & k^{ff}\lambda^2 & 0 & 0 \end{pmatrix} \tag{6}$$

Because **K** is skew-symmetric, local energy conservation law is satisfied exactly. **M** isn't skew-symmetric, so the local momentum of this dissipative system isn't preserved exactly. But if $\rho^s c^s$ and $\rho^f c^f$ are small enough with satisfying the inequality

$$|\Delta_j| \le o(\Delta t, \Delta x^2) \tag{7}$$

the system is generalized multi-symplectic, where $o(\Delta t, \Delta x^2)$ is the difference truncation error. Δ_j is expressed as

$$|\Delta_j| = \max_i\{|\rho^s c^s d\theta^s \wedge d(\partial_t \theta^s) + \rho^f c^f d\theta^f \wedge d(\partial_t \theta^f)|\} \tag{8}$$

Δ_j is the error value of the discrete generalized multi-symplectic conservation law. The error of generalized multi-symplectic local momentum can be expressed as follows,

$$\Delta_p = \rho^s c^s[\partial_t(\theta^s \partial_x \theta^s) - \partial_x(\theta^s \partial_t \theta^s)]/2 + \rho^f c^f[\partial_t(\theta^f \partial_x \theta^f) - \partial_x(\theta^f \partial_t \theta^f)]/2 \tag{9}$$

Referenced formula can consult reference 20. A centred box difference scheme is constructed below,

$$\mathbf{M}(z_{i+1/2}^{j+1} - z_{i+1/2}^{j})/\Delta t + \mathbf{K}(z_{i+1}^{j+1/2} - z_i^{j+1/2})/\Delta x = \nabla_z S(z_{i+1/2}^{j+1/2}) \tag{10}$$

where $z_i^{j+1/2} = (z_i^{j+1} + z_i^j)/2, z_{i+1/2}^j = (z_{i+1}^j + z_i^j)/2, z_{i+1/2}^{j+1/2} = (z_{i+1}^{j+1} + z_{i+1}^j + z_i^{j+1} + z_i^j)/4$ and specific formulas are

$$\begin{cases} \rho^s c^s \dfrac{\theta^{s\,j+1}_{i+1/2} - \theta^{s\,j}_{i+1/2}}{\Delta t} - k^{ss}\lambda^2 \dfrac{\varphi_{i+1}^{j+1/2} - \varphi_i^{j+1/2}}{\Delta x} = e_\Theta(\theta^{f\,j+1/2}_{i+1/2} - \theta^{s\,j+1/2}_{i+1/2}), \\ \rho^f c^f \dfrac{\theta^{f\,j+1}_{i+1/2} - \theta^{f\,j}_{i+1/2}}{\Delta t} - k^{ff}\lambda^2 \dfrac{\phi_{i+1}^{j+1/2} - \phi_i^{j+1/2}}{\Delta x} = e_\Theta(\theta^{s\,j+1/2}_{i+1/2} - \theta^{f\,j+1/2}_{i+1/2}), \\ k^{ss}\lambda^2 \dfrac{\theta^{s\,j+1/2}_{i+1} - \theta^{s\,j+1/2}_{i}}{\Delta x} = k^{ss}\lambda^2 \varphi_{i+1/2}^{j+1/2}, \quad k^{ff}\lambda^2 \dfrac{\theta^{f\,j+1/2}_{i+1} - \theta^{f\,j+1/2}_{i}}{\Delta x} = k^{ff}\lambda^2 \phi_{i+1/2}^{j+1/2} \end{cases} \tag{11}$$

Moreover, the discrete errors of generalized multi-symplectic conservation law and local momentum in step *j*th can be obtained, respectively,

$$|\Delta_j| = \max_{i,j}\{\left|\rho^s c^s d\theta^{sj}_i \Lambda\left[d(\theta^{sj+1}_i - \theta^{sj}_i)/\Delta t\right] + \rho^f c^f d\theta^{fj}_i \Lambda\left[d(\theta^{fj+1}_i - \theta^{fj}_i)/\Delta t\right]\right|\} \le o(\Delta t, \Delta x^2) \quad (12)$$

$$|(\Delta_p)_j| = \max_i \left| \frac{1}{2}\rho^s c^s [\frac{1}{\Delta t}(\theta^{sj+1}_i \frac{\theta^{sj+1}_{i+1} - \theta^{sj+1}_i}{\Delta x} - \theta^{sj}_i \frac{\theta^{sj}_{i+1} - \theta^{sj}_i}{\Delta x}) - \frac{1}{\Delta x}(\theta^{sj}_{i+1} \frac{\theta^{sj+1}_{i+1} - \theta^{sj}_{i+1}}{\Delta t} - \theta^{sj}_i \frac{\theta^{sj+1}_i - \theta^{sj}_i}{\Delta t}] + \frac{1}{2}\rho^f c^f [\frac{1}{\Delta t}(\theta^{fj+1}_i \frac{\theta^{fj+1}_{i+1} - \theta^{fj+1}_i}{\Delta x} - \theta^{fj}_i \frac{\theta^{fj}_{i+1} - \theta^{fj}_i}{\Delta x}) - \frac{1}{\Delta x}(\theta^{fj}_{i+1} \frac{\theta^{fj+1}_{i+1} - \theta^{fj}_{i+1}}{\Delta t} - \theta^{fj}_i \frac{\theta^{fj+1}_i - \theta^{fj}_i}{\Delta t})] \right| \quad (13)$$

where $o(\Delta t, \Delta x^2) \approx \Delta t^2 + \Delta x^3$.

4. Numerical experiment

In order to simulate the thermal conduction process of one-dimensional local thermal nonequilibrium saturated porous media, the following two cases are considered.

4.1. Case1

$$\theta^s(x,0) = -0.01x^2 + 0.02x, \quad \theta^f(x,0) = 0, \quad x \in [0,1] \quad (14)$$

$$\theta^s\big|_{x=0} = \theta^f\big|_{x=0} = 0, \quad \partial_x \theta^s\big|_{x=1} = \partial_x \theta^f\big|_{x=1} = 0, \quad x \in [0,1], \quad t \ge 0 \quad (15)$$

Letting c^s=0.3, ρ^s=0.35, k^{ss}=0.4, c^f=0.7, ρ^f=0.65, k^{ff}=0.6 and λ^2=0.0001, the numerical solution of equation can be obtained by scheme (11), where the time-step and space-step are Δt=0.03 and Δx=0.1, respectively. The following analytical solution of equation (4) is obtained by separation of variables.

$$\theta^f(x,t) = \sum_{n=1}^{\infty}\left[C_{n1}\exp(\beta_{n1}t) + C_{n2}\exp(\beta_{n2}t)\right]\sin[(2n-1)\pi x/2] \quad (16)$$

$$\theta^s(x,t) = \sum_{n=1}^{\infty}\{\rho^f c^f \xi / e_\Theta + \zeta[k^{ff}\lambda^2(2n-1)^2\pi^2/4 + e_\Theta]\}\sin[(2n-1)\pi x/2] \quad (17)$$

where $\xi = C_{n1}\beta_{n1}\exp(\beta_{n1}t) + C_{n2}\beta_{n2}\exp(\beta_{n2}t)$, $\zeta = C_{n1}\exp(\beta_{n1}t) + C_{n2}\exp(\beta_{n2}t)$, $C_{n2} = -C_{n1}$, $C_{n1} = 32/(2n-1)^3\pi^3\rho^f c^f(\beta_{n1} - \beta_{n2})$,
$\beta_{n1} = \left(-B_n - \sqrt{B_n^2 - 4\rho^f c^f \rho^s c^s D_n}\right)/2\rho^f c^f \rho^s c^s$, $B_n = \rho^s c^s k^{ff}\lambda^2(2n-1)^2\pi^2/4 + e_\Theta\rho^s c^s + \rho^f c^f k^{ss}\lambda^2(2n-1)^2\pi^2/4 + e_\Theta\rho^f c^f$,
$\beta_{n2} = \left(-B_n + \sqrt{B_n^2 - 4\rho^f c^f \rho^s c^s D_n}\right)/2\rho^f c^f \rho^s c^s$, $D_n = k^{ss}\lambda^2(2n-1)^2\pi^2[k^{ff}\lambda^2(2n-1)^2\pi^2/4 + e_\Theta]/4 + e_\Theta k^{ff}\lambda^2(2n-1)^2\pi^2/4$.

According to table 1 and table 2, the relative errors between solid phase and pore fluid are both small. Therefore, the thermal conduction process is simulated effectively by the scheme (11) and the high precision of this method is also reflected well.

Table 1. Relative error between analytical solution and numerical solution of solid phase temperature with time and location (i.e. x) when e_Θ=1.0.

time	x=0.1 relative error (‰)	x=0.2 relative error (‰)	x=0.4 relative error (‰)	x=0.7 relative error (‰)	x=0.9 relative error (‰)
0.45	4.03	2.90	3.11	3.21	3.20
0.9	1.30	0.39	0.06	0.03	0.03
1.2	1.58	0.48	0.08	5.52×10^{-4}	6.65×10^{-3}

Table 2. Relative error between analytical solution and numerical solution of pore fluid temperature with time and location (i.e. x) when e_Θ=1.0.

time	x=0.1 relative error (‰)	x=0.2 relative error (‰)	x=0.4 relative error (‰)	x=0.7 relative error (‰)	x=0.9 relative error (‰)
0.45	0.06	1.02	0.84	0.74	0.75
0.9	1.18	0.40	0.10	1.97×10^{-3}	0.01
1.2	1.47	0.46	0.08	3.55×10^{-4}	7.54×10^{-3}

Table 3. Temperatures of solid phase and pore fluid at x=0.4 and x=0.9 when e_Θ=1.0.

time	x=0.4 solid phase temperature	x=0.4 pore fluid temperature	x=0.9 solid phase temperature	x=0.9 pore fluid temperature

0.81	0.001199783954719	0.00119942836808	0.001856134233953	0.001855536475685
0.84	0.001199657769257	0.001199432759522	0.001855951090021	0.001855555163308
0.87	0.001199563182989	0.001199429767285	0.001855816879380	0.001855562481376
0.9	0.001199490728570	0.001199421580124	0.001855716953053	0.001855561813531

Table 4. Relative error between local thermal equilibrium temperatures of two phases.

e_Θ	relative error (‰)	
	x=0.4	x=0.9
1.0	0.11	0.14
0.5	0.19	0.24

Table 5. Relative error between local thermal equilibrium temperatures with e_Θ=1.0 and e_Θ=0.5.

x	local thermal equilibrium temperature		relative error of temperatures (‰)
	e_Θ=1.0	e_Θ=0.5	
0.4	0.001199429767285	0.001198801816009	0.52
0.9	0.001855562481376	0.001854945906886	0.33

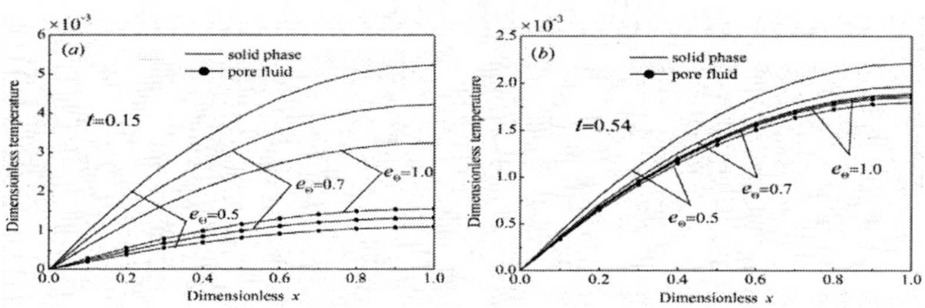

Figure 2. Temperature profiles of solid phase and pore fluid at different dimensionless times with different e_Θ for case 1.

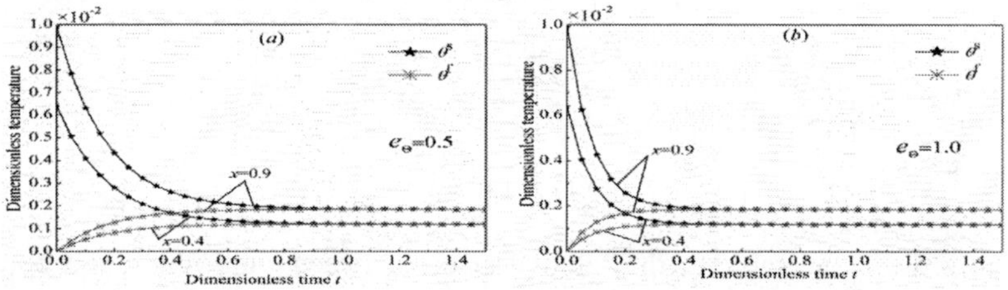

Figure 3. Temperatures of solid phase and pore fluid change with time at location x=0.4 and x=0.9 for different e_Θ in case 1.

From figure 2, the time from local thermal nonequilibrium to local thermal equilibrium decreases with the value of e_Θ increasing. When t=0.54, the system with e_Θ=1.0 has been already close to local thermal equilibrium state. The reason is that larger value of e_Θ leads to more adequate heat transfer between two phases. In figure 3, the effect of heat exchange constant e_Θ on the time from local thermal nonequilibrium to local thermal equilibrium is also indicated exactly. Table 3 gives the specific values of temperatures of solid phase and pore fluid when the time is close to local thermal equilibrium state. Because the temperature of pore fluid is lower than the temperature of solid phase in the initial state, temperature of pore fluid increases until reach the local thermal equilibrium state. Then with conducting heat to the location with lower temperature, the local thermal equilibrium temperature decreases, so the moment with maximum temperature of pore fluid is the time reaching local thermal equilibrium state. When e_Θ=1.0, the moments reaching the local thermal equilibrium state are t=0.84 and t=0.87 at x=0.4 and x=0.9, respectively. But the relative error between these two moments is small (3.45%), so it can be considered that different spatial locations reach the local thermal equilibrium state at the same time (t=0.87). In the same way, above conclusion can also be obtained when e_Θ=0.5. Moreover, the temperatures of local thermal equilibrium are independent on the value of e_Θ at both spatial positions. It is attributed to the independence of local thermal equilibrium on the heat exchange coefficient between two phases.

According to a series of small relative errors in table 4 and table 5, the temperatures of local thermal equilibrium are 0.001199 and 0.001856 at x=0.4 and x=0.9, respectively.

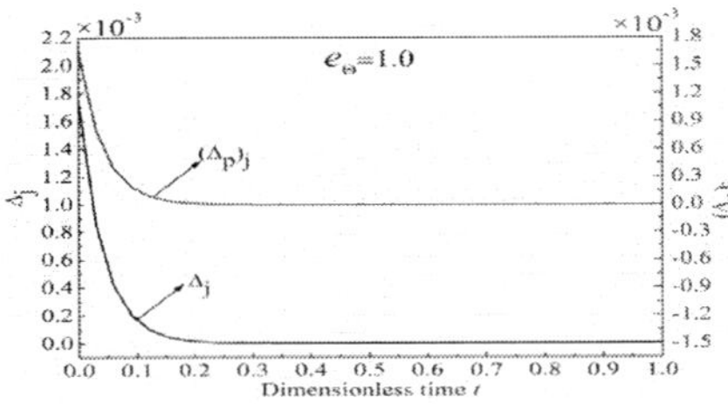

Figure 4. Numerical errors of generalized multi-symplectic conservation law and local momentum with e_Θ=1.0 for case 1.

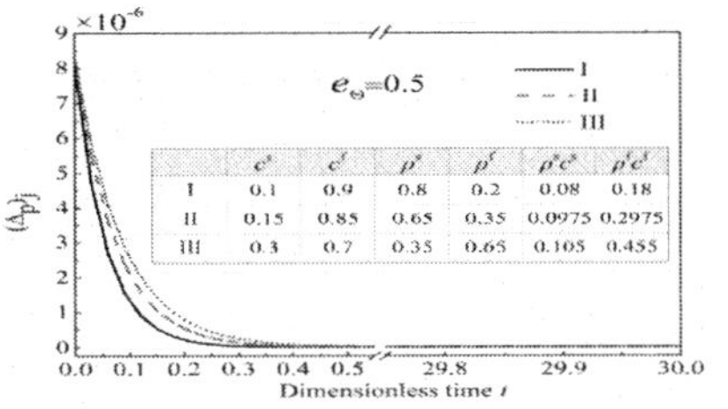

Figure 5. Numerical errors of generalized multi-symplectic local momentum with different parameters for case 1.

In figure 4, order of errors are 10^{-3} at the beginning, and it is already less than 10^{-9} which tends to zero after t=0.5. In figure 5, the order of error obtained by formulation (13) is 10^{-6} and all errors tend to zero after a short time. Moreover, error value and the time both decrease with the values of $\rho^s c^s$ and $\rho^f c^f$ decreasing.

4.2. Case2

$$\theta^s(x,0) = -0.01x^2 + 0.01x, \quad \theta^f(x,0) = 0, \quad x \in [0,1] \tag{18}$$

$$\theta^s\big|_{x=0} = \theta^f\big|_{x=0} = 0, \quad \theta^s\big|_{x=1} = \theta^f\big|_{x=1} = 0, \quad x \in [0,1], \quad t \geq 0 \tag{19}$$

In case2, letting Δt=0.01 and Δx=0.1, the numerical solutions are obtained by scheme (11).

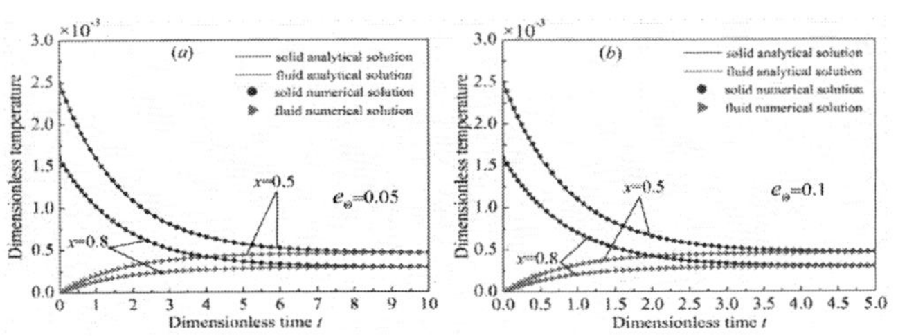

Figure 6. Temperatures of solid phase and pore fluid change with time at location x=0.5 and x=0.8 for different e_Θ in case 2.

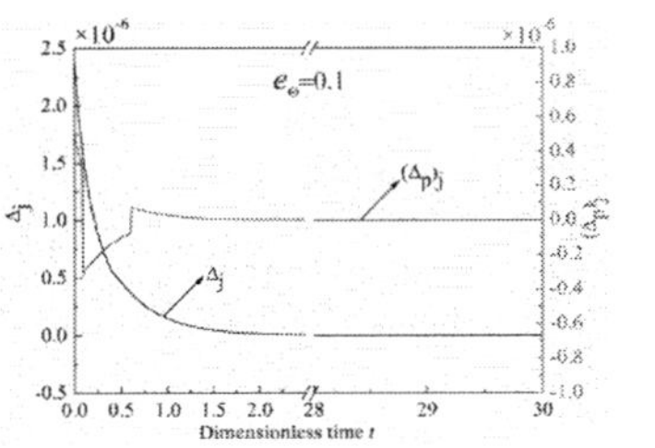

Figure 7. Numerical errors of generalized multi-symplectic conservation law and local momentum with e_Θ=0.1 for case 2.

In figure 6, the numerical solutions coincide with the analytical solution very well and the process of thermal conduction is also indicated clearly. According to the numerical errors in figure 7, the excellent long-time numerical stability of scheme (11) is demonstrated once more.

5. Conclusions

Based on multi-symplectic idea, a generalized multi-symplectic structure-preserving method is proposed to solve the thermal conduction equations in local thermal nonequilibrium saturated porous media. Then, a centred box difference scheme is constructed, which satisfies the condition of the generalized multi-symplectic. In addition, the errors of the generalized multi-symplectic conservation law and local momentum are also obtained. Two typical types of the initial-boundary value problems are surveyed. Comparing the numerical solutions with the analytical solutions, it is concluded that the multi-symplectic structure-preserving scheme constructed in this paper can solve the thermal conduction equations accurately. Moreover, the time from local thermal nonequilibrium to local thermal equilibrium decreases with the value of e_Θ increasing, while the temperature of local thermal equilibrium is independent on the value of e_{Θ}. Finally, according to the numerical errors of the generalized multi-symplectic conservation law and local momentum with two typical cases, the excellent local structure-preserving property and long-time numerical stability of this generalized multi-symplectic structure-preserving algorithm are presented well. These conclusions may provide some useful information and a new numerical method for the temperature analysis of porous medium.

References

[1] Simacek P and Advani S G 2001 Heat Mass Transfer **38** 25
[2] Simacek P and Advani S G 2002 Heat Mass Transfer **38** 497
[3] Hang X D, Sun W and Ye C 2012 Comput. Fluids **57** 25
[4] Liu B C and Liu W 2005 *Int. J. Heat and Mass Transfer* **48** 4579
[5] Sajjadi M and Azaiez J 2015 *Int. J. Heat Mass Transfer* **88** 926
[6] Belmiloudi A 2010 *Nonlinear Anal. Real World Appl.* **11** 1345
[7] Zhu G M, Liu W, Zeng T F and Yang K 2002 *IMEC2002 (2002 International Mechanical Engineering Congress and Exposition)* (New Orleans:USA) p 93
[8] Chen W 2011 *Energy Convers. Manage.* **52** 2217
[9] Hayat T, Mustafa M and Pop I 2010 *Commun. Nonlinear Sci. Numer. Simul.* **15** 1183
[10] Nield D A and Kuznetsov A V 2010 *Transp. Porous Media.* **83** 425
[11] Dehghan M, Jamal-Abad M T and Rashidi S 2014 *Energy Convers.Manage.* **85** 264
[12] Yang X 2005 *Acta Mech. Solida Sin.* **18** 37
[13] Aydın A and Karasözen B 2007 *Comput. Phys. Commun.* **177** 566
[14] Wang Y Sh, Wang B and Qin M Zh 2008 *Sci. China, Ser. A Math.* **51** 2115
[15] Zhang S Y and Deng Z C 2005 *Chinese Journal of Computational Mechanics* **22** 47
[16] Qin Y Y, Deng Z C and Hu W P 2014 *J. Appl. Math. Mech.-Engl. Ed.* **35** 783
[17] McDonald F, McLachlan R I, Moore B E and Quispel G R W 2016 *Journal of difference equations and applications* **22** 913
[18] Hu W P and Deng Z C 2015 *Nonlinear Dyn.* **79** 325
[19] Hu W P, Han S M and Deng Z C 2012 *Int. J. Non Linear Mech.* **47** 1111
[20] Hu W P, Deng Z C, Han S M and Zhang W R 2013 *J. Comput. Phys.* **235** 394
[21] Liu X M, Deng Z C and Hu W P 2015 *J. Nw. Polytech. Univ.* **33** 265

Study on Numerical Simulation of Wind Turbine

S Yan, Y T Guo, J Y Zhao, S P Shi and X Liu

Huaneng Clean Energy Research Institute, Beijing102209, China
E-mail: yanshu@hnceri.com

Abstract. Wake effect may decrease the overall power output of the wind farm. It is necessary to study the wind turbine wake, which would benefit wind turbine micro-sitting selection optimization. In order to improve the accuracy of wind turbine wake simulation, dynamic mesh method based on CFD (computational fluid dynamics) is proposed. Simulation comparison tests between dynamic meshing method and actuator disk method have been done with a 1.5MW wind turbine. The result shows that with the effect of the rotating blades being considered, dynamic meshing method can capture the characteristics such as vortexes, velocity distribution, flow recovery rate, and the wake radius along the axis direction of wind turbine wake more accurately.

1. Introduction

The rotor deserves the most concern in a wind turbine, as it can largely influence the power output and efficiency. As wind passes through the rotating blades, its direction and speed can both be changed, and this phenomenon is called the wake effect of wind turbines. In most cases, the wake effect of one wind turbine may cast negative impacts on another one located down the stream, including fatigue load, service life, structural reliability and power output, which will further reduce the total output of the whole wind farm. As a result, more intensive studies of wake effect are in need, with the intention of investigating the wind flow filed, predicting power output, making better use of wind source, reducing the land scale of wind farms, and optimizing the location of wind turbines.

The study of wind turbine wake has a long history. Early works mainly focused on the theoretical analyze of hydromechanics. As the application of CFD became more and more common, researchers began to use aerodynamic theories and computational simulation in their works. As a result, studies on more complicated cases, with larger and more turbines, were carried out. Risø National Laboratory [1] combined finite volume theory with characteristics of wind farms, in order obtain the aeromechanic performance of a number of turbines when they are arranged tandem or parallel, as well as the velocity filed under different cases. Barthelemie R J [2] introduced a disk, in replacement of the rotor, to simulate the whole wind farm, and got the contour plots of momentum, pressure and velocity. Alexandros Makridis [3] simulated two rows of wind turbines in complex terrains with RSM model and k-wSSt model, finding out the contour of wind velocity when two turbines are placed over mountain slopes. Zhu Chong [4] did numerical simulation on NH1500 models with general actuator method, discussed the feasibility and got the proper distance of two tandem arranged wind turbines, in order to avoid wake effect. Based on Simplec method and SST *k-w* turbulence model, Zhou Yunlong [5] simulated the three-dimensional aeromechanic performance of wind turbines and found out that as the number of blades increases, power output and efficiency also see an increase; turbulence becomes more violent, while the total stability gets worse.

As is list above, recent researches only focus on the numerical modeling of individual or multiple wind turbines in different dimensional scales, with static meshes, intrinsically. However, there's a lack of enough studies on flow characteristics themselves, especially with dynamic meshes. As a result, this paper, using dynamic mesh method, carries out simulation on 1.5MW wind turbines and compares the result with that of actuator disk model.

2. Numerical Simulation Method

2.1. Dynamic mesh model

2.1.1. Creating single *turbine model and dynamic meshes*
This paper carries on numerical simulations of a 1.5MW wind turbine on FLUENT15.0, while the blade design and optimization is based on Matlab [6]. The computing model is a cuboid area, with enough inner space for the wake to develop. The length of the area is calculated along Y-axis to represent the flow direction. The width is measured along Z-axis, representing the horizontal direction. The X-axis is along the vertical direction, and which means XOZ is the rotating plane and ZOY is the horizontal plane. The computing area can be divided into two parts, the inner space of the cuboid and the extended area of the disk [7]. Based on the center of the disk, the extended area has a length of 4D(D is the diameter of the disk) up the stream and 10D down the stream, while the radial diameter is 3D, shown Figure 1. Three blades with the same scale are included in the disk, with a diameter of 70 m. Moreover, the hub height is 65 m, hub radius is 2m, and the angle between main axis of the turbine and horizontal surface is 5°. Leaving the body of the turbine out, the disk now has a cavity inside and is calculated as rotating area with dynamic meshes. The computing model is meshed in GAMBIT, using line mesh method on blades and surface mesh method on all certain faces, as is shown in Figure 2. The disk and the cuboid have 3 interfaces, which are the joints of static and dynamic meshes, so the "interface" function is applied. As the cuboid has a series of bodies inside, and proper mesh quality can both reduce computing time and increase accuracy, this paper divides the cuboid into several parts. For the disk, hybrid-tetrahedral meshes are applied and get 1.79 million meshes in total. While for the other parts, we use hexahedron meshing and get 3.08 million meshes.

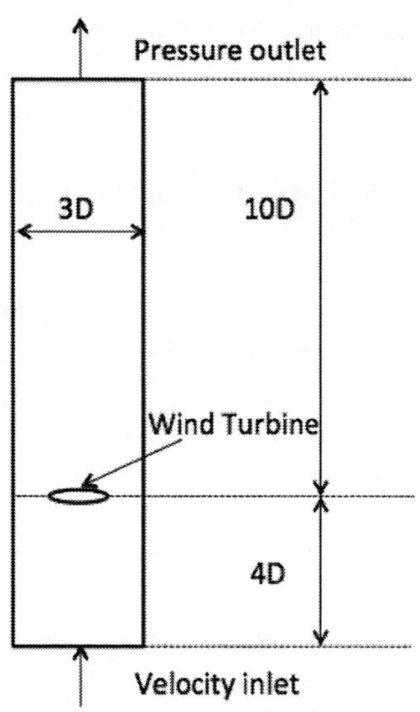

Figure 1. Dynamic mesh model of plane graph

Figure 2. Computational grid domain of Dynamic mesh model

2.1.2. Mesh-preview and computing results of Dynamic mesh model
After the meshing work in GAMBIT, some definitions are in need before the preview [8]. First, define Y-axis as the flow direction, and the inlet is uniform velocity entrance. The outlet is defined as a uniform pressure one. The bottom is treated as solid, smooth wall without slipping. The other surfaces

are defined as symmetry walls. The flow area is defined as well. Second, import the model into FLUENT15.0 to adjust the parameters, apply "mesh interface" for the joints of static and dynamic meshes, write UFD, and assign the value of rotating speed and time. Third, import UFD profile, and assign the mesh condition together with rotating vector. Have a preview of the moving area and dynamic meshes to make sure the dynamic meshes rotate in the right direction. Last, adopt three-dimensional solver with double precision, along with k-epsilon model. For pressure and velocity coupling, use SIMPLE method. Other assumptions include the intensity of air ρ is 1.225kg/m3, the dynamic viscosity μ is $1.7894\times10^{-5}\text{Pa}\cdot\text{s}$, the inlet velocity is 12m/s, and the angular velocity of the disk is 17.32r/min.

2.2. *Actuator disk model (ADM)*

In order to verify the property of using dynamic mesh method and its accuracy, this paper also adopts numerical simulation with actuator disk model.

When air passes though the rotor disk, the velocity within the flow tube is changed, as a result of the resistant effect [9]. According to the law of mass conservation, if we assume the intensity of air keeps constant, the area of the cross-section in flow tube changes contrarily with velocity. That means as velocity goes down, the cross-section becomes larger . As a result, we named the plane of rotor as actuator disk.

From classic Rankine-Froude theory, when the flow passes through the actuator disk, there's a gap between frontier pressure and rear one, and this gap can be calculated by:

$$\Delta \text{p} = 2\rho v_0^2 a(1-a) \tag{1}$$

Δp is the pressure difference when air passes actuator disk, ρ is air intensity, v_0 is wind velocity, a is axial induction factor and defined as $1-\sqrt{1-C_T}/2$. Simplify wind turbines as actuator disks and adopt the function for porous cells:

$$\Delta \text{p}=-\left(\frac{\mu}{\alpha}v_0+\frac{1}{2}C_2\rho v_0^2\right)\Delta m \tag{2}$$

μ is laminar viscosity, α is penetration rate, C_2 is pressure coefficient, v_0 is normal velocity, Δm is the thickness of the medium. Combining Equation (1) and (2), α and C_2 is obtained.

3. Simulation results and analysis

3.1. *Wake velocity distribution: wake flow velocity contours*

In order to come up with a common rule of wake effect, we compare the cross-section of velocity at the distance of D, 2D, 3D, 4D, 5D and 6D(where D is the diameter of the rotor, 70m), for both AD model and dynamic mesh model, as is shown is Figure 3.

IOP Conf. Series: Journal of Physics: Conf. Series **1039** (2018) 012039 doi:10.1088/1742-6596/1039/1/012039

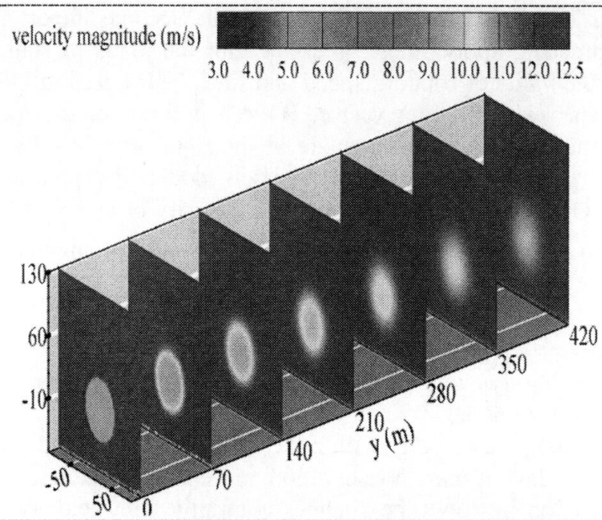

(a). Velocity distribution for ADM

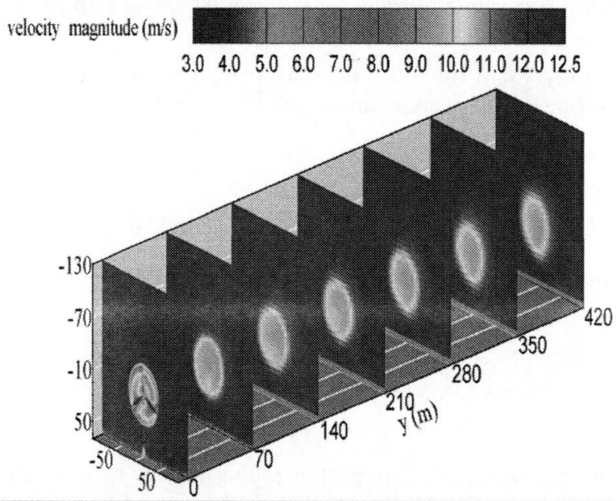

(b). Velocity distribution for DMM

Figure 3. Comparison of wake flow velocity contours between ADM and DMM

We know from Figure 3(a) and Figure 3(b) that the distributions of velocity in the wake flow for both models are quite alike. As the distance after the turbine grows the gradient of velocity becomes smaller, and the influenced area gets larger. This regulation is in good agreement with wake theory. The result of numerical modeling is also used to investigate the average velocity of each section and the thrust factor [10]. In Figure 3(a), the area with a velocity less than 11m/s shows the shape of a circle, more or less. As a result, the influenced area can be measure by radius and the center of the circle. In Figure 3(b), the influenced area in each cross-section resembles the outline of the turbine, and develops spirally along the wake stream. Precisely, the velocity near the turbine sees a little bit increase for the tip part, while shows a decrease for the inner part. Although the influenced area becomes larger as wake developed, this regulation still works. All these show that the distribution of velocity is quite different from the uniform one at entrance.

3.2. Wake velocity distribution: axial velocity contours

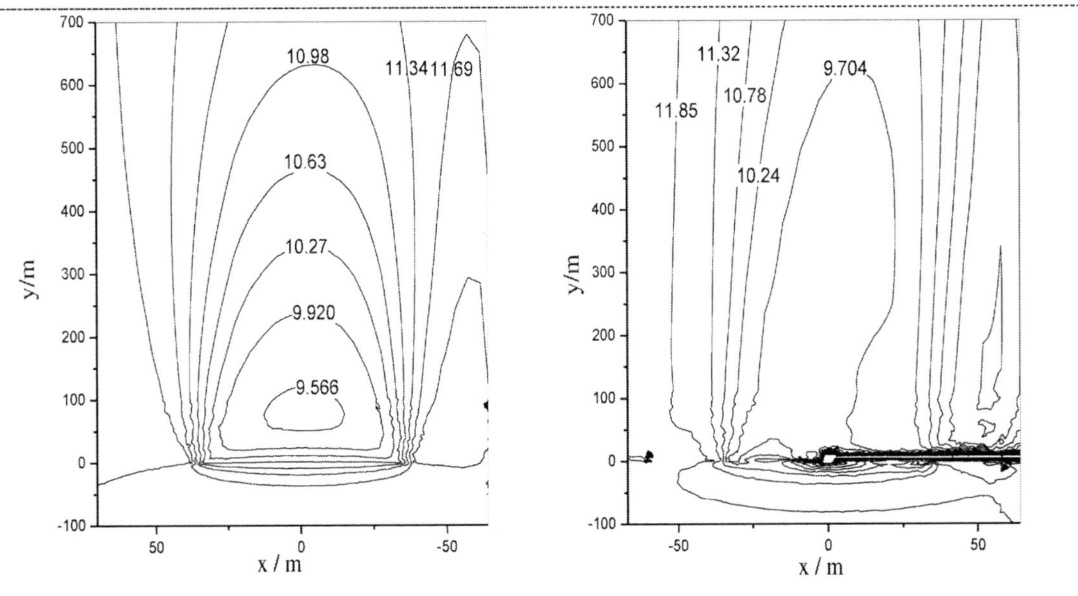

(a). Velocity distribution for ADM **(b).** Velocity distribution for DMM

Figure 4. Comparison of the axial velocity contours between ADM and DMM

Figure 4(a) shows that the velocity sees a significant decrease after the actuator disk. In the nearer down stream, the outlines of vortexes in the wake is divided into two asymmetrical parts because of gravity. The velocity gradient near the edge of the disk is far larger than the other parts, indicating that the flow in the stream tube is mainly powered by the pressure difference before and after the actuator disk. In the nearer down stream, the central velocity is larger than the surroundings. This is because the normal and tangential forces are both relatively small in the center, so the decrease of velocity in this area is not so significant due to the law of momentum [11]. As the wake develops down the stream, it keeps having momentum exchange with flow around, as a result of turbulence effect. In the end, the velocity recovers to the initial value, and shows symmetrical distribution around the rotor center. Figure 4(b) indicates that the axial velocity in the nearer down stream has dropped a lot, showing a flame-shaped area with low velocity ($u \leq 9.7\mathrm{m/s}$). The wake keeps expending to the outside. Circle fluid appears around the tip, forms closed regions with varied gradient, and expends towards outside. Around the center of hub, the wake appears an ellipse area, with a major axis of 70m. Near the bottom of tower, circle flow happens and keeps breaking away from the surfaces. The phenomenon indicates that the inlet flow forms two main vortex areas with different velocity after the turbine. From eddy theory, as the distance increases down the stream, the vortex center deviates from the rotor center, while the kinetic energy of vortex keeps changing. In the further stream area, as vortex expends and develops, the velocity recovers gradually, until it equals to the value in free flow. However, because of the influence of hub and ground, when applying dynamic mesh model, a larger and more complicated wake area is generated.

3.3. Velocity Vectors

In order to analysis the distribution of vortex and velocity vector after wind turbine, we generate three cross-sections at the distance of 2D, 6D and 10D (D represents the diameter of rotor), with both AD model and DM model, shown in Figure 5.

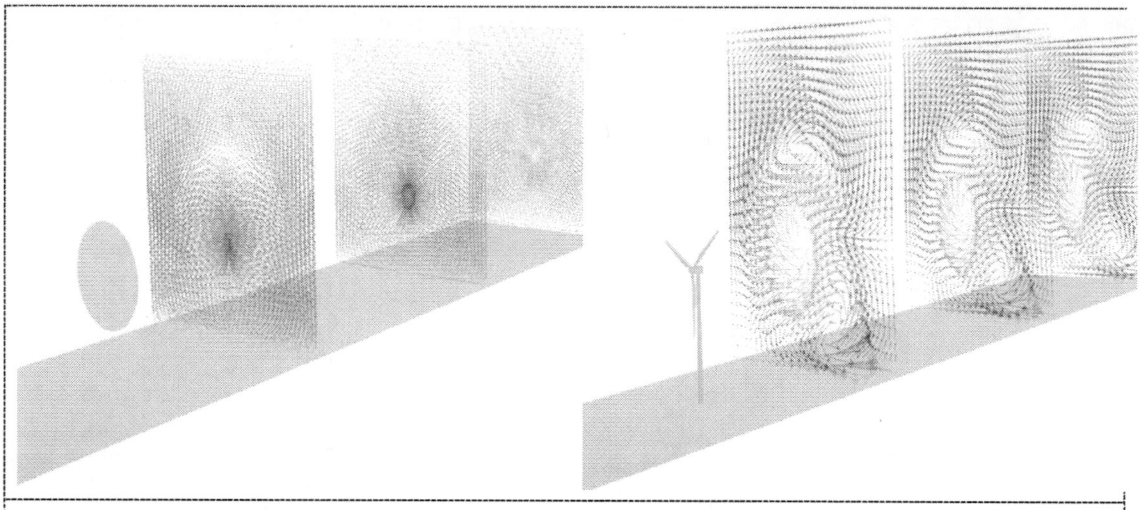

(a). Velocity distribution for ADM **(b).** Velocity distribution for DMM

Figure 5. Comparison of the velocity vectors between ADM and DMM

From Figure 5(a), we know that the inlet flow forms a round wake area after it passed the actuator disk. The vectors all have the direction towards the center. This is because the outside pressure is higher than inside, so the outside air keeps flowing into the wake area to recover the velocity and kinetic energy. Figure 5(b) shows that, the wake of a single turbine develops in a helix way, and the rotating center is no longer coincident with the symmetrical center. In cross-sections at all distances (in XOZ plane), two vortexes with different velocity always appear, the one with higher velocity happens near the tip and the one with lower velocity happens near the interface of hub center and ground (whose rotating direction is in agreement with that of rotor). As the distance grows, the centers of the two vortexes move downwards; the two centers get closer; the shapes of vortexes also become variable.

When air flows through the turbine, its velocity drops dramatically and recovers gradually down the stream, the recovery rate is defined as

$$\eta=\frac{v_n}{v_0} \tag{3}$$

v_0 is the inlet velocity; v_n is the average wind speed of the cross-section at the distance of nD. In order to come up with a common rule of developing wake, the velocities of both models are analyzed and compared, shown in Figure 6.

Figure 6 shows that, as wake goes down stream, the velocities of the two models both shows some recovery, but at different rates. The computing length (10D) is not long enough for velocity to recover entirely. For distance $y \leq 4D$, the recovery rates of DM model and AD model are almost the same, and the velocity difference changes quite slow with distance. For distance $y > 4D$, the recovery rate of AD model are larger than that of DM model, the difference of velocity changes dramatically, with a peak value of 16.77%. That is because when wake flows a certain distance down the stream, the pressure gap is no longer the main factor, turbulence intensity and the variance of vortexes become more important instead. In the true practice, the vortexes will dissipate gradually, leading to a decrease in total energy and velocity.

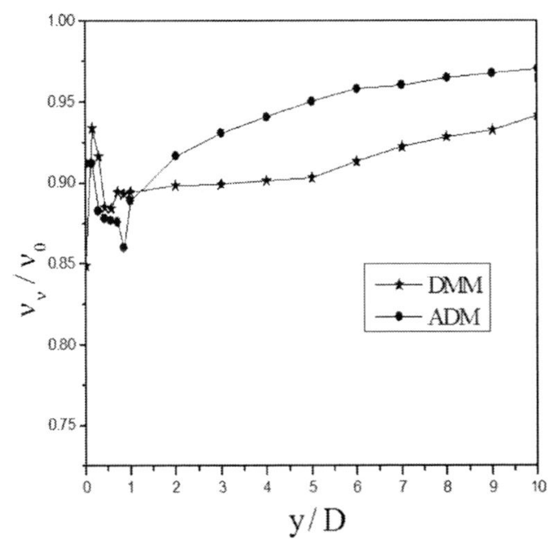

Figure 6. Comparison of the wake flow recovery rate between two different models

The above shows that the recovery rate of DM model is smaller than that of AD model and there exists a big disparity. There are mainly two possible reasons. First, in an AD model, it is mass conservation law and momentum conservation law that determine the force on the disk, which will change the local velocity on and near the disk. At the same time, it consumes energy for wake to rotate at certain angular velocity, leading to a static pressure loss. As the wake develops, this loss becomes smaller and that's why the recovery rate gets larger. While for the true rotor, the force is determined by axial and tangential energy functions. Second, unlike AD model, DM model take the rotating blades into consideration, and this rotation will lead to large scales of circle flow, turbulence and vortex. The dissipation of vortex will cost a lot of energy, resulting in lower velocity. Based on what is stated above, DM model resembles the true practice most, as it considers both rotating effect and precise geometric blade information. In future works, we will also adopt the information of real wind farms to confirm and improve the conclusion.

4. Conclusion

In this paper, we use Fluent15.0 to simulate 1.5MW wind turbines with dynamic meshes. The velocity of wake is plotted and made comparison with that of AD models. Combining with true practice, it is confirmed that DM method has high validity and accuracy. Some conclusions are made as following:

- Build dynamic mesh model, simulate the flow field, get the velocity distribution within the wake of a single turbine. The result can be helpful in investigating the velocity filed in the design process of wind farm.
- Simplify the actuator disk model with theory study and CFD. Use the model to solve the wake flow field, get the distribution of axial velocity and fluid at different distance down the stream.
- With the same wind speed, analysis and compare the wake flow distributions at multiple cross-sections for both two methods. For DM model, two vortexes appear in the down stream and dissipate gradually as distance increases. As for AD model, there's no such phenomenon. That indicates that dynamic mesh modeling is more precise and convincing, and can capture fluid details in larger scale.
- Comparing the contour plot of velocity at different cross-sections of the two models, we find that the smallest value appears at the distance of 1D after the hub, rather than right after it. As the distance increases, velocity recovers gradually in a linear way for both models.
- Using the recovery rate of velocity, power output, and wind utilizing factor as indicators, there's a deviation between DM model and AD model. Moreover, as DM model takes the rotating of blades into consideration, its result is quite close to the true practice. In conclusion, DM model is a new method to simulate wind farms, and in terms of reliability and accuracy,

DM method performs better than any other methods.Footnotes should be avoided whenever possible. If required they should be used only for brief notes that do not fit conveniently into the text.

5. References

[1] Sten F, Rebecca B and Sara P 2004 *The necessary distance between large wind farms offshore study* (Roskilde：Risø National Laboratory) p 2-11.

[2] Barthelemie R J, Rathmann O and Frandsen S T 2007 Modelling and measurements of wakes in large wind farms *J. Phys. Conf. Ser.* **75** 12-49.

[3] Alexandros M and John C 2009 *CFD modeling of the wake interactions of two wind turbines on a Gaussian hill* (Florence: EACW) p 9-13.

[4] Zhu C, Wang T G and Zhong W 2013 Investigation of wake interaction between wind turbines in tandem *J. Mech. Eng.* ***35*** *17-22.*

[5] Zhou Y L, Yang C Z and Li L 2012 3D Flow field numerical simulation on aerodynamic characteristics of new double-rotor wind turbines *J. Chin. Soc. Pow. Eng.* **32** 698-705.

[6] Li R N, Guo X W and Yang R 2012 The design of 1.5 MW wind turbine blade and its optimization based on the Matlab *J. Gans. Sci.* **24** 73-76.

[7] Li S H, Wang D H and Yue W P 2011 Numerical simulation on wake interaction and array of double wind turbine at varying wind directions *J. Chin. Soc. Pow. Eng.* **31** 768-78.

[8] Li S H, Yue W P and Kuang Q F 2011 Numerical of simulation of wake interaction and array of double wind turbine *J. Proc. CSEE* **31** 101-7.

[9] Tian L L, Zhao N and Zhong W 2012 Numerical simulation of wake interactions of wind turbine *J. Acta Ener. Solar. Sin.* **33** 1315-20.

[10] Qian Y R and Wang T G 2016 Large-Eddy Simulation of Wind Turbine Wake and Aerodynamic Performance with Actuator Line Method *J. Trans. Nanj. Univ. Aero. Astro.* **33** 26-36.

[11] Yuan W, Tian W, Ozbay A and Hu H 2014 An experimental study on the effects of relative rotation direction on the wake interferences among tandem wind turbines *J. Sci. Chin.* **57** 935-49.

Acknowledgement

This work was supported by China Huaneng Group Science and Technology Fund (HNKJ16-H25)

IOP Conf. Series: Journal of Physics: Conf. Series **1039** (2018) 012040 doi:10.1088/1742-6596/1039/1/012040

On the free surface motion with a vertical vortex sheet under water

Jing Li

Shi-jun Liao

State Key Laboratory of Ocean Engineering, School of Naval Architecture, Ocean and Civil Engineering, Shanghai Jiao Tong University, Shanghai 200240, China

E-mail: lijing_@sjtu.edu.cn

Abstract. The vortex sheet always represents a thin shear layer across which there is a discontinuity in fluid velocity. We show a vertical one in a 2D water flow and simulate the interaction between its evolution and the free surface. It is quite stimulating that this phenomenon may result from a section of the seafloor lifting up along the fault. Since it also reflects the tsunami generation, the significance for this study is highlighted. We numerically solve the Navier-Stokes equations to tackle this issue. By analyzing the data from computation we attempt to discover the nature of this system and to derive some rule controlling the hydrodynamic process for our problem. The identity of this flow structure and the way the free surface moves are explored as well.

1. Introduction

Helmholtz [1] for the first time noticed that an interface of discontinuity is highly unstable. The most remarkable work is the built of Birkhoff-Rott equation [2, 3] which is an integro-differential equation and indicates the temporally evolutionary profile. Moore [4] contributed much theoretical work in this area . Some key results have been verified by Krasny [5], Shelley [6] via numerical approaches and by Calflish and Semmes [7] via analytical means. However, the equivalence between this mathematical model and the 2D Euler system is still a weak form. Hence more researchers choose Direct Numerical Simulation (DNS) to deal with this problem. Hoepffner et. al. [8] reported their numerical experiment that local perturbation accounts for a self-similar growth. Lecoanet et. al. [9] computed vortex sheet in the scenario of continuous density distribution as a test for the spectral method code. However, only a few publications discussed the effect of vortex sheet upon the free boundary. Evans [10] introduced the propagation of plane surface wave across a vortex sheet constructed by two opposite flows. McKee and Tesoriero [11] extended the study by moving the background into finite depth water. These studies do not analyze the direct interaction of vortex sheet with the free surface. Tryggvason [12] performed the calculation numerically for a free surface with a horizontal submerged vortex sheet and concluded free surface steeply deforms due to the roll-up of the vortex sheet. Tsai and Yue [13] considered the wake of a vertical surface-piercing plate as the shed vortex sheet which interacts with the free surface.

Although the previous work provides sufficient data for supporting the nature of interactions between free surfaces and vortex sheets, few of them ever talked about the scenario that consists of the top free surface and a vertical vortex sheet excited by the discontinuous tangential components of the flow velocity. The plate motions in the ocean may exert significant influence on the flow. We assume that some cases akin to a sharp upwards movement would for somehow result in a vertical flux right over the active sea bottom and finally forms a steep shear layer across the fault. The motivation for the present manuscript is to discover and predict the deformation of presumed quiet ocean surface with a vertical vortex sheet under the water. In addition, what kind of features this sheet may possess is another angle of sight for observation. When KHI occurs the vortex sheet turns to roll up into a spiral, which may benefit water mixing in ocean meteorology, yet the downside may take a powerful impact on submerged structures.

2. Mathematical Model

The actual nature is rather complex. We simplify our model into a 2D plane and in a dimensionless context. We carefully choose d, ρ and V as characteristic length, density and velocity scales to nondimensionalize Navier-Stokes equations:

$$\nabla \cdot \boldsymbol{u} = 0 \tag{1}$$

$$\boldsymbol{u}_t + \boldsymbol{u} \cdot \nabla \boldsymbol{u} = -\frac{1}{\rho}\nabla p - g + \frac{1}{Re}\nabla^2 \boldsymbol{u} + \frac{1}{We}\kappa \delta_s \boldsymbol{n} \tag{2}$$

where Re and We are the Reynolds number and the Weber number. Accordingly the initial vertical velocity filed turns into

$$v(x, y, 0) = \begin{cases} \sin\left(\frac{\pi}{2}y\right) e^{-x}, & x \geq 0, \quad 0 < y < 1 \\ -\sin\left(\frac{\pi}{2}y\right) e^{x}, & x < 0, \quad 0 < y < 1 \end{cases} \tag{3}$$

3. Numerics

The Gerris Flow Solver [14, 15] is employed in our research for its accurate interface capturing and reconstructing capability of a modified VOF approach. A good thing is that it is an open source code based on the quad-tree (octree for 3D) grids. Here we directly solve the incompressible Navier-Stokes equations (Eq. 1-2). The computational domain consists of 15 2D square boxes and the length of one side each is 2 thus the domain is a rectangle with the length of 30 and the height of 2 in a nondimensionalized frame. As mentioned before, the dimensionless water depth is 1 which in the real circumstance is d. Given the disturbance which water flow may lead to on the free surface, the domain height is high enough to act as the far field. Therefore, the upper boundary is set to be symmetric and the bottom as well. To avoid wave reflecting and absorbing, periodic boundary conditions are imposed on the left and the right sides.

4. Results and discussions

First, we would like to examine the vortex sheet. We deliberately put a scalar so called tracer to mark the two parts on both sides, which is shown in Fig.1. It is hard to measure the evolution of this structure quantitatively, thus We define L_{vs} as the absolute distance between the horizontal positions of the upper tip and the bottom end of this thin layer (Fig.1b) to uncover the vortex sheet motion.

Fig.2 shows that during the initial stage this vortex sheet experiences a gradually bending process. Since these curves fall into a narrow range and have almost nothing to do with V and d, they share a similar developing trend. Let us move on to Fig.2a for some details. For this series, V is the major factor affecting the rise. If we look at the curves, the vortex sheet is expanding

IOP Conf. Series: Journal of Physics: Conf. Series **1039** (2018) 012040 doi:10.1088/1742-6596/1039/1/012040

(a) (b)

Figure 1: Map of tracer for case of $V = 0.3, d = 1$. (a) $t = 0$; (b) $t = 1$

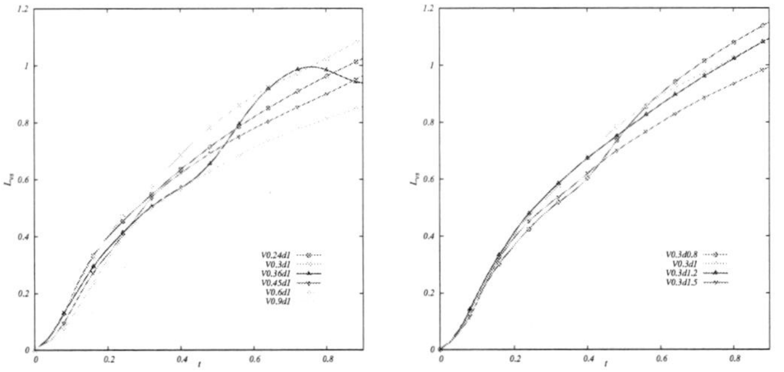

(a) (b)

Figure 2: History of L_{vs} during the initial stage. (a) V varies; (b) d varies.

slightly faster when V is bigger, yet after the curve of $V0.36d1$ shows up this tendency is shaken and later on reversed. The tracer map gives us some hints to find the reason. In Fig.3 we may observe the vortex sheet at $t = 0.8$ in different cases. When V is lower than 0.36, it is easy to have the concept that greater the kinetic energy is further the distance expands. However, so far we ignore the KHI. When this singularity turns to be strong enough and the perturbation focuses on the lower part, a dominated spiral will be yielded locally. This structure seriously hinders the expansion, thus as V increases L_{vs} stretches slower. When V approximates 0.6 the free surface breaks. This dissipative procedure may also impede L_{vs} developing. Back to Fig.2b, we may find some twists of the curves but still scattered in a narrow range. A better understanding of Fig.3g, Fig.3b, Fig.3h and Fig.3i may help us know the reason. The change of d in fact shrinks or expands the vertical room, which leads to the front of this layer being confined or fully developed. The first spirals have a quite organized law behind if we zoom in and rearrange them in Fig.4, which is the evidence.

So far we find that vortex sheets move in a similar trend at the beginning. Therefore it would be an important feature for us to judge how much extent this thin layer would have an impact on the free surface.

The most direct way to analyze free surface motion is to follow the water wave evolution. Fig.5a shows directly how the free surface develops. The flat water around the center part indicates that the wave generation procedure has a similar mechanism with the drop waves, which means this wave propagation only depends on the initial disturbance. The two purple lines outline the tail positions of the wave yielded, which corresponds with the wave energy transport. This phenomenon may disappoint us somehow since the underneath singularity seems to have no obvious impact on the free surface evolution. If the continuous influence does not exist, there should be some connection between the vortex sheet and the free surface at the beginning. We examine the maximum value of the free surface elevation during this stage. A strong connection is found in Fig.5b. This linear relation technically reveals, at the beginning,

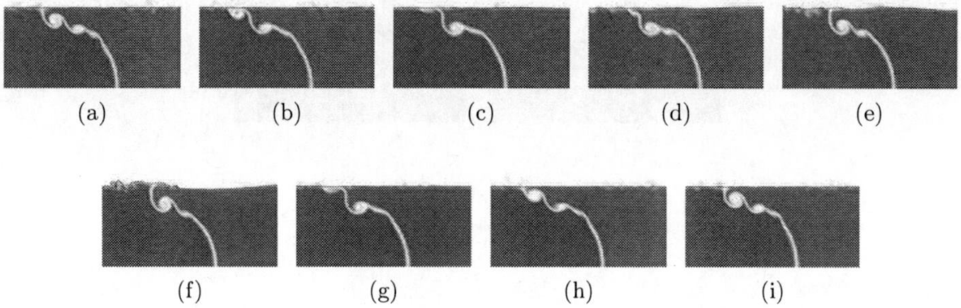

(a) (b) (c) (d) (e) (f) (g) (h) (i)

Figure 3: Map of tracer for different cases at $t = 0.8$. (a) $V = 0.24$ $d = 1$; (b) $V = 0.3$ $d = 1$; (c) $V = 0.36$ $d = 1$; (d) $V = 0.45$ $d = 1$; (e) $V = 0.6$ $d = 1$; (f) $V = 0.9$ $d = 1$; (g) $V = 0.3$ $d = 0.8$; (h) $V = 0.3$ $d = 1.2$; (i) $V = 0.3d = 1.5$.

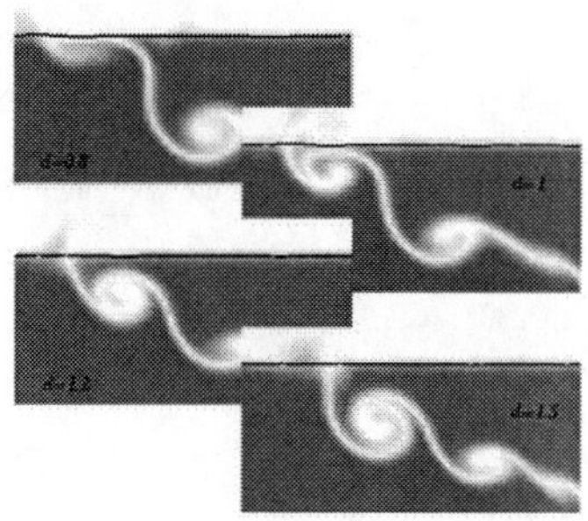

Figure 4: Spirals along the vortex sheet for cases of $V = 0.3$ and different d.

that the wave rising depends on the initial velocity field under water, and V represents the intensity of the vortex sheet. The potential needed by wave rising procedure is provided by the fluid kinetic power. Fig.6a shows that. All kinetic curves reach the lowest point at the same time and the water wave is on its first crest at that moment as well. This fact tells us at least two things. First, when d is fixed, whatever V is, the real time consuming is equal for kinetic power converting into potential energy. Second, the well agreement of curves implies that the energy magnitude fits the dimensionless scaling rule, which is also verified by Fig.6b. Compared with Fig.2, we find their interaction is quite rapid or instantaneous, since the energy transport follows a dimensional time scale yet as we know L_{vs} goes the other way. If the change of the two items cannot be in the same pace, we hardly see any continuous interaction for both of them although they do have a strong connection illustrated in Fig.5b.

5. Conclusion

We directly solve Navier-Stokes equations by numerical method to study the free surface motion with a vortex sheet under water. By analysis, we find that the vortex sheet bending process does not rely on the choice of characteristic scalars: V and d, thus it has a similar trend under dimensionless frame. The expansion procedure of this thin layer has nothing to do with the water wave propagation, but the vortex sheet does have a strong connection with the free surface elevation at the initial stage, and this relation is linear between V and H_{max}. In addition, during the initial energy transport, time consuming is equal while kinetic energy is converting into the potential one. We argue that these results may help people understand the interaction between the vertical vortex sheet and the free surface.

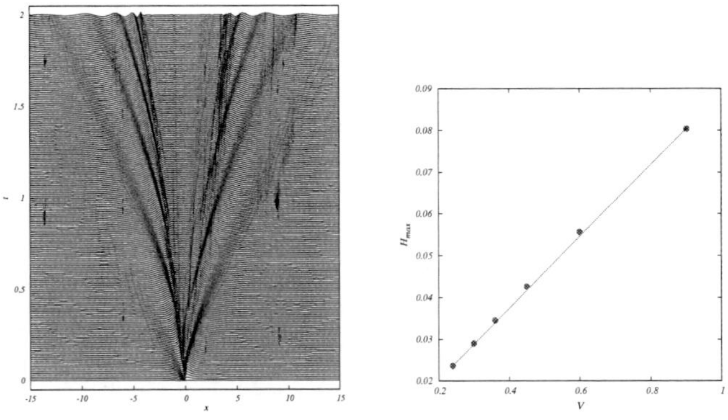

(a) (b)

Figure 5: (a) Wave profiles with time elapse for case of $V = 0.3$ and $d = 1$. (b) The maximum free surface elevation H_{max} at the beginning versus V when d is fixed.

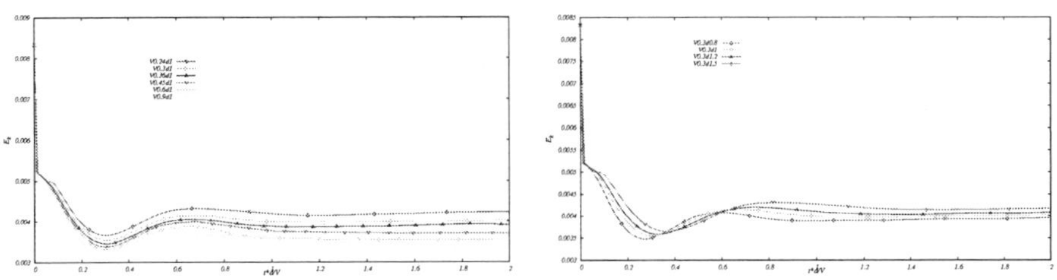

(a) (b)

Figure 6: Kinetic energy for different cases versus dimensional time. (a) V varies; (b) d varies.

Acknowledgment
This work is partly supported by National Natural Science Foundation of China (Approval No. 11272209 and 11432009).

References
[1] Helmholtz H 1868 *Über discontinuierliche Flüssigkeits-Bewegungen* (Akademie der Wissenschaften zu Berlin)
[2] Birkhoff G 1962 *Proceedings of a Symposium in Applied Mathematics of the American Mathematical Society* **13** 55–76
[3] Rott N 1956 *Journal of Fluid Mechanics* **1** 111–128
[4] Moore D W 1979 *Proceedings of the Royal Society of London. Series A, Mathematical and Physical Sciences* **365** 105–119
[5] Krasny R 1986 *Journal of Fluid Mechanics* **167** 65–93
[6] Shelley M J 1992 *Journal of Fluid Mechanics* **244** 493–526
[7] Caflisch R E and Semmes S 1990 *Physica D* **41** 197–207
[8] Hoepffner J, Blumenthal R and Zaleski S 2011 *Physical Review Letters* **106** 104502
[9] Lecoanet D, McCourt M, Quataert E, Burns K J, Vasil G M, Oishi J S, Brown B P, Stone J M and O'Leary R M 2016 *Monthly Notices of the Royal Astronomical Society* **455** 4274–4288
[10] Evans D V 1975 *Journal of Fluid Mechanics* **68** 389–401
[11] McKee W D and Tesoriero F 1987 *Journal of the Australian Mathematical Society - Series B* **29** 127–141
[12] Tryggvason G 1988 *Physics of Fluids* **31** 955–957
[13] Tsai W t and Yue D K P 1993 *Journal of Fluid Mechanics* **257** 691–721
[14] Popinet S 2003 *Journal of Computational Physics* **190** 572–600
[15] Popinet S 2009 *Journal of Computational Physics* **228** 5838–5866

Research on electromagnetic propulsion system based on reusable space vehicle

Fu Hui[1,2], Wang Pei[1,2], Zhang Ke[1,2], Lv Mei Bo[1,2],Ge Zhi Lei[1,2]

[1] College of Astronautics, Northwestern Polytechnical University, Xi'an 710072China
[2] National Key Laboratory of Aerospace Flight Dynamics, Northwestern Polytechnical University, Xi'an 710072 China

Email: jimmy5579@163.com

Abstract. It is inefficient and uneconomical to complete space launch mission with the assistance of unusable rockets which is the typically used approach today, and finding a more economical way has been the pursuit of scholars all over the world. According to the purpose above, this paper proposed a new design to lift a payload to the desired orbit. In this design, the electromagnetic propulsion system (EMPS) is applied to replace the first rocket, leaving the usable rocket to take care of the rest job. Based on this design, the structure of the propulsion system is constructed and the electromagnetic properties are analyzed by finite element analysis tool*Maxwell*, then the mathematic model of EMPS is simulated and calculated. The result shows that EMPS is capable of accelerating the vehicle, which weighs hundreds of tons, up to one hundred meters per second. Therefore, it is foreseeable that EMPS has a certain potential in the future space launch mission.

1. Introduction

Although technology is developing swiftly after entering the 21st century, the mature aerospace launch technology is still traditional one-time launch rocket. But the traditional rocket that relies on chemical energy has a critical weakness, that is, the ratio of the payload to the total take-off weight is quite small[1]. In the case of a Low-Earth Orbit (LEO) mission, the ratio mentioned above is usually only a few percent, while the implementation of the geostationary orbit (GTO) task corresponds to a lower ratio, or even less than one percent. Therefore, the completion of one launch mission requires a huge amount of money, transporting a kilogram of load into the LEO need to spend over four thousandAmerican dollar[2]. And the traditional rocket releases harmful exhausts in the combustion process because of relying on the chemical energy, it contains pollution risk to the atmosphere. However, humans have never stopped thinking, scholars have been committed to finding a low-cost space launch technology, propose a variety of launching ideas in order to reduce the expenses, such as Skyhook[3], space elevator[4], reusable launch vehicle and electromagnetic propulsion system. These ideas have been carried out on different sizes of prototype construction, except space elevator. Among them, the electromagnetic propulsion system is one of the most practical concepts[5].

Electromagnetic launch technology developed rapidly and comprehensively for the past 20 years. Researchers not only concerned about the exit velocity of the emitter but also pay great attention to the weight of the emitter. Balikci A. and Zabar Z.[6]verified the possibility to launch missiles with large weight using electromagnetic force. Li Weibo and Cao Yanjie[7]also proposed a similar simulation model which launches missiles vertically using electromagnetic launch system. The United States Lake Hacker Naval Air Force Engineering Station successfully launched the T-45 Goshawk Trainer, C-2 Greyhound Transport and F/A-18 Hornet Fighter and other spacecraft, which makes the electromagnetic launch of large weight spacecraft into a new stage. From another point of view, it is

IOP Conf. Series: Journal of Physics: Conf. Series **1039** (2018) 012041 doi:10.1088/1742-6596/1039/1/012041

also feasible to launch large scale spacecraft with the application of electromagnetic propulsion technology.

2. Overall concept of EMPS

In this paper, the mathematical model of the coil electromagnetic propulsion system is established by using the equivalent current loop method. The usable launch vehicle is economically beneficial in that it can be recycled and will get ready for the next mission with necessary maintenance. Besides that, massive researches have been carried out by scholars[8] about reusable rockets and the conclusion that reusable vehicle is technically advisable can be drawn[9]. This paper takes the reusable vehicle as an object to be accelerated by EMPS based on the truth above.

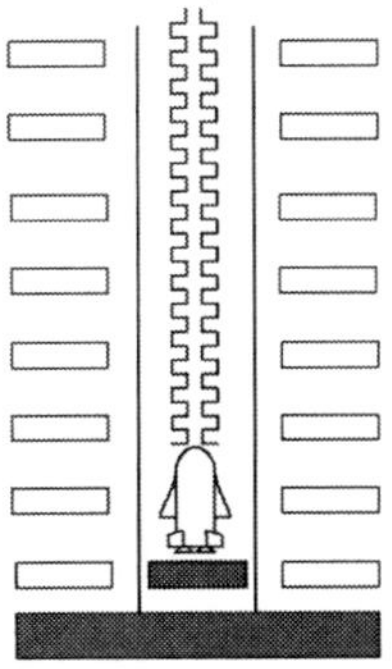

Figure1. Overall concept of electromagnetic propulsion system.

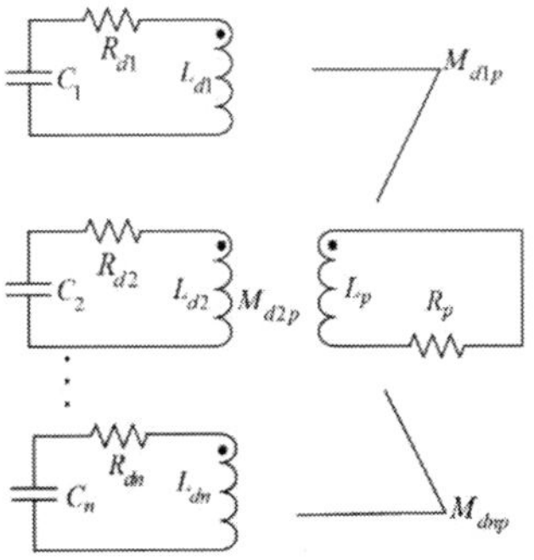

Figure 2. Schematic diagram of the equivalent current loop method for ignoring inter - stage coupling.

As it shows in figure1, this paper presents a new approach to launch spacecraft to the given orbit. The concept consists of two parts, one is the spacecraft, and the other is the electromagnetic launch.

3. Electromagnetic launch system

Multi-stage coil electromagnetic launch concept is applied to electromagnetic launch device. Drive part of the system consists of 3000 coils and armature coil radial size is slightly smaller than the drive coil. In fact, the entire launch system can be considered as the composition of multi-stage induction coil linear motor. The drive coil is equipped with a guide rail, which can restrain the armature and vehicle combination, make them run smoothly on the propulsion line. The following is a mathematical model of the coil-type electromagnetic launch system.

The coil-type electromagnetic launch system is simple in structure, high in energy utilization, capable of propelling large-quality objects, and suitable for the propulsion task of spacecraft.

Assuming that the drive coil has n-stage, the capacitance, initial inductance, self-inductance, and resistance of each stage are C_i , u_{ci} , L_{di} and R_{di} $(i=1,2,\cdots,n)$, as it shows in figure 2. The self-inductance and resistance of the armature are L_p, R_p and the mutual inductance of armature and drive coil is M_{dip} .

The drive coil power supply in each stage triggers the discharge in turn, so assume that the armature moves to the trigger point of the stage i drive coil at t , and the power supply of the stage i drive coil starts to trigger the discharge. Since the trigger switches of $i+1, i+2, \cdots, n$ stage drive coils is in off state, the current in these coils is zero and the voltage across the capacitor is still the initial voltage. Then,

$$u_{ck}(t) = u_{ck}(0) - \frac{1}{C_k}\int_0^t i_{dk}\,\mathrm{d}t \qquad (k=1,2,\cdots,i) \tag{1}$$

$$i_{dk} = 0 \qquad (k=i+1, i+2, \cdots, n) \tag{2}$$

The velocity of the armature can be acquired by solving equation (3),

$$\mathbf{RI} + \mathbf{L\dot{I}} - \mathbf{M_I\dot{I}} - v_p \frac{\mathbf{dM_I}}{\mathrm{d}x}\mathbf{I} = \mathbf{U} \tag{3}$$

among them,

$$\mathbf{R} = diag(R_p, R_{d1}, \cdots, R_{di}, \cdots, R_{dn}), \mathbf{L} = diag(L_p, L_{d1}, \cdots, L_{di}, \cdots, L_{dn})$$

$$\mathbf{I} = \begin{bmatrix} I_p \\ I_{d1} \\ \vdots \\ I_{di} \\ \vdots \\ I_{ds} \end{bmatrix}, \dot{\mathbf{I}} = \begin{bmatrix} \dot{i}_p \\ \dot{i}_{d1} \\ \vdots \\ \dot{i}_{di} \\ \vdots \\ \dot{i}_{ds} \end{bmatrix}, \mathbf{M}_\mathrm{I} = \begin{bmatrix} 0 & M_{d1,p} & \cdots & M_{di,p} & \cdots & M_{dn,p} \\ M_{d1,p} & & & & & \\ \vdots & & & \vdots & & \\ M_{di,p} & & & & & \\ \vdots & & & & \vdots & \\ M_{dn,p} & & & & & \end{bmatrix}, \mathbf{U} = \begin{bmatrix} u_{c1} \\ u_{c2} \\ \vdots \\ u_{ci} \\ \vdots \\ 0 \end{bmatrix}$$

The constant transformation of equation (13) yields:

$$\dot{\mathbf{I}} = (\mathbf{L} - \mathbf{M}_\mathrm{I})^{-1}\left(\mathbf{U} + v_p \frac{\mathrm{d}\mathbf{M}_\mathrm{I}}{\mathrm{d}x}\mathbf{I} - \mathbf{RI}\right) \tag{4}$$

The initial velocity and current of the armature are zero, so the initial current derivative of the moment is

$$\dot{\mathbf{I}}_0 = (\mathbf{L}_0 - \mathbf{M}_{\mathrm{I}0})^{-1}\mathbf{U}_0 \tag{5}$$

The time step is Δt, then the time series will be $t_n (n = 0,1,\cdots)$, so the current derivative at t_{n-1} is:

$$\dot{\mathbf{I}}_{n-1} = (\mathbf{L}_{n-1} - \mathbf{M}_{I(n-1)})^{-1}\left(\mathbf{U}_{n-1} + v_{p(n-1)} \frac{\mathrm{d}\mathbf{M}_{I(n-1)}}{\mathrm{d}x}\mathbf{I}_{n-1} - \mathbf{RI}_{n-1}\right) \tag{6}$$

Where, the component of $\mathbf{U}_{n-1}$ is $u_{ci(n-1)} = u_{ci(n-2)} - \frac{\Delta t}{C_i} I_{di(n-2)}$, then the current $\mathbf{I}_n$ at t_n is:

$$\mathbf{I}_n = \mathbf{I}_{n-1} + \Delta t \dot{\mathbf{I}}_{n-1} \tag{7}$$

The axial electromagnetic force F_n received by the armature at t_n can be expressed as:

$$F_n = \sum_{i=1}^{n} \left(\frac{\mathrm{d}M_{di,p}}{\mathrm{d}x}\right)_n I_{din} I_{pn} \tag{8}$$

$\left(\frac{\mathrm{d}M_{di,p}}{\mathrm{d}x}\right)_n$ is the interaction gradient between the drive coil of stage i and the armature at t_n,

I_{din} is the current of the stage I drive coil at t_n,

I_{pn} is the momentary current of the armature at t_n.

In this paper, the drive coils at all stages are identical in structural parameters. Therefore, if the axial midpoint of the first stage drive coil is used as the reference origin, the mutual inductance function between the drive coils and the armature coils at each stage can be regarded as a function of the reciprocal function relative to the reference origin translation. If it is assumed that the position of i-th drive coil is x_{di} and the position of the $(i+1)$-th drive coil is $x_{d(i+1)}$,

$$M_{d(i+1),p}(x) = M_{di,p}(x - (x_{d(i+1)} - x_{di})) \tag{9}$$

4. Simulation andanalysis of EMPS

4.1. Simulation parameters set-up

The simulation parameters are shown in Table 1 and Table 2.

Table 1. Parameters of the coils.

Coil Type	Material	Outside Diameter(m)	Inside Diameter(m)	Axial Length(m)	Number of Turns
Drive Coil	Copper	5.5	4.5	2	1000
Armature Coil	Copper	4.4	3.5	2	1000

IOP Conf. Series: Journal of Physics: Conf. Series **1039** (2018) 012041 doi:10.1088/1742-6596/1039/1/012041

Table 2. Parameters of the track and the vehicle.

Track Length (km)	Series number	Weight of armature and vehicle(ton)
6	3000	350

4.2. Mutual inductance simulation of armature and drive coil

It can be seen from equation (18) that the magnitude of the electromagnetic force is related to the current in the armature and the drive coil, and the mutual inductance between the armature and the drive coil. The former is related to the discharge process of the discharge circuit. In the discharge circuit, the other circuit parameters other than the mutual inductance can be considered as constant. If the armature and the driving coil are coaxial, the mutual inductance is due to the axial relative position variety. So in the entire simulation process, how to accurately get mutual inductance is essential.

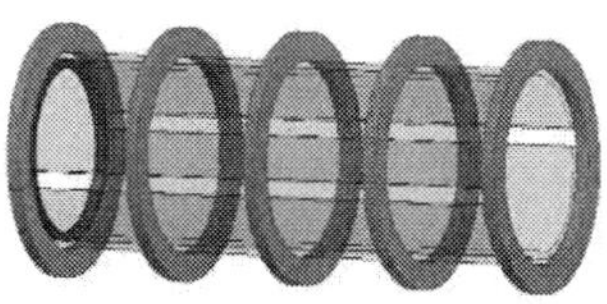

Figure3. Brief description of configuration of coil launchsystem.

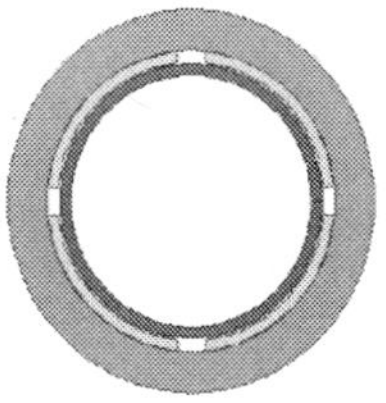

Figure4. Side view of coil launch system.

The parameters of the coils have been shown in table 1 and table 2.Based on these parameters, Maxwell finite element model was established as figure 3 and figure 4 and analysis method is performed to analyze the mutual inductance of the armature and the drive coil in different positions.

The simulation results of the electromagnetic launch are shown in Figure 5 toFigure8.Figure 5 to Figure 7is the simulation results of the single stage EML. Figure 5 shows the maximum thrust of the single stage device reaches 1500 kN, reaches the peak at 1 second, and then decreases gradually until it is close to zero at nearly 2 second. The armature acceleration coincides with the armature thrust curve, which makes the single-stage acceleration process reach a stable high value at 2 second as it shows in Figure 6. After this time point, the electromagnetic thrust is small and is not sufficient for the armature acceleration, and because of the existence of inertia, armature continuingly performs uniform motion, which can be seen from Figure 7.

As it shows in Figure 8, the acceleration of the assembly in the whole multi-stage launch is gradually decreasing. This is due to the increase in the speed of movement of the assembly, so the effective acceleration effect of each the drive coil on the armature is decreasing.

5. Conclusion

Based on the principle of coil electromagnetic propelling, the equivalent current loop model is constructed and the mutual inductance and static magnetic field between the driving coil and the armature coil are simulated by Maxwell finite element analysis software. Combining with the model and simulation Data, the calculation of the electromagnetic propulsion can accelerate the spacecraft weight 133.3 tons to a velocity of 207 meters per second. Given that the weight of typical reusable vehicle is hundreds of tons, EMPS is doable to propel such spacecraft.

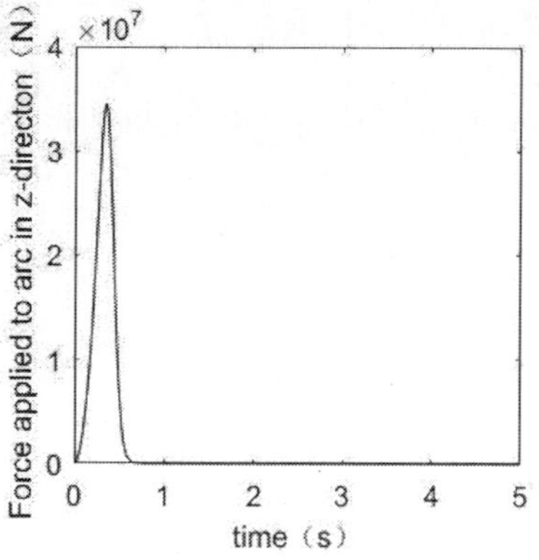

Figure5. Electromagnetic forces applied to armature in z-direction.

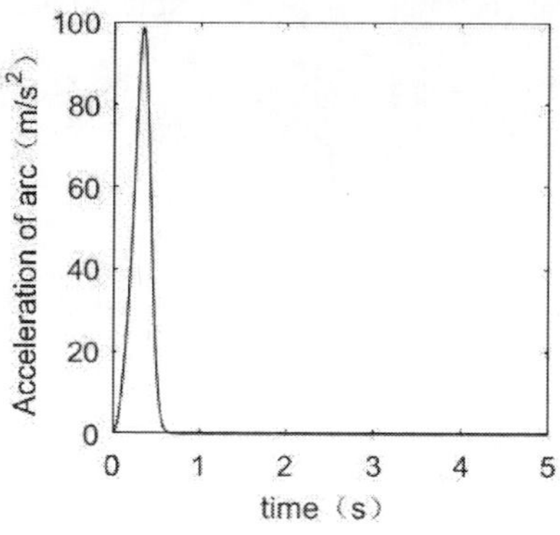

Figure6. Acceleration of armature.

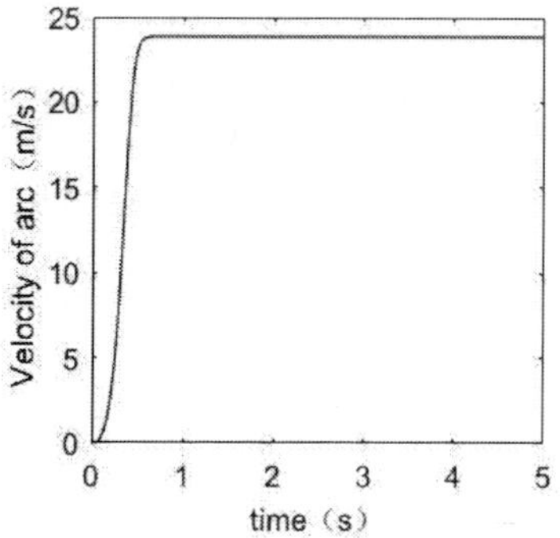

Figure7. Velocity of armature Velocity of armature.

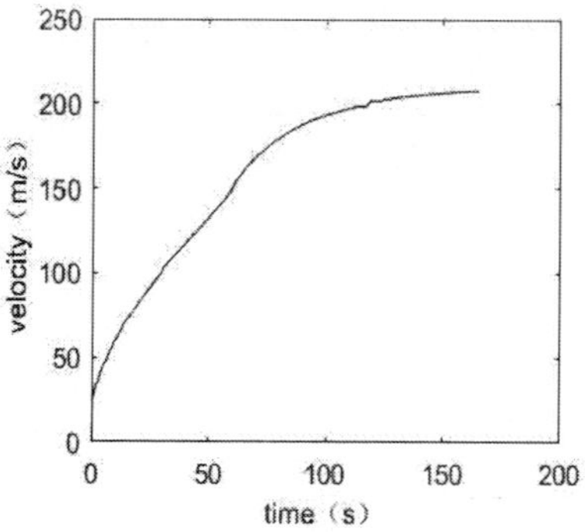

Figure8. Acceleration of multi-stage electromagnetic launch.

References

[1] Zhusheng L,Xiaobo P2014*Space Transportation System* (Beijing: China Science Press) p 6
[2] Bhase PS2015Magnetically Launching of Space Vehicle in Earth's Lower Orbit. (Shivajinagar:College of Engineering Shivajinagar Pune) pp 1-2
[3] Thomas J B, Michal E B, and Mark J L1999*Int.Conf.9th Space Planes and Hypersonic Systems and Technologies*pp 1-14
[4] Bradley C E2003 The space elevator NIAC phase II final report*NASA Institute for Advanced Concepts Report* p 2
[5] SchultzJH, RadovinskyA, Thome RJ2001*J. IEEE Transactions on Applied Superconducttivity***11**1749-1752
[6] Balikci A, Zabar Z, Birenbaum L2007*J. IEEE Trans. on Magn.***43**107-110
[7] Weibo L, Yanjie C,LiangpengZ2015*J. Ordnance Industry Automation***34**18-21
[8] Heppeenheimer T A. 1987*The national aerospace plane*(Maryland: University Press of America)
[9] S Weingertner1993 Sanger-The reference concept of the German hypersonic technology program *AIAA report* pp 1-9

Acknowledgments

This work is supported by the Astronautical Supporting Technology Foundation of China under grant 2015-HT-XGD.

Three-dimensional Trajectory Tracking Guidance Law Based on Linear Quadratic Regulator

Zhilei Ge, Yanni Wang and Meibo Lv

School of Astronautics, Northwestern Polytechnical University, Xi'an 710072, China

Email: 13991336193@139.com

Abstract. A three-dimensional trajectory tracking guidance law is designed based on the theory of linear quadratic regulator(LQR), which can help to solve the problem of long range trajectory tracking for air-defense missile in high altitude. Firstly, the missile mass model is linearized by using the coordinate variable of X axial. Secondly, the guidance parameters and mathematical model are obtained by designing state variables and control variables. At last, the three-dimensional trajectory tracking guidance law were proposed with theory of LQR. When the trajectory deviation is given, the simulation results show that the guidance law designed in this paper can suppress the disturbance and track ideal trajectory accurately.
Keywords: Three-dimensional trajectory; Trajectory tracking; LQR; Guidance law

1. Introduction

A two-dimensional trajectory tracking guidance law were designed based on the theory of LQR [1]. The literature [2] studied model of hypersonic vehicle and attitude control based on feedback linearization method. A roll angle control method based on LQR control algorithm was put forward for controlling roll angle on the two-dimensional course correction [3]. The trajectory tracking device was designed by LQR, which was used to track the optimal trajectory based on the theory of Gauss pseudospectral [4]. A trajectory tracking guidance law was proposed based on the theory of sliding mode variable structure control, which can help to solve the problem of trajectory tracking for air-defense missile [5]. An aircraft trajectory tracking guidance law were designed by nonlinear spatial inversion [6]. The literatures [7-9] established a nonlinear inverse control law based on distance to land to make the aircraft follow accurately a vertical profile and a desired airspeed. The literature [10] established the control law by using nonlinear dynamic inversion to make the aircraft accurately follow 3D+T desired trajectories. The literature [11] tackled the nonlinear control problem of aircraft trajectory tracking in the framework of multiple linear time-varying constrained control. The literature [12] discussed the aircraft trajectory tracking in three-dimensional space by six-degree-of-freedom nonlinear flight control scheme for unmanned aerial vehicles (UAVs). The literature [13] studied the problem of converting a trajectory tracking controller to a path tracking controller for a nonlinear non-minimum phase longitudinal aircraft model based on LQR. The literature [14] studied a flight path tracking problem for underactuated UAV in 3D space. A decoupling trajectory tracking method for gliding reentry vehicles was presented to improve the reliability of the guidance system in literature [15], and compared with the linear quadratic regulator (LQR) approach in some specific conditions.

The trajectory deviations of the tracking guidance law are basically tens of meters, up to several hundred meters. In this paper, long range tracking of air-defense missile and initial error up to 1 km are considered. The missile mass model is linearized by using the coordinate variable of X axial. A three-dimensional trajectory tracking guidance law is designed based on the theory of linear quadratic regulator(LQR) by designing normal overload and lateral overload as control variables and selecting y, z, θ, ψ as state variables. A simulation example is given to show the effectiveness of the law at last.

2. Mathematical model of missile

Considering a missile as a particle, the mathematical model is as follows:

$$m\dot{v} = P\cos\alpha\cos\beta - C_x qS - mg\sin\theta \tag{1-a}$$

$$mv\dot{\theta} = P\sin\alpha + C_y qS - mg\cos\theta \tag{1-b}$$

$$-mv\cos\theta\dot{\psi} = -P\cos\alpha\sin\beta + C_z qS \tag{1-c}$$

$$\dot{x} = v\cos\theta\cos\psi \tag{1-d}$$

$$\dot{y} = v\sin\theta \tag{1-e}$$

$$\dot{z} = -v\cos\theta\sin\psi \tag{1-f}$$

Here, m and S are quality and reference area of missile, respectively. v is missile velocity. P represents engine thrust. θ and Ψ are tilt angle and deflection angle of trajectory, respectively. q represents dynamic pressure calculated by $q = \frac{1}{2}\rho v^2$. ρ is the atmospheric density calculated by $\rho = \rho_0 e^{-h/h_p}$, $\rho_0 = 1.225 kg/m^3 . h_p = 7254.3m$. h, that is, y represents the altitude of missile. C_x, C_y, C_z are drag coefficient, lift coefficient and lateral force coefficient. $g = 9.8m/s^2$.

3. Design of three-dimensional trajectory tracking guidance law

3.1. Description of optimal control for trajectory tracking

Suppose that a linear system can be represented by state vectors:

$$\dot{x} = Ax + Bu \tag{2}$$

A state feedback controller is designed to satisfy the desired performance of the system:

$$\dot{x} = (A - BK)x = A_0 x \tag{3}$$

Performance index:

$$J = \frac{1}{2}\int_{t_0}^{t} x^T Qx + u^T Rudt \tag{4}$$

Here, Q and R are weight matrices, Q is positive semidefinite matrix, R is positive definite matrix. When t tends to infinity, x approaches 0, which guarantees the stability of the system. The minimum control variable u means the optimal control is obtained with the minimum control energy.

The optimal control variable is $u = -Kx$, equation (4) can be converted into:

$$J = \frac{1}{2}\int_{t_0}^{t} x^T(Q + K^T RK)xdt \tag{5}$$

Suppose there is a constant matrix P so that equation (6) is established:

$$\frac{d}{dt}(x^T Px) = -x^T(Q + K^T RK)x \Rightarrow \dot{x}^T Px + x^T P\dot{x} + x^T Qx + x^T K^T RKx = 0 \tag{6}$$

Combing equation (3) and equation (6):

$$x^T A_0^T Px + x^T PA_0 x + x^T Qx + x^T K^T RKx = x^T((A - BK)^T P + P(A - BK) + Q + K^T RK)x = 0 \tag{7}$$

$K = R^{-1}B^T P$, equation (7) can be converted into:

$$PA + A^T P - PBR^{-1}B^T P + Q = 0 \tag{8}$$

Thus, the optimal control variable matrix is $u = -Kx$. The optimal feedback gain matrix is $K = R^{-1}B^T P$. P is the solution of equation (8).

3.2. Design of three-dimensional trajectory tracking guidance law

The three-dimensional trajectory tracking guidance law is designed to make parameters deviations (position, tilt angle and deflection angle of trajectory, etc.) between actual trajectory and ideal trajectory minimum by establishing the linearized state-space equation with the state variable of trajectory parameters deviation. The equation (5) is the performance index. The optimal feedback gain matrix is designed based on the theory of LQR, which can make the least trajectory deviation.

It is necessary to linearize the nonlinear mathematical model of missile to design trajectory tracking guidance law based on the theory of LQR. The speed of missile is uncontrollable. Combing equation (1), the missile mass model is linearized by using the coordinate variable of X axial.

$$\frac{dy}{dx} = \frac{\tan\theta}{\cos\psi} \tag{9-a}$$

$$\frac{dz}{dx} = -\tan\psi \tag{9-b}$$

$$\frac{d\theta}{dx} = \frac{P\sin\alpha + C_y qS - mg\cos\theta}{mv^2\cos\theta\cos\psi} = \frac{gn_y}{v^2\cos\theta\cos\psi} - \frac{g}{v^2\cos\psi} \tag{9-c}$$

$$\frac{d\psi}{dx} = -\frac{-P\cos\alpha\sin\beta + C_z qS}{mv^2(\cos\theta)^2\cos\psi} = -\frac{gn_z}{v^2(\cos\theta)^2\cos\psi} \tag{9-d}$$

State variable matrix:

$$x = [\Delta y \quad \Delta z \quad \Delta\theta \quad \Delta\psi]^T = \left[y - y_b \quad z - z_b \quad \theta - \theta_b \quad \psi - \psi_b\right]^T \tag{10}$$

Subscript b represents the data of ideal trajectory。

The optimal control variable matrix is u=$\left[\Delta n_y \quad \Delta n_z\right]^T$. The linearized state-space equation is:

$$\frac{d}{dt}[\Delta y \quad \Delta z \quad \Delta\theta \quad \Delta\psi]^T = A[\Delta y \quad \Delta z \quad \Delta\theta \quad \Delta\psi]^T + B\left[\Delta n_y \quad \Delta n_z\right]^T \tag{11}$$

$$A_{11} = \partial\frac{dy/dx}{\partial y} = 0; A_{12} = \partial\frac{dy/dx}{\partial z} = 0; A_{13} = \partial\frac{dy/dx}{\partial\theta} = \frac{1}{(\cos\theta)^2\cos\psi}; A_{14} = \partial\frac{dy/dx}{\partial\psi} = \frac{\tan\theta\tan\psi}{\cos\psi}; A_{21} = \partial\frac{dz/dx}{\partial y} = 0;$$

$$A_{22} = \partial\frac{dz/dx}{\partial z} = 0; A_{23} = \partial\frac{dz/dx}{\partial\theta} = 0; A_{24} = \partial\frac{dz/dx}{\partial\psi} = -\frac{1}{(\cos\theta)^2}; A_{31} = \partial\frac{d\theta/dx}{\partial y} = \frac{-C_y\rho_0 e^{(-h/h_p)}S}{2mh_p\cos\theta\cos\psi}; A_{32} = \partial\frac{d\theta/dx}{\partial z} = 0;$$

$$A_{33} = \partial\frac{d\theta/dx}{\partial\theta} = \frac{gn_y\tan\theta}{v^2\cos\theta\cos\psi}; A_{34} = \partial\frac{d\theta/dx}{\partial\psi} = \frac{g\tan\psi}{v^2\cos\psi}(\frac{n_y}{\cos\theta} - 1); A_{41} = \partial\frac{d\psi/dx}{\partial y} = \frac{C_z\rho_0 e^{(-h/h_p)}S}{2mh_p(\cos\theta)^2\cos\psi}; A_{42} = \partial\frac{d\psi/dx}{\partial z} = 0;$$

$$A_{43} = \partial\frac{d\psi/dx}{\partial\theta} = \frac{-2gn_z\sin\theta}{v^2(\cos\theta)^3\cos\psi}; A_{44} = \partial\frac{d\psi/dx}{\partial\psi} = \frac{-gn_z\tan\psi}{v^2(\cos\theta)^2\cos\psi}; B_{11} = \partial\frac{dy/dx}{\partial n_y} = 0; B_{12} = \partial\frac{dy/dx}{\partial n_z} = 0; B_{21} = \partial\frac{dz/dx}{\partial n_y} = 0;$$

$$B_{22} = \partial\frac{dz/dx}{\partial n_z} = 0; B_{31} = \partial\frac{d\theta/dx}{\partial n_y} = \frac{g}{v^2\cos\psi\cos\theta}; B_{32} = \partial\frac{d\theta/dx}{\partial n_z} = 0; B_{41} = \partial\frac{d\psi/dx}{\partial n_y} = 0; B_{42} = \partial\frac{d\psi/dx}{\partial n_z} = \frac{-g}{v^2\cos\psi(\cos\theta)^2};$$

Performance index:

$$J = \frac{1}{2}\int_{t_0}^{t}[\Delta y \quad \Delta z \quad \Delta\theta \quad \Delta\psi]Q[\Delta y \quad \Delta z \quad \Delta\theta \quad \Delta\psi]^T + \left[\Delta n_y \quad \Delta n_z\right]R\left[\Delta n_y \quad \Delta n_z\right]^T dt \tag{12}$$

The optimal control variable matrix $u = -Kx = -K[\Delta y \quad \Delta z \quad \Delta\theta \quad \Delta\psi]^T$ can make J minimum. The optimal feedback gain matrix:

$$K = R^{-1}B^T P \tag{13}$$

P is the solution of equation (14).

$$PA + A^T P - PBR^{-1}B^T P + Q = 0 \tag{14}$$

Normal overload and lateral overload of actual trajectory required:

$$\left[\Delta n_y \quad \Delta n_z\right]^T = -K[\Delta y \quad \Delta z \quad \Delta\theta \quad \Delta\psi]^T \tag{15-a}$$

$$\begin{bmatrix} n_y & n_z \end{bmatrix}^T = \begin{bmatrix} n_{yb} & n_{zb} \end{bmatrix}^T + \begin{bmatrix} \Delta n_y & \Delta n_z \end{bmatrix}^T \quad (15\text{-b})$$

3.3. Work Flow for guidance law

When t tends to infinity, state variable matrix approaches zero matrix. Actual trajectory coincides with ideal trajectory and overloads are optimal. The flow chart and working procedure are as follows:

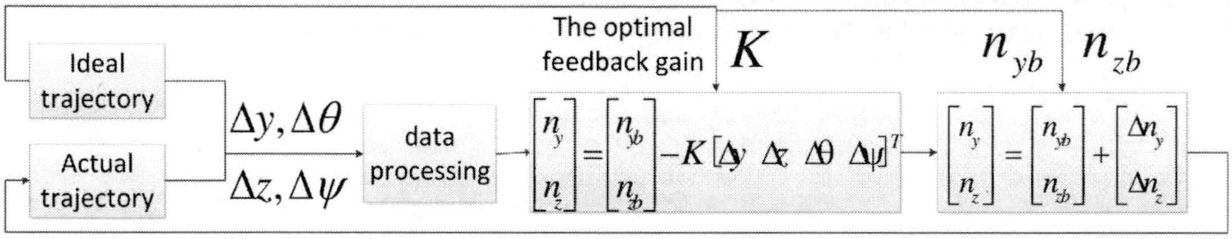

Figure 1. Flow chart of guidance law.

(1) Combing equation (11), (14) and equation (13), the optimal feedback gain matrix is calculated.
(2) Combing equation (10), trajectory data is sampled and optimal control variable matrix is calculated.
(3) According to equation (15) and ideal trajectory, the missile is controlled to fly.

4. Simulation results and analysis

The initial conditions of simulation: $v_0 = 3000m/s, \theta_0 = 15^\circ, \psi_0 = 60^\circ, m = 3000kg$, $h = 0.001$. The ideal trajectory is 500km long. The initial height of missile is 26km, and the height of hitting target is 20km. Trajectory deviation is $\Delta y = 1000$. Set parameters to $R = diag[100 \quad 100], Q = diag[0.0007 \quad 0.01 \quad 1 \quad 1]$.

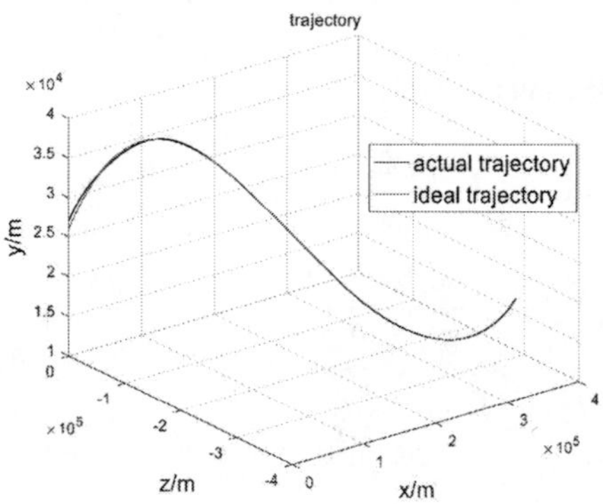

Figure 2. Trajectory curve.

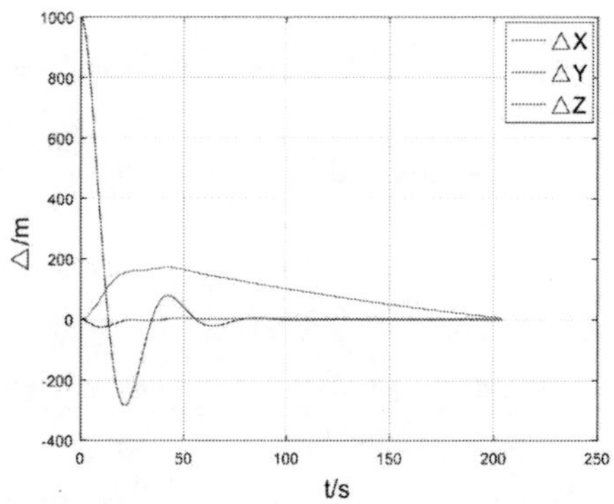

Figure 3. Deviation curve of position.

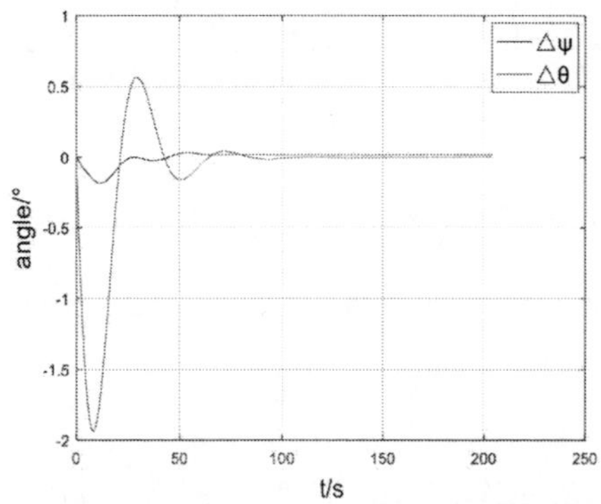

Figure 4. Deviation curve of angles.

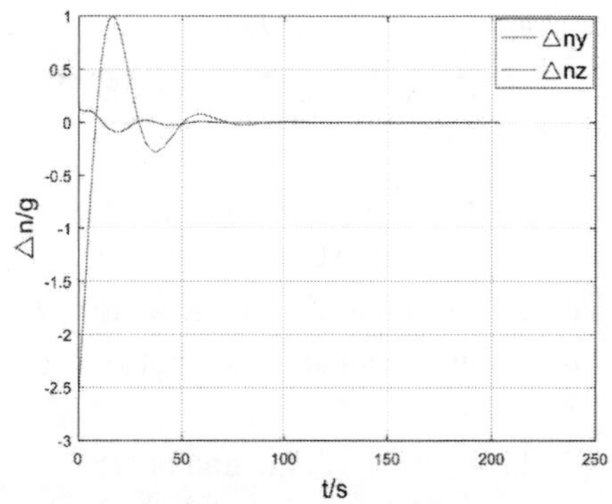

Figure 5. Deviation curve of overload.

Table 1. Statistical data of simulation results.

Parameters	$\Delta x(m)$	$\Delta y(m)$	$\Delta z(m)$	$\Delta\psi(\circ)$	$\Delta\theta(\circ)$	$\Delta n_y(g)$	$\Delta n_z(g)$
Deviation	4.64	-1.07	2.61	0.0199	0.0029	3.10×10^{-5}	-2.79×10^{-4}

Simulation results analysis: Simulation results is shown in figure 2-5 and table 1. As can be seen, the guidance law can suppress the disturbance and track ideal trajectory in altitude, overloads, tilt angle and deflection angle of trajectory accurately. The actual curves of deviation vary smoothly. The last tilt angle and the ideal angle are no more than 0.001 deg. The last deflection angle and the ideal angle are no more than 0.02 deg. Deviations of normal overload and lateral overload approaches 0. The last position to the target is no more than 6 m, thus performs high precision.

5. Conclusions

Aiming at the problem of long range trajectory tracking for air-defense missile in high altitude, the missile mass model is linearized by using the coordinate variable of X axial. Trajectory deviation can be obtained by comparing trajectory data between actual trajectory and ideal trajectory. According to this, optimal control variable is calculated. Then the optimal control variable is added to the overload of the ideal trajectory to obtain the overload of the actual trajectory required. In the simulation process, the controller parameters of LQR are designed. When the trajectory deviation is given, the simulation results show that the three-dimensional trajectory tracking guidance law designed in this paper based on the theory of LQR can suppress the disturbance and track the reference trajectory accurately.

References

[1] Zhang Dayuan, Lei Humin, et al. A Trajectory Tracking guidance law based on feedback linearization method[J]. *Journal of Ballistics*, 2014, **4(26)**:7-12.
[2] Liu Xiaoyun, Wang Jing. The Hypersonic vehicle attitude control based on feedback linearization/LQR method[J]. *Aerospace Control*, 2014, **4(32)**:37-41.
[3] Lei Longjie, Chen Ruihua, et al. Linear quadratic regulator algorithm for roll angle control[J]. *Journal of Detection & Control*, 2016, **38(01)**:33-36.
[4] Leng Shan, Wang Lixin, Cui Minghai. Trajectory optimization and guidance for transition phase of missile[J]. *Tactical Missile Technology*, 2013, **(01)**:82-86+110.
[5] Yang Rongjun, Ye Yao, et al. Application of sliding mode control on trajectory following for guided munitions[J]. *Journal of Ballistics*, 2015, **27(04)**:24-29.
[6] Bouadi H, Moracamino F. Aircraft trajectory tracking by nonlinear spatial inversion[C]. *AIAA Guidance, Navigation, and Control Conference*. 2012:1-17.
[7] Bouadi H, Mora-Camino F, Choukroun D. Aircraft time-2d longitudinal guidance based on spatial inversion of flight dynamics[C]. *Ieee/aiaa, Digital Avionics Systems Conference.*, 2012:3C4-1-3C4-14.
[8] Bouadi H, Choukroun D, Mora-Camino F. Aircraft longitudinal guidance based on a spatial reference[M]. *Springer Berlin Heidelberg*, 2013:427-442.
[9] Bouadi H, Moracamino F, Choukroun D. Space-indexed control for aircraft vertical guidance with time constraint[J]. *Journal of Guidance Control & Dynamics*, 2014, **37(37)**:1103-13.
[10] Wahid M A, Panomruttanarug B, Drouin A, et al. Space-indexed aircraft trajectory tracking[C]. *Control and Decision Conference. IEEE*, 2016:5076-81.
[11] Hameduddin I, Bajodah A H. Generalized dynamic inversion control for aircraft constrained trajectory tracking applications[C]. *American Control Conference. IEEE*, 2012:4599-4606.
[12] Xie Z, Xia Y, Fu M. Robust trajectory tracking method for UAV using nonlinear dynamic inversion[C]. *IEEE, International Conference on Cybernetics and Intelligent Systems*, 2011:93-98.
[13] Al-Hiddabi S A, Mcclamroch N H. Aggressive longitudinal aircraft path tracking using nonlinear control[J]. *Asian Journal of Control*, 2010, **3(4)**:280-288.
[14] Guan J, Wen Jun Y I, Chang S J, et al. Study of flight path tracking and control of an UAV in 3D space[J]. *Acta Armamentarii*, 2016, **37(1)**:64-70.

[15] Liang Z, Ren Z, Shao X. Decoupling trajectory tracking for gliding reentry vehicles[J]. IEEE/CAA *Journal of Automatica Sinica*, 2015, **2(1)**:115-120.

Acknowledgments
We sincerely thank the support given by the National Natural Science Foundation of China (no. 61374209).

IOP Conf. Series: Journal of Physics: Conf. Series **1039** (2018) 012043 doi:10.1088/1742-6596/1039/1/012043

Measuring on Sugar Content of Sugarcane Based on Phase Locked Loop with Capacitive Sensor

Siriporn Nonta[1], Anucha keawphusuk[1] and Paramote Wardkein[2]

[1] Department of Physics, Faculty of Science, Naresuan University , Phitsanulok, Thailand
[2] Department of Telecommunication Engineering, Faculty of Engineering, King Mongkut's Institute of Technology Ladkrabang , Ladkrabang, Bangkok ,Thailand

Email: sirinnta@gmail.com

Abstract. The article proposes instrument construction for measuring the sugar content in sugarcanes juice. A interdigital capacitor acting as a sensing element is connected with the sinusoidal generator. Whose output frequency depends on changing the capacitance of interdigital capacitor .To convert the frequency to the direct current (DC) voltage. A sin wave from the generator is converted square wave by comparator circuit before feeding to the phase locked loop circuit. Finally, the DC voltage is changed to concentration of the sugar content in sugarcanes juice

1.Introduction

Currently, technologies for determining the sugar content in agricultural products can be divided in three groups which consist of the optical property, specific gravity and electrical property. The optical properties are divided in near-infrared and visible light. The near-infrared light is used to examine the sugar content in fruit. The reflectance, transmittance and interactance are the properties of light that have been applied to literatures. The reflectance of banana skin surfaces was measured by UV-VIS-NIS spectrometer. The result indicate that NIR spectra responded directly to pulp sugar and the correlation coefficients is 0.96 Thus, NIR spectroscopy in banana provides valuable information on internal fruit maturity and fruit quality[1]. A spectrum measurement instrument was designed to obtain the transmitted energy spectra. It consisted of two halogen lamps and tray having a small hole at the center was used to hold the pear sample. The results showed the pre-processed by second derivative have a good performance with 0.82 so, the transmittance method is possible for measurement sugar content in pear[2]. Measuring the content of intact carambola by interactance mode ,where the light source and detector are positioned next to each other so that the light due to specular reflection cannot enter the detector. The results showed the pre-processed by second derivative with correlation coefficients is 0.82 Thus, the results can be applied to larger frits and broader sugar content ranges [3].The use of natural light on the refraction. When light through the solution of different concentrations. As the refractive index varies with concentration [4]. But solution must be clear to detect the sugar content accurately. In the wine industry ,Specific gravity values are used to measure sweetness of grape juice for calculating the alcohol content in wine[5].But examination requires a lot of solution. For the more, the electrical properties of the sugar solution were also investigated using time-domain reflectometer.The decreasing of resistance depends on the concentration of the solution but the dielectric constant is increased [6]. The dielectric constant was applied to determine of sugar

IOP Conf. Series: Journal of Physics: Conf. Series **1039** (2018) 012043 doi:10.1088/1742-6596/1039/1/012043

content in food and juice industry by the interdigital capacitor sensor[7] .The frequency range for measurement is investigated and the amplitude value of sinusoidal output signal inversely proportional to the sugar content. However the generators ,oscilloscopes and other devices are required in this system. The research is just a study not make system to work on its own, also requires the generator and oscilloscope etc. In this article presents measurement the sugar content in sugarcane juice the by phase locked loop circuit with interdigital capacitor sensor, it consists of a oscillator generating waveform that whose frequency is variable by the capacitance of the interdigital capacitor and the sin wave from the generator is converted square wave by comparator circuit before feeding to the phase locked loop circuit, After that the output frequency is converted to the DC output voltage which relates with the concentration of sugar content in sugarcane juice.

2. Principles of measuring sugar content in sugarcane juice

Principles of sugar content measurement in sugarcane juice includes interdigital capacitor sensor, quadrature oscillator, comparator circuit and phase locked loop circuit.

2.1. Interdigital capacitor sensor

The interdigital capacitor is conductor plate placed alternate between anode and cathode. The interdigital capacitor, show in figure 1.is used to measure the sugar content in solution. The total capacitance of the interdigital capacitor has been revealed[8].

$$C_0 = 2\varepsilon_0\varepsilon_1\frac{t_h}{S} + \varepsilon_0(\varepsilon_1+\varepsilon_2)\frac{K\left(\sqrt{1-k^2}\right)}{K(k)}(n-1)L \quad (1)$$

Where ε_0 is the free space permittivity (8.854x10^{-12}F/m), ε_1 is the relative permittivity of material between electrodes, ε_2 is the relative permittivity of the substrate, t_h is thickness of finger, S is space between finger, $K[x]$ is a complete elliptic integral of the first kind, n is number of finger, L is length finger and k is equal to S/W.

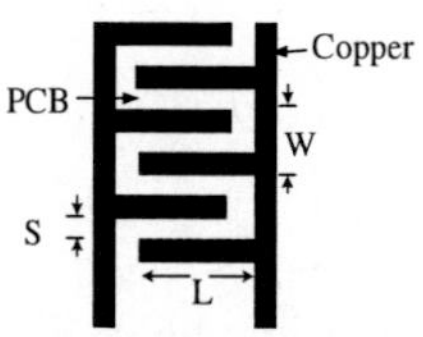

Figure 1. Characteristics of the interdigital capacitor.

From (1) we define $k_x = \varepsilon_0\varepsilon_2\left(\frac{K\left(\sqrt{1-k^2}\right)}{K(k)}\right)(n-1)L$ and $k_y = \frac{2t_h}{S} + \frac{k_x}{\varepsilon_o}$ .The total capacitance of interdigital capacitor is rewritten as

$$C_0 = \varepsilon_o\left[\left(\varepsilon_1 k_y\right) + k_x\right] \quad (2)$$

2.2. Quadrature oscillator

The quadrature oscillator is another type of phase-shift oscillator. The outputs are labelled as sine and cosine because the cosine is actually the same sine shifted to 90°degrees. The frequency is as follow [9]

$$f_i = \frac{1}{2\pi\sqrt{R_1C_1R_2C_2}} \quad (3)$$

In (3), capacitance (C_2) is interdigital capacitor (C_0) paralleled with one capacitor constant (C). We define Where $R_1 = R_2 = R_x$ so the frequency in quadrature oscillator can be stated as

$$f_i = 1\Big/2\pi R_x\sqrt{C_1\left[\varepsilon_o\left(k_x + \left(\varepsilon_1 k_y\right)\right)\right] + C} \quad (4)$$

2.3. Phase locked loop for frequency to DC voltage conversion

The conventional phase locked loop (PLL) is used to converts frequency to DC voltage. Figure 2. Shows the block diagram of the PLL which consists of phase detector (PD), a loop filter or a low pass filter (LPF), voltage controlled oscillator (VCO) and an integrator [10].The parameters $\phi_i(s)$, $\phi_o(s)$, $\phi_e(s)$, $V_d(s)$ and $V_e(s)$ are analyzed by applying the Laplace transform.The parameters $\omega_o(s)$ and $\omega_r(s)$ are the output frequency and the running frequency of VCO A_o, K_d, K_o and K_I are gain of loop filter, phase detector ,VCO and integrator,respectively. The $F(s)$ is transfer function of a loop filter.

In figure 1.the relationship of the system can be written as $\phi_d(s)=\phi_i(s)-\phi_o(s)$,$V_d(s)=K_d(\phi_d(s))$, $V_e(s)=F(s)K_d(\phi_d(s))$,$F(s)=A_o/(1+\tau s)$ and $\omega_o(s)=\omega_r(s)+K_oV_e(s)$.

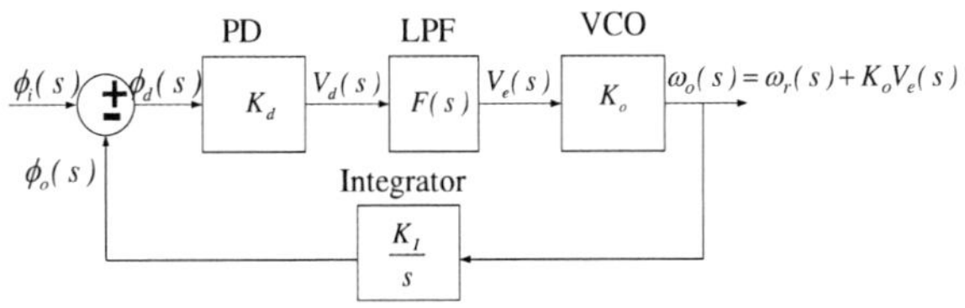

Figure 2. Block diagram of a conventional PLL.

From $(\phi_d(s))$,$(V_d(s))$,$(V_e(s))$, $(F(s))$ and $\omega_o(s)$,the relationship of system can be written as

$$\tau s^2\phi_o(s)+s\phi_o(s)+K_T\phi_o(s)=K_T\phi_i(s)+K_I\omega_r(s)\tau s+K_I\omega_r(s) \tag{5}$$

From (5) we let $K_T=K_IK_oK_dA_o$ and take inverse Laplace transform to both sides of (5),which can be stated as

$$\tau\frac{d^2\phi_o(t)}{dt^2}+\frac{d}{dt}\phi_o(t)+K_T\phi_o(t)=K_T\phi_i(t)+K_I\tau\frac{d}{dt}\omega_r(t)+K_I\omega_r(t) \tag{6}$$

Generally, $\phi_o(s)$ is the composed of two parts which are natural response and forced response. For this work, a natural response is zero when the system is in a steady state. Only steady state forced response is of interest. To determine the forced response, the $\phi_i(t)$ can be assumed as $\phi_i=\omega_it+\theta_i$,hence

$$\phi_o(t)=\omega_it+\theta_i+\frac{K_I\omega_r}{K_T}-\frac{\omega_i}{K_T}$$

(7)

The phase error $\phi_d(t)$ can be calculated by

$$\phi_d(t)=\phi_i(t)-\phi_o(t)=-\frac{K_I\omega_r(t)}{K_T}+\frac{\omega_i(t)}{K_T} \tag{8}$$

When the system is locked state, the phase error $\phi_d(t)$ depends on $\omega_i(t)$ because $\omega_r(t)$ is constant time increases. From (8), it is seen that the phase error $\phi_d(t)$ is directly proportional to input signal. $V_e(s)=F(s)K_d(\phi_d(s))$ and take inverse Laplace transform. Thus we can written as

$$V_e(t)=K_vA_oe^{t/\tau}+K_dA_o\phi_d(t) \tag{9}$$

From (9), time increases $K_ve^{-t/\tau}$ is zero convergence. $\phi_d(t)$ from (8) is achieved. When $\omega_r=2\pi f_r$, $\omega_i=2\pi f_i$ and $K_R=2\pi K_dA_0/K_T$ is constant, the $(V_e(t))$ can be expressed as

$$V_e(t)=K_R(-K_If_r+f_i) \tag{10}$$

From (10), when f_r is constant, the output voltage $(V_e(t))$ of phase locked loop system depends on f_i only. The frequency of input signal is obtained from quadrature oscillator circuit. So output voltage of phase locked loop circuit is

$$V_e(t)=K_R\left(-K_If_r+1\big/2\pi R_x\sqrt{C_1\left[\varepsilon_o\left(k_x+\left(\varepsilon_f k_y\right)\right)\right]+C}\right) \tag{11}$$

Let $C_t=C+C_1\varepsilon_0k_x$,the equation (11) is rewritten to be

$$V_e(t)=K_R\left[-K_I f_r + 1\Big/2\pi R_x\sqrt{1+\frac{C_1}{C_t}\varepsilon_o\varepsilon_1 k_y}\right] \tag{12}$$

Distribution of term $1\Big/\sqrt{1+\frac{C_1}{C_t}\varepsilon_o\varepsilon_1 k_y}$ with binomial series of $1/\sqrt{1+x}$ [11] is applied in (12).The output voltage can be given by

$$V_e(t) \approx K_R\left[-K_I f_r - \frac{1}{2\pi R_x\sqrt{C_t}} - \frac{C_1\varepsilon_o\varepsilon_1 k_y}{4\pi R_x C_t\sqrt{C_t}}\right] \tag{13}$$

We can observe that, the output voltage of the vary with the dielectric constant (ε_1) in sugarcane juice.

3. Experimental setup

The proposed principle can be compound in the circuit as shown in figure 3.

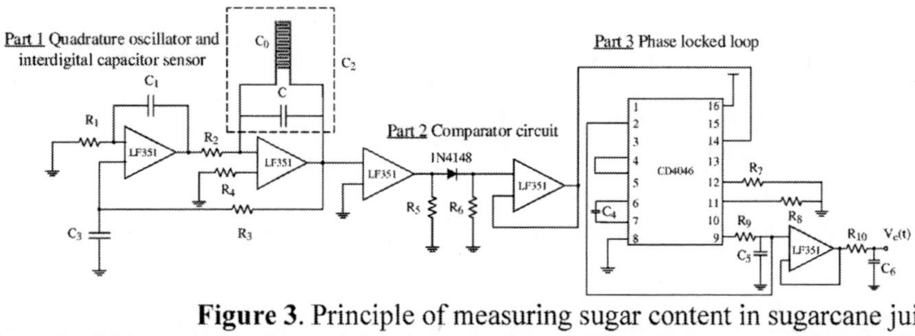

Figure 3. Principle of measuring sugar content in sugarcane juice.

A total of 80 sugarcane is used for the experiment. 50 samples are used for calibration model and 30 samples are used for prediction model. The sugar content in sugarcane juice is detected by a refractometer.The average of measurement results is recorded. And used in later research. After measured sugar content by a refractometer.After that 20 ml of the sugarcane juice is poured on surface of interdigital capacitor sensor. The sample is scanned and the average DC voltage data is saved.

4. Results and discussions

The results of measuring sugar content of part 1 is demonstrated in figure 4.and 5 by using the proposed principle.

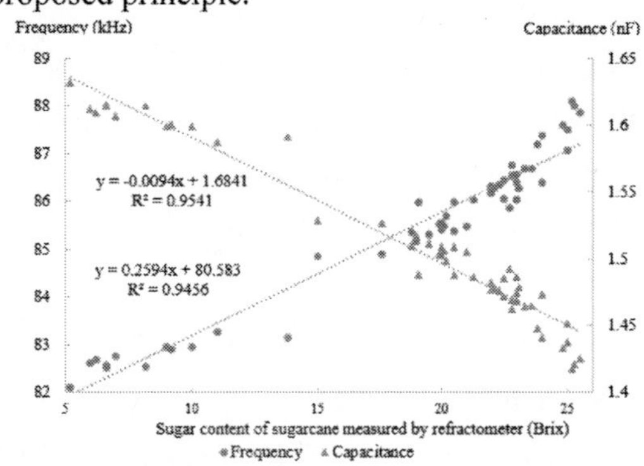

Figure 4. Relationship between frequency and capacitance with sugar content in sugarcane juice.

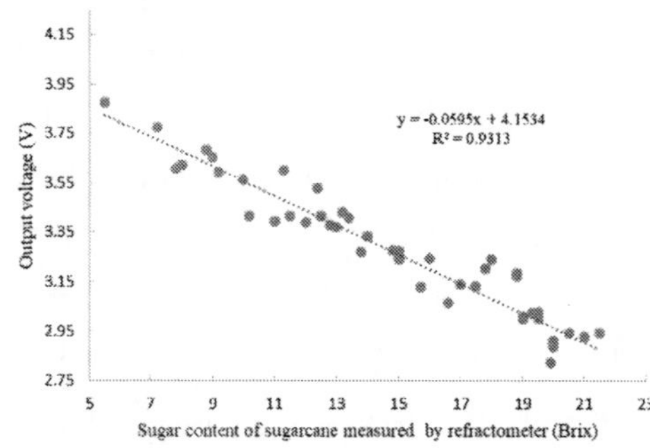

Figure 5. The results of output voltage with sugar content of sugarcane juice.

The interdigital capacitor sensor is used as a sensing element of the quadrature oscillator circuit. The output signal is a sinusoidal waveform which its frequency can be calculated to capacitance of interdigital capacitor sensor. The frequency, capacitance and sugar content of sugarcane juice measured are plotted in figure 4. The experimental results show that the frequency of the sugarcane juice is increased as the sugar concentrations are increased. On the other hand, the capacitance of the solutions is decreased as the sugar concentration are increases. In Part 2, the comparison circuit is used to convert the sine wave from quadrature oscillator circuit to square wave for signal to phase locked loop in part 3. The DC output voltage of sugarcance juice are decreased as the sugar concentration increases as shown in figure 5. The comparison between the sugar content measured by the proposed principle the refractometer is shown in figure 6.The correlation coefficient (R^2) of the our result is 0.9805.

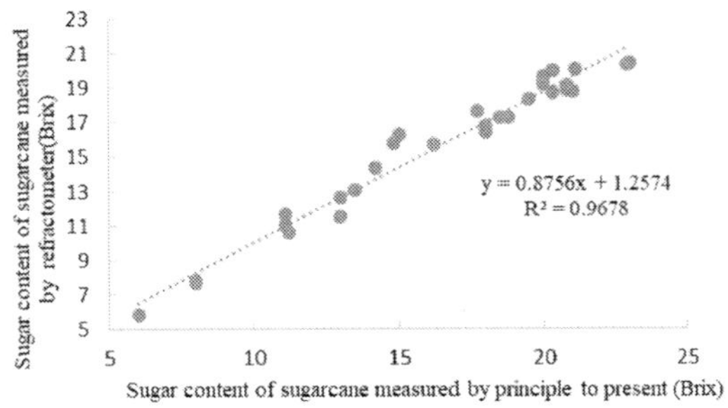

Figure 6. Sugar content of sugarcane juice measured by principle to present with measured from refractometer.

5. Conclusion

This article reports the principle of measurement sugar content in sugarcane juice. The capacitor sensor is used as sensing element of the phase locked loop. The results show that the values of frequency and out voltage increase as the sugar content in sugarcane juice increases, whereas capacitance decreases .The results of the sugar content measured by the proposed principle are in good agreement with the sugar content measured by the refractometer.The proposed for measuring in sugarcane juice can be applied to measure the different concentration of different solution.

References

[1] Zude, M. (2003). Non-destructive prediction of banana fruit quality using VIS/NIR spectroscopy. Fruits, **58**(3), 135-142.

[2] Wenqiang Wu(2011). Nondestructive Detection of Sugar Content in Early Season Maturity Pear based on Half Transmittance Technique.IEEE 978-1-4577-0321

[3] Omar, A. F. and Matjafri, M. F. (2013). Near Infrared Spectral Linearisation in Quantifying Soluble Solids Content of Intact Carambola. *Sensors*, **13**(1), 4876-4883.

[4] Surasit pligbua (2004). National Metal and Materials Technology Center Thailand. , 60-62

[5] Murli Dharmadhikari. Composition of Grapes. Iowa State University Extension and Outreach Midwest Grape and Wine Industry Institute.

[6] Badhe, S. G. and Helambe, S. N. (2013). Electrical impedance analysis of commonly used preservatives, Nacl (salt) and C12H22O11 (Sugar). *Science Research Reporter*, **3**(2), 239-242.

[7] Angkawisittpan, N. and Manasri, T. (2012). Determination of Sugar Content in Sugar Solutions using Interdigital Capacitor Sensor. *Measurement Science Review*, **12**(1), 8-13.

[8] Alexander, C.K., Sadiku, M.N.O. (1999).Fundamentals of Electric Circuits. McGraw Hill, 608-610.

[9] Ron Mancin. (2000) Design of op amp sine wave oscillators, *Analog Applications Journal.*

[10] Sitthichai Pokai-udom (2006) *Negative feedback circuits and oscillators* Mahanakorn University of Technology

[11] Pappas, T. "Pascal's Triangle, the Fibonacci Sequence & Binomial Formula." The Joy of Mathematics. San Carlos, CA: Wide World Publ.Tetra, pp. 40-41, 1989.

Acknowledgments

This article is based on work supported by the Research Center for Academic Excellence in Applied Physics of Naresuan University.

IOP Conf. Series: Journal of Physics: Conf. Series **1039** (2018) 012044 doi:10.1088/1742-6596/1039/1/012044

A Study Thermal Efficiency of Solar Air Heater with Wire Mesh Stainless Installation: Using Solar Simulator

Ratinun Luampon and Bundit Krittakom

Department of Mechanical Engineering, Faculty of Engineering and Architecture, Rajamangala University of Technology Isan, Nakhonratchasima, Thailand.

Email: ratinun.lu@rmuti.ac.th

Abstract This research was purposed to study thermal efficiency of solar air heater with wire mesh stainless installation. In experimental, solar irradiance for solar air heater testing was used solar simulator in the laboratory. This research consists of two main parts. First part: solar simulator testing. Solar irradiance used 4 halogen lamps with 1,500 W. The EN-12975-2 standard was evaluated the solar simulator performance for steady-state efficiency test. Experimental result, solar irradiance was good homogeneity at distance 70 cm with solar irradiated 980.4 W/m^2 and the uniformity value was lowest as 6.84%. It was suitable to test the solar air heater. Second part: a study thermal efficiency of solar air heater. Solar air heater performance was evaluated with the ASHRAE 93-86 standard. In experimental, wire mesh stainless was varied pores per inch (PPI) at 4, 8 and 12 PPI and constant layer at 11 layers. Air mass flow rate was constant at 0.037 kg/s and average solar irradiance about 980.4 W/m^2. A comparison between solar air heater with and without wire mesh stainless was conducted. It was found that, thermal efficiency of solar air heater with installing wire mesh stainless was increased and greater than without wire mesh stainless about 5.10 %, 16.75 % and 34.81 %, respectively.

1. Introduction

Solar air heater is a kind of heat exchangers that transform solar energy into heat. Usually, they are used for heating air in drying agricultural products and as an air heater in combination with auxiliary heaters for air conditioning of buildings [1]. Flat plate solar air heaters are non-adiabatic radiative heat exchangers; they are essentially used at low temperature levels (T<375 K) in air heating and drying systems [2]. Solar air heaters are simple devices to heat air by utilizing solar energy. Such heaters are implemented in many applications which require low to moderate temperature below 60 °C [3]. The main part of flat plate collector consists of: a case which holds a back insulation, an absorber and a transparent cover [4]. The transparent cover reduces heat losses towards the front, meanwhile the air flows in between the absorber plate and the thermal insulation. To increase thermal efficiencies, heat has to be transferred efficiently from the absorber to the flowing air [5]. Therefore, many configurations of the absorber plate have been designed to improve the heat transfer for the air flow in the passage. Velmurugan and Ramesh [6] was evaluated thermal performance of wire mesh solar air heater. The matrix solar air heater with a wire mesh produces higher thermal efficiency over the conventional flat plate solar air heater. Chabane, et al. [7] was studied heat transfer and thermal performance with longitudinal fins of solar air heater. Longitudinal fins were used interior the absorber plate to increase the heat exchange and render the flow fluid in the channel uniform. The maximum efficiency values obtained for the 0.012 and 0.016 kg/s with and without fins were 40.02%,

51.50% and 34.92%, 43.94%, respectively. Anuradha and Oommen [8] were studied a v-groove solar air collector. The results showed that the collector efficiency is about 35%. Alvarez [2] was developed and evaluated efficiency of single-glass air solar collector with an absorber plate made of recyclable aluminum cans. The result was shown that, the efficiency was increased with using recyclable aluminum cans. The maximum efficiency reached was 74%. Rajendra Kava [9] was augmented heat transfer and friction in asymmetrically heated rectangular ducts with ribs on the heated wall in transverse, inclined, v-continuous and v-discrete pattern. It was found that, v-discrete pattern was highest thermal efficiency and lowest friction factor, transverse ribs pattern was lowest thermal efficiency, v-continuous ribs pattern was highest friction factor. Tanda [10] was studied heat transfer coefficient in rectangular channels with transverse and v-shaped broken ribs. Result was shown, heat transfer coefficient was increased with increased ribs high and heat transfer coefficient was higher than smooth flat plat solar air heater about 300 %. Benlu and Jiang [11] was studied heat transfer coefficient and friction factor of rectangular solar air heater with inclined ribs between 0 to 90 degrees. The result was, inclined ribs with 60 degrees was highest heat transfer coefficient and pressure drop. And inclined ribs with 20 degrees were highest thermal efficiency.

From investigations, solar air heater is high thermal efficiency with increase heat transfer area but pressure drop and friction factor will be increase. For these reasons, the researchers will apply wire mesh stainless to increase heat transfer area. Because wire mesh stainless is a kind of porous media, it can absorb, emit heat and enhance heat transfer with absorber plate and air [12-15], and can reduce pressure drop and friction factor. Therefore, this research is proposing to install wire mesh stainless in solar air heater for increase thermal efficiency. The results of this experiment will be useful and guide development the solar air heater in further.

2. Thermal efficiency of solar air heater

The efficiency of a solar air heater is defined as the ratio of the amount of useful heat collected to the total amount of radiation striking the collector surface during any period of time.

$$\eta = \frac{SolarEnergyCollected}{TotalSolarStrikingCollectorSurface} = \frac{Q_c}{I \times A_c} \tag{1}$$

Useful heat collected for an air-type solar collector (solar air heater) can be expressed as:

$$Q_c = \dot{m}C_p\left(T_{air,out} - T_{air,in}\right) \tag{2}$$

So, thermal efficiency is calculated by Eq. (3)

$$\eta = \frac{\dot{m}C_p\left(T_{air,out} - T_{air,in}\right)}{I \times A_c} \tag{3}$$

where C_p = Specific heat of air (kJ/kg K)
$T_{air,out}$ = Air temperature at outlet solar air heater (K)
$T_{air,in}$ = Air temperature at inlet solar air heater (K)
$\dot{m}$ = Mass flow rate of air (kg/s)
I = Solar irradiance (W/m^2)
A_c = Solar absorber area (m^2)

3. Equipment and methodology

3.1 Equipment

3.1.1 Solar simulator design.

In experiment, solar irradiance used solar simulator in laboratory for solar air heater testing because some parameter can control such as solar irradiance, ambient temperature and ambient air velocity. Solar simulator frame made from box steel, solar irradiance used 4 halogen lamps with 1,500 W, shown in Figure 1. The EN-12975-2 standard was evaluated the solar simulator performance for steady-state efficiency test, because light field is the most important part of the solar simulator, an acceptable uniformity of irradiance in the irradiated surface and having stable irradiance over the time of test, are design factor of an authority light source for scientific aims [16]. The EN-12975-2 standard details were following [17]: The lamps shall be capable of producing a mean irradiance over the

IOP Conf. Series: Journal of Physics: Conf. Series **1039** (2018) 012044 doi:10.1088/1742-6596/1039/1/012044

collector aperture of at least 700 W/m^2. At any time, the irradiance at a point on the collector aperture shall not differ from the mean irradiance over the aperture by more than ±15 % (uniformity value). The irradiation map is usually measured with an automated X-Y system that moves the reference pyranometer every 150 mm in both directions. The pyranometer is maintained about 20-30 seconds in each position.

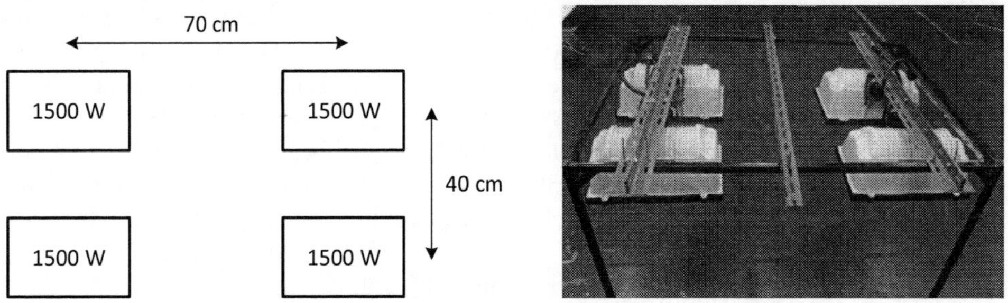

a) b)

Figure 1. a) Halogen lamps arrangement, b) 4 Halogen lamps with 1,500 W.

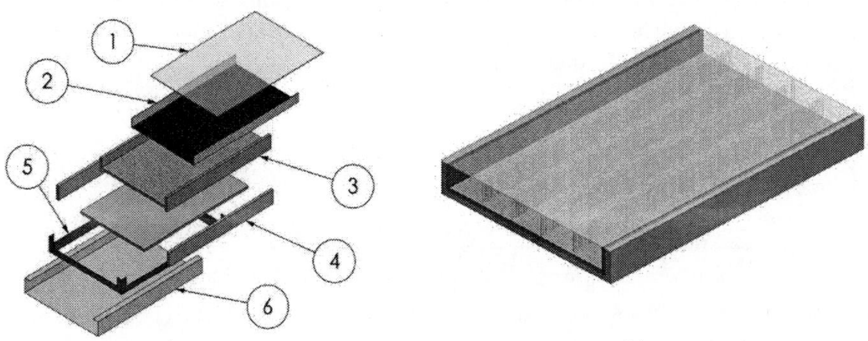

a) b)

Figure 2. a) Consists of solar air heater: 1. Glass, 2. Aluminum plate (absorber plate), 3. Wood, 4. Fiber Glass, 5. Steel frames and 6. Zinc plate, b) Solar air heater with wire mesh stainless.

3.1.2 Solar air heater design. Solar air heater made from angle bar 1 m length, 0.7 m width and 0.1 m height. Solar absorber made from aluminum plate 1 m length, 0.6 width and black painted. On top of solar air heater was cover with transparent glass. Side and bottom was insulated with wood and fiberglass 0.02 m thickness, and cover outside with zinc plate, was shown in Figure 2a. Wire mesh stainless was installed inside of solar air heater, was shown in Figure 2b.

4. Methodology

Solar simulator in laboratory used to evaluate thermal efficiency of solar air heater. The solar simulator performance was evaluated with the EN-12975-2 standard and solar air heater performance was evaluated with the ASHRAE 93-86 standard. The schematic diagram of solar collector testing with solar simulator was shown in Figure 3. Before the thermal efficiency evaluated, solar air heater was exposed in solar simulator about 30 minutes and collected data at 5 hours [5]. Air was flow through the flow straightener at solar air heater inlet with blower for reduce air swirl. Air mass flow rate was measured with the hot wire anemometer, air temperature was measured with type K thermocouple at 15 positions of air flow through solar air heater, 3 positions of solar absorber plate, shown in Figure 4 and solar irradiance (I) was measured by pyranometer. Experimental data were collected with data logger at interval 10 minute for 5 hours. Solar air heater with wire mesh stainless was varied pores per inch (PPI) at 4, 8 and 12 PPI and wire mesh stainless layers constant at 11 layers. Air mass flow rate was constant at 0.037 kg/s each experimental. The experimental result would analyze and compare between solar air heater with and without (PPI=0) wire mesh stainless.

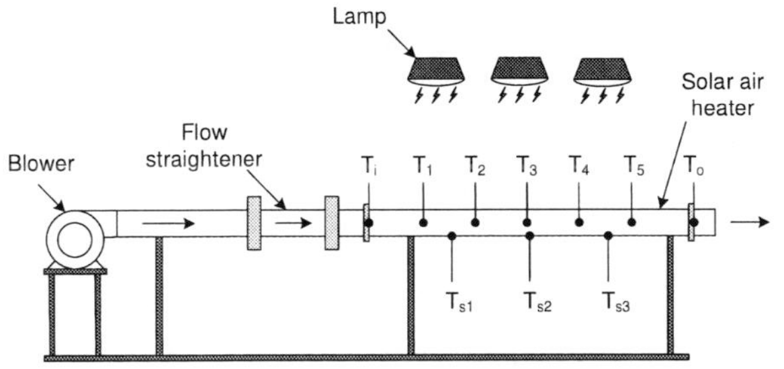

Figure 3. Schematic diagram of solar air heater testing with solar simulator.

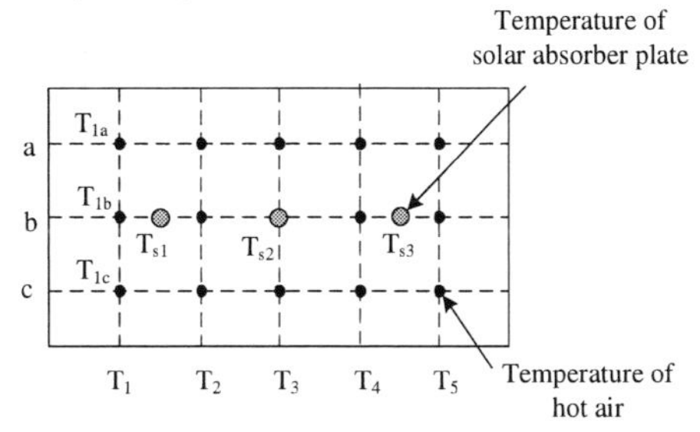

Figure 4. Temperature measurement positions on solar air heater.

5. Experimental result

5.1 Solar simulator

Solar simulator used 4 halogen lamps with 1,500 W to generated solar irradiance. The EN-12975-2 standard was evaluated the solar simulator performance for steady-state solar air heater efficiency test. From experimental result, it was found that solar irradiance was increased with decreased distance between halogen lamp and solar air heater. The solar irradiance was good homogeneity at distance 70 cm and the uniformity value was lowest as 6.84%. The solar simulator was provided maximum and minimum solar irradiance as 1,036.5 W/m^2 and 918 W/m^2, respectively and average solar irradiance was 980.4 W/m^2, shown in Figure 5. Therefore, this solar simulator was suitable for evaluate solar air heater thermal efficiency.

937.8	1006.9	918.0	947.6	937.8
1006.9	1036.5	1016.7	987.1	967.4
927.9	1016.7	947.6	977.2	947.6
1006.9	1036.5	987.1	1036.5	1006.9
927.9	1006.9	947.6	947.6	1036.5

Figure 5. Solar simulator irradiated distributions (I, W/m^2).

5.2 Solar air heater thermal efficiency

A comparison between solar air heater with and without (PPI=0) wire mesh stainless with varied pores per inch (PPI) at 4, 8 and 12 PPI and constant layer at 11 layers. Air mass flow rate was constant at 0.037 kg/s and average solar irradiance at 980.4 W/m^2**.** From result, it was found that solar air heater without wire mesh stainless was provided thermal efficiency at 48.24% and hot air outlet average temperature at 40.37 °C, shown in Fig 6a.

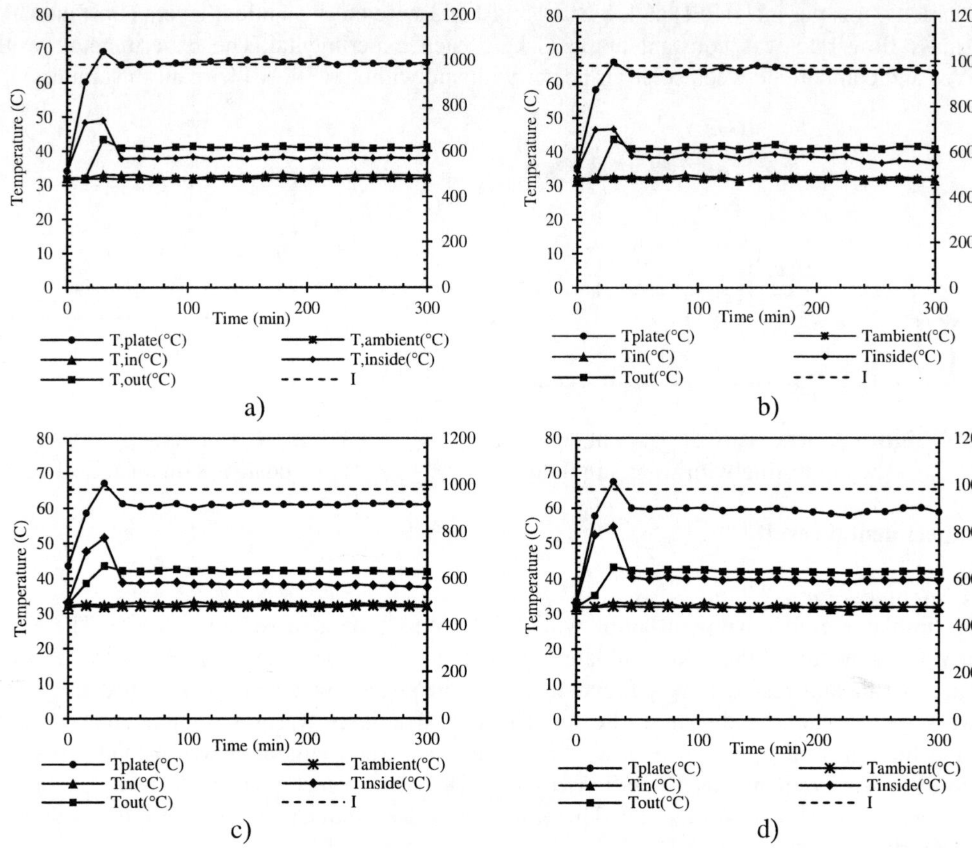

Figure 6. Temperature of air and absorber plate with I=980.4 W/m^2
a) PPI= 0, b) PPI=4, c) PPI=8 and d) PPI=12.

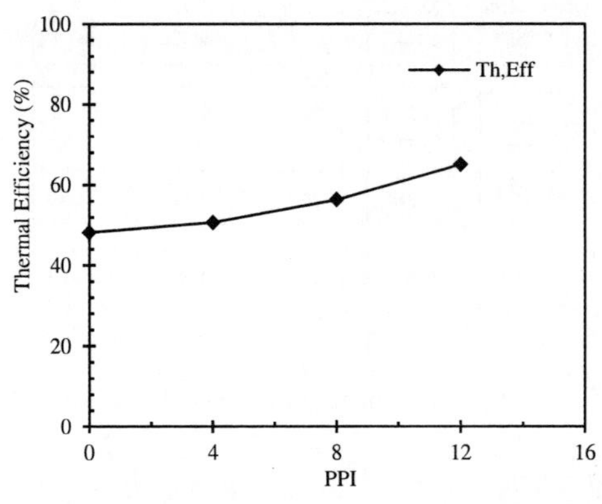

Figure 7. Thermal efficiency of solar air heater with (PPI=4, 8, 12) and without (PPI=0) wire mesh stainless.

Experimental result of solar air heater with wire mesh stainless installation, it was found that thermal efficiency was higher than without wire mesh stainless. Thermal efficiency and air temperature was increased with increased wire mesh stainless PPI. Because an installing wire mesh stainless inside solar air heater was an increasing heat transfer area and wire mesh stainless was a kind of porous media, it was absorber and emitter and can enhanced heat transfer from solar irradiance to the air flow through wire mesh stainless. Therefore, air temperature and thermal efficiency was increased. A result was shown in Figure 6 and 7.

From Figure 7, solar air heater with installing wire mesh stainless at PPI=12 was provided highest thermal efficiency about 65.04% and air outlet average temperature about 41.40 °C. A comparison between with and without wire mesh stainless it was found that, wire mesh stainless at PPI=4, 8 and 12 can increased thermal efficiency about 5.10 %, 16.75 % and 34.81 % greater than without wire mesh stainless, respectively.

6. Conclusions

A study thermal efficiency of solar air heater with wire mesh stainless installation. Solar simulator was constructed and tested in laboratory with EN-12975-2 standard. The solar irradiance was good homogeneity at distance 70 cm and the uniformity value was lowest about 6.84% and provided average solar irradiance about 980.4 W/m^2.

An evaluated thermal efficiency of solar air heater it was found that, the solar air heater with wire mesh stainless was greater than without wire mesh stainless and at PPI=12, 11 layers was provided highest thermal efficiency about 65.04% and greater than without wire mesh stainless about 34.81%.

References

[1] Alvarez G, Arce J, Lira L and Heras MR 2004 *Sol. Energy.* **77** 107
[2] Ajam H, Farahat S and Sarhaddi F 2005 *Int. J. of thermodynamics* **8(4)** 183
[3] Alta D, Bilgili D, Ertekin C and Yaldiz O 2010 *J. of Applied Energy* **87** 2953
[4] Kurtbas I and Durmus S 2004 *Renew. Energ.* **29** 1489
[5] Pottler K, Sippel CM, Beck A and Fricke J 1999 *Sol. Energy.* **67(1–3)** 35
[6] Velmurugan P and Ramesh P 2001 *Indian. J. Sci. Technol.* **4(1)** 12
[7] Chabane F, Moummi N and Benramache S 2014 *J. Adv. Res.* **5** 183
[8] Anuradha A and Oommen R 2013 *J. Sci. Eng. Res.* **4(6)** 2072
[9] Karwa R 2003 Expermental, *Int. Comm. Heat. Mass. Transfer.* **30(2)** 241
[10] Tanda G 2004 *Int. J. Heat. Mass. Transfer.* **47** 229
[11] Benlu and Jiang PX 2006 *Exp. Therm. Fluid.* Sci. **30(6)** 513
[12] Tongtem P and Jakjai S 2009 *Conf. of the mechanical engineering network of Thailand 22 (Thailand)*
[13] Mital R, Gore JP and Viskanta R 1997 *Combust. Flame.* **111** 175
[14] Boonmak S, Chavaratsinthon N and Krittakom B 2013 *Thailand Renewable Energy Community Association Conference 9 (Thailand: Chaingmai)*
[15] Srisuwan P, Chavaratsinthon N and Krittakom B 2013 2013 *Thailand Renewable Energy Community Association Conference 9 (Thailand: Chaingmai)*
[16] Sabahi H, Tofigh AA, Kakhki IM and Bungypoor-Fard H 2016 *Sustainable. Energy. Technol. Assess.* **15** 35
[17] Kovacs P 2012 *A guide to the standard EN 12975* (Technical Research Institute of Sweden)

Acknowledgements

The authors gratefully acknowledge Department of Mechanical Engineering, Faculty of Engineering and Architecture, Rajamangala University of Technology Isan and Development in Technology of Porous Material Research Laboratory (DiTo-Lab) for their support in this study under research projects.

Testing change points and its application in urban transportation

H Y Wang[1], S ZShi[1*], Z X Jiang[1] and H F Jia[2]

1 Science School, Changchun University of Science and technology. Changchun, China
2 School of Transportation, JiLin University. Changchun, China

Email: whywanghongyu@yeah.net

Abstract. In this paper, we propose the method of testing change points with the exponential family distribution. Under the condition of the same variance, we test whether exist change point in the mean value, we give the test statistics under some conditions when the distribution is unknown, and prove that the asymptotic distribution is the extreme value distribution. The result is same as many literatures. This method is extended to testing multiple change points. By simulating Gamma distribution, Weibull distribution, Log-normal distribution, two-parameter exponential distribution and their mixture distributions, we found the appropriate segment interval size and give the algorithm of the single change point and the multiple change points. We applied this method to testing change points with the pedestrian flow data in Shanghai Railway Station.
Key Words: Change point; Exponential family distribution; Mean test; Asymptotic distribution Extreme value distribution; Transportation; Simulation Theory

1. Introduction

Studies of change-point detection date back to 1950s. In the past half century, the topic has attracted a great deal of attention in such fields as statistics, engineering, economics and bioscience. For example Miao (1988) proposed the inference in a model with at most one slope-change point; Hsu (1979) proposed the method of detecting shifts of parameter in Gamma sequences with applications to stock price and air traffic flow; Cobb (1978) applied the change point model to hydrology; Braun et al. (2000) solved DNA sequence segmentation by using the multiple change point fitting via quasilikelihood. Fong et al. (2015) employed the change point model-based test in the study of immune responses that are associated with the risk of mother to child transmission of HIV-1.

Depending on the goal of the data that are collected, detection of change points can be crucial for decision making or necessary for understanding certain scientific issues. Many frequentist's and Bayesian methods have been introduced to detect the change of the mean, variance, slope of regression line, hazard rate, or nonparametric distribution for various models. We refer to the books by Chen (1988); Brodsky and Darkhovsky (1993); Csörgő and Horváth (1997); Chen and Gupta (2011) for various aspects for classical change-point analysis, and to an article Lee (2010) for a list of comprehensive bibliography of books and research papers on this topic.

While this body of work constitutes a rich literature, it mainly deals with the inference of change points when the distributions are known. In fact, the distribution of sequence cannot be judged in some applications. This resulted in some sequence, we couldn't use these methods of change point test to analyze, such as traffic flow, pedestrian traffic problems etc. But for detecting traffic abnormal incident, analysis and research for the phenomenon of the change of traffic flow is very important. So

regardless of the distribution function, we hope to find a quick way to check the number of change points and their locations. Then analysis the reason. So in this article, we consider the change point test problem when the distribution is unknown and find the minimum data size of the change point by some known distribution, such as Gamma distribution, Weibull distribution, Log-normal distribution, two-parameter Exponential distribution and their mixture
distribution. Then quickly find the change point according to the minimum data size.

Let $X_1,\cdots,X_n$ be a sequence of independent random variables with a change point τ, $\tau = k/n$. The distribution function of $X_1,X_2,\cdots,X_k$ is $F(x;\mu_1,\sigma^2)$. The distribution function of $X_{k+1},X_{k+2},\cdots,X_n$ is $F(x;\mu_2,\sigma^2)$ μ_1,μ_2,σ^2 are the mean and variance, where μ_1,μ_2,σ^2 are all unknown.

In section 2, we intend to test the location of change points by using the mean test. We give the statistic of mean test. At the same time, asymptotic distribution of estimator of change point is present when the non-negative random variables X_i is a common exponential family distribution. Furthermore, the estimator of change point $\hat{\tau}$ is also present. Section 3 focuses on the simulation results of segmentation of the multiple change point. By simulating the change point test of the common exponential family distribution function to find the minimum data which suitable the change point test. As a practical application of change point problem, in section 4, we research the pedestrian flow at subway gates by change point analysis methods in the common exponential family distribution.

2. The test of change point

Let $X_1,\cdots,X_n$ be a sequence of independent random variables with a common exponential family distribution. The value of mean μ, variance σ^2 are assumed unknown. The hypotheses can be described in more formal terms as

$$H_0:\mu_1=\mu_2 \tag{2.1}$$

Where the mean of $X_1,X_2,\cdots,X_k$ is μ_1 and the mean of $X_{k+1},X_{k+2},\cdots,X_n$ is μ_2, k is unknown. For convenience, we use some notations:

$$\bar{X}_1=\frac{1}{k}\sum_{i=1}^{k}X_i,\quad \bar{X}_2=\frac{1}{n-k}\sum_{i=k+1}^{n}X_i,\quad \bar{X}=\frac{1}{n}\sum_{i=1}^{n}X_i,\quad S^2=\frac{1}{n}\sum_{i=1}^{n}(X_i-\bar{X})^2 .$$

Denoting that $\overline{X_1}$ and $\overline{X_2}$ is the estimator of μ_1 and μ_2, respectively. S^2 is the estimator of σ^2. The statistic T_k defined in (2.2) can test whether there was a change point along the sequence.

$$T_k=\sqrt{\left(\frac{k(n-k)}{n}\right)}\left(\frac{1}{k}\sum_{i=1}^{k}\frac{X_i}{S}-\frac{1}{n-k}\sum_{i=1+k}^{n}\frac{X_i}{S}\right) \tag{2.2}$$

The following theorem give the asymptotic distribution of $\max_{1\le k\le n}|T_k|$, we give the lemma of the domain of attraction of the normal law shown in appendix and the following theorem.

Theorem2.1 Let the random variables $X_1,\cdots,X_n$ be a sequence of non-degenerate and nonnegative i.i.d. random variables. The distribution of X_i is an exponential family distribution, $Ee^{tX_i}<+\infty$, $0\le t\le 1$, the four order moments of X_i exists, then

$$\lim_{n\to\infty}p\{\max_{1\le k<n}|T_k|\le a(x,\log n)\}=\exp\{-2e^{-x}\}\quad -\infty<x<+\infty \tag{2.3}$$

$a(x,\log n)$ is given for

$$a(x,T)=[2\log T]^{-\frac{1}{2}}[x+2\log T+\frac{1}{2}\log\log T-\frac{1}{2}\log\pi] \tag{2.4}$$

The proof of the Theorem 2.1 is shown in the appendix. The method borrows an idea advanced by Tan and Miao (2005). Theorem 2.1 provides a method to test the null-hypothesis $H_0:\mu_1=\mu_2$. Given the test level α, from the equation

$$\exp\{-2e^{-x}\}=1-\alpha .$$

We can get $x(\alpha) = -\log[-\frac{1}{2}\log(1-\alpha)]$ if and only if when $\max_{1\le k<n} |T_k| > a(x(\alpha), \log n)$, we refused H_0, which means there is a change point exists. If there is a change point in the test results, the change point is estimated to be

$$\hat{\tau} = \frac{1}{n}\min\{k : |T_k| = \max_{1\le j\le n} |T_j|\} \tag{2.5}$$

The problem of making inference in this model is important in practical applications and of much theoretical interest. Many authors have contributed to it. To name a few among others, Chen(1988) use this method in Normal distribution.

3. Simulation

When the sequence have multiple change points, we need to segment the sequence. In order to verify the feasibility of the statistics given in Section 2 for testing the location of change points, and find out the smallest sample size of the common variable point test in the exponential family as the segmentation basis of the multiple point test, we give the following simulation.

3.1. Simulation test of a single change point

By using the mean change point test, the observation data are searched and tested as follow. First, simulating exponential family distribution. Each distribution is divided into two parts which have different means and same variance. The sample size of two parts is N_1 and N_2 respectively, total sample size is $N = N_1 + N_2$. Then, we use the mean test proposed in section 2 to find the location of change point $\hat{k}$, and then compared with the actual change point k_0. Finally, repeat experiment 500 times, when the probability of $P > 0.9$, output N, otherwise, improve the number of samples to continue to repeating the experiment.

In the simulation, we find that when the mean and the variance are more than 10, the probability of finding the change point is about 50%, but in reality, the mean and variance in the field of traffic exceed this number. In order to solve this problem, we normalize the data to carry out the change point test. The normalization method is carried out by the equation (3.1). Kao (2010)

$$\frac{X_i - \min_{1\le k\le n}\{X_i\}}{\max_{1\le k\le n}\{X_i\} - \min_{1\le k\le n}\{X_i\}} \tag{3.1}$$

According to the simulation results, we find that when the sample size of each sequence is 59, the probability of finding the position of the change point is more than 90%. Therefore, we choose 59 for segmented data. The results of the four distributions (Gamma distribution, Weibull distribution, Log-normal distribution, two-parameter Exponential distribution) of exponential families and their mutual combination are shown in Table 1,

Table1 Common exponential family distribution simulation test results

	μ_1	μ_2	σ^2	n	k0	$\hat{k}$	P	μ_1	μ_2	σ^2	k0	$\hat{k}$	P
Gamma	8	4	2	59	20	20	0.9999	120	100	20	30	30	0.9999
Weibull	2.7	11	2	59	20	20	0.9999	14.3	27.8	29.5	30	30	0.9999
Log-normal	2	5	1	59	20	20	0.9878	30	70	400	30	30	0.9200
Exponential	3	6	10	59	20	20	0.9999	20	60	10	30	30	0.9999
Gamma,Weibull	8	4	2	59	30	30	0.9800	40	20	10	30	30	0.9900
Gamma,Log-normal	10	15	5	59	30	30	0.9600	10	15	25	30	30	0.9900
Gamma,Exponential	12	4	4	59	30	30	0.9999	72	26	36	30	30	0.9999
Weibull,Log-normal	6	14	9	59	30	30	0.9999	40	15	10	30	30	0.9999
Weibull,Exponential	3	7	2	59	30	30	0.9999	30	15	10	30	30	0.9999
Exponential,Log-normal	2	10	4	59	30	30	0.9999	19	50	81	30	30	0.9999

3.2. Simulation test of multiple change points

According to the simulation results in section 3.1, this paper select the change points of distribution to repeat 150 times the simulation of multiple change points. We divide the sample into several groups, each group has 59 samples, and test the change point of each group. When we find the change point, recording the position. Choosing the next 59 data from this position (not include this data) to find the next position of change points, and so on. When there is no change point in group n, increase the size of sample to several times of 59, repeat the steps above until find the change point. The sample size of the next change point analysis is still 59, and finally find all the change points.

We select 310 data for the Gamma distribution, Weibull distribution, Log-normal distribution, Two-parameter Exponential distribution and mixed distribution, respectively. The mixed distribution is the mixture of four distributions. The order is Gamma distribution, Weibull distribution, Log-normal distribution, Two-parameter Exponential distribution, Gamma distribution, Weibull distribution, Log-normal distribution and Two-parameter Exponential distribution They have 8 different means and same variance. Their mean are 20, 40, 60, 80, 100, 80, 50, 30; Their variance are 5, 1, 4, 1, 1 and 10, 10, 64, 81, 10. The size of data are 30, 30, 50, 20, 40, 50, 50, 50. Their shape are stepped. Through MATLAB simulation, one of the results is shown in Figure 2

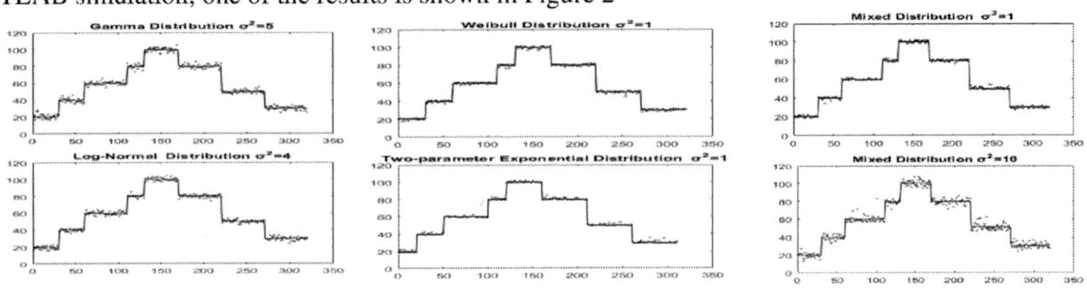

Fifure2. Simulation of four Exponential Family distribution and Mixed Distribution by multivariate change point test estimated mean, the variance is 5,1,1,4,1,10, in which the dotted line is the true value and the solid line is the estimated mean line chart. Where the abscissa indicates the position of the change point, the ordinate indicates the size of the change point.

4. Application in transportation

In this section, we select April 1, 2015 Shanghai Metro Line 1 pedestrian flow data. We analyze the data of gate entrance and exit obtained by every 5 minutes in Shanghai Metro Line 1 pedestrian flow. The number of pedestrian flow at the entrance of the subway gates is 212 and the number of the exit is 210. Firstly, the data were normalized. Then, the method proposed in Section 2 and the simulating process in Section 3 were used to test the multiple change points. Finally, the results are shown in Figure 3.

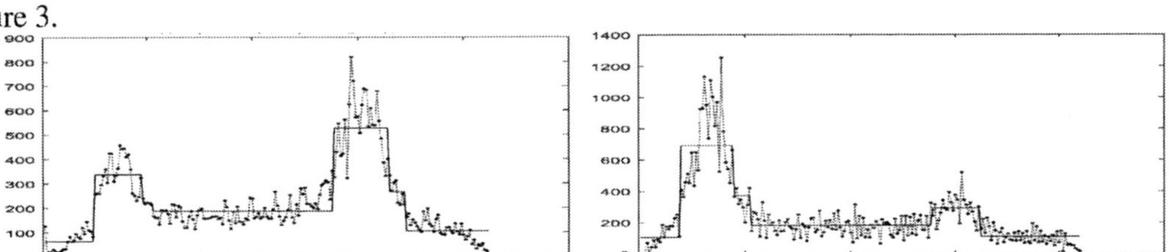

Figure 3. The pedestrian flow of the entrance and exit gates of the Shanghai Railway Station subway, in which the dotted line is the true value and the solid line is the line graph of the estimated mean. Where the abscissa indicates the time at which the change point occurred, the ordinate indicates the size of the change point.

From Figure 3 we can see that 7:30-9:20 and 17:00-19:05 are the peaks of inbound stations, the means are 337 and 527 respectively; 7:15-9:15 is the peak of outbound stations, the mean is

690 people, a larger flow of people. According to the Shanghai Railway Station timetable we can know that in the morning 7:00-9:30, 20 trains reach the Shanghai Railway Station, and this period of time is the morning peak, we can verify Shanghai is the most interprovincial office workers city as reported in news.

We can get the following conclusions from the results of the application: this method is suitable for the change points whose distance is large enough. The estimation location of change points is basically consistent with the actual situation. The distribution of samples is the common exponential distribution. The size of sample is more than 59.Conclusion

In this paper, we propose a fast testing method to find the change points when the distribution is unknown. The simulation results show that the test results are accurate under certain conditions, but when the number of data between two change points is less than 10, this method is difficult to detect the change points. Much work remains to be done for this problem.

References

[1] Braun J V, Braun R K, and Muller H G. Multiple change point fitting via quasilikelihood, with application to dna sequence segmentation. Biometrika, 87(2):301–314(14), 2000.
[2] Brodsky B E and Darkhovsky B S. Nonparametric Methods in Change-Point Problems. Kluwer Academic Publishers, 1993.
[3] Chen J and Gupta A K. Parametric Statistical Change Point Analysis. Springer Basel AG, 2011.
[4] Chen X R. Inference in a sample change point model. Scientia Sinica, 6:654–667, 1988.
[5] Cobb G W. The problem of the nile: Conditional solution to a change point problem. Biometrika, 65(2): 243-251, 1978.
[6] Csörgő M and Horváth L. 20 nonparametric methods for change point problems. Handbook of Statistics, 7(88):403-425, 1988.
[7] Csörgő M and Horváth L. Limit theorems in change-point analysis. John Wiley and Sons Ltd Chichester, 1997.
[8] Csörgő M, Szyszkowicz B, and Wang Q Y. Donsker's theorem for self-normalized partial sums processes. Annals of Probability, 31(3):1228-1240, 2003.
[9] Fong Y, Di C., and Permar S. Change point testing in logistic regression models with interaction term. Statistics in Medicine, 34(9):1483-94, 2015.
[10] Giné E, Götze Friedrich, and Mason David M. When is the student t-statistic asymptotically standard normal? Annals of Probability, 25(3):1514-1531, 1997.
[11] Hsu D A. Detecting shifts of parameter in gamma sequences with applications to stock price and air traffic flow analysis. Journal of the American Statistical Association, 74(365):31-40, 1979.
[12] Kao C. Fuzzy data standardization. IEEE Transactions on Fuzzy Systems, 18(4):745-754, 2010.
[13] Lee T S. Change-point problems: Bibliography and review. Journal of Statistical Theory and Practice, 4(4):643-662, 2010.
[14] Miao B. Q. Inference in a model with at most one slope-change point. Journal of Multivariate Analysis, 27(2):375-391, 1988.
[15] Tan C C and Miao B Q. Inference in γ-distribution with at most one change-point. J. univ. sci. technol. china, (1):51-58, 2005.
[16] Zhu. H L. Statistical inference in the jump change-point estimator. Master's thesis, Annals of Probability, 2014.

Acknowledgement

This work was supported by NSF grant 51378076 from the National Natural Science Found in China and grant 20140101199JC from the Natural Science Foundation of Jilin Province

Disturbance Rejection Attitude Controller Design for Launch Vehicle based on Orthogonal Function Neural Network

Wang Pei[1,2], Lv Mei Bo[1,2],Ge Zhi Lei[1,2], Zhang Zi Qi[1]

[1] College of Astronautics, Northwestern Polytechnical University, Xi'an 710072China
[2] National Key Laboratory of Aerospace Flight Dynamics, Northwestern Polytechnical University, Xi'an 710072 China

E-mail: jimmy5579@163.com

Abstract. A disturbance rejection attitude controller of launch vehicle based on chebyshev orthogonal function (COF) neural network and proportional-differential (PD) control is proposed to attenuate effect of parameter variation and disturbances of launch vehicle on attitude control accuracy and stability. The chebyshev orthogonal function (COF) neural network is used to estimate and compensate impact of parametric uncertainties and disturbances by taking full advantage of information of gyros and attitude sensors. The designed controller applies proportional-differential control to realize attitude control. Simulation results show that the controller exhibits good dynamic performance, high ability and strong robustness against external disturbance and parameter uncertainty.

1. Introduction

The structure of launch vehicle is more and more complex, launch vehicle has become a highly complex nonlinear system and its attributes are aerodynamically unstable. Meantime the mass properties and aerodynamic parameters vary substantially throughout the flight regime that makes a challenging task to design its control systems, so numerous nonlinear control methods have been used to design the attitude control system of launch vehicle.

$H\infty$ synthesis has been providing efficient scheduled controllers for launcher vehicle in atmospheric ascent [1].But depending on the augmented model, the controller order is commonly very high and has to be reduced before implementation. Nonlinear Dynamic Inversion (NDI) control law is formulated to eliminate system's nonlinearities by means of feedback. To make the NDI robust against model uncertainties and unknown but bounded external disturbances, dynamic constraint is defined and an additional Sliding Mode Control based loop is augmented [2].Due to the ground test limitation and the complexity of hypersonic aerodynamics, the aerodynamics coefficients of Reusable Launch Vehicle may have large uncertainty. Thus an in-flight parameter identification method for lateral control departure parameter is proposed utilizing the retrospective cost optimization and an adaptive controller is applied to control the lateral motion of launch vehicle. But the real-time character is not discussed [3]. By design the extended state observer to compensate model disturbances and uncertainty, active disturbance rejection control (ADRC) is well suited for design launch vehicle attitude control system and obtains good control effect [4]. How to tuning parameters quickly and effectively has become a very important issue of the application of ADRC.

Considering the internal and external interference and parameters uncertainties imposed on launch vehicle attitude dynamic model, a disturbance rejection controller was proposed for launch vehicle attitude control. This new controller scheme contains two components, which were based on PD control and COF neural network respectively. Simulation results show that the proposed controller is effective and robust in launch vehicle attitude control.

2. Attitude dynamic model of launch vehicle

Given the symmetry properties of launch vehicle, the longitudinal motion is adopted in subsequent sections. The longitudinal model is represented by three parts, including rigid body model, flexible body model and attitude information detect model [5].

The rigid body model of launch vehicle is expressed as

$$\begin{cases} \Delta\dot{\theta} = c_1\Delta\alpha + c_2\Delta\theta + c_3\delta_\vartheta + c_{30}\ddot{\delta}_\vartheta + \sum_i (c_{1i}\dot{q}_i + c_{2i}q_i) + c_{10}\alpha_w - \bar{F}_{by} \\ \Delta\ddot{\vartheta} + b_1\Delta\vartheta + b_2\Delta\alpha + b_3\delta_\vartheta + b_{30}\ddot{\delta}_\vartheta + \sum_i (b_{1i}\dot{q}_i + b_{2i}q_i) + b_{20}\alpha_w = \bar{M}_{bz} \\ \Delta\vartheta = \Delta\alpha + \Delta\theta \end{cases} \tag{1}$$

The flexible-body model of launch vehicle is given by the following expressions:

$$\ddot{q}_i + 2\zeta_i\omega_i\dot{q}_i + \omega_i^2 q_i = D_{1i}\Delta\vartheta + D_{2i}\Delta\alpha + D_{3i}\Delta\delta_\vartheta + D_{30i}\Delta\ddot{\delta}_\vartheta + \bar{Q}_{iy} \tag{2}$$

The attitude information measurement model is the following:

$$\begin{cases} \Delta\vartheta_{gz} = \Delta\vartheta - \sum_i W_z(X_{gz})q_i \\ \Delta\dot{\vartheta}_{st} = \Delta\dot{\vartheta} - \sum_i W_z(X_{st})\dot{q}_i \end{cases} \tag{3}$$

Where, $\Delta\theta$ is the trajectory inclination angle deviation of launch vehicle. $\Delta\alpha$ is the attack of angle deviation of launch vehicle. $\Delta\delta_\vartheta$ is swing angle of longitudinal pitch channel motor. $\Delta\ddot{\delta}_\vartheta$ is swing angle acceleration of longitudinal pitch channel motor. α_w is the angle of attack caused by wind. $\bar{F}_{by}$ is the linearized disturbance force of launch vehicle. $\bar{M}_{bz}$ is the linearized disturbance moment of launch vehicle. $\Delta\vartheta$ is pitch angle deviation. $\Delta\vartheta_{gz}$ and $\Delta\dot{\vartheta}_{st}$ are pitch angle and angle velocity which are measured by inertial integrated navigation platform under the influence of elastic vibration respectively. The definition of the rest parameters and expression can be found in [5].

3. Compensator design based on orthogonal function neural network

Chebyshev orthogonal function neural network has strong capability of nonlinear approximation with high reconstructing accuracy and fast training rate. So the adaptive chebyshev orthogonal function neural network is design to approach the disturbance and uncertainties of the system.

The chebyshev orthogonal function neural network is augmented as

$$\hat{f} = \sum_{i=1}^{N} w_i\phi_i + \varepsilon = W^T\vec{\Phi}(X) + \varepsilon \tag{4}$$

Where, $x = (x_1,\cdots,x_n)^T$ is the input of network, N is the number of network input. $\phi_i(X) = P_{1i}(x_1)P_{2i}(x_2)\cdots P_{ni}(x_n) = \prod_{j=1}^{n} P_{ji}(x_j)$, $P_{ji}(x_j)$ is the chebyshev orthogonal function. The expression of chebyshev orthogonal function are

$$P_{j1}(x_j) = 1,\ P_{j2}(x_j) = x_j, P_{ji}(x_j) = 2x_jP_{j(i-1)}(x_j) - P_{j(i-2)}(x_j)\ , j = 1,2,\cdots,N, i \le 3.$$

$\vec{\Phi} = [\phi_1 \quad \phi_2 \quad \cdots \quad \phi_n]^T$ is the network output vector of chebyshev orthogonal function. W is neural network weight.The network input are selected as $x = [e \quad \dot{e}]^T$.

The chebyshev orthogonal function network can approximate nonlinear continuous system to any degree of accuracy based on the following assumptions [6].

1) Network output $f(x,W)$ is continuous;
2) For a small positive number ε_0 , there is the ideal
3) Approximation network output $\hat{f}(x,W^*)$, Satisfied

$$\max\left\|\hat{f}(x,W^*)-f(x,W)\right\|\le \varepsilon_0 \tag{5}$$

Then, $W^* = \arg\min_{\theta\in\beta(M_\theta)}[\sup_{x\in\phi(M_x)}\left\|\hat{f}(x,W)-f(x)\right\|]$ is the best network weight for approaching $f(x)$. Taking $\delta^* = f(x)-\hat{f}(x,W^*)$ as the optimal approximation error.From the approach ability of chebyshev orthogonal function network, the approximation error is bounded, assuming the bounded is δ_0 .then

$$\delta_0 = \sup\left\|f(x)-\hat{f}(x,W^*)\right\| \tag{6}$$

Where, $\hat{f}(x,W^*) = W^{*T}h(x)$.

4. Launch Vehicle Attitude controller Design

Pitching angle differential equation of launch vehicle longitudinal channel can be represented as

$$\begin{aligned}\Delta\ddot{\vartheta} &= -b_1\Delta\vartheta - b_2\Delta\alpha - b_{30}'\delta_\vartheta - b_3''\ddot{\delta}_\vartheta - \\ &\sum_i(b_{1i}\dot{q}_i + b_{2i}q_i) - b_2'\alpha_w + \bar{M}_{bz} - b_{30}\delta_\vartheta\end{aligned} \tag{7}$$

The swing angle coefficient can be described as $b_3 = b_{30}' + b_{30}$. b_{30} is swing angle coefficient in nominal model, b_{30}' is the uncertainty of swing angle coefficient. By treating $f(x) = -b_1\Delta\vartheta - b_2\Delta\alpha - b_{30}'\delta_\vartheta - b_3''\ddot{\delta}_\vartheta - \sum_i(b_{1i}\dot{q}_i + b_{2i}q_i) - b_2'\alpha_w + \bar{M}_{bz}$ as unknown part of equation, and $g(x) = -b_{30}$. So the launch vehicle longitudinal channel can be represented as

$$\begin{cases}\dot{x}_1 = x_2 \\ \dot{x}_2 = f(x) + b_{30}u\end{cases} \tag{8}$$

Where， $x_1 = \Delta\vartheta$, $x_2 = \Delta\dot{\vartheta}$, $u\in R$ are control input.

Suppose control command is y_m, have $e = y_m - \Delta\vartheta$, $\vec{e} = (e,\dot{e})^T$ .We use the chebyshev orthogonal function network to estimate the uncertainty part, the control law is designed as

$$u = -\frac{1}{b}(-\hat{f}(x,W) + \ddot{y}_m + \vec{K}^T\vec{e}) \tag{9}$$

Where, selecting $\vec{K} = (k_2,k_1)$ can make the polynomial $s^2 + k_1s^1 + k_2 = 0$ satisfy Hurwitz criterion.

Substituting (9) into (8), we obtain (10) as follows

$$\dot{\vec{e}} = A\vec{e} + b\{(W-W^*)^T\vec{\Phi}(X) + \delta\} \tag{10}$$

Where, $A = \begin{bmatrix}0 & 0\\ -k_2 & -k_1\end{bmatrix}, b = \begin{bmatrix}0\\1\end{bmatrix}$.

We define Lyapunov function as,

$$V = \frac{1}{2}\vec{e}^TP\vec{e} + \frac{1}{2\gamma_1}(W-W^*)^T(W-W^*) \tag{11}$$

Where, γ_1 is a positive constant， P is a positive definite matrix and meet Lyapunov equation

$$A^TP + PA = -Q \tag{12}$$

Where, Q is a arbitrary 2*2 positive definite matrix. Calculating the derivative of the Lyapunov function, so

$$\dot{V} = \frac{1}{2}\dot{\vec{e}}^TP\vec{e} + \frac{1}{2}\vec{e}^TP\dot{\vec{e}} + \frac{1}{\gamma_1}(W-W^*)^T\dot{W} \tag{13}$$

Substituting (10) into (11),

$$\dot{V} = -\frac{1}{2}\vec{e}^TQ\vec{e} + \frac{1}{\gamma_1}(W-W^*)^T(\gamma_1\vec{e}^TPb\vec{\Phi}(X) + \dot{W}) + \vec{e}^TPb\delta \tag{14}$$

If we select chebyshev orthogonal function network weight adaptive law as,

$$\dot{W} = -\gamma_1\vec{\Phi}(X)\vec{e}^TPb \tag{15}$$

The derivative of the Lyapunov function is obtained as

$$\dot{V} = -\frac{1}{2}\vec{e}^T Q\vec{e} + \vec{e}^T Pb\delta \tag{16}$$

Because the approximation error can be sufficiently small, so $\dot{V} \le 0$ is approximately satisfied. We can build a compensation control law for uncertainty based on COF network. The control signal of controller is

$$\begin{aligned} u_{control} &= -\frac{1}{b}(-\hat{f}(x,W) + \ddot{y}_m + u_{pd}) \\ u_{pd} &= K_2 e + K_1 \dot{e} \\ \hat{f}(x,W) &= W^T \vec{\Phi}(X) \quad \dot{W} = -\gamma_1 \vec{\Phi}(X)\vec{e}^T Pb \end{aligned} \tag{17}$$

Where, K_2 and K_1 are proportional and differential gains, respectively. They are the adjustable parameters. $\hat{f}(x,W)$ is the estimation of the unknown disturbance item, The adjustable parameters include γ_1 and Q .Commonly the proposed controller parameters include K_1、K_2、q_{11}、q_{22}、γ_1 . Compared with ARDC method, the number of controller parameters dramatically reduced. It is convenient to design parameter.

5. Simulation

In order to validate the proposed controller, simulations of attitude control for the launch vehicle in some flight second, are presented. Structural parameters and aerodynamic coefficients for the nominal case are from [4].Considering the nominal situation without disturbance and uncertainty, the PD controller ($K_1 = 30, K_2 = 16$) is designed based on Hurwitz criterion.The chebyshev orthogonal function network parameters are set as

$$, Q = \begin{bmatrix} q_{11} & 0 \\ 0 & q_{22} \end{bmatrix} = \begin{bmatrix} 300 & 0 \\ 0 & 100 \end{bmatrix}, \gamma_1 = 100, N = 10.$$

The performance of proposed controller was analyzed from response time, steady-state error and ITAE index.

Case 1: In nominal situation without disturbance and parameter perturbation, initial pitch angle deviation is 5°.Simulation result is shown as Figures 1.

In nominal situation, the designed controller eliminates the initial attitude deviation fast and accurately. The rise time index is about 0.57 second, and the adjusting time index is about 0.92 second. The stable error is 0.005 deg, and the ITAE index is 1.08.The Table 1.shows that the proposed controller performance significantly better than PD controller.

Table 1. Control performance of different controller in Case 1

Controller	Rise time	Adjusting time	Stable error	ITAE index
PD+COF	0.57s	1.08s	0.005°	1.08
PD	4.56s	6.91s	0.008°	23.6

Case 2: To validate the robustness of the controller under time-varying disturbance, which is mainly caused by inaccurate modeling of the aerodynamic data and external perturbations, in the case 2, all of the aerodynamic parameters used in modeling the plant are increase of 30%. The wind attack of angle is $a_W = 50^\circ \sin(t)$, pitch channel linearized force and torque disturbance are respectively $\overline{F}_{by} = 0.4s^{-1}\sin(2t)$ and $\overline{M}_{bz} = 0.3s^{-2}\sin(3t)$. Initial condition was the same as the case 1. Simulation result is shown as Fig. 4.

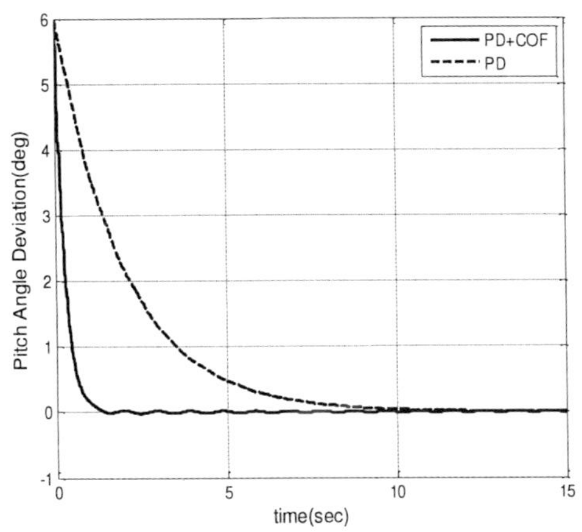

Figure 1. The curve of pitch angle varied with time in case 1.

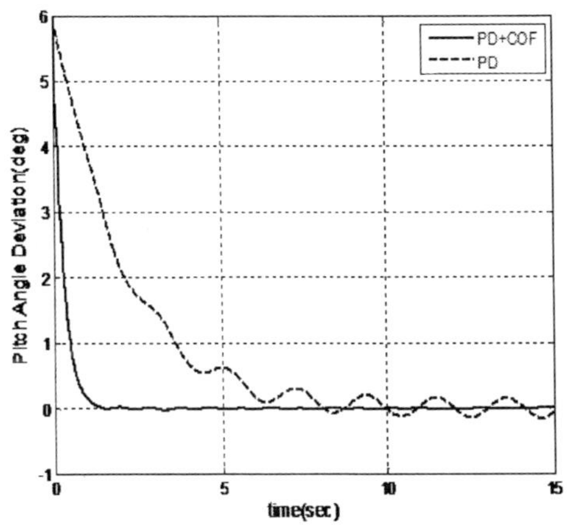

Figure 2. The curve of Pitch angle varied with time in case 2.

Table 2. Control performance of different controller in Case 2

Controller	Rise time	Adjusting time	Max Stable error	ITAE index
PD+ COF	0.59s	0.99s	0.0078°	0.91
PD	4.38s	7.7s	0.151°	30.3

Under the time-varying disturbance situation, the response time index of PD controller is obvious degenerate. The control error presents the similar periodicity in changes of disturbance. Stable error peak value of PD controller is 0.15° .However, the proposed controller has a good response. Compared with case 1, the performance has slightly declines as shown in Table II. The dynamic performance and control precision are still much better than PD controller result. The simulation results show the strong robustness and interference resistance of proposed controller.

6. Conclusion

In view of the launch vehicle attitude control problem, the main controller based on PD control was designed for control attitude of launch vehicle and the compensator based on COF network was used to estimate and compensate impact of parametric uncertainties and disturbances. Simulations of attitude control for the launch vehicle in some flight second are presented. Adding the COF network compensator, the rapidity and precision of attitude control were significantly improved.Under the disturbance case, the designed controller still has good control performance. The robustness of the proposed control scheme is embodied.

References

[1] M. Knoblauch, D. Saussie, C. Berard 2012 American Control Conference pp 967-972

[2] Uzair Ansari, Abdulrahman H. Bajodah 2016 Proceedings of 2016 4th International Conference on Control Engineering & Information Technology pp 1-6

[3] Shicong Dai,Quanjun Liu, Ying Wang, Zhang Ren, Yudong Wang, Changwan Min 2017 Proceedings of the 36th Chinese Control Conference pp:3565-3569

[4] Haoyu C,Chaoyang D,Qing W 2015 *J. Systems Engineering and Electronics* **37** 2109-2114

[5] Yanwang X 1999*Design and analysis on control system of the ballistic missile and launch vehicle* (Beijing: Astronautic Publishing House) pp 63-65

[6] *Bhargav Y. Vyas, Biswarup Das, Rudra Prakash Maheshwari 2016 J. IEEE Transactions on Neural Networks and Learning Systems* **8** 1631-1642.

Institute of Physics
Dirac House, Temple Back
Bristol BS1 6BE UK

ISSN: 1742-6588
ISBN 978-1-5108-6589-1
9 781510 865891

CURRAN ASSOCIATES INC.
proceedings.com